Neurotransmitter Release

Frontiers in Molecular Biology

SERIES EDITORS

TITLES IN THE SERIES

1. **Human Retroviruses**
Bryan R. Cullen

2. **Steroid Hormone Action**
Malcolm G. Parker

3. **Mechanisms of Protein Folding**
Roger H. Pain

4. **Molecular Glycobiology**
Minoru Fukuda and Ole Hindsgaul

5. **Protein Kinases**
Jim Woodgett

6. **RNA–Protein Interactions**
Kyoshi Nagai and Iain W. Mattaj

7. **DNA–Protein: Structural Interactions**
David M. J. Lilley

8. **Mobile Genetic Elements**
David J. Sherratt

9. **Chromatin Structure and Gene Expression**
Sarah C. R. Elgin

10. **Cell Cycle Control**
Chris Hutchinson and D. M. Glover

11. **Molecular Immunology (Second Edition)**
B. David Hames and David M. Glover

12. **Eukaryotic Gene Transcription**
Stephen Goodbourn

13. **Molecular Biology of Parasitic Protozoa**
Deborah F. Smith and Marilyn Parsons

14. **Molecular Genetics of Photosynthesis**
Bertil Andersson, A. Hugh Salter, and James Barber

15. **Eukaryotic DNA Replication**
J. Julian Blow

16. **Protein Targeting**
Stella M. Hurtley

17. **Eukaryotic mRNA Processing**
Adrian Krainer

18. **Genomic Imprinting**
Wolf Reik and Azim Surani

19. **Oncogenes and Tumour Suppressors**
Gordon Peters and Karen Vousden

20. **Dynamics of Cell Division**
Sharyn A. Endow and David M. Glover

21. **Prokaryotic Gene Expression**
Simon Baumberg

22. **DNA Recombination and Repair**
Paul J. Smith and Christopher J. Jones

23. **Neurotransmitter Release**
Hugo J. Bellen

Neurotransmitter Release

EDITED BY

Hugo J. Bellen

Howard Hughes Medical Institute
Departments of Molecular and Human Genetics
Cell Biology, Division of Neuroscience
and Program in Developmental Biology
Baylor Collge of Medicine
Houston, TX

UNIVERSITY PRESS

OXFORD
UNIVERSITY PRESS

Great Clarendon Street, Oxford OX2 6DP

Oxford University Press is a department of the University of Oxford
and furthers the University's aim of excellence in research, scholarship,
and education by publishing worldwide in

Oxford New York

Athens Auckland Bangkok Bogotá Buenos Aires Calcutta
Cape Town Chennai Dar es Salaam Delhi Florence Hong Kong Istanbul
Karachi Kuala Lumpur Madrid Melbourne Mexico City Mumbai
Nairobi Paris São Paulo Singapore Taipei Tokyo Toronto Warsaw

with associated companies in Berlin Ibadan

Oxford is a registered trade mark of Oxford University Press
in the UK and in certain other countries

Published in the United States
by Oxford University Press Inc., New York

First published 1999

A catalogue record for this book is available from the British Library

Library of Congress Cataloging in Publication Data

Neurotransmitter release / edited by Hugo J. Bellen.
p. cm.—(Frontiers in molecular biology ; 23)
Includes bibliographical references and index.
1. Neurotransmitters. 2. Synapses. 3. Neural transmission. I.
Bellen, Hugo J. II. Series.
QP364.7 .N4745 1999 573.8'54—dc21 99-16240

ISBN 0 19 963767 9 (Hbk)
ISBN 0 19 963766 0 (Pbk)

Typeset by Footnote Graphics, Warminster, Wilts

Printed in Great Britain
on acid-free paper by
The Bath Press, Avon

Preface

Understanding the mechanisms that underlie brain activity and function remains one of the major frontiers of biology. One must marvel at the speed and precision by which we co-ordinate our movements, sense our surroundings, react to stimuli, and learn and retain information. All these processes rely on complicated networks of neurons that communicate with each other and their targets. This intercellular signalling mostly occurs at synapses, specialized processes of neurons that release chemical signals, called neurotransmitters. This form of chemical intercellular communication is incredibly fast and accurate, and it is precisely the fascination with this process of release that has driven much of the research presented in this book.

This book is intended to provide the reader with an extensive but not necessarily comprehensive background on aspects of neurotransmitter release. First and foremost, it is aimed at generating excitement about the field for prospective graduate or postgraduate researchers. None of the chapters assumes extensive knowledge, and basic introductions on most topics are provided such that the material that is covered in more detail can be understood easily. The book also covers most aspects of neurotransmitter release in enough depth to interest many experts in the field. Most chapters emphasize the thought process by which proteins are assigned specific functions rather than providing facts about function. Given that neuroscientists, biochemists, molecular biologists, structural biologists, geneticists, and physiologists often combine forces to tackle questions of protein function, a multidisciplinary approach is often presented and favoured when possible.

Our knowledge of the molecular mechanisms underlying neurotransmitter release has expanded tremendously in the past ten years. The first chapter introduces the basic features and properties of the synapse. It is followed by a chapter which details several important methods used to measure and visualize various aspects of the release process. Chapter 3 describes many of the biochemical approaches that are used to identify proteins involved in neurotransmitter release, and Chapters 4 and 5 focus on more specific aspects of synaptic transmission: the proteins that transport neurotransmitters and the role of phospholipids in the process. Given that the role of many proteins implicated in the release process remains to be determined, the next five chapters focus on approaches to unravel the function of many proteins *in vivo*. Chapter 6 and 7 focus on the use of toxins, and giant squid synapses, respectively, to understand protein function. The next three chapters (8, 9, and 10) provide a review of the knowledge that we have acquired using genetic approaches in worms (*C. elegans*), fruitflies (*Drosophila*), and mice. Finally, Chapter 11 summarizes the burgeoning field of endocytosis. Although many, and possibly most proteins required or implicated in the process have been isolated, determining their functions will remain a major challenge.

I would like to sincerely thank the people who helped me with this endeavour: Susan Amara, Jay Bhave, Vytas Bankaitis, Phillis Hanson, Dan Johnston, Regis Kelly, Harvey McMahon, Stan Misler, Jim Rand, David Sweat, Konrad Zinsmaier, and the members of my laboratory, Jay Bhave, Bassem Hassan, Tom Lloyd, Giusy Pennetta, Mark Wu, and Bing Zhang for their help and suggestions. Salpy Sarikhanian provided invaluable help with the preparation of this book.

Houston, Texas, USA H. J. B
January 1999

Contents

List of contributors *xv*

Abbreviations *xvii*

1 An introduction to the nerve terminal 1

REGIS B. KELLY

1. Introduction 1

2. Structure of the synapse 2

2.1 Endocrine cell morphology 3

2.2 Varicosities are non-contacting synapses 3

2.3 The neuromuscular junction 4

2.4 The central nervous system synapse 5

3. Electrophysiological analysis of synaptic transmission 7

3.1 Calcium entry triggers neurotransmitter release 9

3.2 Mobilization, docking, and fusion of synaptic vesicles 10

3.3 Activity-dependent modification of release 12

4. The biochemistry, pharmacology, and genetics of synaptic transmission 13

4.1 The SNARE complex and exocytosis 14

4.2 Recycling of synaptic vesicle membranes 15

4.3 Adhesion molecules in the synapse 20

4.4 Molecular organization of the synapses 22

5. Development of the nerve terminal 24

6. Plasticity of the synapse 25

References 27

2 Techniques for elucidating the secretory process: from Ca^{2+} entry to exocytosis 34

THOMAS E. FISHER AND ANDRES F. OBERHAUSER

1. Introduction 34

2. Membrane capacitance measurements with the patch-clamp technique 36

2.1 Capturing the initial stages of the exocytotic fusion pore with the patch-clamp technique 42

3. Amperometric measurements of exocytosis 46

3.1 Monitoring exocytosis using simultaneous measurements of cell membrane capacitance and amperometry 49

4. Ca^{2+} imaging 51

4.1 Fluorescence 51

4.2 Optical measurement 52

4.3 Ca^{2+} indicator dyes 56

4.4 Caged Ca^{2+} compounds 60

4.5 Proteinaceous Ca^{2+} indicators 61

4.6 New imaging techniques 63

5. Imaging of vesicle dynamics 68

5.1 Activity-dependent labelling; the FM dyes 68

5.2 Total internal reflection fluorescence microscopy 71

Acknowledgements 73

References 73

3 Nerve terminal membrane trafficking proteins: from discovery to function 81

CHRISTOPHER D. HAZUKA, DAVIDE L. FOLETTI, AND RICHARD H. SCHELLER

1. Introduction 81

2. Discovery of candidate proteins: methodology 83

2.1 Electric organ and rat brain synaptic vesicle purification 83

2.2 Isolation of cDNAs encoding synaptic vesicle proteins 85

2.3 Protein interaction screens for protein discovery 89

2.4 Clues from intracellular trafficking assays 90

2.5 Insights from yeast secretory mutants 92

3. Ascribing possible function to synaptic vesicle-associated proteins 94

4. A model of possible protein interactions underlying synaptic vesicle exocytosis based on protein binding properties 96

5. Insights into the structures of proteins underlying exocytosis 98

5.1 Structures of NSF in the presence and absence of ATP 98

5.2 Conformation of the VAMP/syntaxin/SNAP-25 complex 100

5.3 The structure of the core complex 102

5.4 Structures of the C_2 domain of synaptotagmin when bound to Ca^{2+} and/or syntaxin 103

6. Current models of protein interaction cascades underlying synaptic vesicle exocytosis 104
6.1 Roles of NSF, αSNAP, and the SNAREs 104
6.2 The Rab3a cycle 107
6.3 Ca^{2+} regulation of neurotransmitter release 110
6.4 Current model of molecular mechanisms of membrane fusion 111
7. Proteins involved in the organization of synaptic vesicle trafficking in the nerve terminal 111
8. Conclusions 113
Acknowledgements 114
References 114

4 The role of phospholipids in neurosecretion 126

THOMAS F. J. MARTIN

1. Introduction 126
2. Phospholipid composition of biological membranes 126
3. Unique diversity of inositol phospholipids achieved by phosphorylation 128
4. Lipid kinases and transport proteins 128
5. Receptor signalling via phosphoinositide hydrolysis 130
6. Receptor signalling via PI 3-kinase 133
7. Phosphoinositide-binding proteins as effectors for membrane-based signalling 134
8. Inositol phospholipids in membrane traffic 136
9. Perspectives 140
References 141

5 Neurotransmitter transporters 145

DAVID E. KRANTZ, FARRUKH A. CHAUDHRY, AND ROBERT H. EDWARDS

1. Introduction 145
2. Na^{+}- and Cl^{-}-dependent plasma membrane transporters 147
2.1 GABA transport 147
2.2 Monoamine transport 149
2.3 Glycine and proline transport 150
2.4 Orphan transporters 151

2.5 Transport mechanism 152
2.6 Regulation 157
3. Excitatory amino acid transporters 158
3.1 Identification and isolation 158
3.2 Pharmacology 159
3.3 Mechanism 160
3.4 Physiological role 164
4. Vesicular neurotransmitter transporters 169
4.1 Identification and characterization of transport activity 169
4.2 Vesicular monoamine transport 170
4.3 Vesicular ACh transport 175
4.4 Vesicular amino acid transport 176
4.5 Additional vesicular transport activities 177
4.6 Mechanisms of vesicular neurotransmitter transport 178
4.7 Regulation of vesicular neurotransmitter transport 180
5. Summary 185
Acknowledgements 185
References 185

6 Toxins that affect neurotransmitter release 208

CAHIR J. O'KANE, GIAMPIETRO SCHIAVO, AND SEAN T. SWEENEY

1. Introduction 208
2. Clostridial neurotoxins 208
2.1 The phenomenology of toxicity 209
2.2 Heavy and light chains 209
2.3 Mechanism of light chain action 212
2.4 Factors affecting toxin–target recognition 213
2.5 Do the toxins have other targets or other activities? 214
2.6 Using clostridial neurotoxins to study the function of their targets 217
3. Latrotoxins and excitatory toxins 220
3.1 α-Latrotoxin 221
4. Calcium channel toxins 223
4.1 Distinguishing the roles of different calcium channels 224
4.2 Using calcium channel toxins to probe synaptic structure 224
5. Phospholipase toxins 225
5.1 Structure–function relationships 225
5.2 Presynaptic activity 226

6. Conclusions 226

References 227

7 Functional studies of presynaptic proteins at the squid giant synapse 237

MARIE E. BURNS AND GEORGE J. AUGUSTINE

1. Introduction 237

2. Overview of the squid giant synapse 238

3. Functional studies of presynaptic proteins 240

3.1 Using the squid giant synapse for microinjection studies 240

3.2 Molecular probes of presynaptic proteins 242

3.3 Electron microscopy defines protein action at the level of synaptic vesicle trafficking 254

4. Summary and future prospects 257

References 259

8 Studying mutants that affect neurotransmitter release in *C. elegans* 265

MICHAEL L. NONET

1. Introduction 265

2. *C. elegans* as a model system 265

2.1 Overview of the nervous system and characteristics of synapses 266

2.2 Identification of genes regulating synaptic transmission 268

2.3 Molecular manipulation of genes 271

2.4 Behavioural and pharmacological analysis 274

2.5 Biochemical and cell biological analysis 275

2.6 Physiological tools 277

3. An overview of *C. elegans* synaptic mutants 281

3.1 The fusion machinery 281

3.2 Calcium signalling at the nerve terminal 284

3.3 Components of the rab3 pathway 287

3.4 G-protein signalling pathways 287

3.5 Vesicular transporters and ATPases 288

3.6 Endocytosis genes 288

3.7 Other components 290
3.8 Are there additional synaptic mutants to be identified? 290
4. A brief comparison of phenotypic consequences of mutations in different organisms 291
5. Future contributions from the analysis of *C. elegans* 292
Acknowledgements 293
References 293

9 Dissecting the molecular mechanisms of neurotransmitter release in *Drosophila* 304

GIUSEPPA PENNETTA, MARK N. WU, AND HUGO J. BELLEN

1. Introduction 304
2. *Drosophila* as a model system 305
2.1 Introduction 305
2.2 Reverse genetics in *Drosophila* 307
2.3 Overexpression of proteins 307
2.4 The morphology of the neuromuscular junction or NMJ 308
2.5 Electrophysiological paradigms 310
3. Determining the function of *Drosophila* proteins in neurotransmitter release 312
3.1 Introduction 312
3.2 The calcium signalling machinery 312
3.3 The core complex 325
3.4 Proteins that regulate the core complex 332
3.5 Other proteins required for proper neurotransmitter release 338
4. Summary and conclusions 339
Acknowledgements 340
References 341

10 Genetic analysis of neurotransmitter release in mice and humans 352

THOMAS E. LLOYD AND HUGO J. BELLEN

1. Introduction 352
2. Generating mutant mice and phenotypic analyses 353
2.1 'Knock-out' technology 353

2.2 Preliminary phenotypic analysis 354
2.3 Electrophysiological analysis of knock-out mice 355
2.4 Are mouse knock-outs informative? 358

3. Mutations in synaptotagmin and rab3A: implications for a role in the final steps of exocytosis 360
3.1 Synaptotagmin 360
3.2 Rab3A 363
3.3 Guanine nucleotide dissociation inhibitor 1 368

4. Mutations in synaptic vesicle phosphoproteins 368
4.1 Synapsins 368
4.2 Synaptophysin 372

5. Other mutations in proteins implicated in neurotransmitter release 373
5.1 Neurexin Iα 373
5.2 Sec8 374

6. Mutations in the core complex of proteins implicated in neurotransmitter release 374
6.1 Coloboma mice are deficient in SNAP-25 375
6.2 Williams syndrome patients are haploinsufficient for syntaxin1A 377

7. Conclusions 378

Acknowledgements 380

References 380

11 Synaptic vesicle endocytosis and recycling 389

BING ZHANG AND MANI RAMASWAMI

1. Introduction 389

2. Cell biology of synaptic vesicle recycling 390
2.1 Synaptic vesicle recycling: an overview and an alternative view 390
2.2 Kinetics of synaptic vesicle recycling 395
2.3 Distinct pathways of synaptic vesicle endocytosis 397
2.4 Distinct pools of synaptic vesicles 400
2.5 The role of Ca^{2+} in endocytosis 401

3. Molecular mechanisms of synaptic vesicle recycling 403
3.1 Recovery of vesicle proteins and reassembly of endocytic vesicles 407
3.2 Vesicle fission 412
3.3 Post-fission events in synaptic vesicle recycling 417
3.4 Protein–protein interactions in synaptic vesicle endocytosis 418
3.5 Molecular regulation during synaptic vesicle endocytosis 419

4. Conclusion, remarks and perspectives 420
Acknowledgements 421
References 421

Index 433

Contributors

GEORGE J. AUGUSTINE
Department of Neurobiology, Duke University Medical Center, Box 3209, Durham, NC 27710, USA.

HUGO J. BELLEN
Howard Hughes Medical Institute, Department of Molecular and Human Genetics, Baylor College of Medicine, One Baylor Plaza, Houston, TX 77030, USA.

MARIE E. BURNS
Department of Neurobiology, Stanford University Medical School, 299 Campus Drive, Stanford, CA 94305, USA.

FARRUKH A. CHAUDHRY
Department of Neurology and Physiology, University of California, San Francisco, 3rd & Parnassus Streets, San Francisco, CA 94143–0435, USA.

ROBERT H. EDWARDS
Department of Neurology and Physiology, University of California, San Francisco, 3rd & Parnassus Streets, San Francisco, CA 94143–0435, USA.

THOMAS E. FISHER
Department of Physiology and Biophysics, Mayo Foundation, Medical Sciences Building 1–117, Rochester, MN 55905, USA.

DAVIDE L. FOLETTI
Department of Molecular and Cellular Physiology, Howard Hughes Medical Institute, Stanford University Medical School, Beckman Center B155, Stanford, CA 94305, USA.

CHRISTOPHER D. HAZUKA
Department of Molecular and Cellular Physiology, Howard Hughes Medical Institute, Stanford University Medical School, Beckman Center B155, Stanford, CA 94305, USA.

REGIS B. KELLY
Department of Biochemistry and Biophysics, Hormone Research Institute, University of California, San Francisco, 513 Parnassus Avenue, 1090 HSW, San Francisco, CA 94143–0534, USA.

DAVID E. KRANTZ
Department of Neurology and Physiology, University of California, San Francisco, 3rd & Parnassus Streets, San Francisco, CA 94143–0435, USA.

THOMAS E. LLOYD
Department of Cell Biology, Howard Hughes Medical Institute, Baylor College of Medicine, One Baylor Plaza, Houston, TX 77030, USA.

THOMAS F. J. MARTIN
Department of Biochemistry, University of Wisconsin, 433 Babcock Drive, Madison, WI 53706, USA.

MICHAEL L. NONET
Department of Anatomy and Neurobiology, Washington University School of Medicine, 660 South Euclid Avenue, Saint Louis, MO 63110, USA.

ANDRES F. OBERHAUSER
Department of Physiology and Biophysics, Mayo Foundation, Medical Sciences Building 1–117, Rochester, MN 55905, USA.

CAHIR J. O'KANE
Department of Genetics, University of Cambridge, Downing Street, Cambridge CB2 3EH, UK.

GIUSEPPA PENNETTA
Howard Hughes Medical Institute, Department of Molecular and Human Genetics, Baylor College of Medicine, One Baylor Plaza, Houston, TX 77030, USA.

MANI RAMASWAMI
Department of Molecular and Cellular Biology and Arizona Research Laboratories Division of Neurobiology, University of Arizona, Tucson, AZ 85721–210106, USA.

RICHARD H. SCHELLER
Howard Hughes Medical Institute, Department of Molecular and Cellular Physiology, Stanford University Medical School, Beckman Center B155, Stanford, CA 94305, USA.

GIAMPIETRO SCHIAVO
Molecular NeuroPathoBiology Laboratory, 44 Lincoln's Inn Fields, Room 614–5, London WC2A 3PX, UK.

SEAN T. SWEENEY
Department of Genetics, University of Cambridge, Downing Street, Cambridge CB2 3EH, UK.

MARK N. WU
Department of Cell Biology, Howard Hughes Medical Institute, Baylor College of Medicine, One Baylor Plaza, Houston, TX 77030, USA.

BING ZHANG
Department of Molecular and Human Genetics, Howard Hughes Medical Institute, Baylor College of Medicine, One Baylor Plaza, Houston, TX 77030, USA.

Abbreviations

BoNT	botulinum neurotoxin
CAM	cell adhesion molecule
CCD	charged-coupled device
CSP	Cysteine String Protein
DAG	diacylglycerol
DKO	double knock-out
EJC	excitatory junctional current
EJP	excitatory junctional potential
EM	electron microscopy
EPR	electron paramagnetic resonance
EPSC	evoked postsynaptic current
EST	expressed sequence tag
FRET	fluorescence resonance energy transfer
GFP	green fluorescent protein
GST	glutathione *S*-transferase
HC	heavy chain
HRP	horseradish peroxidase
LC	light chain
NA	numerical aperture
NEM	*N*-ethylmaleimide
NMJ	neuromuscular junction
NSF	NEM-sensitive factor
ORF	open reading frame
PC	phosphatidylcholine
PE	phosphatidylethanolamine
PI	phosphatidylinositol
PITP	PI transfer protein
PKC	protein kinase C
PPF	paired-pulse facilitation
PS	phosphatidylserine
PSP	postsynaptic potential
SM	sphingomyelin
SNAP	soluble NSF attachment protein
SNARE	SNAP receptor
SNR	signal-to-noise ratio
SV	synaptic vesicle
TeNT	tetanus neurotoxin

TGN	*trans*-Golgi network
TIRFM	total internal reflection fluorescence microscopy
TMD	transmembrane domain
VAMP	vesicle-associated membrane protein

1 An introduction to the nerve terminal

REGIS B. KELLY

1. Introduction

Synaptic transmission between nerve cells can usefully be viewed as a specialized example of cell-to-cell communication. A cell in a metazoan organism communicates most simply by releasing a signal which diffuses until it is detected by receptors on a neighbouring cell, a process called paracrine communication. If the signalling cell is very far away from the target cell simple diffusion will not work. In such cases the signalling cells can release the signal into the bloodstream, which carries it by the blood flow into the vicinity of the target cell. This is called endocrine communication. We can immediately see the problem facing an endocrine signalling cell. If enough signal is to reach the target, the endocrine cell must release into the body vast quantities of signal compared with what is needed for paracrine signalling. This requirement is reflected in the morphology of endocrine cells. Their cytoplasm is packed with secretory vesicles that contain the signalling molecule at concentrations approaching those that cause crystallization.

An alternative strategy to communicate over long distances is used by neurons. Neurons send out long processes that make contact with the target cell at synapses. The distances separating signalling and target cells are very small, about 10–30 nanometres. Although the distance that the signal has to diffuse is small, neurons retain a greater resemblance to endocrine than to paracrine cells. Neurons also package their signals, the neurotransmitters, in specialized secretory organelles, called synaptic vesicles, to concentrations that can be greater than 0.1 M. Both endocrine cells and neurons release their signalling molecules in response to a calcium influx. The protein composition of synaptic vesicle and endocrine secretory vesicle membranes is also remarkably similar. Large dense core vesicles reminiscent of endocrine secretory vesicles are found in many classes of neurons. Thus neuronal cells are believed to be evolutionarily and developmentally related to endocrine cells. In at least one aspect, however, the two cell types differ. Endocrine secretory vesicles are made in the juxtanuclear Golgi complex. Synaptic vesicles are made and filled at the nerve terminal which can be at a considerable physical distance from the Golgi complex in the cell body.

Although the endocrine-like dense core secretory vesicles do participate in communication between neurons, by far the most important mode of neuronal communication is via release of transmitters at synapses. Morphologically, the primary defining characteristic of synapses is the accumulation of synaptic vesicles at the presynaptic side of the synapse. The electrophysiologist can detect the fusion of a single synaptic vesicle with the presynaptic plasma membrane because so much neurotransmitter is released that a voltage change can be detected in the postsynaptic cell using an intracellular microelectrode. The postsynaptic signal caused by the contents of a single synaptic vesicle is called a 'quantum' event. The ability to count quanta has led to a remarkably precise ability to determine whether a change in synaptic efficiency is due to a change in the number of transmitter molecules per vesicle, the number of vesicles that fuse when calcium enters, or the sensitivity of the postsynaptic receptors to transmitter. Our knowledge of the molecular events that occur during transmitter release is due in large part to the abundance and size-uniformity of the synaptic vesicles. This led to the purification of synaptic vesicles to homogeneity, which in turn allowed antibodies to be raised to vesicle proteins, the DNA of which were then cloned and sequenced. From our knowledge of synaptic vesicle proteins has come our relatively detailed understanding of the molecular events that take place during transmitter release.

In addition to being a site of cell–cell communication the synapse is a site of cell–cell contact. We are also beginning to understand what is the molecular basis for the specific adhesion between pre- and postsynaptic membranes. The ability of the brain to process data is circumscribed by its ability to make precise and defined synaptic connections between the correct neuronal pairs. Thus it is not enough to know how the synapse works, we need to know how it is formed. We are only beginning to learn how any cell, neuronal or non-neuronal, binds specifically to its correct neighbour. Fortunately, it looks as if the same basic adhesion mechanisms are used in the synapse and in non-neuronal cells which should accelerate our understanding of synaptic development. This is a welcome development because it now looks as if the process of learning is going to closely resemble the late stages of synaptic development.

The goal in this introductory chapter is to review, for those new to neurobiology, the structure, function, and composition of synapses, how they form, and the link between synaptic development and synaptic plasticity. It is hoped that appetites whetted on this introductory chapter will seek more meat in the in-depth chapters that follow.

2. Structure of the synapse

Synapses all share one morphological feature, the clustering of synaptic vesicles near the plasma membrane of the nerve terminal. In other morphological aspects, however, synapses vary greatly. The detailed structure and the different types of synapses are well described in textbooks (1). Some aspects of synaptic structure are summarized here, starting from what appear to be the simplest synapses and progressing to the more complex. The morphology of endocrine cells is discussed first because they can be viewed as the most primitive form of nerve cell, or neuron.

2.1 Endocrine cell morphology

Endocrine cells and neurons both can release signals. In the case of endocrine cells, the signalling molecule is packed to high concentration into large (80–150 nm) vesicles that accumulate in the cytoplasm. Because these vesicles are filled with densely staining material in electron micrographs they are often referred to as dense core secretory granules or just secretory granules. Like synaptic vesicles, the secretory granules release their contents by exocytosis. Unlike synaptic vesicles, secretory granules are made in the *trans*-Golgi network. After exocytosis the secretory granule membrane is vesiculated and returned to the Golgi to be reutilized to form new secretory granules. Some neuroendocrine cells also have small vesicles with the same dimensions and composition as synaptic vesicles. Their physiological function is poorly understood, except in the case of insulin-secreting beta cells, where the small synaptic-like vesicles have been shown to store the transmitter GABA (2).

2.2 Varicosities are non-contacting synapses

To become a neuron, an endocrine cell has to extend a long process, then target the synaptic-like vesicles to the processes, where they can fuse and release their contents. The simplest neurons, those of the peripheral autonomic system for example, resemble modified endocrine cells. They do not make synapses but send out long processes that release neurotransmitter in the vicinity of a target organ such as a blood vessel. In such cases, the signal is released as closely as 400 nm from the target, which improves considerably the neuron's efficiency of chemical communication compared with endocrine cells. The signal is released from specialized regions along the process called varicosities or boutons (Fig. 1). Varicosities contain both dense core secretory granules and synaptic vesicles, but it is the latter that are the major source of signalling molecules. Unlike secretory granules, synaptic vesicle membranes are recycled and refilled with transmitter locally in the varicosity. Thus when signalling cells improve the selectivity of their targeting by extending processes, they also switch away from Golgi-synthesized secretory granules, to the locally-made synaptic vesicles, gaining efficiency through decentralization.

The varicosity allows transmitter to be released close to its target organ but does not make direct physical contact with the cells of the target organ. Such varicosities are found in slowly responding peripheral tissues and in some catecholamine-releasing neurons in the brain (3). The varicosity are a simple synapse in terms of morphology and communicates with its target much more slowly than most other neuronal types.

Neurons growing in culture also have varicosities along the length of their neurites (4–6). These varicosities are enriched in synaptic vesicles, which undergo exocytosis on stimulation. Why such varicosities exist and are so regularly spaced is still unknown.

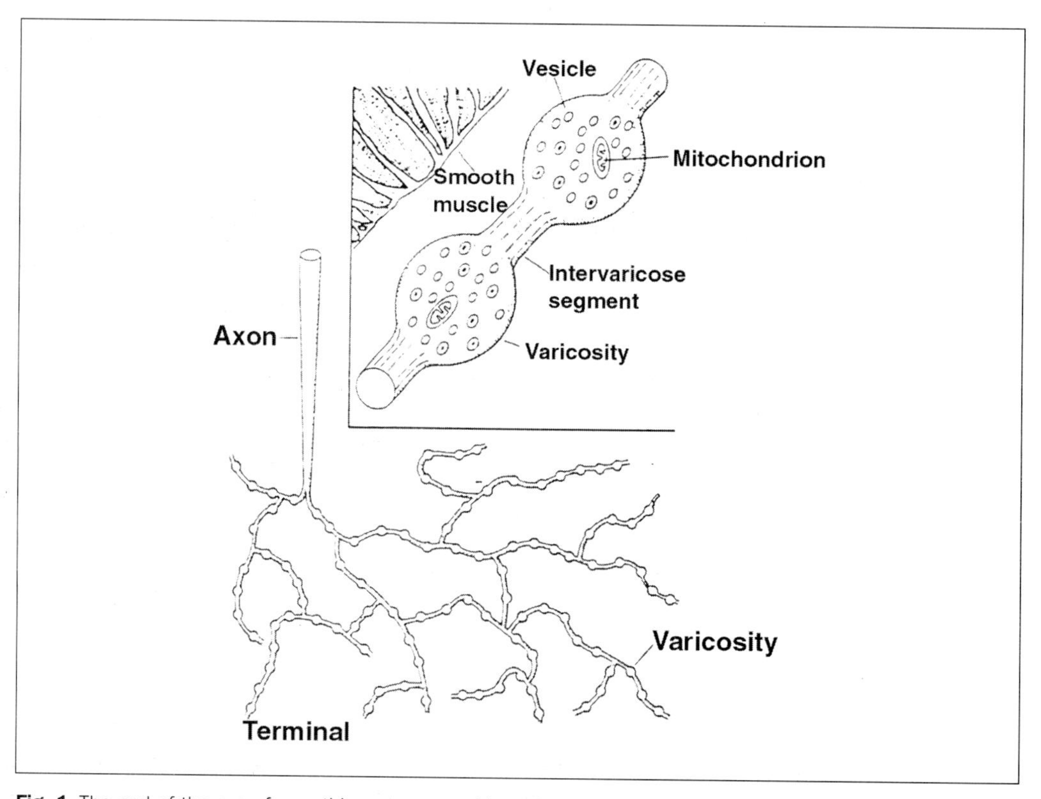

Fig. 1 The end of the axon forms thin processes with periodic swellings called 'boutons' or 'varicosities' along their length, in the autonomic nervous system. Synaptic vesicles accumulate in varicosities (inset). Varicosities are usually a significant distance from the target organ, in this case smooth muscle. (From ref. 1, Fig. 12–9. © Appleton and Lange (1985), with permission.)

2.3 The neuromuscular junction

A motor neuron that stimulates a striated muscle must evoke a much quicker response than the varicosity regulating a blood vessel. To achieve a quicker response the ending of the nerve comes to within 20–30 nm of the target muscle, which reduces diffusion time to 100 μsec or less. The synaptic vesicles are not distributed uniformly throughout the motor nerve ending but are localized to small zones or clusters. For example, at the frog neuromuscular junction, the nerve terminal is a tube with a hundred or more clusters of synaptic vesicles about 2 μm apart (Fig. 2). The plasma membrane under the synaptic vesicle cluster is enriched in calcium channels and dense bodies. The dense bodies have different morphologies in different species, sometimes called a bar, sometimes a ribbon. The composition and function of these presynaptic dense bodies are unknown but they mark the release site or *active zone*, the preferred location for synaptic vesicle fusion. Exactly opposed to the presynaptic release site is a cluster of postsynaptic receptors, held together by a densely-staining band of intracellular material. Understanding the molecular

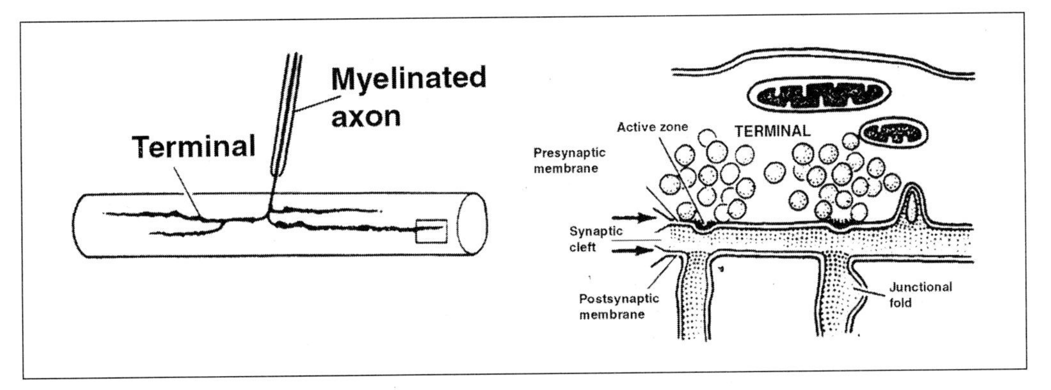

Fig. 2 The organization of the frog neuromuscular junction. The nerve terminals are long tubes along which are clusters of synaptic vesicles above active zones of plasma membrane. An adult neuromuscular junction has about 100–200 such active zones. (From ref. 112, p. 268, with permission).

composition and function of active zones, and how they are linked to postsynaptic receptor clusters are major goals of molecular neurobiologists.

2.4 The central nervous system synapse

Usually each muscle cell receives input from only a single synapse. In contrast, a neuron in the brain is festooned with thousands of synaptic inputs. Almost all parts of the neuron can receive synaptic input, the dendrites, the cell body, the axon hillock where action potentials are initiated, and even the nerve terminal itself. Usually the presynaptic endings are specializations of the axon, but not always. A dendritic region receiving input from an axon can extend a short process and release neurotransmitter on the nerve terminal that innervates it, to form a reciprocal synapse. The sizes of the nerve terminals can vary enormously from the tiny synaptic boutons (nerve endings) on dendritic spines (Fig. 3) to the massive, 5 μm synapses found on mossy fibres in the hippocampus or the glomeruli in the cerebellum (7). All the synapses contain synaptic vesicles 30–50 nm in diameter. The number of synaptic vesicles varies with the size of the terminal. The dendritic spine boutons, for example, contain only about 100 synaptic vesicles and one active zone, compared to more than 100 000 synaptic vesicles and over 100 active zones in frog neuromuscular junctions.

Almost all synapses also contain the mysterious presynaptic density or dense bar that characterizes active zones of exocytosis. In central nervous system neurons these densities are often spots laid out as a triangular grid on the presynaptic membrane (8). The composition of these presynaptic densities is unknown.

The morphology of the synapse has been used to advantage by biochemists. Because nerve terminals are much smaller than cell bodies they are resistant to the shearing forces commonly used to break open cells. Thus homogenates of brains are enriched in intact nerve terminals that have been pinched-off from their axons

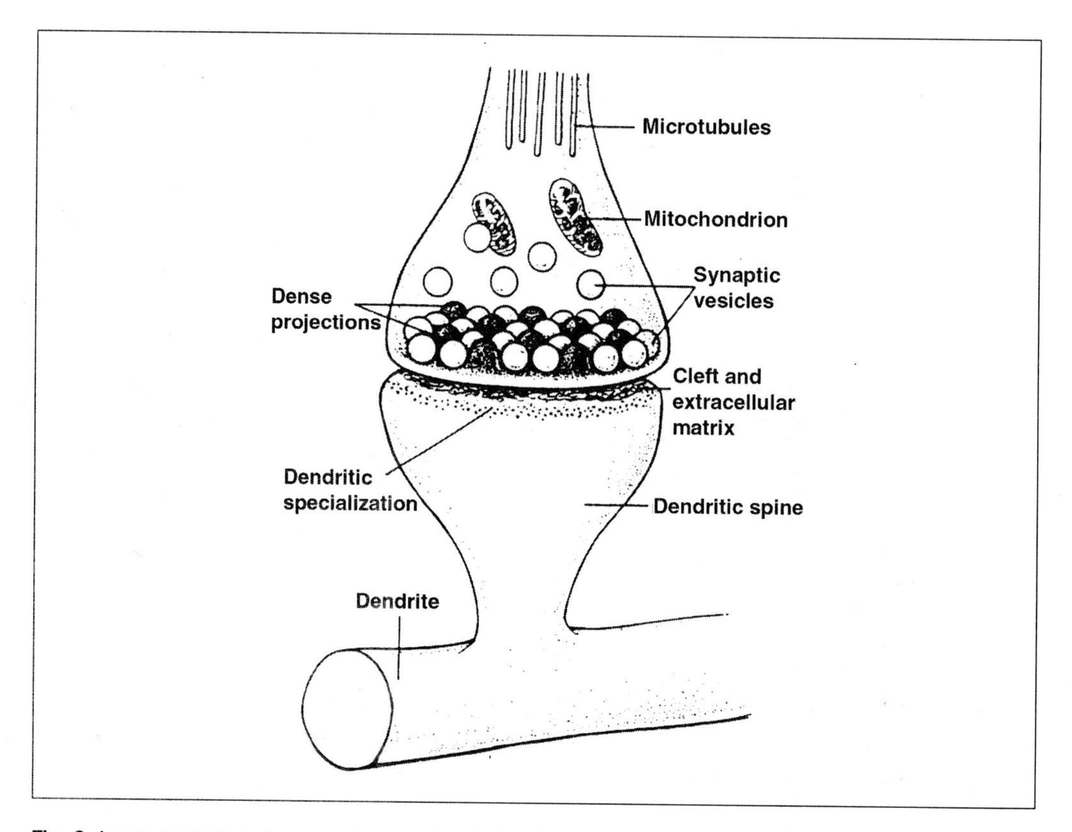

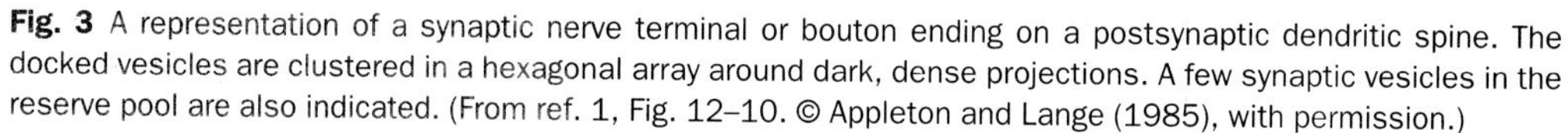
Fig. 3 A representation of a synaptic nerve terminal or bouton ending on a postsynaptic dendritic spine. The docked vesicles are clustered in a hexagonal array around dark, dense projections. A few synaptic vesicles in the reserve pool are also indicated. (From ref. 1, Fig. 12–10. © Appleton and Lange (1985), with permission.)

during homogenization. These 'synaptosomes' are extensively used to study neurotransmitter release and as a source from which synaptic vesicles can be purified. A second useful preparation derives from the dense band of material that lies under the postsynaptic receptors, and crosslinks them into an array. Because detergents dissolve membranes but do not break up protein–protein interactions it is possible to isolate from brain large 'postsynaptic densities' which are enriched in the proteins of the synaptic contact site (9).

A final morphological issue is the 'docking' of synaptic vesicles. About 1–10% of the synaptic vesicles in the synapse appear morphologically to be firmly attached to the plasma membrane in the region of the active zone (10). Because vesicles have to fuse with the plasma membrane a fraction of a millisecond after the nerve terminal is stimulated, there is no time to move a vesicle to the plasma membrane using any known mechanism of organelle motility. Thus 'docked vesicles', already touching the plasma membrane, are the only ones that can undergo exocytosis. The remainder are in a reserve pool that can be mobilized to fill empty sites at the active zone. Even the reserve pool of vesicles appears to be anchored in the vicinity of the docked

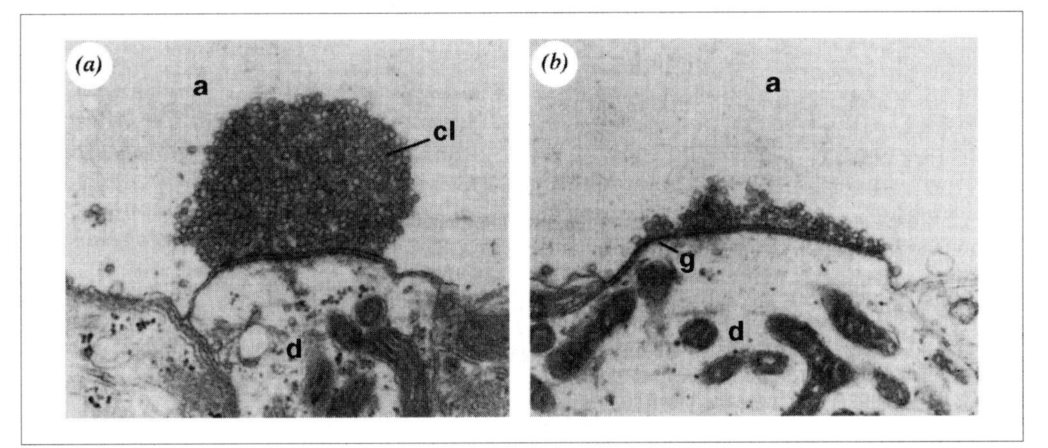

Fig. 4 Synaptic vesicle clustering in electron micrographs of the lamprey reticulospinal axons. (a) In normal axons, massive vesicle clustering can be seen. (b) After removing the reserve pool of vesicles, using anti-synapsin antibodies, a second pool of more tightly docked vesicles is revealed. a, Axon; cl, synaptic vesicle cluster; d, postsynaptic dendrite; g, gap junction. (Reprinted by permission from *Nature*, ref. 11, Fig. 2. © with permisssion from Macmillan Magazines Ltd.)

vesicle. What might be an extreme case of such clustering can be clearly seen in electron micrographs of lamprey synapses (11) (Fig. 4).

3. Electrophysiological analysis of synaptic transmission

It is difficult to think of a physiological event that is better described than synaptic transmission. Measurements can be made of the number of vesicles that are released per stimulus, the transmitter content of the vesicle, the latencies between stimulus and transmitter release, the opening of transmitter-gated channels in the postsynaptic membrane, and the number of ions that flow per transmitter bound. Further we can discover how these properties change as a function of the number of stimuli, the time between stimuli, and the time after the stimuli have stopped. This treasure trove of quantitative information inspires the experimentalist to develop quantitative models to explain them.

It is rare that a scientific field of such magnitude can be traced back to a single scientist. One such case is the development of the physiology of synaptic transmission by Sir Bernard Katz. Katz was working in London in the 1950s impaling frog muscle cells with the then recently developed glass microelectrodes that allowed precise measurement of the voltage across the plasma membrane. He and his collaborators discovered that when their electrodes were in a region of the muscle cell near a nerve terminal they saw tiny blips of a constant size (0.5–1 millivolts) occurring randomly in time. If they stimulated the nerve while recording, the muscle twitched due to a large 50–100 millivolt change in the transmembrane potential of the muscle. Such large changes in voltage caused the muscle to twitch, which frequently broke the electrode! They found that if they reduced the calcium in the bath, the muscle did not twitch, but stimulation produced voltage changes that were the same size as the spontaneous one, or exactly two or three or four times larger (Fig. 5). Given the importance of developments in physics to scientists of the time and to Katz in particular, he called these unit events of 0.5–1 millivolts quanta, and coined the term 'quantal' release of neurotransmitter.

The series of papers from Katz's laboratory that analysed quantal transmission has few equals in terms of precision, intuition, and clarity of thought (12). The calcium effect was shown to be due to the need for calcium to enter the nerve terminal, triggering transmitter release. Quantal events were linked to the release of the contents of a synaptic vesicle. A second stimulus often gave a stronger response, called facilitation, and was an early clue to the plasticity of synapses. When the nerve terminal was stimulated at high rates for long periods, it could not sustain the same levels of transmitter release. The amount released per stimulus showed a gradual

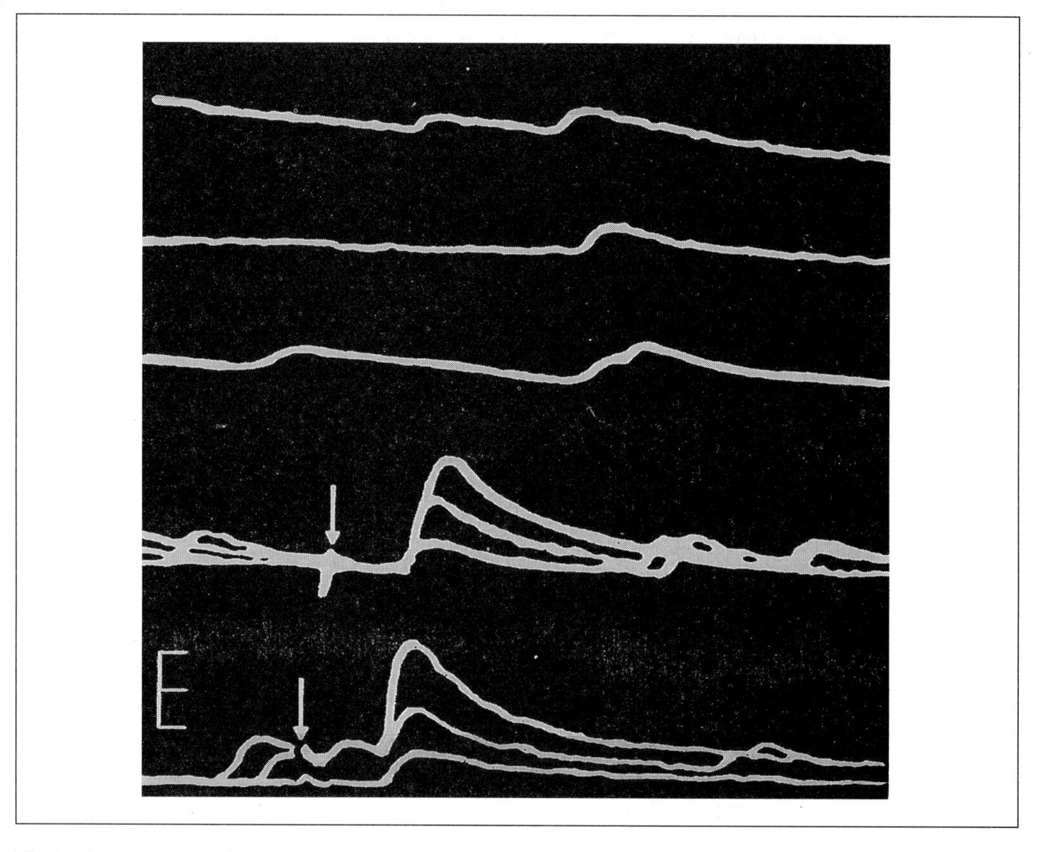

Fig. 5 Spontaneous and evoked transmitter release at the frog neuromuscular junction. The upwards deflections on the next three traces are spontaneous events of about 1 millivolt. The lower two traces show evoked release occurring about 2 msec after the stimulus (arrow). In low calcium the evoked events are small multiples of the spontaneous events. The voltage scale at the lower right hand corner is in millivolts. (From ref. 12, p. 14, with permission.)

reduction, or depression as it is more usually called. Despite frequent and prolonged challenges the key elements of Katz's quantal hypothesis remain the fundamental explanation of neurotransmitter release at peripheral and central synapses.

3.1 Calcium entry triggers neurotransmitter release

The random quantal events are now called spontaneous release or, more colloquially, 'minis'. When an electrical signal travels along an axon and reaches the nerve terminal it evokes release for about one millisecond, during which time the probability of neurotransmitter can be five orders of magnitude greater than during spontaneous release. The intense but brief period of transmitter release is due to calcium entry through voltage-sensitive channels that are very close to the docked synaptic vesicles. The calcium channel opens when the voltage change across the membrane is sufficiently large and closes again about a millisecond later. Because the intracellular free calcium concentration is very low, about 100 nM and the outside concentration high (2 mM or more), the calcium that pours into the nerve terminal through the opened calcium channels causes a very large transient change in free calcium concentration. In regions close to the channel, the intracellular calcium concentration will rise to very high levels for a brief period of time. More distant from the channel the calcium change will be smaller but of longer duration (Fig. 6). Synaptic vesicle exocytosis is triggered at calcium concentrations of 100 μM and higher which are only reached very close to the open calcium channel (13). Furthermore, the probability of release is proportional to the fourth power of the calcium concentration, which also favours a sharp increase in the probability of release for those vesicles close to the calcium channel. Thus the current concept of neurotransmitter release envisions a docked vesicle with associated fusion machinery that is physically linked to the calcium channel which evokes the fusion event.

A unit of exocytosis consisting of a channel bound to a docked vesicle with a calcium-sensitive fusion mechanism provides an elegant explanation of how

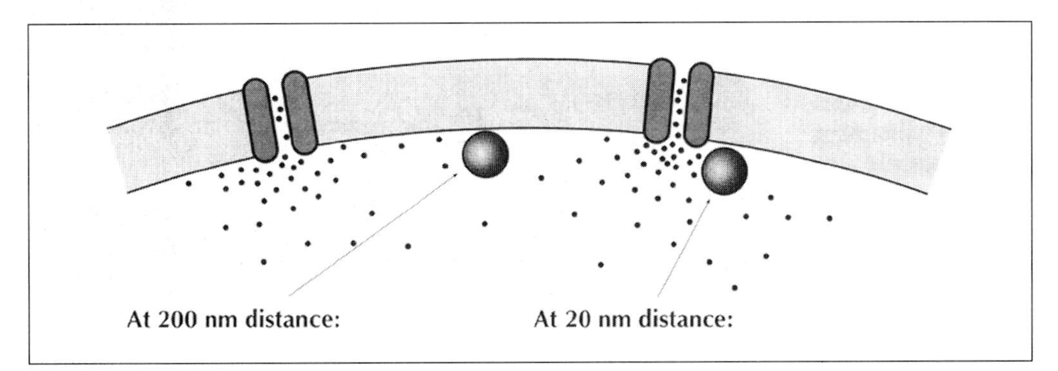

Fig. 6 Fusion only occurs near a calcium channel. A vesicle at 20 nm distance will experience concentrations of calcium that will reach 100 μM but only for a few μsec. A vesicle at ten times the distance experiences calcium concentrations of 5–10 μM and these take about 10 msec to rise and fall. Fusion of synaptic vesicle requires calcium concentrations in the region of 100 μM. (From ref. 13, p. 393 with permission. © Cell Press.)

exocytosis can be triggered so quickly after stimulation and also be turned off equally quickly. Nerve terminals can undergo exocytosis every four milliseconds or so. Returning calcium to 100 nM is next to impossible in such a short time period by using pumps or transporters of conventional efficiency. Rapid shut off of exocytosis is achieved not by pumps and transporters but by the rapid diffusion of calcium away from its entry site and the dependence of exocytosis on the fourth power of calcium concentrations. Since pumps and transporters are ineffective over a four millisecond time interval, we would expect that a second stimulus could be influenced by the 'residual calcium' left in the terminal by the first. There is in fact considerable evidence that residual calcium gives rise to an increase or 'facilitation' of the second pulse if it closely follows a first. This protocol is often called 'paired pulse' facilitation.

Since nerve terminals are usually small, measuring calcium concentrations inside them has been challenging to the experimentalist. The first solution was to find a nerve terminal big enough to impale with the microelectrodes. Katz and Miledi did much of their early microelectrode recording work using the giant stellate ganglion of the squid. To take advantage of this preparation, however, each summer experimentalists had to move their laboratories and their families to sites by the sea either in Naples, Woods Hole, or Friday Harbor. Fortunately, it is now possible to patch-clamp nerve terminals, and so such pilgrimages may no longer be necessary. As described in detail in Chapter 2, calcium influx can be measured by measuring calcium current, or by using a calcium-sensitive dye. A useful trick is to use caged calcium, which can give a uniform increase in calcium concentration when the calcium in the nerve terminal cytoplasm is uncaged by a flash of UV light (13).

A particularly intriguing property of nerve terminals is their plasticity, the changes in the probability of transmitter release per stimulus due to their recent activity, or activity a long time earlier. Short-term synaptic plasticity can occur if there is a change in the amount of calcium that enters per stimulus, or in the calcium sensitivity of the release apparatus. Examples of both have been found.

3.2 Mobilization, docking, and fusion of synaptic vesicles

Calcium entry activates transmitter release. Since release occurs a hundred microseconds or so after calcium ion entry, the synaptic vesicle and the plasma membrane must already be close to physical contact. Two models are currently in favour to explain the fusion event itself (Fig. 7). One is that calcium entry triggers a protein channel to form, linking the lumen of the synaptic vesicle to the synaptic cleft (14). The other postulates protein-catalysed formation of a lipidic pore, formed when the lipids of the synaptic vesicle fuse with those of the nerve terminal plasma membrane (15). Since there are few data that are directly relevant to synaptic vesicle fusion, much of the argument is by extrapolation from 'model' systems (16). Since detailed discussion of these 'model' systems would take us far field from synaptic transmission only the flavour of the arguments is considered here.

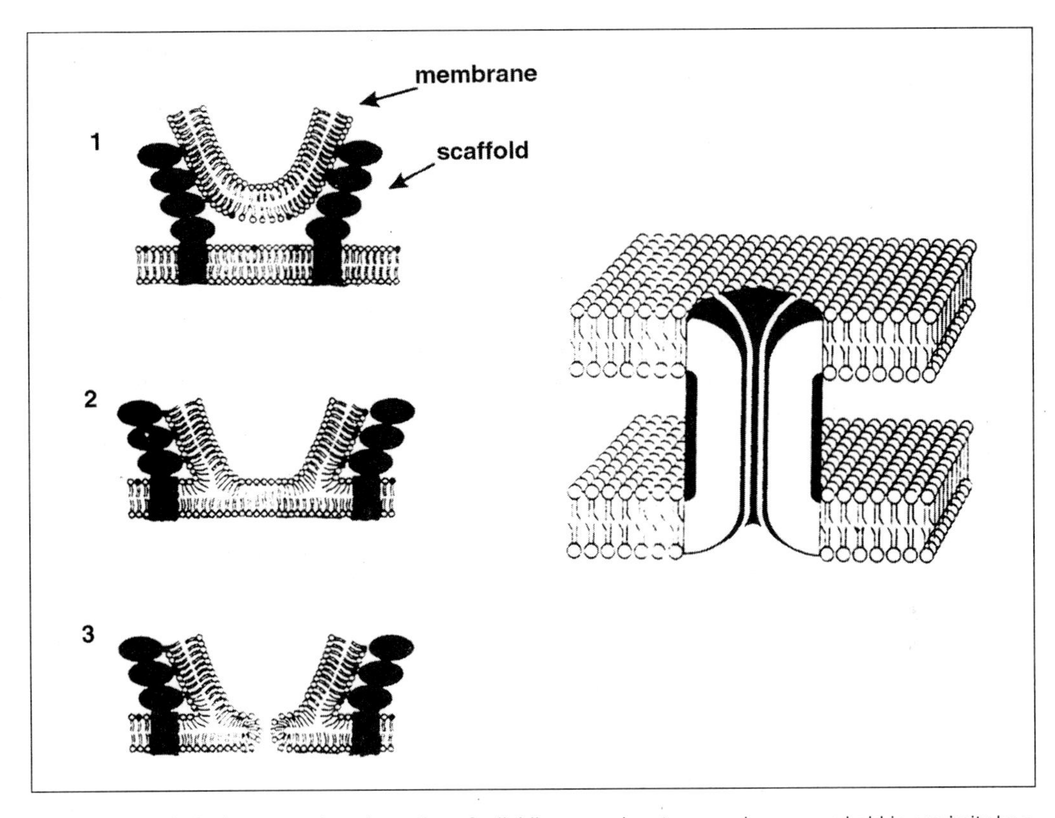

Fig. 7 On the left, fusion occurs by a formation of a lipidic pore, when two membranes are held in proximity by a protein 'scaffold'. (Adapted from Monck and Fernandez. (1994). *Neuron*, **12**, 709.) The diagram on the right illustrates the formation of a proteinaceous pore that could connect the inside of a vesicle to the extracellular space. (Adapted from ref. 14, p. 512, with kind permission from Elsevier Science.)

The best evidence for a fusion pore comes from studies of regulated secretion of secretory granules. When such granules contain biogenic amines, the secretion can be measured quantitatively and with excellent kinetics by placing a carbon fibre electrode on the surface of the cell and using amperometry (17), as described in detail in Chapter 2. A consistent finding is that the release of the bulk of secretory granule content is preceded by the formation of a narrow channel with dimensions comparable to those of an ion channel (18). This channel is referred to as a 'fusion pore'. Exocytosis also increases the surface area of the plasma membrane, which can be accurately measured as an increase in surface capacitance (13) (Chapter 2). Since the charge per unit area of the secretory granule membrane is not the same as that on the plasma membrane, membrane fusion also results in a flow of charge through the neck where the vesicle fuses to the plasma membrane. Measurements of the resistance to ion flow also suggest that the neck is electrically similar to an ion channel. Thus both electrophysiological measurements are consistent with a fusion pore that has, at least transiently, the properties of an ion channel. Since ion channels are proteins, it is thus reasonable to suggest that the fusion pore is, at last transiently, proteinaceous.

Synaptic vesicles are too small for such electrophysiological measurements but, if a similar fusion pore existed, its dimensions are such that almost all the vesicle content could leak out in a millisecond without conventional exocytosis occurring. If the fusion pore closed in a millisecond or two the empty synaptic vesicle could then detach from the active zone, without ever mixing its membrane with that of the plasma membrane. This mechanism is frequently referred to as the 'kiss-and-run' hypothesis (19).

Although the proteinaceous fusion pore and kiss-and-run hypotheses are very elegant and certainly plausible, direct evidence is scarce. No candidate pore-forming proteins have yet been characterized, even in yeast where exocytosis has been extensively studied. Although the kiss-and-run hypothesis cannot be eliminated as a possible mechanism of synaptic vesicle recycling, it is clearly not the only mechanism as is described later.

The alternative model, that proteins catalyse lipid fusion events, has been popular for almost two decades (20). The bases for these models were studies on lipid rearrangements in phospholipid micelles and the extensive literature on fusion between pure lipid bilayers. Measurements of bilayer fusion can be misleading. In some cases they can be due to rupture, followed by resealing, which is presumably not a valid model of synaptic vesicle exocytosis. For this reason, most studies of lipid fusion now measure both lipid mixing and content mixing. Direct experimental evidence for lipid rearrangements during synaptic vesicle fusion does not exist.

3.3 Activity-dependent modification of release

In the simplest vesicular model of transmitter release there are N synaptic vesicles per active zone and the probability of release increases to a value p during an action potential. The number of vesicles released per stimulus is thus N x p. From synapses as varied as the frog neuromuscular junction with 200 active zones, to the hippocampal synaptic bouton with only one, about 0.1 to one vesicle is released per active zone per action potential. Since the hippocampal synapse can have as few as 100 synaptic vesicles, it would release half the number of vesicles after only 50 stimuli in this simple model, and so would show a 50% reduction in transmitter release after 50 stimuli during repetitive or tetanic stimulation. This is not what is seen. For example synapses in the mouse auditory system (Fig. 8) show a 50% reduction in transmitter release after five stimuli (21). This phenomenon, called synaptic depression, can be explained by postulating that the probability of release dropped rapidly during repetitive stimulation. The other explanation is that there are two pools of vesicles, a 'readily releasable' pool pre-docked at the plasma membrane and a reserve pool. The readily releasable pool is about one-tenth the total (10), which could explain the rapid rates of depression. Because of synaptic depression, transmitter release does not drop to zero during prolonged stimulation but is controlled by a new rate-limiting step, the rate of transfer from the reserve pool to the readily releasable pool. This rate has a half-time of about six seconds (22), which brings the rate of release much closer to the

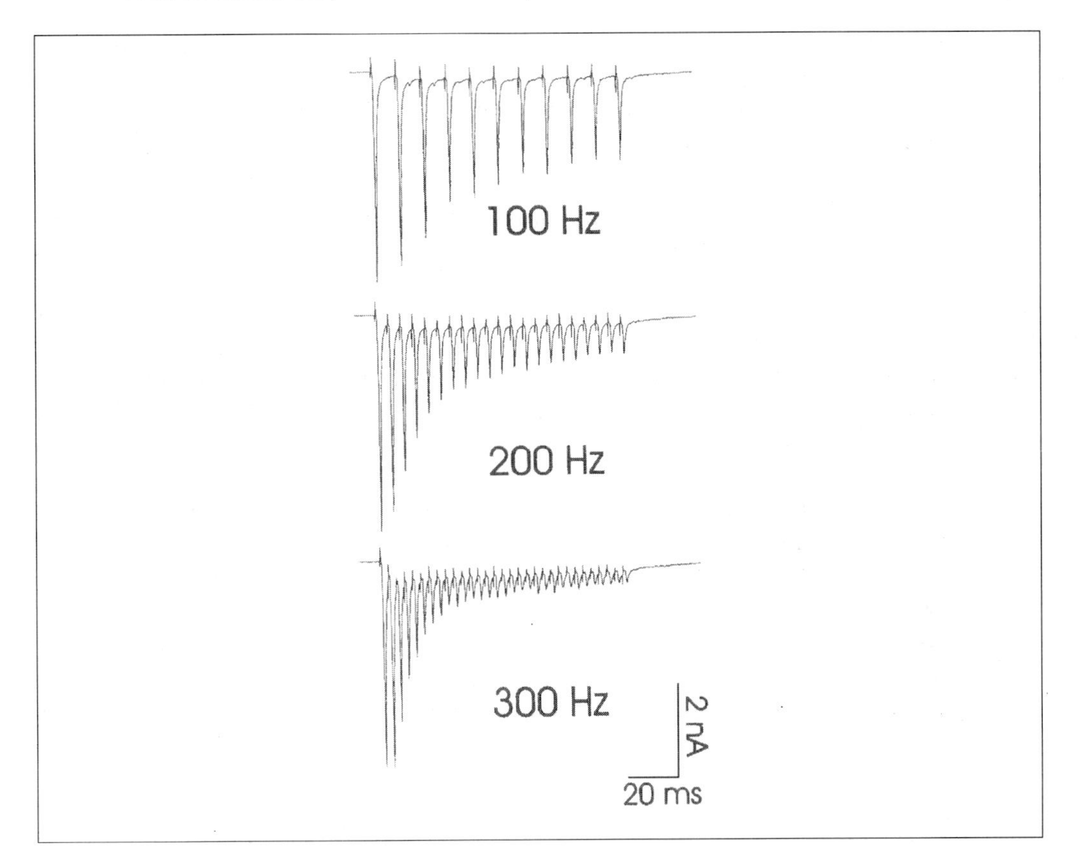

Fig. 8 Depression of neurotransmitter release is very fast. Synapses in the mouse auditory brainstem were stimulated for 100 msec at the frequencies shown. (Reprinted from *Nature*, ref. 21, © 1998 with permission from Macmillan Magazines Ltd.)

synaptic vesicles' recycling rate of 5–10 vesicles/second (23). Thus the synapse can sustain very high frequencies of neurotransmitter release for many seconds, without depleting the vesicle pool.

When a synapse is subjected to several periods of intense, tetanic stimulation, its response many hours later can be potentiated (long-term potentiation) or depressed (long-term depression). These long-term changes, which can require protein synthesis, are frequently used as models of synaptic plasticity. Their mechanisms are very different from the types of short-term depression and facilitation discussed in this section.

4. The biochemistry, pharmacology, and genetics of synaptic transmission

With such a detailed knowledge of what synapses look like and how they work, why do we need to know what proteins are involved? The biochemists' response is that it is through knowing the proteins, and of course the genes encoding them, that we learn how to perturb the synapse pharmacologically and genetically, and therefore how to comprehend genetic abnormalities in synaptic transmission. Since the molecular basis of exo- and endocytosis, transmitter transport, and calcium regulation are considered in much more detail later in this book, our goal here is to define the major questions and to outline their solutions, where known. (Pharmacology of the synaptic transmission is covered in Chapter 6, genetics in Chapters 8, 9, and 10, and *in vitro* reconstitution in Chapter 3.)

A simple but provocative question to begin with is why certain proteins are so abundant in brain. One guesses that proteins that are unique to brain perform some brain-specific function. Secondly, we assume that proteins which are not unique to brain but are abundant in brain perform a function that is much more important to brain tissue than to the rest of the body. Thus tubulin is abundant because of the amount of microtubule-mediated organelle traffic. Proteins required for endocytosis, such as clathrin and AP2, are abundant because, although synapses make up only about 1% of the volume of the brain, they need to be exceedingly efficient at endocytotic recycling of vesicle membranes. Brain is a rich source of proteins involved in membrane fusion, since efficient exocytosis is very important in brain function.

The abundance of certain proteins in the brain has been of major significance to biologists, contributing greatly to our understanding of membrane traffic in general. To gain insight into the molecular basis of synaptic transmission the first step was to purify to homogeneity synaptic vesicles, first from the electric organ (24) and then from rat brain (25). When the proteins of synaptic vesicles had been identified, the powerful technologies for studying protein associations led quickly to the identification of new proteins that interacted with known synaptic vesicle proteins. In particular, the plasma membrane protein syntaxin was discovered because of its interactions with the synaptic vesicle protein synaptotagmin (26). Syntaxin and SNAP-25 were identified by their interaction with synaptobrevin (VAMP), an interaction that was regulated by the two proteins alpha-SNAP and NSF (27). Protein–protein associations have been used successfully in a similar way to identify proteins that interact with dynamin, a protein involved in endocytosis (28). At this juncture, major nerve terminal-specific proteins are being identified at a prodigious rate using molecular technologies. The next goal is to determine the physiological significance of these new proteins. For testing biological significance four basic techniques are available: pharmacology, genetics, *in vitro* reconstitution, and morphology.

4.1 The SNARE complex and exocytosis

An interaction between a synaptic vesicle protein, VAMP, and plasma membrane proteins, syntaxin and SNAP-25 suggested a possible role in either docking the synaptic vesicles, or allowing fusion. Pharmacological evidence that these three proteins were required for exocytosis but not for docking came from studies with bacterial neurotoxins. These endopeptidase toxins simultaneously cleaved one of the three proteins and inhibited exocytosis without affecting vesicle docking (29)

(Chapter 6). Genetic data confirmed these findings. When *Drosophila* mutants are defective in VAMP transmitter release was inhibited but not docking (30) (Chapter 9). Mutations in the yeast equivalents of the three proteins blocked exocytosis and caused accumulation of secretory vesicles (31). This genetic evidence can now be backed up by reconstitution *in vitro* of fusion between liposomes that contain VAMP and liposomes that contain both syntaxin and SNAP-25, albeit at extremely low efficiency compared with physiological fusion (32). Taken together these data implicate VAMP, referred to as a vesicle or v-SNARE (*SN*AP *R*ECEPTOR) and syntaxin and SNAP-25, two target or t-SNAREs, in some aspect of membrane fusion. The exact mechanism whereby the SNAREs facilitate fusion is likely to involve the formation of four parallel bundles of amphipathic helices (33). The nature of these helical bundles and their formation is covered in detail in Chapter 3.

The SNAREs are unlikely to be the only major players in synaptic vesicle fusion. Other suggested participants include the small rab3 GTPases (34), the exocyst complex (35), the calcium-binding protein synaptotagmin (36), the nsec-1 class of proteins (37), the CAPS protein (38), and phosphatidyl inositol phosphates (39) (Chapter 4). It seems likely that the SNAREs will be central to fusion, but not the only players.

The best candidate for a protein that confers calcium sensitivity to the fusion apparatus is synaptotagmin. Here the evidence is almost exclusively genetic. As discussed later (Chapters 8, 9, and 10) worms, flies, and mice that lack synaptotagmin all have deficiencies in their calcium-dependent secretion (40). The secretion mechanism is intact in such mutants because secretion can still show calcium-independent stimulation. Two such stimuli are hyperosmolarity and exposure to alpha-latrotoxin.

The convergence of pharmacological, reconstitution, and genetic studies suggest that the basic molecular elements of the fusion apparatus and its calcium regulation are within our grasp. Still elusive is the mechanism that holds the vesicles docked to the plasma membrane in the 'readily releasable pool'.

4.2 Recycling of synaptic vesicle membranes

After a synaptic vesicle fuses with the plasma membrane and releases its content of transmitter, the membrane is recycled to form a new vesicle, which is refilled with neurotransmitter. Since synaptic vesicle components turn over slowly it is likely that each synaptic vesicle recycles about a thousand times or so per lifetime. This must be a precise process, for a synaptic vesicle that was made without fusion machinery would be incapable of another fusion cycle. Eventually the nerve terminal would fill up with such inert vesicles.

Recycling is covered extensively in Chapter 11. Here we summarize current models of vesicle recycling (Fig. 10) and the experiments on which such models are based.

4.2.1 The three models of vesicle recycling

Vesicle recycling was first detected using an extracellular marker, horseradish peroxidase (HRP), to label recycled vesicles (41, 42). Such experiments implicated clathrin-coated vesicles and intraterminal vacuolar structures, usually called endosomes, in vesicle recycling. Subsequent experiments, particularly using the low density lipoprotein receptor, showed that clathrin-mediated endocytosis into endosomes was characteristic of membrane internalization in almost all cells. In this universal system, membrane proteins that execute a wide array of cellular functions can all be concentrated and internalized non-selectively in the same coated pit. After internalization, the membrane proteins are sorted to their correct intracellular destination from the endosome, which resembles the Golgi complex in its capacity to sort proteins to different intracellular destinations. Since this is the classical mode of membrane recycling in all cells, recycling of synaptic vesicles in nerve terminals by such a mechanism will be referred to as the 'classical' mode of vesicle formation.

From the beginning, the classical mode of vesicle formation was questioned by a major group of neurobiologists, headed by Bruno Ceccarelli (43). They were concerned that they could see very little HRP uptake, depletion of vesicles on stimulation, or clathrin-coated vesicles in the nerve terminal. An alternative hypothesis, described earlier, is the kiss-and-run model (19). In this model the synaptic vesicle makes a fleeting connection with the plasma membrane at the active zone sufficient to allow transmitter to leak out. In this model the membrane of the vesicle does not flatten out, its membrane components do not mix with those of the plasma membrane, and the transmitter leaks from the cell via a small lipidic pore.

The third model resembles the classical model but drops the postulate that membrane protein sorting occurs at endosomes after internalization. Instead it postulates that sorting can take place directly at the plasma membrane. The internalization mechanism is still clathrin-mediated, but clathrin and its accessory proteins are postulated to select specifically the synaptic vesicle proteins for internalization and exclude the others.

There are several variations on each of these models and they are not mutually exclusive. They serve, however, to illustrate the issues with which neurobiologists have been struggling over the years. Fortunately, the answers are now coming quickly, in part due to two technological advances that allow direct measurement of the processes of endocytosis. The first procedure, pioneered by Dr William Betz and described extensively in Chapter 2, uses a membrane-impermeable, fluorescent dye to label the insides of recycling synaptic vesicles (44). After washing the nerve terminal to remove extracellular dye, the amount of dye remaining is proportional to the extent of recycling and the number of synaptic vesicles per terminal. Stimulating exocytosis a second time exposes the inside of labelled synaptic vesicles to the external milieu. Thus loss of dye is a quantitative measure of exocytosis. Since about 30 dye molecules are taken up per synaptic vesicle it is possible to measure the endocytosis (45) as the exocytosis (46) of a single synaptic vesicle. Dye uptake measurements in hippocampal neurons reveal that the half-time to label the synaptic vesicle pool is about 20 seconds, and the half-time before a vesicle is ready for a second round of exocytosis 40 seconds (23). Such speeds of internalization are about ten times faster than measured for classical endocytosis in non-neuronal cells. A second direct measure of the recovery of membrane after exocytosis, a process commonly referred to as compensatory endocytosis, is membrane capacitance (Chapter 2). By patch-clamping the nerve terminals of large retinal bipolar neurons exocytosis can be measured as an increase of capacitance and compensatory endocytosis can be measured as the return of membrane capacitance to baseline activities. Half-times of the order of seconds can be observed at low rates of stimulation (47), consistent with dye measurements.

4.2.2 The 'classical' model of endocytosis

It is usually wisest to assume initially that the nerve cell uses the same basic machinery as non-neuronal cells, which is part of the attraction of the 'classic' model. In classic endocytosis in non-neuronal cells, the membrane proteins are segregated from resident plasma membrane proteins by clathrin and its associated adaptor protein, AP2. Clathrin and AP2 also impart curvature to the membrane, while the final pinching-off process requires the GTPase dynamin, and accessory proteins. Both the sorting proteins, AP2 and clathrin, are highly concentrated in nerve terminals, as well as neuronal dynamin (1), and many of its accessory proteins. It is reasonable to predict that the abundance of these proteins in nerve terminals contributes to highly efficient membrane recycling. This prediction is borne out by morphological (48), genetic (48–51), reconstitution (51), and pharmacological (52, 53) studies.

The evidence. reviewed in Chapter 11, seems overwhelming that synaptic vesicle proteins can be recovered from the plasma membrane by a clathrin-mediated process. The evidence for the second phase, synaptic vesicle protein sorting in endosomes, is less clear-cut. Clathrin-mediated budding of synaptic vesicles from endosomes has been observed (48). Endosomes isolated from the pheochromocytoma line PC12 can generate synaptic vesicles *in vitro* when incubated with ATP and cytosol (54). In this case however clathrin and AP2 are not required for vesicle budding; instead the cytosol requirement for budding can be replaced using a novel adaptor, AP3 and a small GTPase ARF1 (55). Thus there may be two ways to form vesicles from intraterminal vacuoles or endosomes, one using clathrin and the other AP3.

Formation of synaptic vesicles via the AP3 route from endosomes has been characterized in some detail. One satisfying feature of this mode of vesicle formation is that AP3 coat recruitment and vesicle budding is inhibited by tetanus toxin and so requires an intact v-SNARE (56). Initiating a coating interaction by binding to a v-SNARE provides an excellent quality control step to ensure that all vesicles that are made contain fusion machinery.

The endosomal mode of vesicle formation cannot be the major one since dye taken up by a single vesicle is not distributed over many vesicles (46) as would be required if an endosomal intermediate were involved. The AP3 pathway is clearly not essential since mutant mice lacking AP3 are viable and have synaptic vesicles, although they lack zinc (57). The AP3 appears to be necessary for the formation of a subclass of vesicles with a particular protein composition. In the absence of such vesicles, mutant mice have unmistakable neurological symptoms and an unusual theta rhythm in their electrocorticograms (58).

4.2.3 Direct formation of synaptic vesicles from the plasma membrane

Although forming synaptic vesicles directly from the plasma membrane rather than from endosomes is theoretically a more efficient process, it requires that membrane protein sorting takes place at the plasma membrane. Only in recent years, has evidence appeared that supports such sorting (59–61). If we drop the hypothesis that sorting occurs at the endosome after internalization, then the evidence for direct production of synaptic vesicles from the plasma membrane becomes quite impressive (46, 52, 62, 63). As discussed more extensively in Chapter 11, the most plausible scenario at present is that the dominant mode of synaptic vesicle biogenesis is by direct formation from the plasma membrane.

4.2.4 Kiss-and-run recycling

Genetic, morphological, pharmacological, and biochemical data demonstrate unequivocally that recycling of vesicles can be by conventional, clathrin-mediated endocytosis. This does not, however, eliminate the possibility that some transmitter release and vesicle recycling could be by the kiss-and-run mechanism. The only trouble with the very popular kiss-and-run conjecture is the paucity of direct experimental evidence. If the fusion pore has the dimensions of a conductance channel it would not allow HRP or antibodies to enter and so vesicles formed by kiss-and-run would never be measured. Similarly if a proteinaceous fusion pore is only open for a millisecond, it is barely possible that dyes such as FM1-43 might enter during vesicle formation, but inconceivable that they could be lost during kiss-and-run exocytosis, since the off-times of the dyes are measured in the range of seconds (64). If dye could be introduced into a vesicle undergoing kiss-and-run recycling then, under appropriate experimental conditions, it should be possible to see transmitter release without dye release. This has been observed when frog neuromuscular junctions are incubated in the drug stauroporine (65). To date, this remains the best experimental evidence suggestive of kiss-and-run.

4.2.5 Two pathways of recycling

Although the first goal is always to fit available data to a single explanation, circumstantial evidence forces us to seriously consider the possibility of more than one pathway of vesicle recycling. Miller and Heuser (66) suggested that the importance of the endosomal pathway of vesicle recycling observed in earlier electron micrographs might have been exaggerated by the intense stimulation conditions that were then used. Under more physiological conditions synaptic vesicle formation was observed directly from the plasma membranes via clathrin-coated pits. Electron microscopy of *Drosophila* neuromuscular junctions also revealed two pathways, with a rapid internalization near the active zone and slower internalization at more distant sites (67). Internalization of FM1-43 confirmed these observations and suggested that the pool that was slower to fill was essential to maintain transmitter release during prolonged periods of stimulation (68). Vesicle formation by the slow pool was sensitive to disruption of the actin cytoskeleton with cytochalasin D. An appealing explanation of these results is that vesicle recycling can be either directly from the plasma membrane or via the endosomal path (Fig. 9). The latter pathway is slower and is preferentially utilized under periods of intense stimulation. This conjecture fits with the discovery of two parallel pathways of synaptic vesicle biogenesis in PC12 cells (52).

Although the different lines of evidence suggest two pathways of vesicle biogenesis, it is premature to link the AP3 pathway from endosomes to the cytochalasin-

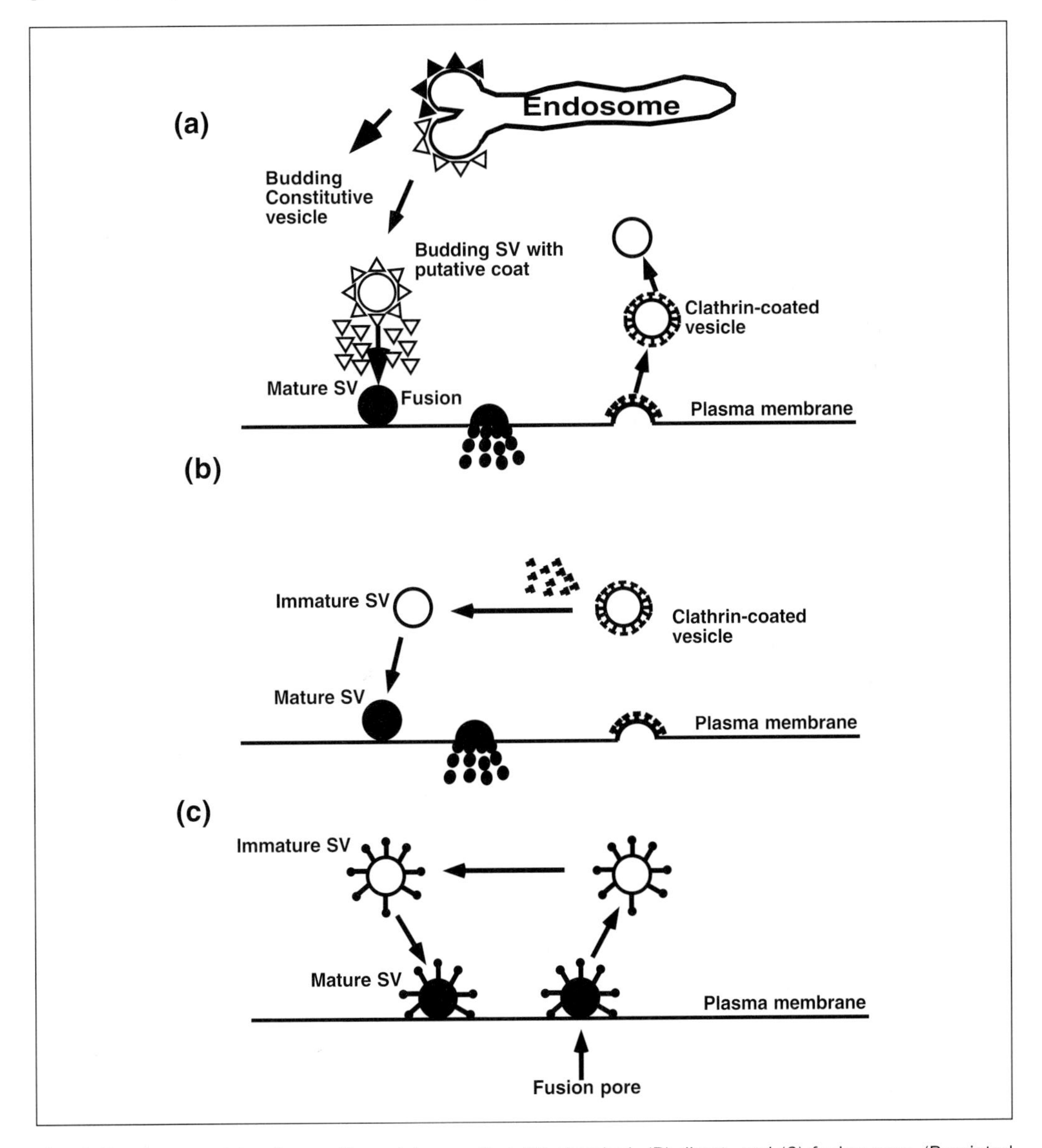

Fig. 9 The three models of synaptic vesicle recycling: (A) classical, (B) direct, and (C) fusion pore. (Reprinted from Mundig and De Camilli. (1994) *Curr. Opin. Cell Biol.*, **6**, 564 with kind permission from Elsevier Science.)

sensitive pathway used at extreme levels of stimulation. We also do not know how to relate these pathways of vesicle recycling to the pathway for *de novo* synaptic vesicle biogenesis.

4.3 Adhesion molecules in the synapse

In addition to being a site of cell-to-cell communication the synapse is a site of cell–cell adhesion. The proteins implicated so far in synaptic adhesion do not appear to be a novel, brain-specific class of proteins. The adhesion molecules are familiar old friends, adapted slightly to the specialized needs of the synapse.

One frequently found class of cell–cell contact is the zonula adherens or adherens junction. At adherens junctions two cells are held together by homotypic interactions between proteins of the cadherin family. Cellular adhesion requires that the cadherins be linked to the actin cytoskeletons of the partnered cells. To do this, beta-catenin links the cytoplasmic tail of the cadherin to alpha-catenin, which in turn connects to actin microfilaments. In brain synapses there are about ten cadherins, beta-catenin, and a neuronal form of alpha-catenin (N-alpha-catenin) (69). Cadherin adhesion sites can be recognized morphologically by the symmetrical pre- and postsynaptic membrane thickening and the associated actin filament bundles. In mature neurons they are most frequently found adjacent to active zones of exocytosis (Fig. 10), particularly at synaptic connections to dendritic spines (70).

A second class of connections found between epithelial cells is the tight junction, which is similar to the septate junction in *Drosophila*. The tight junctions in mammalian cells are held together by the transmembrane protein occludin, which interacts with a tripartite complex of proteins, Z0-1, Z0-1, Z-03. The neurexin family of transmembrane proteins is found in septate junctions (71) and synapses (72) and is also associated with a tripartite complex (73). One of the components of the tripartite complex that binds neurexin is CASK, a large calmodulin-dependent protein kinase. All three components of the complex have PDZ domains which are known to play a major role in clustering ion channels both pre- and postsynaptically (74). Thus the presynaptic neurexins might help link together presynaptic proteins such as the calcium channels. Presynaptic neurexin also binds to neuroligin, a postsynaptic membrane protein. Neuroligin can also associate with postsynaptic proteins that use PDZ domains to cluster postsynaptic receptors. Thus the neurexin–neuroligin complex is ideally located to keep the clusters of postsynaptic receptors in register with the presynaptic calcium channels (Fig. 11).

Exocytosis at the nerve terminal only comes from docked vesicles located close to calcium channels. The neurexin complex could also help keep fusion sites and calcium channels in register. One of the PDZ-containing proteins in the tripartite structure is mint 1, a protein known to interact with nsec1 (munc18) (75), which itself binds very tightly to the t-SNARE, syntaxin. There are two other possible mechanisms for clustering calcium channels with the presynaptic exocytotic machinery. One is a direct interaction between the calcium channel and syntaxin (76, 77). Another is a direct interaction between neurexin and the synaptic vesicle protein,

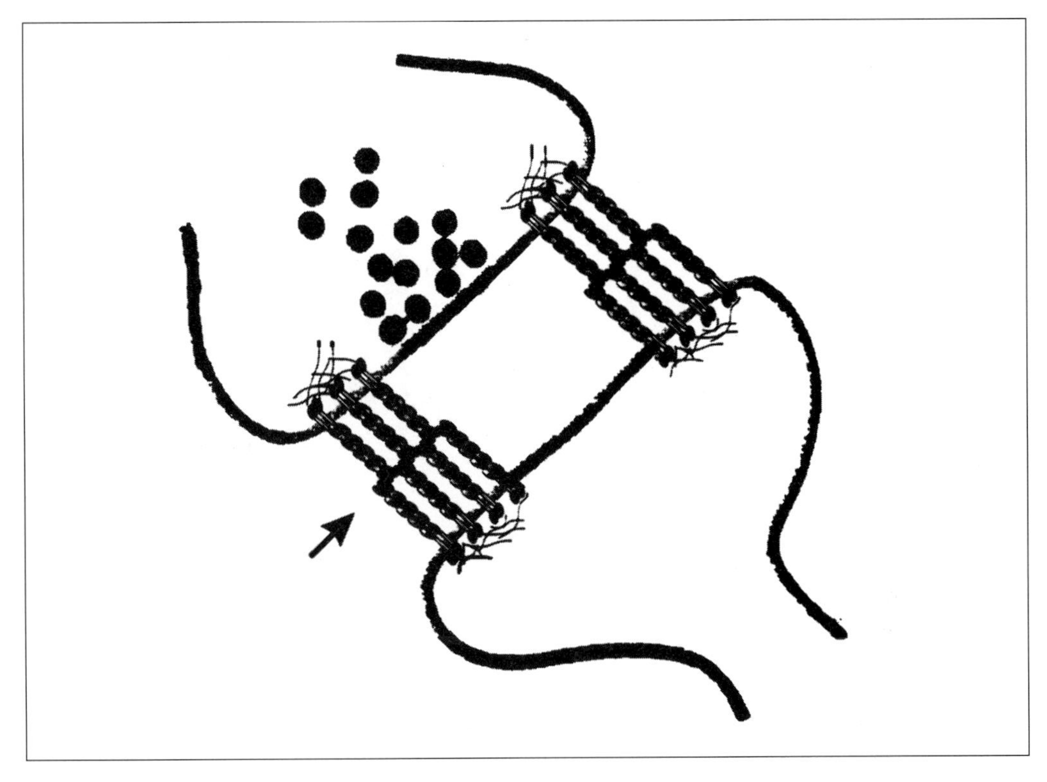

Fig. 10 Cadherin-mediated synaptic adhesion. Cadherins have multiple extracellular domains (ovals) and form homotypic interactions between their most distal ones. Intracellularly, cadherins attach to the actin cytoskeleton via catenins. The adhesion sites lie immediately adjacent to the active zones in many cases.

synaptotagmin (78). Whether these are all physiologically relevant interactions, or reflect irrelevant interactions between purified proteins in a test-tube will be decided by future genetic and pharmacological experiments.

A third family of cell adhesion molecules (CAMs) belongs to the immunoglobulin superfamily. Although CAMs can show homotypic interactions they also interact heterotypically with members of the integrin family of adhesion molecules. Intracellularly, CAMs interact with elements of the cortical actin cytoskeleton. Perhaps the best characterized member of the family is the *Drosophila* CAM, fasciclin II (79). Fasciclin is present both pre- and postsynaptically in larval neuromuscular junctions, and interacts with the Discs-Large protein (dlg), via a PDZ domain. Thus the PDZ-containing proteins can in principle cluster adhesion molecules as well as channels at the synapse.

Non-neuronal cells attach themselves to extracellular matrix at specialized sites called focal contacts. The transmembrane receptors for the extracellular matrix, the integrins, interact via their cytoplasmic tail with parallel bundles of actin microfilaments. In the nervous sytem, extracellular matrix is only found in the synaptic clefts of neuromuscular junctions. Some components of the synaptic cleft matrix are

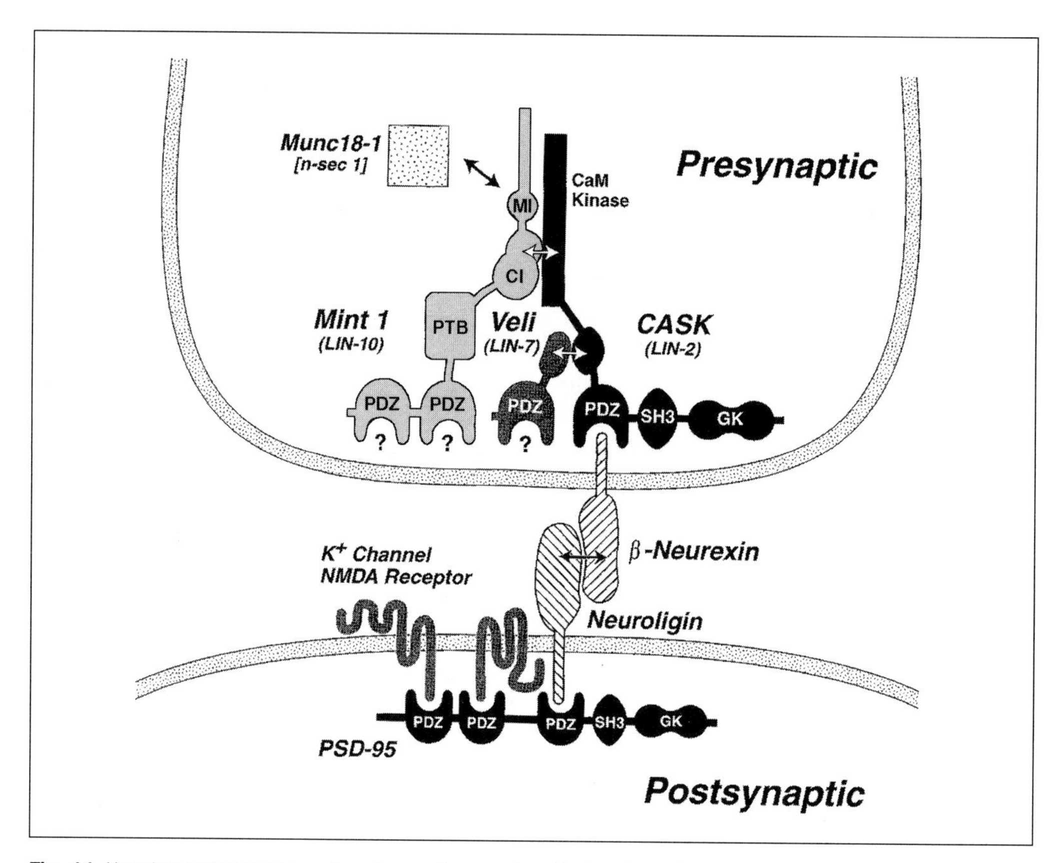

Fig. 11 Keeping active zones and postsynaptic receptor clusters in register. Both neuroligin and postsynaptic receptors bind PSD-95, which has multiple PDZ domains. Neuroligin binds the presynaptic membrane protein neurexin. Neurexins bind a tripartite complex rich in PDZ domains, which could cluster presynaptic channels, and capable of binding to a component of the exocytotic machinery, munc18-1. (From ref. 73, p. 780 with permission, © Cell Press.)

synapse-specific, e.g. synaptic laminin (80). The neuromuscular junctions have specialized integrins to bind these specialized molecules of the extracellular matrix (81). Integrins are also essential in central nervous system synapses since a learning mutation in *Drosophila*, *volado*, is created by a defective integrin (82), and integrin antagonists perturb synaptic function (83). Since extracellular matrix is absent in the central nervous system, integrins may be interacting heterotypically with CAMs.

Synaptic adhesion is mediated, therefore, by at least four classes of adhesion molecules, one of which, the asymmetric neurexin–neuroligin pair, may hold the transmitter release mechanisms and the transmitter detection mechanism in register.

4.4 Molecular organization of the synapses

The nerve terminal is a site of exocytosis, endocytosis, and adhesion. The three functions of the nerve terminal are not independent. The interdependence of the three functions is gradually being revealed as we learn more about the molecular organization of the synapse. While many nerve terminal-specific proteins may be required for exocytosis, endocytosis, and adhesion, others might help organize the architecture of the nerve terminal, organizing and linking domains of endocytosis, exocytosis, and adhesion.

The postsynaptic cluster of receptors is attached to the presynaptic active zone by ill-defined molecular mechanisms. To this fusogenic region of plasma membrane is attached the docked or readily releasable pool of synaptic vesicles. In addition to the readily releasable pool of docked vesicles in the vicinity of presynaptic calcium channels, there is a reserve pool of vesicles retained in close vicinity to the docked pool (Fig. 4). Two very large nerve terminal proteins, piccolo and bassoon, originally isolated using antibodies to the 'postsynaptic' density proteins, co-localize to active zones at mossy fibre terminals in the hippocampus (84, 85) and have been implicated in co-localizing reserve and readily releasable pools. The endocytosis machinery in the nerve terminal clusters at endocytotic zones or 'hotspots' in *Drosophila* neuromuscular junctions (86) and presumably elsewhere. A scaffolding protein that might localize the machinery to the hotspot is the dynamin-binding protein DAP160 (86). The number of such hot spots approximately equals the number of active zones, defined by clusters of postsynaptic receptors. Close examination suggests that the sites of exocytosis and endocytosis do not superimpose but are adjacent to each other (Roos and Kelly, in preparation). The concept of zones of exocytosis surrounded by zones of endocytosis is consistent with recent morphological studies of lamprey synapses (87).

If synapses have active zones of exocytosis, adjacent hotspots of endocytosis and adjacent adhesion sites (Fig. 12) then there must be linking proteins that hold them

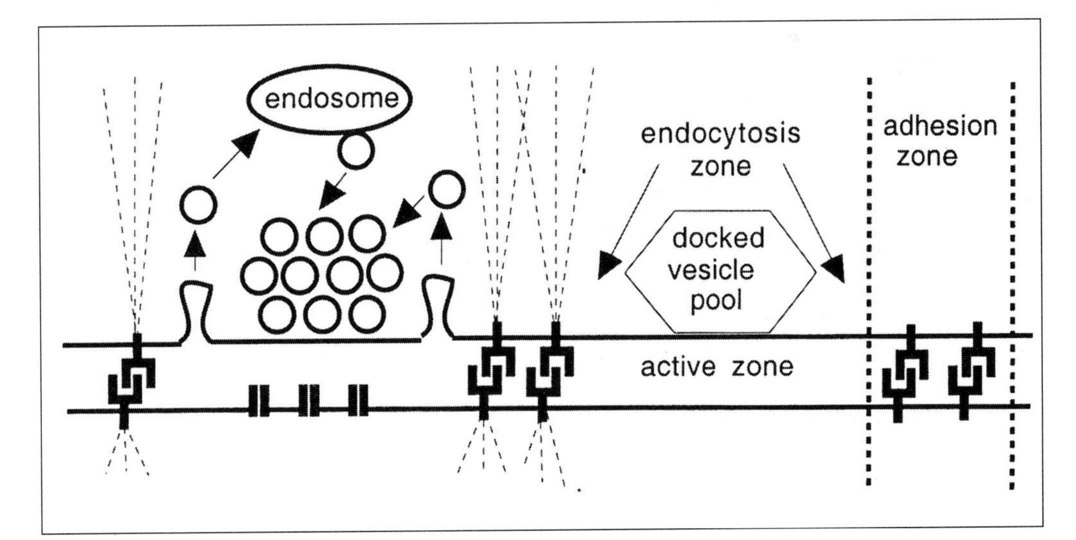

Fig. 12 The three domains of the nerve terminal, the exocytosis (active) zone, the endocytotic recycling zone, and the adhesion zone, are not random, but are organized as shown.

together. Candidates for such linking proteins have recently emerged. Syndapin is a protein that binds both dynamin and the actin-associated N-WASP protein (88). A human homologue of DAP160 may link elements of the endocytotic machinery, dynamin and epsin, to the fusion machinery (Sudhof, personal communication). It may be time to view the synapse as an integrated machine for fusing and recycling vesicle membrane efficiently at sites close to and perhaps surrounded by adhesion domains.

5. Development of the nerve terminal

Making a presynaptic nerve ending involves five steps, the extension of an axonal process, correct sorting of axonal components in the cell body, the generation of sites of calcium-dependent exocytosis in the axon, contact with postsynaptic elements, and modulation of immature synapses by electrical activity.

Extension of processes has been extensively studied in dissociated hippocampal neurons. Several neurites are extended simultaneously. The first to reach a threshold length becomes the axon (for review see ref. 89). The newly formed axon is the target for synaptic vesicle proteins. This implies that newly synthesized axonal membrane proteins leaving the Golgi complex are sorted away from other newly synthesized ones into vesicles for transport along axonal microtubules.

Little is known about the sorting process except that the information resides in the cytoplasmic tails of synaptic vesicle membrane proteins (90). The evidence suggests that the various components of the synaptic vesicle are transported down the axon in different organelles, implying that final vesicle assembly occurs at the nerve terminal (91, 92). Vesicles are transported down the axonal microtubules by the appropriate member of the kinesin protein family (93). The family seems to include dozens of members, with each kinesin being specific for the transport of a given organelle type (94).

A surprising and unexpected feature of synaptic development is that even in the absence of synaptic contact, the proteins of the nerve terminal accumulate at swellings or varicosities along the axon, which are a few microns apart. These varicosities contain the synaptic vesicle proteins, endocytotic, exocytotic, and adhesion machinery, and can show calcium-induced neurotransmitter release and vesicle recycling (4, 5, 95–97). In *Drosophila* larvae defective in muscle formation, motoneurons can have terminals with active zones, again without synaptic contact (98). The nature of the varicosities is obscure. Perhaps immature neurons recapitulate during development the structure of an evolutionarily more primitive form of neuron, similar to the non-contacting synapses described earlier (Section 2.2). Alternatively, the wiring of the brain may be set up so that the postsynaptic target cells select from a large population of immature nerve terminals, or the varicosities could be boluses of cytoplasm being transported to the nerve terminal.

One advantage of differentiating a nerve terminal before contact may be that it becomes primed to release signals that induce postsynaptic specializations such as clustering. An immature motor nerve terminal secretes a protein, agrin, that induces clustering of the postsynaptic receptors via a muscle-specific tyrosine kinase (MuSk) (99). The motor nerve terminal also releases a neuregulin that, via a postsynaptic receptor, controls transcription in muscle nuclei near the motor nerve terminal (100). It is likely that a similar mechanism induces most postsynaptic specialization. In response to signals from the nerve, the target can release signals that regulate the differentiation of the presynaptic nerve terminal. For example, at the neuromuscular junction, the muscle secretes synaptic laminin, which is necessary for proper clustering of synaptic vesicles around the active zone (101).

Contact with the postsynaptic target also induces changes in the calcium sensitivity of the exocytotic machinery, making it more dependent on calcium influx. Before synapse formation, N-type calcium channels are distributed diffusely in the plasma membrane. Synaptic contact is associated with clustering of the calcium channels (102). This correlates with the appearance of a regulatory machinery that can only trigger synaptic vesicle exocytosis at very high levels of cytoplasmic calcium (103).

The last stage of synaptic development occurs after the synaptic contact has been established. The previous four steps in synapse formation have been activity-independent. In the last step, the pre- and postsynaptic properties are fine-tuned to match each other by a mechanism that depends on synaptic activity. The matching process can be followed with great precision in several developmental systems, including the larval *Drosophila* neuromuscular junctions (104). In this system, little of the modification is in the size of the nerve terminal. Instead, reduction in post-synaptic receptors enhances the amount of neurotransmitter released per synaptic bouton, and vice versa (Fig. 13). Such malleability is essential since during development the volume of the larval muscle grows 50-fold and the synaptic input needs to compensate for such growth. Fine-tuning as a result of electrical activity is found for almost all synapses, including those in the visual system (105, 106).

Understanding the molecular basis of synapse regulation by synaptic activity is one of the key goals of the molecular neurobiologist. In the case of *Drosophila*, the cell adhesion molecule, fasciclin II, brings about activity-dependent maturation of synapses. Thus activity-dependent maturation may involve the integrin/CAM class of adhesion, whereas early development of the synapse may need cadherins.

6. Plasticity of the synapse

The urge to decipher the rules of neurotransmitter release, synaptic adhesion, and synaptic development is driven in large part by our need to understand learning, or more precisely, long-term synaptic plasticity, in the brain. For many years it has been known that protein synthesis is required to lay down long-term memory traces (107). Protein synthesis appears to be triggered by an elevation of cAMP, leading to the phosphorylation of the transcriptional regulator CREB (cAMP response element binding protein) (108). Enhanced protein synthesis in the cell body is only part of the story, however. Learning is a synaptic, not a cellular event. It is possible to induce long-term potentiation in some synaptic connections of a hippocampal cell without

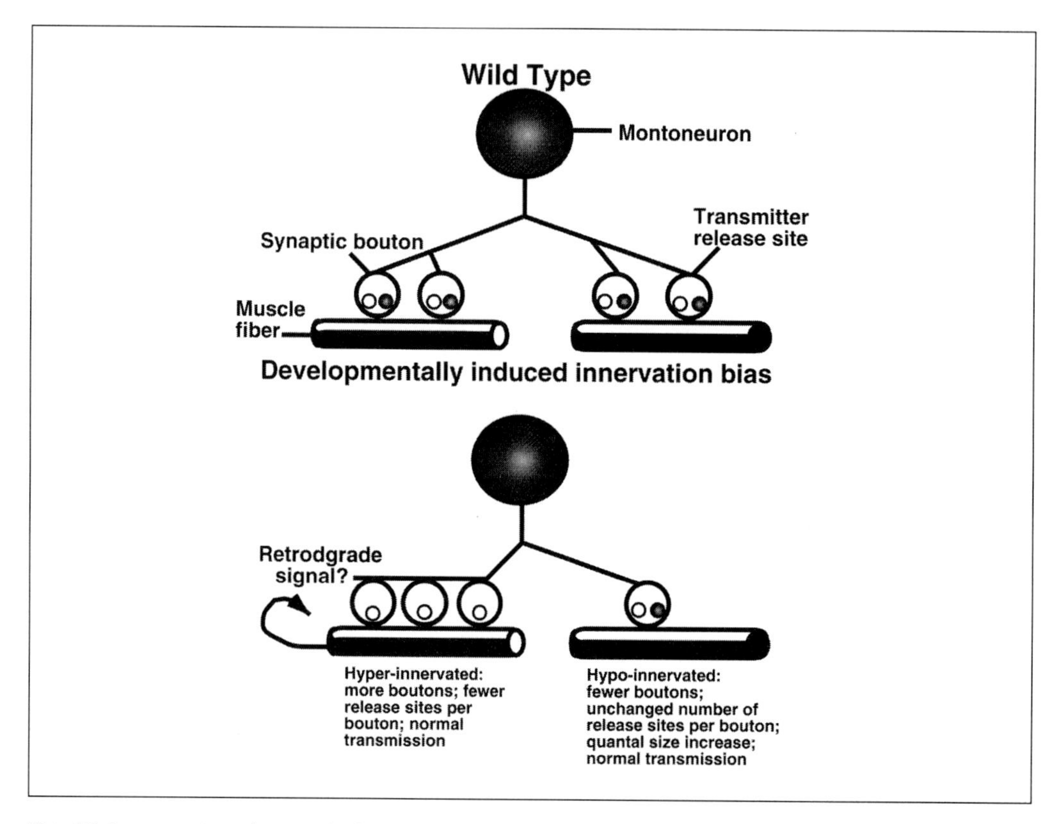

Fig. 13 Compensatory changes in *Drosophila* neuromuscular junctions. A single motoneuron innervates two muscle fibres (top). If the CAM-family molecule fasciclin II is overexpressed in one muscle (bottom) more synapses are made, but the amount released per synapse is reduced. (Adapted from Landmesser. (1998) *Curr. Biol.*, **8**, R565.)

inducing it in others (Fig. 14). Somehow the newly synthesized proteins need to be directed to the synapse that has been 'educated' and not to the rest. Frey and Morris (109) suggested that activity in synapses marks them in some way that ensures that the newly synthesized proteins are targeted to the correct synapse. The cell biology of this remarkable process needs to be worked out. The strongest evidence to date implicates the activation of local protein synthesis in the nerve terminal, at least in Aplysia (110).

One of the most exciting insights comes from analysis of a learning mutation in *Drosophila*. In this mutant, *volado*, the defective gene product is normally expressed in the mushroom body, the site of neural plasticity (82). Learning mutations can be rescued transiently by expressing normal Volado protein, an integrin, suggesting that subtle changes in synaptic adhesivity might underlie learning. Such a conclusion is also suggested by the involvement of neuronal CAMs (NCAMs), and integrins in long-term potentiation (111).

A reasonable expectation is that in the next decade the basic mechanisms of long-

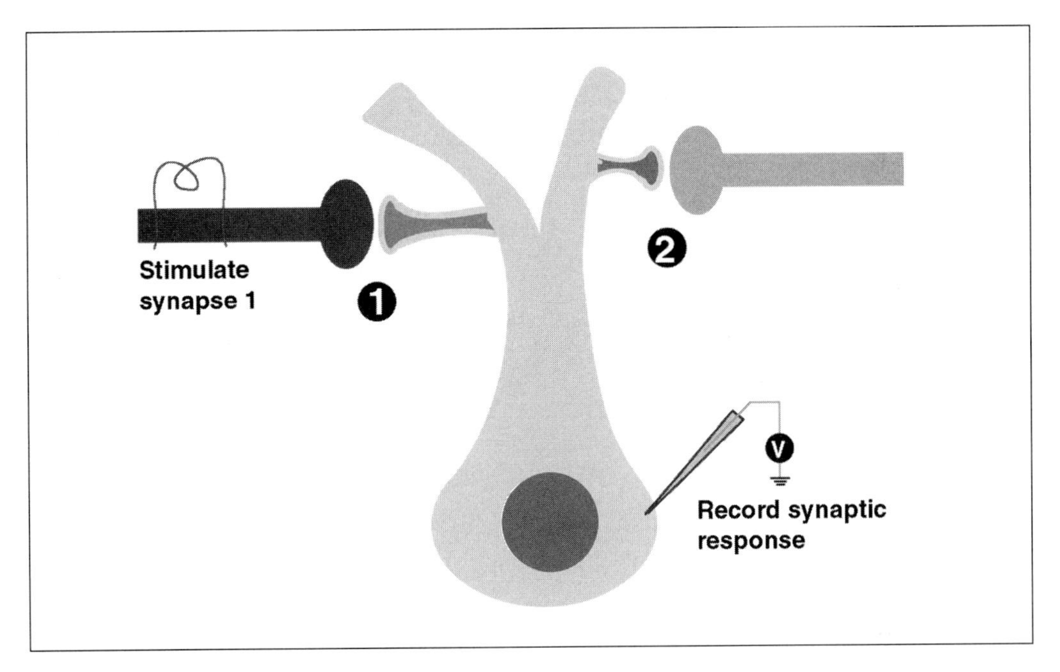

Fig. 14 Synapses learn not neurons. When synapse 1 is exposed to repetitive stimulation to induce long-term potentiation, synapse 2 is unaffected. Potentiation requires protein synthesis in the cell body and 'marking' the stimulated synapse to give selective protein deposition or retention. (Adapted from Bear. (1997) *Nature*, **385**, 481. Reprinted with permission of Macmillan Magazines Ltd.)

term synaptic plasticity will be revealed to us. One major reason for this optimism is the powerful similarities that have appeared between activity-dependent synaptic development and learning. The CAM molecule, fascicilin II, is essential in *Drosophila* synaptic maturation and NCAM is required for plasticity. Trophic factors such as BDNF (brain-derived trophic factor) and other receptors such as the NMDA channel have been implicated in both development and plasticity. A second reason for optimism is the accessibility of systems where long-term synaptic modifications are induced by exposure to drugs rather than behavioural manipulations. Finally, the technologies of molecular biology, sequencing, transgenics, and DNA arrays are increasingly the handmaidens of the neurobiologist.

References

1. Gershon, M. D., Schwartz, J. H., and Kandel, E. R. (1985) Morphology of chemical synapses and patterns of interconnection. In *Principles of neural science* (2nd edn) (ed. E. R. Kandel and J. H. Schwartz), p. 979. Elsevier.
2. Thomas-Reetz, A. C. and De Camilli, P. (1994) A role for synaptic vesicles in non-neuronal cells: clues form pancreatic beta cells and from chromaffin cells. *FASEB J.*, **8**, 209.
3. Pothos, E. N., Davila, V., and Sulzer, D. (1998) Presynaptic recording of quanta from midbrain dopamine neurons and modulation of the quantal size. *J. Neurosci.*, **18**, 4106.

4. Dai, Z. and Peng, H. B. (1996) Dynamics of synaptic vesicles in cultured spinal cord neurons in relationship to synaptogenesis. *Mol. Cell. Neurosci.*, **7**, 443.
5. Kraszewski, K., Mundigl, O., Daniell, L., Verderio, C., Matteoli, M., and De Camilli, P. (1995) Synaptic vesicle dynamics in living cultured hippocampal neurons visualized with CY3-conjugated antibodies directed against the lumenal domain of synaptotagmin. *J. Neurosci.*, **15**, 4328.
6. Zerby, S. E. and Ewing, A. G. (1996) Electrochemical monitoring of individual exocytotic events from the varicosities of differentiated PC12 cells. *Brain Res.*, **712**, 1.
7. Walmsley, B., Alvarez, F. J., and Fyffe, R. E. (1998) Diversity of structure and function at mammalian central synapses. *Trends Neurosci.*, **21**, 81.
8. Pfenninger, K., Sandri, C., Akert, K., and Eugster, C. H. (1996) Contribution to the problem of structural organization of the presynaptic area. *Brain Res.*, **12**, 10.
9. Kennedy, M. B. (1997) The postsynaptic density at glutamatergic synapses. *Trends Neurosci.*, **20**, 264.
10. Zucker, R. S. (1996) Exocytosis: a molecular and physiological perspective. *Neuron*, **17**, 1049.
11. Pieribone, V. A., Shupliakov, O., Brodin, L., Hilfiker-Rothenfluh, S., Czernik, A. J., and Greengard, P. (1995) Distinct pools of synaptic vesicles in neurotransmitter release. *Nature*, **375**, 493.
12. Katz, B. (1969) *The release of neural transmitter substances*, p. 60. Charles C. Thomas, Publisher.
13. Neher, E. (1998) Vesicle pools and Ca^{2+} microdomains: new tools for understanding their roles in neurotransmitter release. *Neuron*, **20**, 389.
14. Lindau, M. and Almers, W. (1995) Structure and function of fusion pores in exocytosis and ectoplasmic membrane fusion. *Curr. Opin. Cell Biol.*, **7**, 509.
15. Monck, J. R. and Fernandez, J. M. (1996) The fusion pore and mechanisms of biological membrane fusion. *Curr. Opin. Cell Biol.*, **8**, 524.
16. Yeagle, P. L. (1997) Membrane fusion intermediates. *Curr. Top. Membr.*, **44**, 375.
17. Leszczyszyn, D. J., Jankowski, J. A., Viveros, O. H., Diliberto, E. J., Near, J. A., and Wightman, R. M. (1991) Secretion of catcholamines from individual adrenal medullary chromaffin cells. *J. Neurochem.*, **56**, 1866.
18. Chow, R. H., von Rueden, L., and Neher, E. (1992) Delay in vesicle fusion revealed by electrochemical monitoring of single secretory events in adrenal chromaffin cells. *Nature*, **356**, 60.
19. Fesce, R., Grohovaz, F., Valtorta, F., and Meldolesi, J. (1994) Neurotransmitter release: fusion or 'kiss-and-run'? *Trends Cell Biol.*, **4**, 1.
20. Kelly, R. B., Deutsch, J. W., Carlson, S. S., and Wagner, J. A. (1979) Biochemistry of neurotransmitter release. *Annu. Rev. Neurosci.*, **2**, 399.
21. Wang, L.-Y. and Kaczmarek, L. K. (1998) High-frequency firing helps replenish the readily releasable pool of synaptic vesicles. *Nature*, **394**, 384.
22. Stevens, C. F. and Tsujimoto, T. (1995) Estimates for the pool size of releasable quanta at a single central synapse and for the time required to refill the pool. *Proc. Natl. Acad. Sci. USA*, **92**, 846.
23. Ryan, T. A., Smith, S. J., and Reuter, H. (1996) The timing of synaptic vesicle endocytosis. *Proc. Natl. Acad. Sci. USA*, **93**, 5567.
24. Carlson, S. S., Wagner, J. A., and Kelly, R. B. (1978) Purification of synaptic vesicles from elasmobranch electric organ and the use of biophysical criteria to demonstrate purity. *Biochemistry*, **17**, 1188.

25. Huttner, W. B., Schiebler, W., Greengard, P., and De Camilli, P. (1983) Synapsin I (protein I), a nerve terminal-specific phosphoprotein. III. Its association with synaptic vesicles studied in a highly purified synaptic vesicle preparation. *J. Cell Biol.*, **96**, 1374.
26. Bennett, M. K., Calakos, N., and Scheller, R. H. (1992) Syntaxin: a synaptic protein implicated in docking of synaptic vesicles at presynaptic active zones. *Science*, **257**, 255.
27. Sollner, T., Bennett, M. K., Whiteheart, S. W., Scheller, R. H., and Rothman, J. E. (1993) A protein assembly-disassembly pathway in vitro that may correspond to sequential steps of synaptic vesicle docking, activation and fusion. *Cell*, **75**, 409.
28. Schmid, S. L., McNiven, M. A., and De Camilli, P. (1998) Dynamin and its partners: a progress report. *Curr. Opin. Cell Biol.*, **10**, 504.
29. Sudhof, T. C. (1995) The synaptic vesicle cycle: a cascade of protein-protein interactions. *Nature*, **375**, 645.
30. Wu, M. N. and Bellen, H. J. (1997) Genetic dissection of synaptic transmission in *Drosophila*. *Curr. Opin. Neurobiol.*, **7**, 624.
31. Bennett, M. K. and Scheller, R. H. (1993) The molecular machinery for secretion is conserved from yeast to neurons. *Proc. Natl. Acad. Sci. USA*, **90**, 2559.
32. Weber, T., Zemelman, B. V., McNew, J. A., Westermann, B., Gmachl, M., Parlati, F., *et al.* (1998) SNAREpins: minimal machinery for membrane fusion. *Cell*, **92**, 759.
33. Sutton, R. B., Fasshauer, D., Jahn, R., and Brunger, A. T. (1998) Crystal structure of a SNARE complex involved in synaptic exocytosis at 2.4 Å resolution. *Nature*, **395**, 347.
34. Geppert, M. and Sudhof, T. C. (1998) RAB3 and synaptotagmin: the yin and yang of synaptic membrane fusion. *Annu. Rev. Neurosci.*, **21**, 75.
35. Hsu, S. C., Hazuka, C. D., Roth, R., Foletti, D. L., Heuser, J., and Scheller, R. H. (1998) Subunit composition, protein interactions, and structures of the mammalian brain sec6/8 complex and septin filaments. *Neuron*, **20**, 1111.
36. Schiavo, G., Gu, Q. M., Prestwich, G. D., Sollner, T. H., and Rothman, J. E. (1997) Calcium-dependent switching of the specificity of phosphoinositide binding to synaptotagmin. *Proc. Natl. Acad. Sci. USA*, **93**, 13327.
37. Pevsner, J. (1996) The role of Sec1p-related proteins in vesicle trafficking in the nerve terminal. *J. Neurosci. Res.*, **45**, 89.
38. Tandon, A., Bannykh, S., Kowalchyk, J. A., Banerjee, A., Martin, T. F., and Balch, W. E. (1998) Differential regulation of exocytosis by calcium and CAPS in semi-intact synaptosomes. *Neuron*, **21**, 147.
39. Martin, T. F., Loyet, K. M., Barry, V. A., and Kowalchyk, J. A. (1997) The role of PtdIns(4,5)P2 in exocytotic membrane fusion. *Biochem. Soc. Trans.*, **25**, 1137.
40. Kelly, R. B. (1995) Synaptotagmin is just a calcium sensor. *Curr. Biol.*, **5**, 257.
41. Holtzman, E. (1977) The origin and fate of secretory packages, especially synaptic vesicles. *Neuroscience*, **2**, 327.
42. Heuser, J. (1989) The role of coated vesicles in recycling of synaptic vesicle membrane. *Cell Biol. Int. Rep.*, **13**, 1063.
43. Ceccarelli, B. and Hurlbut, W. P. (1980) Vesicle hypothesis of the release of quanta of acetylcholine. *Physiol. Rev.*, **60**, 396.
44. Betz, W. J. and Angleson, J. K. (1998) The synaptic vesicle cycle. *Annu. Rev. Physiol.*, **60**, 347.
45. Ryan, T. A., Reuter, H., and Smith, S. J. (1996) Optical detection of a quantal presynaptic membrane turnover. *Nature*, **388**, 478.
46. Murthy, V. N. and Stevens, C. F. (1998) Synaptic vesicles retain their identity through the endocytic cycle. *Nature*, **392**, 497.

47. Matthews, G. (1996) Synaptic exocytosis and endocytosis: capacitance measurements. *Curr. Opin. Neurobiol.*, **6**, 358.
48. Takei, K., Mundigl, O., Daniell, L., and De Camilli, P. (1996) The synaptic vesicle cycle: a single vesicle budding step involving clathrin and dynamin. *J. Cell Biol.*, **133**, 1237.
49. Gonzalez-Gaitan, M. and Jackle, H. (1997) Role of *Drosophila* alpha-adaptin in presynaptic vesicle recycling. *Cell*, **88**, 767.
50. De Camilli, P. and Takei, K. (1996) Molecular mechanisms in synaptic vesicle endocytosis and recycling. *Neuron*, **16**, 481.
51. Tan, P. K., Waites, C., Liu, Y., Krantz, K. E., and Edwards, R. H. (1998) A leucine-based motif mediates the endocytosis of vesicular monoamine and acetylcholine transporters. *J. Biol. Chem.*, **273**, 17351.
52. Shi, G., Faundez, V., Roos, J., Dell'Angelica, E., and Kelly, R. B. (1998) Neuroendocrine synaptic vesicles are formed *in vitro* by both clatrin-dependent and clathrin-independent pathways. *J. Cell Biol.*, **143**, 947.
53. Shupliakov, O., Low, P., Grabs, D., Gad, H., Chen, H., David, C., *et al.* (1997) Synaptic vesicle endocytosis impaired by disruption of dynamin-SH3 domain interactions. *Science*, **276**, 259.
54. Lichtenstein, Y., Desnos, C., Faundez, V., Kelly, R. B., and Clift-O'Grady, L. (1998) Vesiculation and sorting from PC12-derived endosomes *in vitro*. *Proc. Natl. Acad. Sci. USA*, **95**, 11223.
55. Faundez, V., Horng, J. T., and Kelly, R. B. (1998) A function for the AP3 coat complex in synaptic vesicle formation from endosomes. *Cell*, **93**, 423.
56. Salem, N., Faundez, V., Horng, J.-T., and Kelly, R. B. (1998) A v-SNARE participates in synaptic vesicle formation mediated by the AP3 adaptor complex. *Nature Neurosci.*, **1**, 551.
57. Kantheti, P., Qiao, X., Diaz, M. E., Peden, A. A., Meyer, G. E., Carskadon, S. L., *et al.* (1998) Mutation in AP-3 delta in the mocha mouse links endosomal transport to storage deficiency in platelets, melanosomes, and synaptic vesicles. *Neuron*, **21**, 111.
58. Noebels, J. L. and Sidman, R. L. (1989) Persistent hypersynchronization of neocortical neurons in the mocha mutant of mouse. *J. Neurogenet.*, **6**, 53.
59. Wei, M. L., Bonzelius, F., Scully, R. M., Kelly, R. B., and Herman, G. A. (1998) GLUT4 and transferrin receptor are differentially sorted along the endocytic pathway in CHO cells. *J. Cell Biol.*, **140**, 565.
60. Cao, T. T., Mays, R. W., and von Zastrow, M. (1998) Regulated endocytosis of G-protein-coupled receptors by a biochemically and functionally distinct subpopulation of clathrin-coated pits. *J. Biol. Chem.*, **273**, 24592.
61. Bonzelius, F., Herman, G. A., Cardone, M. H., Mostov, K. E., and Kelly, R. B. (1994) The polymeric immunoglobulin receptor accumulates in specialized endosomes but not synaptic vesicles within the neurites of transfected neuroendocrine PC12 cells. *J. Cell Biol.*, **127**, 1603.
62. Maycox, P. R., Link, E., Reetz, A., Morris, S. A., and Jahn, R. (1992) Clathrin-coated vesicles in nervous tissue are involved primarily in synaptic vesicle recycling. *J. Cell Biol.*, **118**, 1379.
63. Schmidt, A., Hannah, M. J., and Huttner, W. B. (1997) Synaptic-like microvesicles of neuroendocrine cells originate from a novel compartment that is continuous with the plasma membrane and devoid of transferrin receptor. *J. Cell Biol.*, **137**, 445.
64. Klingauf, J., Kavalali, E. T., and Tsien, R. W. (1998) Kinetics and regulation of fast endocytosis at hippocampal synapses. *Nature*, **394**, 581.

65. Henkel, A. W. and Betz, W. J. (1995) Staurosporine blocks evoked release of FM1-43 but not acetylcholine from frog motor nerve terminals. *J. Neurosci.*, **15**, 8246.
66. Miller, T. M. and Heuser, J. E. (1984) Endocytosis of synaptic vesicle membrane at the frog neuromuscular junction. *J. Cell Biol.*, **98**, 685.
67. Koenig, J. H. and Ikeda, K. (1996) Synaptic vesicles have two distinct recycyling pathways. *J. Cell Biol.*, **135**, 797.
68. Kuromi, H. and Kidokoro, Y. (1998) Two distinct pools of synaptic vesicles in single presynaptic boutons in a temperature-sensitive *Drosophila* mutant, shibire. *Neuron*, **20**, 917.
69. Uemura, T. (1998) The cadherin superfamily at the synapse: more members, more missions. *Cell*, **93**, 1095.
70. Uchida, N., Honjo, Y., Johnson, K. R., Wheelock, M. J., and Takeichi, M. (1996) The catenin/cadherin adhesion system is localized in synaptic junctions bordering transmitter release zones. *J. Cell Biol.*, **135**, 767.
71. Baumgartner, S., Littleton, J. T., Broadie, K., Bhat, M. A., Harbecke, R., Lengyel, J. A., *et al.* (1996) A *Drosophila* neurexin is required for septate junction and blood-nerve barrier formation and function. *Cell*, **87**, 1059.
72. Missler, M. and Sudhof, T. C. (1998) Neurexins: three genes and 1001 products. *Trends Genet.*, **14**, 20.
73. Butz, S., Okamoto, M., and Sudhof, T. C. (1998) A tripartite protein complex with the potential to couple synaptic vesicle exocytosis to cell adhesion in brain. *Cell*, **94**, 773.
74. Craven, S. E. and Bredt, D. S. (1998) PDZ proteins organize synaptic signaling pathways. *Cell*, **93**, 495.
75. Okamoto, M. and Sudhof, T. C. (1997) Mints, Munc18-interacting proteins in synaptic vesicle exocytosis. *J. Biol. Chem.*, **272**, 31459.
76. Bezprozvanny, I., Scheller, R. H., and Tsien, R. W. (1995) Functional impact of syntaxin on gating of N-type and Q-type calcium channels. *Nature*, **378**, 623.
77. Sheng, Z. H., Rettig, J., Cook, T., and Catterall, W. A. (1996) Calcium-dependent interaction of N-type calcium channels with the synaptic core complex. *Nature*, **379**, 451.
78. Petrenko, A. G., Perin, M. S., Davletov, B. A., Ushkaryov, Y. A., Geppert, M., and Sudhof, T. C. (1991) Binding of synaptotagmin to the alpha-latrotoxin receptor implicates both in synaptic vesicle exocytosis. *Nature*, **353**, 65.
79. Goodman, C. S., Davis, G. W., and Zito, K. (1997) The many faces of fasciclin II: genetic analysis reveals multiple roles for a cell adhesion molecule during the generation of neuronal specificity. *Cold Spring Harbor Symp. Quant. Biol.*, **62**, 479.
80. Martin, P. T., Ettinger, A. J., and Sanes, J. R. (1995) A synaptic localization domain in the synaptic cleft protein laminin beta 2 (s-laminin). *Science*, **269**, 413.
81. Martin, P. T., Kaufman, S. J., Kramer, R. H., and Sanes, J. R. (1996) Synaptic integrins in developing, adult and mutant muscle: selective association of alpha1, alpha7A and alpha7B integrins with the neuromuscular junction. *Dev. Biol.*, **174**, 125.
82. Grotewiel, M. S., Beck, C. D., Wu, K. H., Zhu, X. R., and Davis, R. L. (1998) Integrin-mediated short-term memory in *Drosophila*. *Nature*, **391**, 455.
83. Staubli, U., Chun, D., and Lynch, G. (1998) Time-dependent reversal of long-term potentiation by an integrin antagonist. *J. Neurosci.*, **18**, 3460.
84. Cases-Langhoff, C., Voss, B., Garner, A. M., Appeltauer, U., Takei, K., Kindler, S., *et al.* (1996) Piccolo, a novel 420 kDa protein associated wtih the presynaptic cytomatrix. *Eur. J. Cell Biol.*, **69**, 214.

85. Tom Dieck, S., Sanmarti-Vila, L., Langnaese, K., Richter, K., Kindler, S., Soyke, A., *et al.* (1998) Bassoon, a novel zinc-finger CAG/glutamine-repeat protein selectively localized at the active zone of presynaptic nerve terminals. *J. Cell Biol.*, **142**, 499.
86. Roos, J. and Kelly, R. B. (1998) Dap160, a neural-specific Eps15 homology and multiple SH3 domain-containing protein that interacts with *Drosophila* dynamin. *J. Biol. Chem.*, **273**, 19108.
87. Gad, H., Low, P., Zotova, E., Brodin, L., and Shupliakov, O. (1998) Dissociation between Ca^{2+}-triggered synaptic vesicle exocytosis and clathrin-mediated endocytosis at a central synapse. *Neuron*, **21**, 607.
88. Qualmann, B., Roos, J., DiGregorio, P. J., and Kelly, R. B. (1999) Syndapin I, a synaptic dynamin-binding protein that associates with the neural Wiskott-Aldrich-syndrome protein. *Mol. Biol. Cell*, **10**, 501.
89. Kelly, R. B. and Grote, E. (1993) Protein targeting in the neuron. *Annu. Rev. Neurosci.*, **16**, 95.
90. West, A. E., Provoda, C., Neve, R. L., and Buckley, K. M. (1998) Protein targeting in neurons and endocrine cells. *Adv. Pharmacol.*, **42**, 247.
91. Okada, Y., Yamazaki, H., Sekine-Aizawa, Y., and Hirokawa, N. (1995) The neuron-specific kinesin superfamily protein KIF1A is a unique monomeric motor for anterograde axonal transport of synaptic vesicle precursors. *Cell*, **81**, 769.
92. Mundigl, O., Matteoli, M., Daniell, L., Thomas-Reetz, A., Metcalf, A., Jahn, R., *et al.* (1993) Synaptic vesicle proteins and early endosomes in cultured hippocampal neurons: differential effects of brefeldin A in axon and dendrites. *J. Cell Biol.*, **122**, 1207.
93. Hirokawa, N. (1998) Kinesin and dynein superfamily proteins and the mechanism of organelle transport. *Science*, **279**, 519.
94. Santama, N., Krijnse-Locker, J., Griffiths, G., Noda, Y., Hirokawa, N., and Dotti, C. G. (1998). KIF2beta, a new kinesin superfamily protein in non-neuronal cells, is associated with lysosomes and may be implicated in their centrifugal translocation. *EMBO J.*, **17**, 5855.
95. Buchanan, J., Sun, Y., and Poo, M. (1989) Studies of nerve-muscle interactions in *Xenopus* cell culture: fine structure of early functional contacts. *J. Neurosci.*, **9**, 1540.
96. Benson, D. L. and Tanaka, H. (1998) N-cadherin redistribution during synaptogenesis in hippocampal neurons. *J. Neurosci.*, **18**, 6892.
97. Matteoli, M., Takei, K., Perin, M. S., Sudhof, T. C., and De Camilli, P. (1992) Exo-endocytotic recycling of synaptic vesicles in developing processes of cultured hippocampal neurons. *J. Cell Biol.*, **117**, 849.
98. Prokop, A., Landgraf, M., Rushton, E., Broadie, K., and Bate, M. (1996) Presynaptic development at the *Drosophila* neuromuscular junction: assembly and localization of presynaptic active zones. *Neuron*, **17**, 617.
99. Glass, D. J. and Yancopoulos, G. D. (1997) Sequential roles of agrin, MuSK and rapsyn during neuromuscular junction formation. *Curr. Opin. Neurobiol.*, **7**, 379.
100. Fischbach, G. D. and Rosen, K. M. (1997) ARIA: a neuromuscular junction neuregulin. *Annu. Rev. Neurosci.*, **20**, 429.
101. Noakes, P. G., Gautam, M., Mudd, J., Sanes, J. R., and Merlie, J. P. (1995) Aberrant differentiation of neuromuscular junctions in mice lacking s-laminin/laminin beta 2. *Nature*, **374**, 258.
102. Bahls, F. H., Lartius, R., Trudeau, L. E., Doyle, R. T., Fang, Y., Witcher, D., *et al.* (1998) Contact-dependent regulation of N-type calcium channel subunits during synaptogenesis. *J. Neurobiol.*, **35**, 198.

103. Coco, S., Verderio, C., De Camilli, P., and Matteoli, M. (1998) Calcium dependence of synaptic vesicle recycling before and after synaptogenesis. *J. Neurochem.*, **71**, 1987.
104. Davis, G. W. and Goodman, C. S. (1998) Synapse-specific control of synaptic efficacy at the terminals of a single neuron. *Nature*, **392**, 82.
105. Shatz, C. J. (1996) Emergence of order in visual system development. *Proc. Natl. Acad. Sci. USA*, **93**, 602.
106. Crair, M. C., Gillespie, D. C., and Stryker, M. P. (1998) The role of visual experience in the development of columns in cat visual cortex. *Science*, **279**, 566.
107. Milner, B., Squire, L. R., and Kandel, E. R. (1998) Cognitive neuroscience and the study of memory. *Neuron*, **20**, 445.
108. Silva, A. J., Kogan, J. H., Frankland, P. W., and Kida, S. (1998) CREB and memory. *Annu. Rev. Neurosci.*, **21**, 127.
109. Frey, U. and Morris, R. G. (1997) Synaptic taggin and long-term potentiation. *Nature*, **385**, 533.
110. Martin, K. C., Casadio, A., Zhu, H. E. Y., Rose, J. C., Chen, M., Bailey, C. H., *et al.* (1997) Synapse-specific, long-term facilitation of aplysia sensory to motor synapses: a function for local protein synthesis in memory storage. *Cell*, **91**, 927.
111. Hagler, D. J., Jr. and Goda, Y. (1998) Synaptic adhesion: the building blocks of memory? *Neuron*, **20**, 1059.
112 Kuffler, S. W., Nicholls, J. G., and Martin, A. R. (1984) *From neuron to brain*. pp. 651. Sinauer Assoc. Inc., publ. Sunderland, MA.

2 Techniques for elucidating the secretory process: from Ca^{2+} entry to exocytosis

THOMAS E. FISHER and ANDRES F. OBERHAUSER

1. Introduction

Technical innovation has always been central to progress in the study of neurotransmission. The first demonstration of chemical neurotransmission was in fact made possible by a novel and ingenious experimental approach. In 1921 Otto Loewi rushed to his laboratory in the middle of the night to perform an experiment that had occurred to him in a dream. It had been known for some time that stimulation of the vagal nerve causes slowing of the frog heart. Loewi's insight was to stimulate the vagus of a perfused frog heart and then add the perfusate to the solution bathing a second frog heart in a separate vessel. This caused the slowing of the second heart, demonstrating that the vagal action was caused by release of a chemical substance, later shown to be acetylcholine. From this simple, elegant beginning the techniques for studying neurotransmission have developed to the point where it is now possible to measure exocytosis at the level of a single neuron, with the precision to detect the fusion of a single synaptic vesicle with the neuronal membrane.

In this chapter we introduce the reader to four of the most important and widely used techniques for the investigation of exocytosis in live cells. Our discussions are intended to describe basic principles in sufficient detail to allow the reader to understand the application of these methods found in the scientific literature, and their limitations. We will also alert the reader to some promising recent advances in each of the fields. The techniques, which are illustrated in Fig. 1, are:

(a) Electrical monitoring of the change in cell capacitance caused by fusion of vesicles with the plasma membrane.

(b) Electrochemical detection of released substances.

(c) Visualization of the influx of Ca^{2+} that evokes secretion.

(d) Specific labelling of vesicles to follow their progress through the exocytotic pathway.

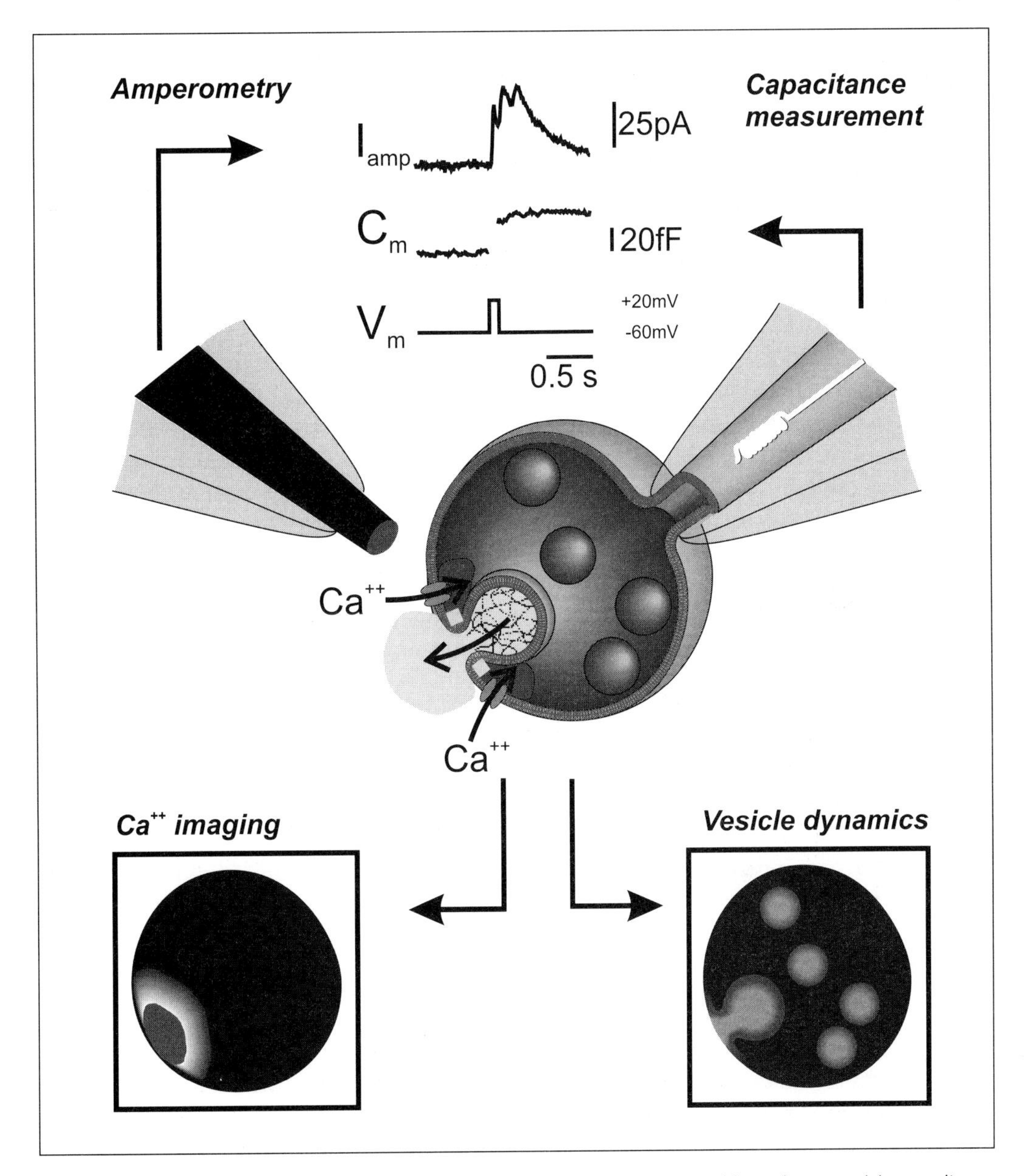

Fig. 1 Techniques for monitoring exocytosis. The techniques discussed in this review are: (a) capacitance measurements, (b) amperometry, (c) Ca^{2+} imaging, and (d) vesicle dynamics. A brief depolarization of a patch-clamped cell (V_M trace) can cause Ca^{2+} influx near release sites (lower left), release of oxidizable substances measured as a spike on the amperometric trace (upper left), an increase in total membrane area detected as a jump in the capacitance trace (upper right), and a transfer of stained vesicular membranes to the cell surface (lower right) see colour plate section between pages 46 and 47.

2. Membrane capacitance measurements with the patch-clamp technique

Membrane capacitance measurements take advantage of the fact that the exocytotic fusion of a vesicle with the plasma membrane causes an increase in the surface area of the cell. This change in area causes an increase in the electrical capacitance that can be measured by observing the response of a cell to an applied voltage. This concept was first demonstrated more than 60 years ago by K. G. Cole when he measured a change in the total membrane capacitance of a population of sea urchin oocytes in response to fertilization (1). The development of patch-clamp recording techniques (2) has now made it possible to monitor changes in cell capacitance continuously with a resolution precise enough to detect the fusion of single granules.

In order to describe membrane capacitance measurements it is necessary first to understand the concept of electrical capacitance. A capacitor can be described as two electrically conductive media separated by a thin insulating layer. Since charges cannot pass through the insulator, a capacitor behaves very differently to an imposed voltage change than does a resistor. When a voltage is applied to a resistor a current flows through it continuously according to Ohm's law (Fig. 2A). When there is a change in the potential applied to a capacitor, however, charge pushes up against one side of the non-conducting layer causing displacement of charge from the other side. This displacement of charge, which is equivalent to a current and may be measured as such, occurs until a new equilibrium is formed between charges on the two sides. At that point there is no charge movement since the insulator prevents charge transfer between the two sides. The 'capacitive current' that occurs whenever there is a step change in voltage across a capacitor is therefore a transient current (see Fig. 2A). If the voltage across the capacitor is then returned to its original value, the charges return to their original distribution resulting in an equal but opposite current. The size of the current evoked depends on the size of the membrane capacitance (C_M), which is defined by the following equation:

$$C_M = \varepsilon \frac{A}{d}$$

where ε is the dielectric constant of the insulator, A is the area of the insulator, and d is the distance between the two conducting media. Capacitance is measured in farads, which may be thought of as a measure of the capacity to build-up, store, and release charges.

The phospholipid bilayers of cellular plasma membranes form thin insulating layers (ignoring for the moment the conductivity of ion channels) between intracellular and extracellular solutions. As such, they act as excellent capacitors and store charge in response to applied voltages. Since vesicular and plasma membranes have similar dielectric constants and thickness, the addition of new membrane to the cell surface by exocytotic fusion can be directly measured as a change in total capacitance. For example, a cell with a diameter of 10 μm will have a membrane capacitance of

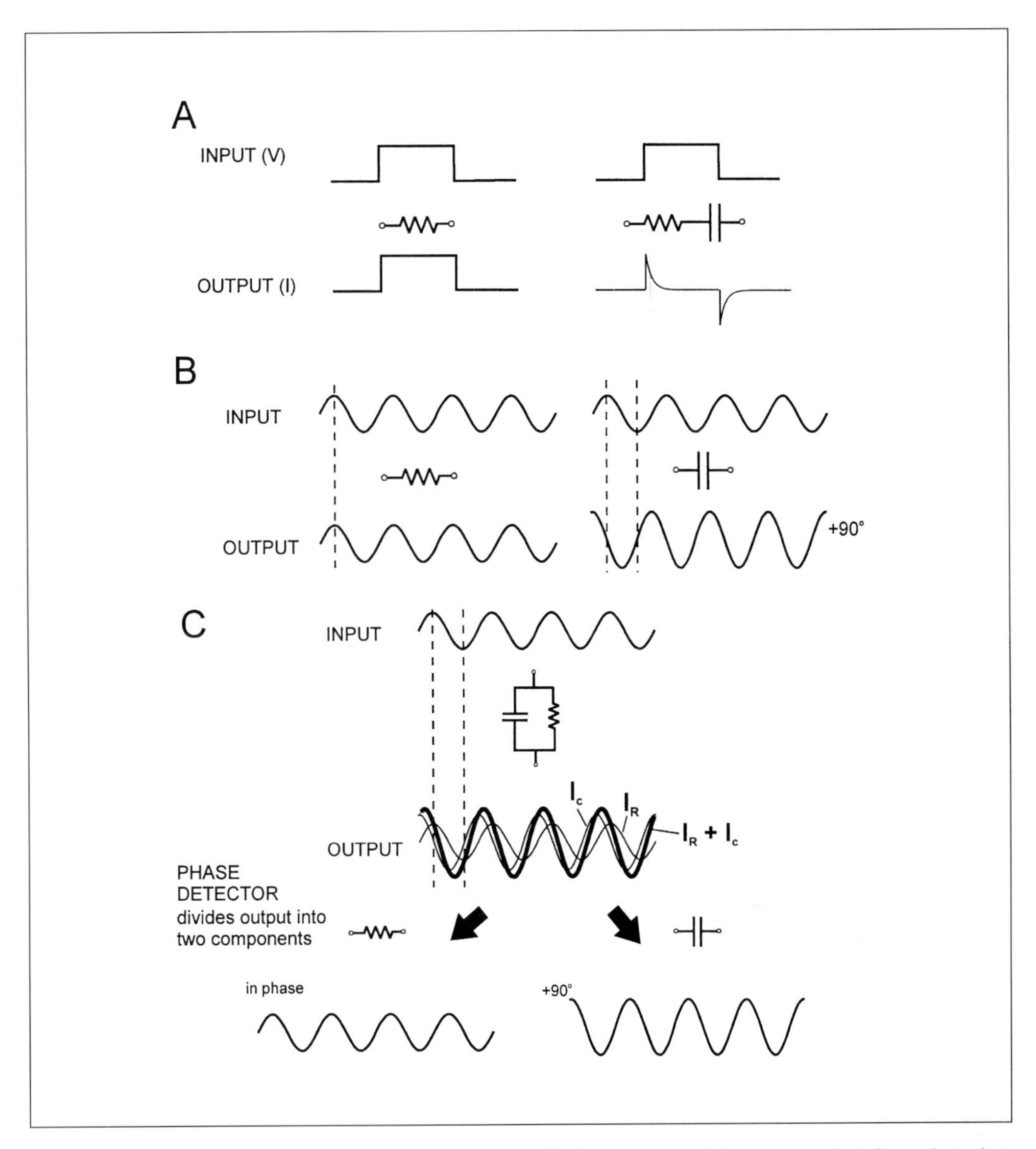

Fig. 2 Principles of capacitance and its measurement. (A) Time course of the current that flows through a resistor (left) or a capacitor (right) after a step change in voltage. The time course of the current across a resistor is identical to that of the change in voltage, according to Ohm's law ($I = V/R$). For an ideal capacitor, a step increase in voltage will produce an infinitely fast charging current since the current depends on the rate of change of the voltage, $I = C\, dV/dt$. Capacitors always have an associated resistance, however, which results in a current output that decays with an exponential time course (lower right). (B) With a sine wave voltage as an input, the current through a resistor will be a sinusoid with the same phase as the input, while in a capacitor the output current is phase shifted by 90° with respect to the input. (C) When a capacitor and a resistor are combined in parallel the resulting current is $I_R + I_C$. The total current will be phase shifted with respect to the stimulus and the size of the phase shift will depend on the relative values of R and C. With a phase detector, the sinusoidal current can be separated into two components: a component in phase with the stimulus voltage and a component 90° out of phase with the voltage.

about 3 picofarads (pF or 10^{-12} farads). The fusion of a secretory granule with a diameter of 300 nm will increase the total capacitance by about 3 femtofarads (fF or 10^{-15} farads). This represents the limit of resolution of most membrane capacitance measurements.

Changes in membrane capacitance can be monitored by repetitively giving voltage steps such as that illustrated in Fig. 2A. This form of measurement is slow, however, as it is limited by the rate at which the capacitive current decays. To measure membrane capacitance continuously, the voltage must be continuously changed. This is accomplished by applying voltage in a sinusoidal pattern (Fig. 2B). When a sinusoidal voltage is applied to a resistor the current output is a sinusoid perfectly in phase with the stimulus. When a sinusoidal voltage is applied to a capacitor, however, the current output is displaced relative to the voltage. As we saw in Fig. 2A the capacitive current is maximal when the rate of change of voltage is maximal and is zero when the rate of change of voltage is zero. Thus, during a sinusoidal stimulus, the measured current output from the capacitor reaches a peak when the voltage is crossing zero and is at zero when the voltage is at a peak. The current output from a capacitor is therefore a sine wave of equal frequency that is shifted 90° relative to the stimulus.

As stated earlier, a cell membrane can be considered both as a capacitor (due to its dielectric property) and as a resistor (due to the presence of ion channels). The equivalent electrical circuit for a cell would therefore include a resistor and a capacitor in parallel. When a sinusoidal stimulus is applied to such a circuit the output is equal to the sum of the output from the two components. This is illustrated in Fig. 2C and can be described using the following equations.

The current (I) that passes through the cell membrane capacitance (C_M) may be expressed as:

$$I_{CM} = C_M \, (dV(t)/dt)$$

while that passing through the membrane resistance is:

$$I_{RM} = V/R_M$$

When a sinusoidal stimulus is applied, the voltage at a given time V(t) is related to that at time zero (V_o) by:

$$V(t) = V_o \sin(\omega t)$$

with ω being equal to 2πf (f being the frequency of the sine wave). The current through the resistive element of the membrane is therefore:

$$I_{RM}(t) = V(t)/R_M = (1/R_M)\, V_o \sin(\omega t) \qquad [1]$$

and the current through the capacitive element is:

$$I_{CM}(t) = C_M \, (dV(t)/dt) = \omega \, C_M \, V_o \cos(\omega t) = \omega \, C_M \, V_o \sin(\omega t + 90°) \qquad [2]$$

The total current across this simple equivalent circuit for a cell will be the sum of the current across these circuit elements, $I_{RM} + I_{CM}$. The total current will be neither in phase nor orthogonal to the applied voltage; the phase shift with respect to the voltage will depend on the relative values of R_M and C_M (Fig. 2C). In order to measure the component of the current that is proportional to the capacitance, we can use a method known as *phase detection*. With the phase detector technique, the sinusoidal current is separated into two components: a component in phase with the stimulus voltage and a component 90° out of phase with the voltage. The phase detector can be implemented in hardware (lock-in amplifier) (3, 4) or software (digital phase detector) (5, 6). The lock-in amplifier has the advantage of better time resolution (as little as 1 msec per capacitance determination). The digital phase detector has the advantage of a high sensitivity over a larger dynamic range. With either method, changes in cell capacitance are tracked simply by measuring the component of the membrane current that is 90° out of phase relative to the command voltage. This would result in an output signal proportional to capacitance and independent of cell conductance (which is equal to $1/R_M$). In a patch-clamped cell (Fig. 3A), however, the equivalent circuit is more complex because additional phase shifts are introduced by the patch-pipette series resistance, R_S, and by the different electronic components of the patch-clamp used to filter and amplify the membrane current. Thus, the correct phase angle for measuring changes in resistance and capacitance measurements are both shifted by some extra phase angle that has to be determined in each experiment.

Neher and Marty in 1982 (3) and then Fernandez *et al.* in 1984 (4) used the whole-cell configuration of the patch-clamp in combination with the phase detector method to measure the small changes in cell membrane area caused by the fusion of single secretory granules in chromaffin and mast cells, respectively. In these two works the authors followed changes in cell *admittance* that occur during exocytosis. In circuits in which the current flow is frequency-dependent (for example those with capacitors) the conductance is frequency-dependent and is called admittance (Y). The admittance is determined from the measured current using Ohm's law ($Y = I/V$). From eqns [1] and [2] it is clear that the changes in admittance (ΔY) of a patch-clamped cell have two components; these may be referred as the 'real' component that is in phase with the membrane voltage ($Re[\Delta Y]$), and the 'imaginary' component that is 90° out of phase with the membrane voltage ($Im[\Delta Y]$). In these first experiments the correct phase angle of the phase detector was found manually. After achieving the whole-cell configuration, a small voltage pulse was applied and the compensation circuit of the patch-clamp amplifier was used to nullify the cell capacitance and series resistance. A sinusoidal voltage was then applied and the correct phase angle was determined by applying small changes in capacitance using the compensation circuit of the patch-clamp amplifier. The phase angle of the phase detector was adjusted until these changes in capacitance were not reflected in the $Re[\Delta Y]$ output. After the correct phase angle for the phase detector was found, the $Im[\Delta Y]$ output was proportional to changes in membrane capacitance. These two works were the first high resolution membrane capacitance recordings of exocytosis in secretory cells. Since this development, the measurement of membrane capacitance has become an important tool for studying exocytosis, endocytosis, phagocytosis, and stimulus-secretion coupling mechanisms in a large variety of cells (7–11).

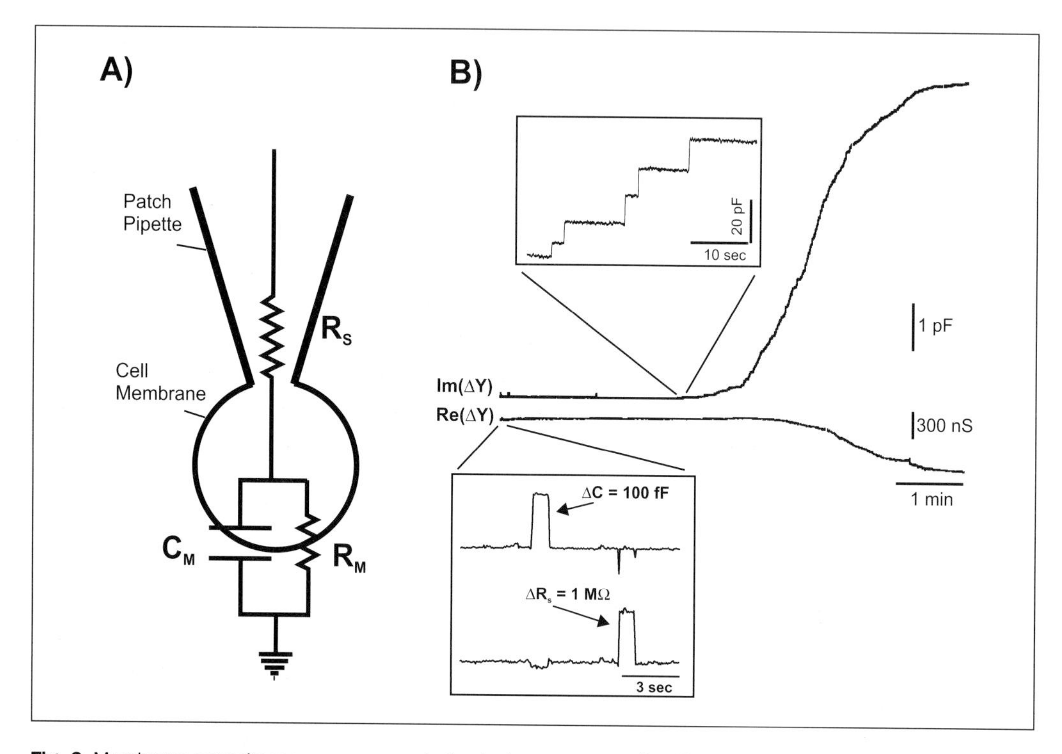

Fig. 3 Membrane capacitance measurements in single secretory cells. (A) Schematic diagram of the equivalent electrical circuit of a patch-clamped cell. (B) Time course of the cell membrane capacitance (represented as Im [ΔY]) increase measured during a complete degranulation in a peritoneal mast cell caused by including 10 μM GTPγS in the pipette solution.. Each capacitance determination was obtained at a rate of one point every 14 msec. After the phase angle was properly aligned using the phase tracking method, the Re [ΔY] output of the phase detector is proportional to changes in resistance, and the Im [ΔY] output is proportional to changes in the cell membrane capacitance. The deflections at the beginning of the traces (see bottom inset) represent calibrations of 100 fF and 1 MΩ, for the capacitance and conductance traces respectively. The upper inset shows an expansion of the membrane capacitance trace near the beginning of the degranulation.

One problem in cell membrane capacitance measurements using the phase detector technique is that the series resistance and membrane capacitance are constantly changing during an experiment, due to clogging of the pipette tip and changes in membrane area due to exocytotic fusion. Since the proper phase angle depends on cell parameters such as R_S, R_M, or C_M, the phase detector must be periodically adjusted. When the phase angle of the cell admittance becomes misaligned from the phase detector, the phase detector outputs will no longer report pure changes in capacitance or resistance (i.e. changes in R_S or R_M will appear as artefactual changes on the C_M trace). In 1989, Fidler and Fernandez (6) presented a simple solution to this problem by introducing the method of *phase tracking*. By inserting a resistor between the cell and ground, the series resistance can be changed by a known amount and the resulting change in admittance can be used to calculate the new phase angle and to realign the phase detector. This method provides a simple and fast way of periodically finding the appropriate phase angle at which to measure current proportional to capacitance changes. In this way, single fusion events can be detected throughout the secretory response of secretory cells.

The sensitivity of cell membrane capacitance recording depends on the amplitude of the applied sinusoidal voltage and on its frequency (because the current through a capacitor depends on the input frequency; see eqn [2]). The amplitude of the sinusoidal voltage is normally kept small (between 10 mV and 50 mV applied on top of the holding voltage) to prevent the activation of voltage-dependent ion channels, but large enough to produce a significant current through the capacitive and resistive elements of the cell. It has been found that the optimal frequency range of the voltage stimulus in patch-clamped cells is usually between 100 Hz and 2000 Hz (5, 12). The value of the optimal stimulus frequency has to be found empirically, however, and depends on the values of C_M, R_M, and R_S. For example, if a cell is voltage-clamped at low frequencies (< 100 Hz) most of the current will flow through the membrane resistance, and step changes in capacitance due to the fusion of single secretory granule will not be resolvable.

Figure 3 illustrates the use of the patch-clamp technique in combination with a software-based phase detector and phase tracking to measure the exocytotic response of a single mast cell. The recording starts shortly after gaining access to the cytosol by disrupting the patch of membrane under the pipette tip with gentle suction. In the whole-cell recording mode the cytosol is dialysed with the solution in the pipette, allowing control of the intracellular environment. Using the whole-cell configuration of the patch-clamp technique, it is possible to perfuse the cell with buffered Ca^{2+} solutions or guanine nucleotides to stimulate a secretory response from a cell. In this example, the pipette solution contained GTPγS, which bypasses some of the normal stimulus–secretion coupling reactions in mast cells to stimulate secretion more directly (4). In this experiment the cell admittance was measured by applying a sinusoidal voltage (833 Hz, 50 mV peak to peak) on top of a holding potential of 0 mV. The resulting current was measured at two different phases, Φ and $\Phi - 90°$, relative to the stimulus with a digital phase detector (5). The phase detector was aligned so that one output (at $\Phi - 90°$) reflected the real part of the changes in cell admittance ($Re[\Delta Y]$) and the second output reflected changes in the imaginary part ($Im[\Delta Y]$, proportional to changes in membrane capacitance). The phase angle was found periodically (every 10 sec) using the phase tracking technique (6). The deflections at the beginning of the traces (see bottom inset) represent calibrations of the two signals. When the phase detector is properly aligned the introduction of a 1 MΩ resistor in series with the cell causes a deflection only in the $Re[\Delta Y]$ trace and not in the C_M trace ($Im[\Delta Y]$). Conversely, unbalancing the patch-clamp capacitance compensation circuit by adding 100 fF causes a deflection only in the C_M trace. As the GTPγS took effect, the cell began to secrete and the capacitance increased in a sigmoidal fashion. The membrane capacitance increased from 4.2 pF to a stable value of 10.7 pF. If this trace is expanded, at any point in time, it can be seen that it is made of step increases in capacitance (top inset in Fig. 3). Fernandez *et al.* (4) demonstrated that each step increase in the membrane capacitance represents the exocytosis of one secretory vesicle. The capacitance trace in Fig. 3 represents the exocytotic fusion of hundreds (about 250 in this cell) of secretory granules each contributing about 25 fF of membrane capacitance (or 2.5 μm^2 of membrane area).

2.1 Capturing the initial stages of the exocytotic fusion pore with the patch-clamp technique

Chandler and Heuser (13) used quick-freezing and freeze-fracture techniques to show that an early event in the exocytotic fusion of a secretory granule is the formation of a pore spanning the granule and plasma membranes. Subsequent patch-clamp measurements revealed that the fusion pore is first apparent as a small pore (1–2 nm in diameter, see below) (14, 15). These measurements also revealed that the exocytotic fusion pore is a remarkably dynamic structure. It opens abruptly and undergoes rapid fluctuations in size, usually culminating in the irreversible expansion and release of the secretory granule contents.

Admittance measurements with the patch-clamp technique can detect the formation of the fusion pore from the instant it conducts ions. Figure 4 shows a high resolution recording of the time course of the formation of an exocytotic fusion pore in a mast cell using the admittance technique. The recording shows the changes in the Im[ΔY] and Re[ΔY] traces during the fusion of a single mast cell granule with the plasma membrane. The granule fuses first transiently (event no. 1) and then irreversibly (event no. 2). If we assume that the resistance of the cell membrane is high and the pipette series resistance is low compared to the changes in membrane capacitance, we can model the fusion of a secretory granule as the granular membrane capacitance (C_g) in series with the conductance of the fusion pore (G_P) (Fig. 4A). When the fusion pore is small, most of the voltage drop occurs across it. As the fusion pore expands, G_p increases and the current through the secretory granule capacitance increases. The contribution of the granule admittance, as G_P progresses from zero to infinity, is thus measured as a change in the total admittance of the cell reflected in the Im[ΔY] and Re[ΔY] traces. On fusion of a granule, the imaginary and real parts of the cell admittance change as (15):

$$\mathrm{Im}(\Delta Y) = \frac{\omega\, C_g\, G_P{}^2}{G_P{}^2 + \omega^2 C_g}; \quad \mathrm{Re}(\Delta Y) = \frac{\omega^2 C_g^2 G_P}{G_P{}^2 + \omega^2 C_g}$$

While the fusion pore is small, G_P is much less than ωC_g, and the imaginary part of the admittance, Im[ΔY], will be not be a simple function of C_g. Rather, it will vary according to the changes in G_P. When the fusion pore is completely expanded, however, G_P is much greater than ωC_g, and Im[ΔY] will be proportional to the capacitance (Im[ΔY] = ωC_g). The fusion pore conductance can be found by simply dividing the imaginary by the real component of the cell admittance:

$$G_p = \omega C_g\, \mathrm{Im}[\Delta Y]/\mathrm{Re}[\Delta Y]$$

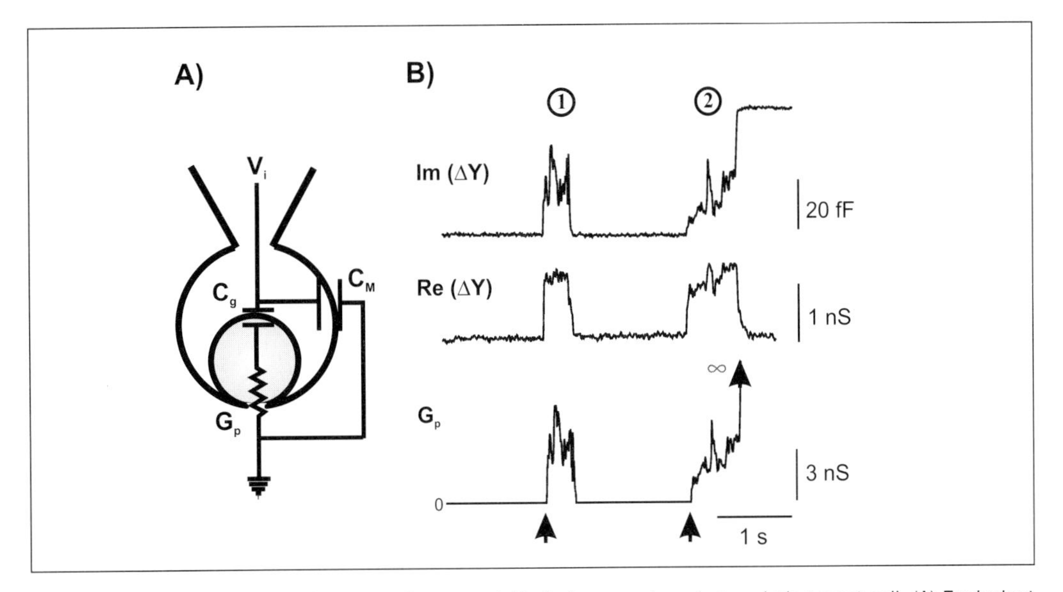

Fig. 4 Time course of the conductance of an exocytotic fusion pore in a degranulating mast cell. (A) Equivalent electrical circuit for a cell at the instant at which the fusion pore between a secretory granule and the plasma membrane is formed. (B) Continuous recordings of the imaginary and real parts of the admittance (Im [ΔY] and Re [ΔY]) and the calculated fusion pore conductance, G_P, during two fusion events. In this experiment the fusion pore opens abruptly and fluctuates for several hundred milliseconds before closing (event no. 1) or opening irreversibly (event no. 2). The final expansion of the fusion pore is marked by a stable increase in C_M. See text for description.

The reconstructed time course of the pore conductance shows that the fusion pore opens abruptly (marked by arrows) to about 3 nanosiemens (nS) within 100 msec. Then the pore enlarges more slowly, but with large and rapid fluctuations in its conductance. The pore remained in this 'flicker' state hundreds of milliseconds before expanding irreversibly (event no. 2) or closing (event no. 1). This latter observation is important because it indicates that the formation of a fusion pore is not an irreversible process. Although the physiological relevance of transient fusion is not yet clear, it has been observed in all secretory cells studied to date (4, 14, 16). It has been suggested that transient fusion events might be the main form of exocytosis at the neuronal synapse (17–19).

One limitation of the admittance technique in measuring the fusion pore conductance is that the time resolution is insufficient (only about 10 msec) to measure the initial values of the fusion pore conductance. An alternative method can be used to capture the first milliseconds of the conductance of the fusion pore (14, 20). When a secretory granule fuses with the plasma membrane a capacitive current is generated (Fig. 5B). This is because the cell membrane potential, V_m, is different from the membrane potential of the secretory granules, V_g. Secretory granules maintain a positive transmembrane potential ($\sim +70$ mV) mainly due the presence of an electrogenic proton pump (21, 22). Thus, when a granule fuses with the cell membrane it is

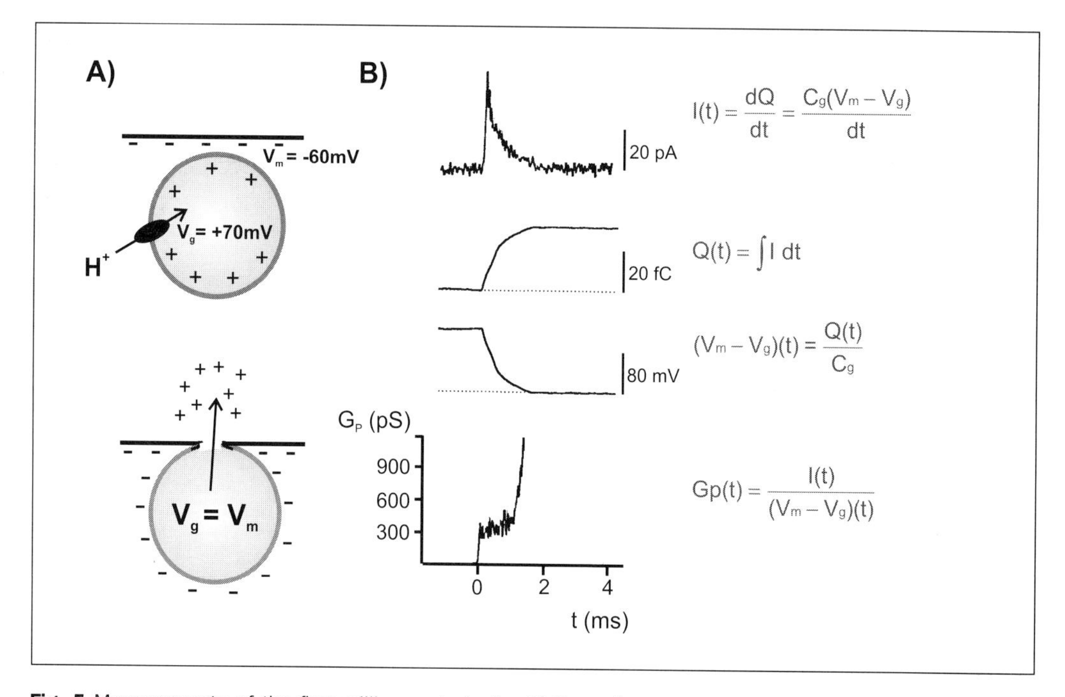

Fig. 5 Measurements of the first milliseconds in the lifetime of an exocytotic fusion pore. (A) Diagram of a secretory granule before and after fusion with the plasma membrane. Note that prior to fusion the granule has a very positive resting potential that is maintained by a proton pump. Upon fusion this potential rapidly equilibrates with the membrane potential of the cell. (B) In order to capture the current transient that marks the opening of the fusion pore, a threshold detector method was used. After establishing the whole-cell configuration, the membrane current was continuously monitored at a holding potential of –60 mV. When the measured current exceeded a threshold (set at two times the noise level), a record of 1000 points (10 msec) was saved (top trace). This was immediately followed by a measurement of the C_M to determine the size of the vesicle that fused (300 fF in this example). The time course of the fusion pore conductance was reconstructed by first calculating the time integral of the current (Q(t); second trace) and then calculating the voltage driving the current (Q(t) divided by the observed change in C_M; third trace). Since this plot is equal to $(V_M - V_g)$ (t) (see equation), the ratio of the recorded current transient and the plot of $Q(t)/C_g$ gives a plot of the $G_p(t)$ (bottom trace).

rapidly charged to the potential of the cell membrane (i.e. V_g becomes equal to V_m). This change in voltage generates a capacitive current, $I_C(t) = C_g\, dV/dt$, which is very fast (only 2–10 msec in duration) and marks the formation of a fusion pore. From this current transient it is possible to reconstruct the time course of the changes in pore conductance.

Figure 5B shows a current transient through an exocytotic fusion pore recorded in a degranulating mast cell. This experiment was similar to that shown in Fig. 3, except that the membrane current was monitored at much higher time resolution (at 10 μsec/point). After detection of a current transient (see Fig. 5 legend) the membrane capacitance was measured using the phase detection method. The fusion pore conductance, G_P is calculated from the discharge current, I, and the potential difference driving the current, $(V_m - V_g)$. The potential difference is calculated from the time integral of I, divided by the capacitance of the granule membrane, C_g. The calculated pore conductance shows that the formation of the fusion pore can be divided into several distinct phases (14, 20). The fusion pore opens abruptly (< 100 μsec) to about 300 pS (range: 80–1000 pS); then it expands to about 5 nS within 20 msec at about 300 pS/msec. This expansion phase is characterized by large, rapid fluctuations in pore conductance (flicker). At this point, the fusion pore can either:

- open to a very large structure (> 20 nm) in less than 100 msec, or
- close completely

What are the initial dimensions of a fusion pore with a conductance of 300 pS? By modelling the fusion pore as a cylinder with a length l and a radius r, we find that the resistance is given by:

$$R_{pore} = \rho \frac{l}{\pi \cdot r^2}$$

where ρ is the resistivity of the solution (about 100 Ωcm for a 100 mM NaCl solution). By assuming that the fusion pore spans the thickness of a lipid bilayer (~ 5 nm), we can calculate that the initial diameter of the fusion pore is about 1 nm. Thus, the initial dimensions of the fusion pore is comparable to that of a large ion channel like a gap junction channel (23). Patch-clamp measurements of the activity of individual fusion pores in mast cells have shown, however, that the fusion pore has some unusual and unexpected properties that are very different from typical ion channels:

(a) There is a large flux of lipid through the lining of the pore (24).

(b) The rate of pore closure has a discontinuous temperature dependency (25).

(c) Comparisons of experimental data with theoretical fusion pores (26) and with breakdown pores (27) supports the view that the fusion pore is initially a pore through a single bilayer, as would be expected for membrane fusion proceeding through a hemifusion mechanism.

Based on these observations, Monck and Fernandez (28) proposed a model in which the fusion pore is initially a pore through a single bilayer. In this model, it is envisioned that the formation of a fusion pore is regulated by a macromolecular scaffold of proteins that is responsible for bringing the plasma membrane into a highly curved dimple very close to a tense secretory granule membrane, creating a condition in which strongly attractive hydrophobic forces cause the membranes to form a 'hemifusion' intermediate. Membrane fusion is completed by the formation of a pore after rupture of the shared bilayer. The microenvironment of the interface when the pore first opens, dominated by charged groups on glycoproteins in the lumen of the secretory vesicle and on phospholipids, might influence the release of secretory products (29, 30). Therefore, the amount of the neurotransmitter released might be finely regulated by the kinetics of opening and closing of the fusion pore.

3. Amperometric measurements of exocytosis

An alternative method of monitoring secretion is to measure the release of oxidizable secretory products by placing a carbon fibre microelectrode (held at a constant potential) in close proximity to a single isolated cell (31–34). Release of secretory granule constituents results in the generation of brief current spikes as secretory products are oxidized on the surface of the carbon fibre. One advantage of the amperometric method of following secretion is that it only monitors the release of secretory components into the extracellular milieu and is therefore not complicated by some of the factors affecting capacitance measurements (like membrane retrieval by endocytosis). Amperometry provides exquisite temporal resolution and allows detailed analysis of the kinetics of the release of neurotransmitters. The sensitivity of the amperometric technique is remarkable: the release of only ~ 5000 serotonin molecules has been detected in leech neurons (35). Therefore, the amperometric technique provides an alternative method of studying the properties of single fusion pores in preparations where the patch-clamp technique does not have enough resolution to resolve the activity of single fusion events. A number of secreted molecules can be detected using this technique. Among the most intensively studied are catecholamines (from chromaffin cells), serotonin (from mast cells and invertebrate neurons), dopamine (from glomus cells), and insulin (from pancreatic beta-cells) (31, 35–40).

By placing an electrode near a secretory cell it is possible to detect an electrochemical current when secretion occurs (Fig. 6). Detection of secreted molecules is based on the electrochemical processes of oxidation and reduction (Fig. 6A, inset). Electrochemically active molecules like catecholamines or serotonin have a characteristic redox potential. The redox potential is the potential at which half of the molecules are oxidized and half are in the reduced form at equilibrium. To favour conversion of molecules to the oxidized form, a voltage must be applied that is higher than the redox potential. Thus, by holding the voltage of an electrode (like a carbon fibre electrode) at a voltage more positive than the redox potential, the molecules will be oxidized and an electrochemical current will be created that is proportional to the number of molecules reacting with the surface of the electrode. The number of molecules that are electrolysed can be found using Faraday's law, which states that the charge (Q, time integral of the current) is directly proportional to the number of molecules detected (M):

$$Q = \int I dt = ezM$$

where z is the number of electrons involved in the electrochemical reaction ($z = 4$ for serotonin) (41) and e is the elementary charge (1.6×10^{-19} coulombs).

The time course of the amperometric signal of single secretory vesicles reflects the time course of transmitter release and the diffusion of molecules from the release site to the carbon fibre detector. The recorded amperometric signal therefore critically depends on the distance between the electrode and the cell, because the diffusional

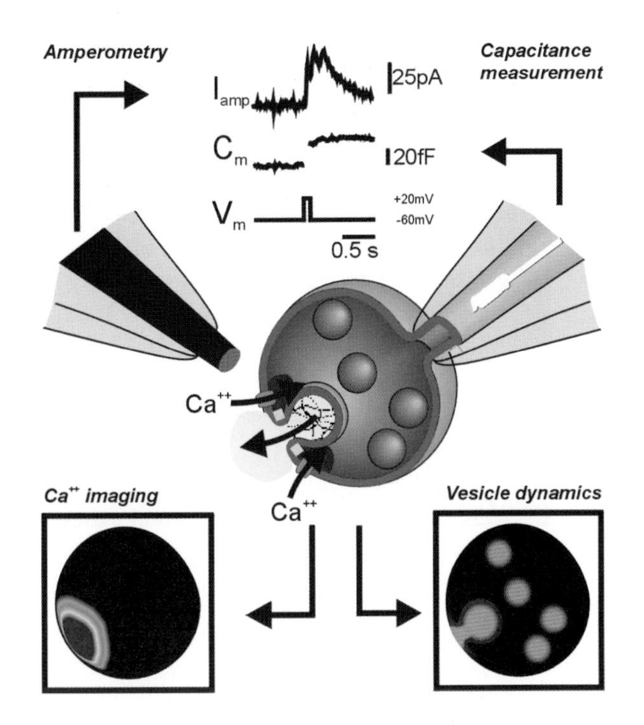

Fig. 1 (Chapter 2, p. 35). Techniques for monitoring exocytosis. The techniques discussed in this review are: (a) capacitance measurements, (b) amperometry, (c) Ca^{2+} imaging, and (d) vesicle dynamics. A brief depolarization of a patch-clamped cell (V_M trace) can cause Ca^{2+} influx near release sites (lower left), release of oxidizable substances measured as a spike on the amperometric trace (upper left), an increase in total membrane area detected as a jump in the capacitance trace (upper right), and a transfer of stained vesicular membranes to the cell surface (lower right).

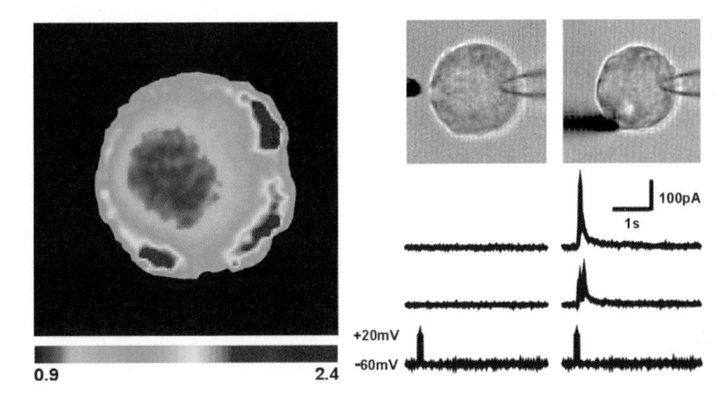

Fig. 11 (Chapter 2, p. 64). Pulsed laser imaging of Ca^{2+} influx combined with amperometric detection of release. The image on the left shows the highly localized Ca^{2+} influx evoked in a patch-clamped chromaffin cell by a 50 msec depolarization from –60 mV to +20 mV. The false colours (see scale below image) indicate the ratio of Ca^{2-} levels compared to that in a control image. The upper two images on the right show the same cell with an amperometric carbon fibre in two different positions. When in the first position, which is not close to one of the hotspots of Ca^{2+} entry, the fibre does not detect release of catecholamines. When in the second position, however, which is close to one of the Ca^{2+} entry sites, depolarization of the cell evokes clear spikes in the amperometric trace. These data demonstrate that the sites of Ca^{2+} influx are associated with sites of preferential release of catecholamines. Reproduced from Robinson *et al.* (1995) *Proc. Natl. Acad. Sci. USA,* **92**, 2474–8, by copyright permission of The National Academy of Sciences, USA.

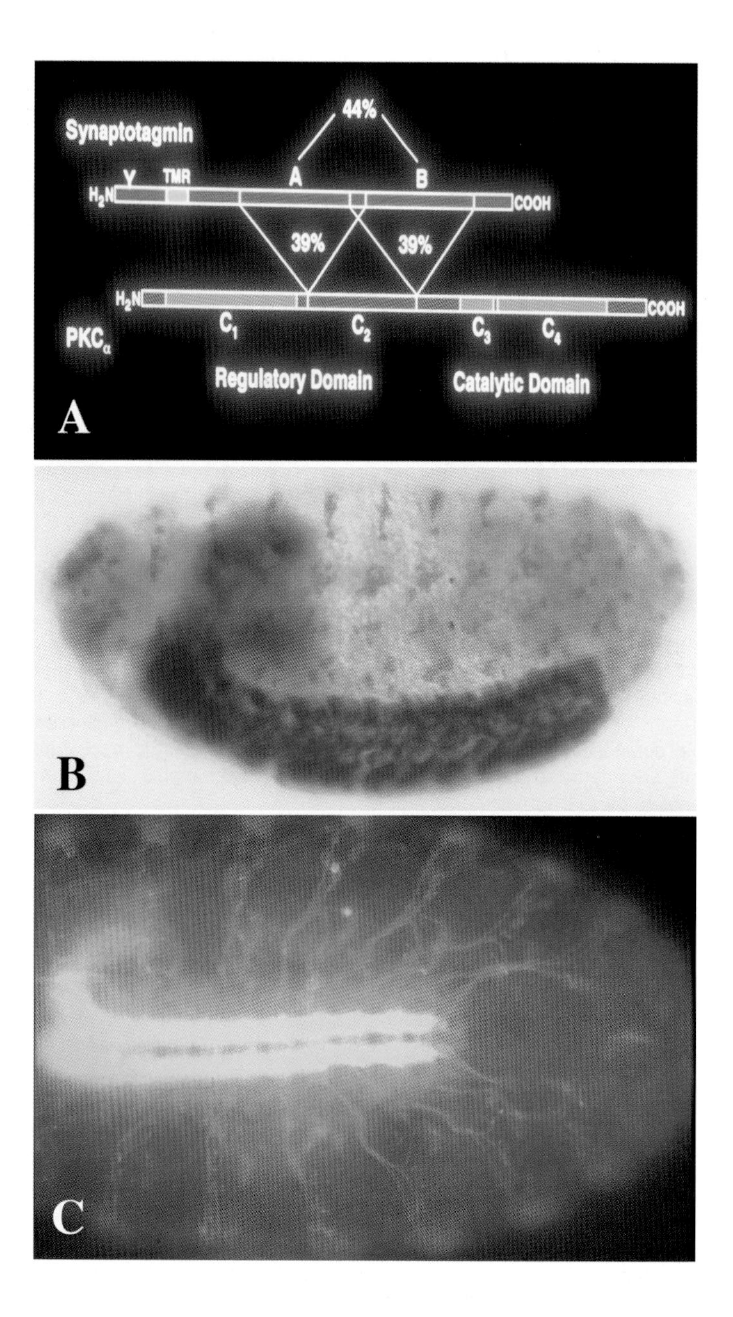

Fig. 3 (Chapter 9, p. 317). Synaptotagmin structure and pattern of expression. (A) Domain structure of Synaptotagmin. TMR, transmembrane region; Y, glycosylation site; A and B, C2 domains. (B) *In situ* hybridization of whole mount embryos using a digoxygenin-labelled *sytnaptotagmin* cDNA as a probe. (C) Immunohistochemical staining of embryos with an anti-Synaptotagmin antibody (green and yellow) and Monoclonal Ab. 22C10 (red and orange). Note the green dots in the periphery. They correspond to NMJs. (From ref. 17, courtesy of the company of Biologists Limited.)

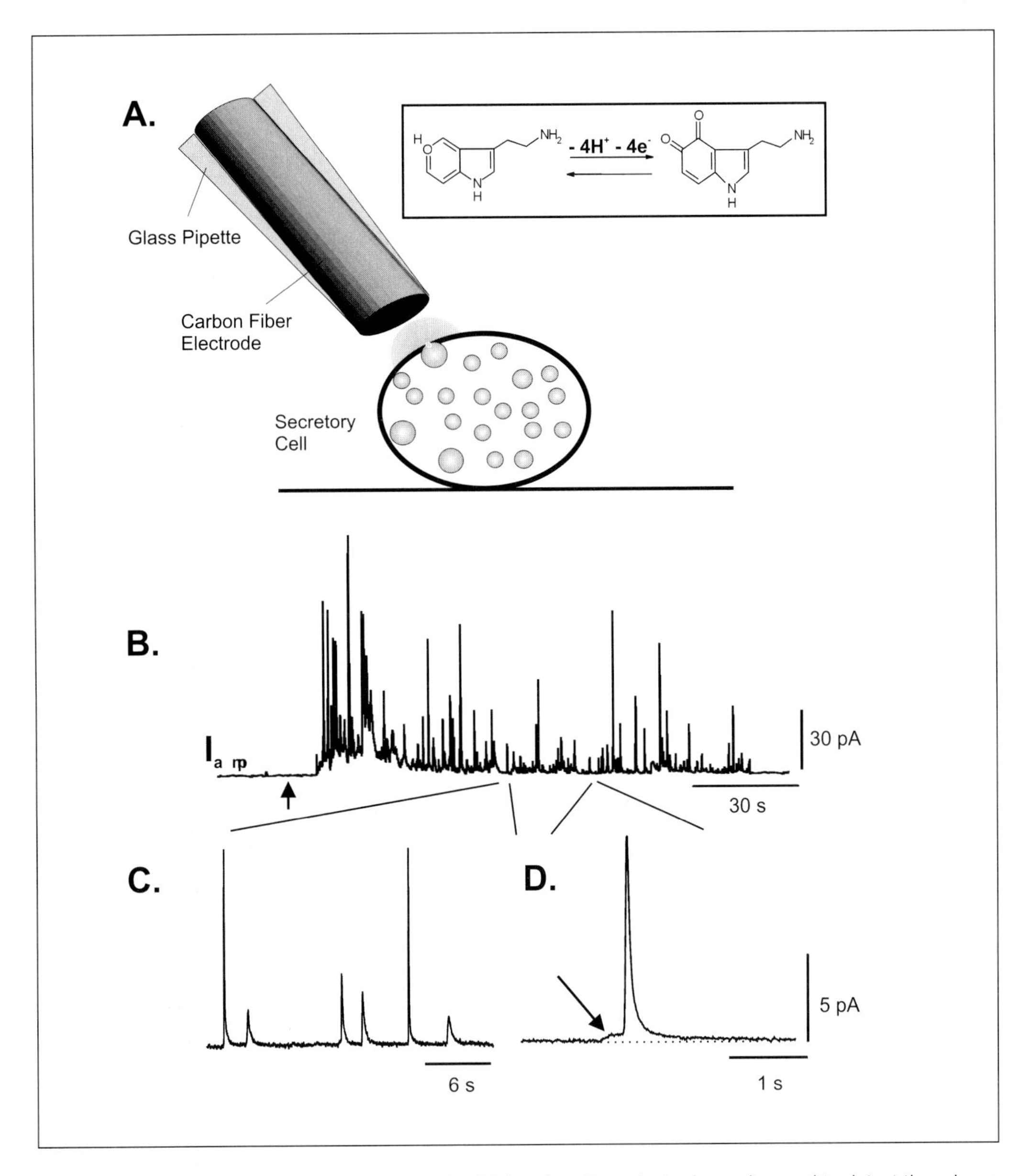

Fig. 6 Amperometric measurement of exocytosis. (A) A carbon fibre electrode can be used to detect the release of secretory products from a single exocytotic fusion event. Application of a constant potential through the electrode causes the oxidation (or reduction) of susceptible molecules that reach the fibre resulting in the generation of small electric currents (I_{amp}) that can be measured with a patch-clamp amplifier. For example, when the fibre is held at a very positive potential (+700 mV) secreted serotonin molecules undergo the electrochemical oxidation reaction shown in the inset. (B) Secretory response recorded from a single mast cell. The trace shows the spikes of current evoked by the oxidation of serotonin released from individual secretory granules. The arrow marks the application of a secretagogue (compound 48/80) to the cell. (C, D). Enlarged examples of single amperometric spikes obtained from (B). The spike shown in (D) displays a 'foot' (arrow), which corresponds to the release of serotonin through the nascent fusion pore prior to its final expansion.

time course depends on the square of the distance. This is described by the Einstein–Smolochowski equation for one-dimensional diffusion:

$$\langle x^2 \rangle = 2Dt$$

where $\langle x^2 \rangle$ is the mean square displacement, D is the diffusion constant, and t the time. Thus, an electrode distant from a cell will respond with a longer delay, a slower rise time, and lower amplitude than one close to the cell (33). This problem is particularly important for the detection of quantal secretory events when high time resolution is required. Thus it is important to use carbon fibre electrodes with a large detecting surface placed in close proximity to the cell to minimize diffusional delays and to maximize the probability of detecting exocytotic events. The electrodes used to monitor the release of neurotransmitters from single cells are typically based on carbon fibres (5–10 μm in diameter) placed inside an insulating material (like a glass capillary or plastic tubing) (see refs 33 and 42 for detailed descriptions of the method). The surface of the carbon fibre electrode is commonly bevelled at a 45° angle to increase the surface area of the tip to maximize the detection of exocytotic events.

Figure 6B shows an amperometric recording obtained from a single degranulating mast cell. The cell was stimulated by adding 10 μg/ml of the secretagogue 48/80 to the cell (marked by the arrow). The exocytotic release of serotonin from the cell causes a large number of overlapping spikes to appear on the recorded amperometric current trace. Some individual spikes are shown in an expanded time scale in Fig. 6C and 6D. Combined amperometric and capacitance measurements have shown that the detection of a single amperometric spike corresponds to the fusion of a single granule with the plasma membrane, observed as a stepwise increase in the membrane capacitance (37, 43). Therefore, each spike in Fig. 6C represents the detection of the release of the contents of a single granule. The charge during the larger amperometric spikes shown in Fig. 6C corresponds to about 2×10^{-12} coulombs, corresponding to the release of 3×10^{6} serotonin molecules (about 5 attomoles).

The development of amperometry to detect the release of secretory products from single cells has yielded exciting new information regarding the mechanisms by which the neurotransmitters are released to the extracellular medium. As shown in Fig. 6D some of the amperometric spikes are preceded by a small 'foot' (arrow). This small amount of release before the main phase of release represents the leakage of serotonin through the nascent fusion pore. A 'foot' followed by a 'spike' indicates a fusion pore expanding slowly and then becoming fully opened. Similar observations have been made in a wide variety of secretory cells including chromaffin cells (36), glomus cells from the carotid body (38), and Retzius neurons from the leech (35). Alvarez de Toledo *et al.* (37) found that the release during the foot or transient fusion events was proportional to the measured fusion pore conductance. These observations led to the proposal that the release of neurotransmitters is rate-limited by the exocytotic fusion pore (36, 44, 45). In beige mouse mast cells, however, the amount of transmitter released during the foot is only 1–2% of the total serotonin detected amperometrically and the rate of release was significantly smaller than predicted if all the serotonin molecules were available for release. The most likely explanation for this surprising observation is that most of the serotonin is trapped in the granule lumen and is unavailable for release through the fusion pore. Secretory granules from many different cell types contain proteoglycans and other acidic proteins which form crosslinked matrices or gels (46, 47). Neurotransmitters bind to this highly charged polymer gel matrix and are therefore osmotically inactive; this mechanism allows the storage of very large amounts of secretory products (up to 800 mM). Recent studies on amperometric detection of serotonin from electroporated secretory granules have demonstrated that the diffusion coefficient of serotonin within the granules is about 10^{-8} cm^2/sec (48–50). This is almost three orders of magnitude lower than bulk diffusion (about 10^{-5} cm^2/sec). Thus, in addition to the geometry of the fusion pore, the diffusivity of neurotransmitters within the secretory vesicle matrix affects the rate of release.

3.1 Monitoring exocytosis using simultaneous measurements of cell membrane capacitance and amperometry

There are certain limitations of membrane capacitance measurements that must be borne in mind in studies of secretion. In mast cells and in other cells with large granules, step increases in cell membrane capacitance observed during exocytosis unambiguously indicate the fusion of single secretory vesicles (4). Membrane capacitance measurements recorded in whole-cell patch-clamp, however, can only resolve, as step increases, the fusion of individual vesicles that are larger than ~ 300 nm in diameter. Fusion of secretory vesicles smaller than this could only be detected as a smooth increase in capacitance. Cell membrane capacitance recordings are based on the assumption that the cell can be modelled as a simple electrical equivalent circuit (see Section 2). Any change in the electrical equivalent circuit of the cell that are unrelated to secretion will affect the capacitance measurements. Changes in the cell shape, surface morphology, or changes in the dielectric constant contributed by membrane proteins can produce changes in the cell membrane capacitance that are not related to the fusion of secretory vesicles. Also, cell membrane capacitance measurements can be complicated by the fact that following a secretory response cells actively retrieve membrane area by endocytosis (3, 43, 51, 52). Therefore, without the use of an independent method to measure exocytosis, capacitance measurements can be very difficult to interpret. Figure 7 illustrates this point.

Figure 7A shows an experiment where guanine nucleotides were used to trigger exocytosis in a patch-clamped mast cell. Exocytosis was independently monitored by measuring the cell membrane capacitance and by measuring the exocytotic release of secretory compounds using the amperometric method. The time integral of the amperometric current trace is shown as a line in Fig. 7A. A time integral of the area under the amperometric trace is a measure of the total amount of secretory products that were released by a cell following its stimulation and as such allows a direct

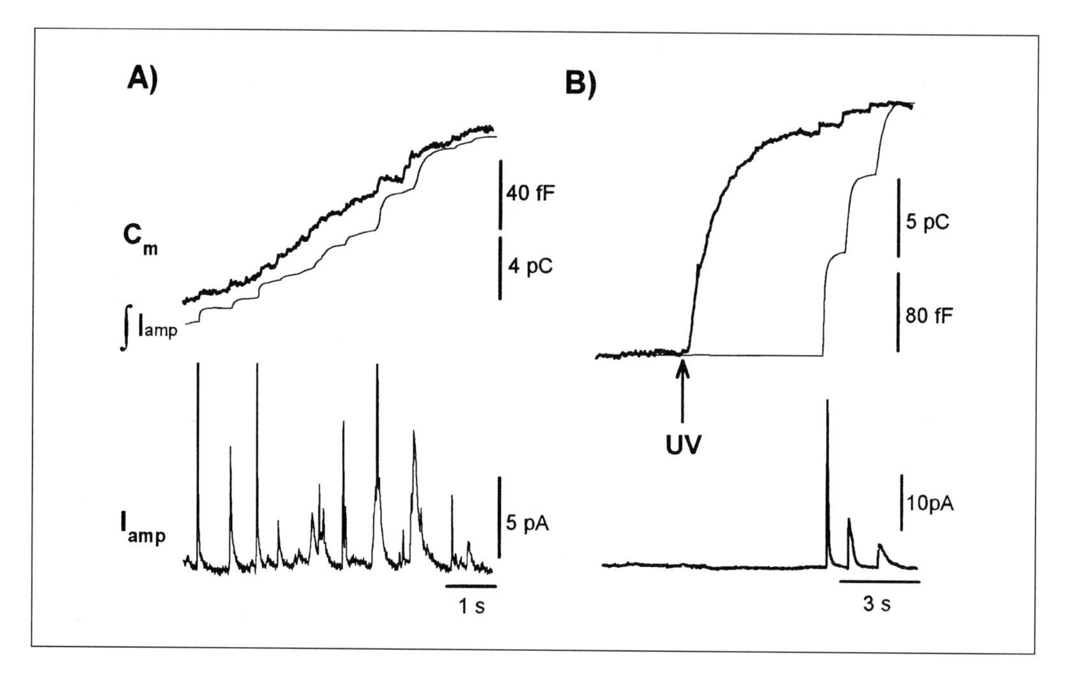

Fig. 7 Measurements of exocytosis by simultaneously monitoring membrane capacitance and the release of vesicular contents by measuring amperometric currents. (A) Secretory response of a mast cell induced by perfusion with GTPγS. The time integral of the amperometric recording shows a very good correlation with the increase in cell membrane area. (B) Secretory response of a mast cell elicited by photolysis of caged Ca^{2+} (arrow). The patch pipette solution contained 1.5 mM GTP to trigger exocytotic secretion. The time integral of the amperometric current shows that there is a long delay (~ 5 sec) between the increase in C_M and the first amperometric spike.

comparison of the events measured using this technique with those measured by capacitance. Notice that the time integral of the amperometric current trace follows a similar time course to the membrane capacitance recording. The close agreement between the capacitance and amperometry recordings demonstrates that the amperometric technique provides an alternative method of assaying the extent and kinetics of the secretory response of a cell.

Figure 7B shows traces similar to those in Fig. 7A, but in this case secretion was induced by rapidly and transiently increasing the $[Ca^{2+}]_i$ by photolysis of the caged Ca^{2+} compound DM-nitrophen (see section 4.4). By exposing the cell briefly to UV light (500 msec, arrow) rapid increases in $[Ca^{2+}]_i$ are triggered. These increases cause an immediate increase in C_M that was smooth at first and then increased in a stepwise fashion (53). Surprisingly, however, the first amperometric spike was seen after a long delay (about 5 sec in this cell) coinciding with the first detectable step increase in the capacitance trace. In order to more readily compare the amperometric response with the capacitance trace, we again calculated the time integral of the amperometric trace (shown as a line superimposed on the C_M trace). It is evident that during the smooth phase there is no detectable release of oxidizable substances and that the increase in the integral of the amperometric current lags the increase in C_M. This lag was seen only following the first stimulus, and was never seen for subsequent stimuli. It is not clear what this smooth increase in C_M represents; it could reflect the fusion of small vesicles with the plasma membrane or a phenomenon unrelated to membrane fusion.

This experiment clearly demonstrates that there can be large changes in membrane capacitance in secretory cells that are not due the fusion of secretory granules containing oxidizable substances. Thus, increases in membrane capacitance that are not resolved as steps cannot be readily interpreted as secretory events unless this is confirmed with independent techniques (such as amperometry or imaging techniques; see sections 4 and 5).

4. Ca^{2+} imaging

Ca^{2+} is the trigger that activates neurotransmission. Although the mechanism by which this occurs is not yet fully understood (hence this book) there are many important questions relating to the action of Ca^{2+} that are independent of the biochemical target. Is exocytosis evoked by a local increase in Ca^{2+} mediated by entry through single Ca^{2+} channels? Through clusters of channels? Or does it depend on a more global increase in cellular Ca^{2+} mediated by multiple, distantly spaced Ca^{2+} channels? What concentration of Ca^{2+} is required? How may Ca^{2+} influence pre-fusion events? Ca^{2+} imaging provides a means to address these questions. Ca^{2+} imaging is based on the use of intracellular indicator dyes whose emission of light depends on the concentration of Ca^{2+} (for reviews see refs 54 and 55). These emissions are translated into an image that reflects the distribution of Ca^{2+} in the cytoplasm. A comparison of images at different times indicates how Ca^{2+} levels changed. Thus, Ca^{2+} imaging can provide information about where an increase occurred, when it occurred, and how much of an increase occurred. These parameters translate into the fundamental requirements for the techniques used to image Ca^{2+} — spatial resolution, temporal resolution, and quantification.

4.1 Fluorescence

To understand the principles behind Ca^{2+} imaging it is necessary first to discuss the principle of fluorescence. Fluorescence refers to the ability of a molecule, a fluorophore, to absorb a photon of a given energy and then release a photon of a diminished energy (a longer wavelength). The absorption of the incident photon elevates the fluorophore into a short-lived excited state, which typically lasts 1–10 nanoseconds. Part of the added energy is immediately dissipated, at which point three things may happen. The molecule might emit a photon of light (i.e. it may fluoresce), thereby returning to the ground state. It might also return to the ground state by releasing the added energy in some other way, without the creation of a photon. After either of these processes the molecule is ready to undergo further cycles of excitation. Thirdly, the molecule might undergo an irreversible change, called photobleaching, such that it is unable to return to the ground state and can no longer contribute to the fluorescent signal. Photobleaching therefore causes a progressive loss of signal from a sample.

The probability that a molecule will fluoresce depends on the intensity of the excitation light and on the intrinsic properties of the molecule. These properties can be expressed as the molar extinction coefficient, which is a measure of the tendency of the fluorophore to absorb a photon at a given wavelength, and the quantum yield, which is a measure of the probability that the excited fluorophore will release a photon. The intensity of the emitted fluorescence is therefore proportional to the product of these two parameters.

4.2 Optical measurement

From a practical point of view, fluorescence imaging depends on the ability to separate light into different wavelengths, allowing the measurement of the fluorescence that is emitted from the fluorophore while preventing interference from the light used to excite it. This is accomplished using filters and mirrors that exclude certain wavelengths of light while allowing others to pass. Consider the example illustrated in Fig. 8, which shows an inverted microscope in epifluorescence mode (an arrangement in which both the excitation and emission light pass through the microscope objective). Polychromatic source light passes through an excitation filter, called a bandpass filter, that transmits light only in a narrow range of wavelengths. This range corresponds to the wavelengths that are optimal for the excitation of the fluorophore. The filtered light then strikes a dichroic mirror, or dichroic beam splitter, placed at a 45 degree angle beneath the microscope objective. Dichroic mirrors have thin optical coatings that allow them to reflect wavelengths shorter than a cut-off value and to transmit wavelengths longer than that value. The cut-off wavelength is chosen such that the excitation beam is reflected up into the objective where it is then focused on the specimen to excite the fluorophore, while the emitted fluorescence light passes through the dichroic to strike the detector. A final emission filter, in this case a longpass filter, is used to exclude any stray excitation light while passing the fluorescent light emitted by the sample.

The most important determinant of the proportion of emitted light that reaches the detector is the choice of microscope objective. Since fluorescence is emitted by a sample in all directions, only a small portion will be picked up by the objective lens. The size of this portion is dependent on the light gathering capacity of the objective, which is described by its numerical aperture (NA). The NA depends on the aperture angle (α, or one-half the maximum angle at which light can enter the lens) and the refractive index of the medium between the lens and the sample (η) according to the following equation:

$$\mathrm{NA} = \eta \sin \alpha$$

Since the refractive index of oil is larger than that of air (about 1.5 compared to 1.0), oil immersion lenses have superior light gathering capability. The dependence of NA

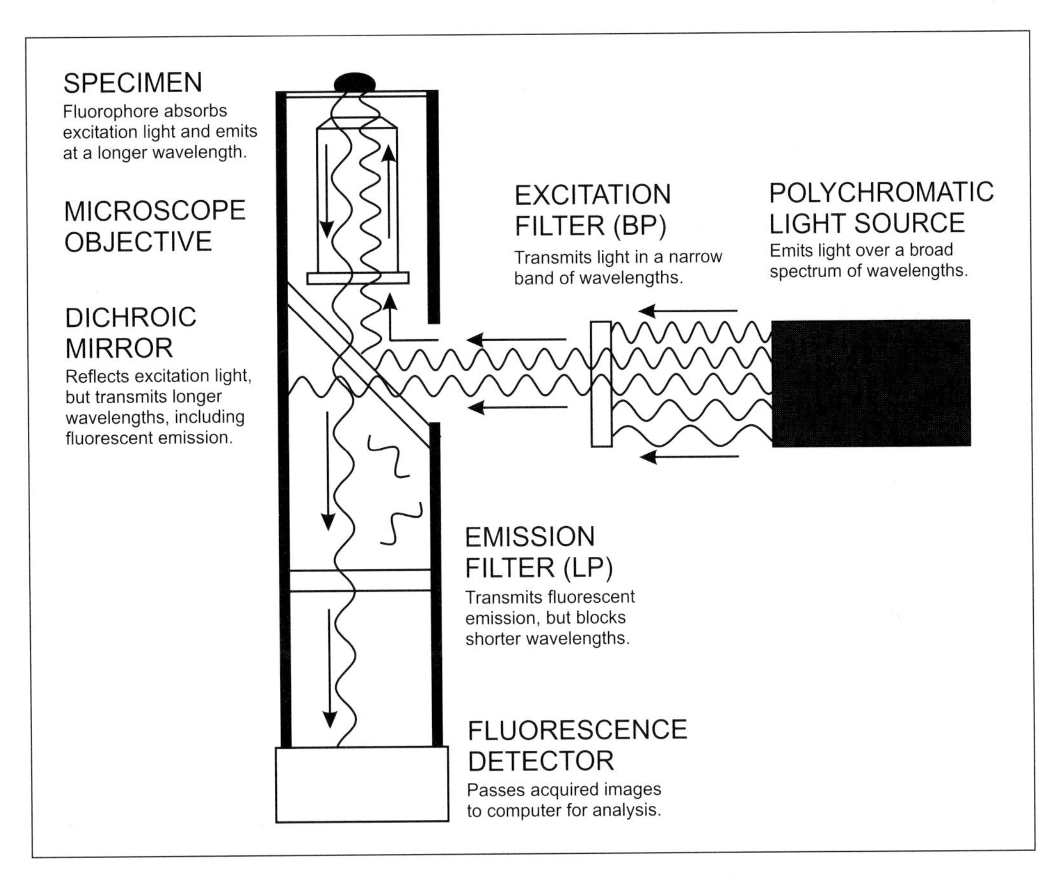

Fig. 8 Excitation of a fluorescent sample and detection of the emitted fluorescence. A typical set-up for measuring fluorescence in an inverted microscope. See text for description.

on α also explains why image quality is enhanced by having a short working distance between objective and specimen. The NA is particularly important for epifluorescence measurements since both excitation and emission light passes through the objective. The relative brightness (B) of the fluorescent image in epifluorescence is given by the equation:

$$B = (NA)^4/(\text{magnification})^2$$

It is therefore desirable to have an objective with the largest possible NA and the smallest magnification capable of resolving necessary detail. The highest NA objectives commercially available are 100× oil immersion lenses with a NA of about 1.40. Lower magnification objectives generally have a lower NA, however, and the optimal objective for a specific application will depend also on the desired magnification and the wavelengths of light used.

4.2.1 Detection with charge-coupled devices

The next important determinant of fluorescence measurement is the detection system. The most commonly used device for the spatially-resolved measurement of emitted fluorescent light is the charge-coupled device or CCD camera. We will therefore concentrate our discussion on properties and limitations of CCD cameras, although much will apply to other types of cameras as well. Several recent books and reviews provide excellent broader discussions of electronic cameras and digital imaging (56–59).

The heart of a charged-coupled device consists of a square or rectangular array of picture elements, or pixels. Each of these pixels, which for fluorescence microscopy are usually from a few to 30 μm across, is designed to accumulate a charge corresponding to the number of photons that strike it. The size of the array varies from a few hundred to a few thousand pixels per side; the total length of the detector is therefore only one or two centimetres. After a specific period of integrating incident light, a programmed series of gate potentials allows the charge in each pixel to be read out sequentially. This information can then be used to reconstruct the pattern and intensity of light that made up the original microscopic image.

i. Temporal resolution

For a given period of integration, therefore, the rate of acquisition is limited by the rate at which the charge collected in the pixel array can be read and digitized for further analysis. The readout rate is a measure of the time required to digitize a single pixel and is usually expressed as kilopixels or megapixels per second (expressed as kHz or MHz). The rate at which images can be acquired, the frame rate, can therefore be estimated by adding the exposure time to the number of pixels divided by the readout rate (although various delays in data handling normally decrease the true rate). For example, for a 1000 × 1000 CCD detector with a readout rate of 5 MHz, it would take 200 msec to digitize the image. If the exposure time were 50 msec, the frame rate would be about one every 250 msec or four per second. The frame rate can be improved by choosing a small area within the total field (a subarray) and digitizing only the input from the pixels in that area. Binning refers to the digitization of multiple adjacent pixels (e.g. a 2 × 2 square of pixels) as single units. These enlarged areas may be counted nearly as quickly as single pixels. The major drawback of binning is a loss of spatial resolution (see discussion below). Data acquisition may also be accelerated by the use of a process called frame transfer. In frame transfer the CCD pixel array is divided into one section that is used for the acquisition of images and one or more other sections that are used for storage. After integration of an image the information is quickly transferred to a storage array. This allows the reading of the pixels in the storage array to occur simultaneously with the acquisition of a second image in the image array.

CCD cameras can be made to acquire images with very short integration times (less than a millisecond). Using frame transfer and rapid digitization methods these images can be accumulated very rapidly. The limiting factor, however, is likely to be the amount of light given off by the sample. The relationship between the level of light and the signal quality will be discussed in detail below, but suffice it to say here that short integration times may yield insufficient light to acquire useful images.

ii. Spatial resolution

What defines the resolution of the image? Remember that the light impinging upon the pixel array is an image of the specimen magnified by the microscope objective. The resolution therefore depends on the resolving power of the microscope and on the sample size of the detector (i.e. the pixel dimensions). The resolving power of the microscope is determined by the NA using the Rayleigh criteria:

$$d_{min} = 0.61\lambda/NA$$

where d_{min} is the distance between two points that can just be resolved, and λ is the wavelength of the illuminating light. For example, using a 63× objective with a NA of 1.4 illuminated with 500 nm light, the d_{min} is equal to 218 nm. In the magnified image that is projected onto the CCD detector, this distance will correspond to a length of 14 μm (218 nm × 63). This value represents the smallest object resolvable by the microscope and defines the limit of resolution of the image produced by the CCD. To accurately record the image data in digital form, the pixel size must be one-half or less the size of the d_{min} of the magnified analogue image. In this case, optimal sampling of the image would require a pixel size of 7 μm. If the pixel size is larger than the maximum allowable size defined by the objective, then this becomes the limiting factor in the resolution and image detail is lost. This phenomenon is called undersampling. It may be corrected by adding additional optical magnification before the focal plane of the CCD detector.

iii. Sensitivity and noise

These considerations define the theoretical limit for the spatial resolution of the system. For this resolution to be meaningful, however, the detector must be able to differentiate the levels of light in different pixels. The ability of the CCD to discriminate levels of light may be defined in several ways. Quantum efficiency (QE) refers to the effectiveness of the detector in generating an electric charge in response to incident photons. This efficiency is wavelength dependent, usually being greatest with visible light, and can vary greatly. Whereas the least expensive CCD chips may have a QE in the range of 10–20%, those that are useful for low light imaging are generally above 20%, and the best chips currently available show efficiencies of about 80%.

An important measure of the detector sensitivity is the signal-to-noise ratio (SNR), which may be defined as the magnitude of the signal divided by the uncertainty of that signal. This uncertainty is due to the presence of three types of noise. The first, called photon or shot noise, is due to the inherent variation in the photon flux of any source of light and is equal to the square root of the amplitude of the signal. This form of noise is therefore unavoidable, and defines the optimal SNR. The other two forms, readout and dark noise, are independent of the size of the signal. As the signal gets larger the photon noise forms a larger part of the total noise, which indicates that the noise level is approaching the theoretical minimum. Other things being equal, therefore, the larger the signal the better the SNR. Readout (or read) noise is due to the error introduced as the signal is transferred from the detector to be quantified and put into a usable electronic format. Dark noise arises from thermally generated electrons known as dark current. The amplitude of the dark noise is equal to the square root of the amplitude of the dark current. Since the dark noise generated is a function of time, it contributes little noise during rapid sampling. It is also strongly dependent on the temperature of the detector and thus cooling of the CCD camera, usually with a Peltier-type thermoelectric cooler, can drastically decrease the dark noise. Cooling from room temperature to –25 °C can reduce the dark current by 99% or more. Cooled CCD cameras are therefore particularly useful for applications in which there are low light levels and the requirement for long integration times. Under those conditions, both the readout noise and the dark noise are minimal, thereby allowing maximal SNR.

Another important parameter defining the sensitivity of a CCD camera is its dynamic range, which is a measure of the ability to quantify dim and bright light within a single image. This is expressed as the range between the read noise at each pixel, which is the minimum achievable noise, and the pixel full-well capacity, which is the total number of photons that each pixel can accept before saturating. For example, if the read noise of a CCD was 10 electrons and the full-well capacity was 40 000, the dynamic range of that chip would be 4000:1. This value would usually be expressed either as 12-bit (since 4000 is close to 2^{12} or 4096) or as 72 decibels (which is given by 20 × the log of the ratio). The dynamic range specifies the optimal precision with which the data should be digitized. For this example, it would be ideal to have a 12-bit digitization. If the data is digitized at a lower precision, say 8-bit, the discrimination of light levels within the image will be dependent on the digitization, rather than on the detector, and resolution will be lost.

4.3 Ca^{2+} indicator dyes

4.3.1 Single wavelength

The application of fluorescence to the measurement of free intracellular Ca^{2+} concentration $[Ca^{2+}]_i$ began with the development of Ca^{2+}-sensitive fluorescent indicator dyes (60). These dyes were developed from the highly selective Ca^{2+} chelator EGTA and its relative BAPTA. Since then, because of the usefulness of Ca^{2+} imaging techniques in answering a host of biological questions, there is now a myriad of Ca^{2+} indicator dyes differing in structure, fluorescence properties, sensitivity to Ca^{2+}, routes through which they are introduced into cells, and how they distribute within different cellular compartments. For reviews on Ca^{2+} indicator dyes see refs 61–64.

Fluorescent Ca^{2+} indicator dyes differ from other fluorophores in that their fluorescence properties change when they bind with free Ca^{2+}. Changes in $[Ca^{2+}]_i$

therefore result in a measurable alteration in the fluorescence, which can then be used to determine the location of changes in free Ca^{2+} level and also to estimate the amount by which it has changed. The types of changes that Ca^{2+} evokes in fluorescence properties are of three main types: a change in quantum yield, a change in the excitation spectrum, and a change in the emission spectrum. The simplest of these is the change in quantum yield; dyes that undergo this type of change are called single wavelength indicators. An example is illustrated in Fig. 9 (left panel), which shows the emission spectrum for rhod-2. This shows the intensity of the fluorescence emission measured at wavelengths from 550–650 nm (excitation at 540 nm) in a series of different Ca^{2+} concentrations. The emission maximum is invariant at about 580 nm. The intensity, however, is strongly dependent on the concentration of Ca^{2+}, being about 100-fold greater in saturating levels of Ca^{2+} than in the absence of Ca^{2+}. It is this property that enables the visualization of increases in $[Ca^{2+}]_i$. When the specimen is excited with light near 540 nm, the fluorescence measured at wavelengths near 580 will be a function $[Ca^{2+}]_i$.

Most of the common single wavelength indicators are based on the structures of one of the two archetypal fluorophores, fluorescein and rhodamine, which are excited by, and emit light in, the visible range. Those based on fluorescein, such as fluo-3 and the Calcium Green dyes (65), are excited by wavelengths of about 505 nm and emit at wavelengths of about 530 nm (green light), whereas those based on rhodamine, such as rhod-2 and Calcium Orange (65), are excited by wavelengths of about 555 nm and emit at wavelengths of about 575 nm (yellow light). Different dyes differ markedly in their quantum yield, fluorescence in the absence of Ca^{2+}, and enhancement of fluorescence at saturating Ca^{2+} (65).

Another important characteristic of indicator dyes is their affinity for Ca^{2+}. The

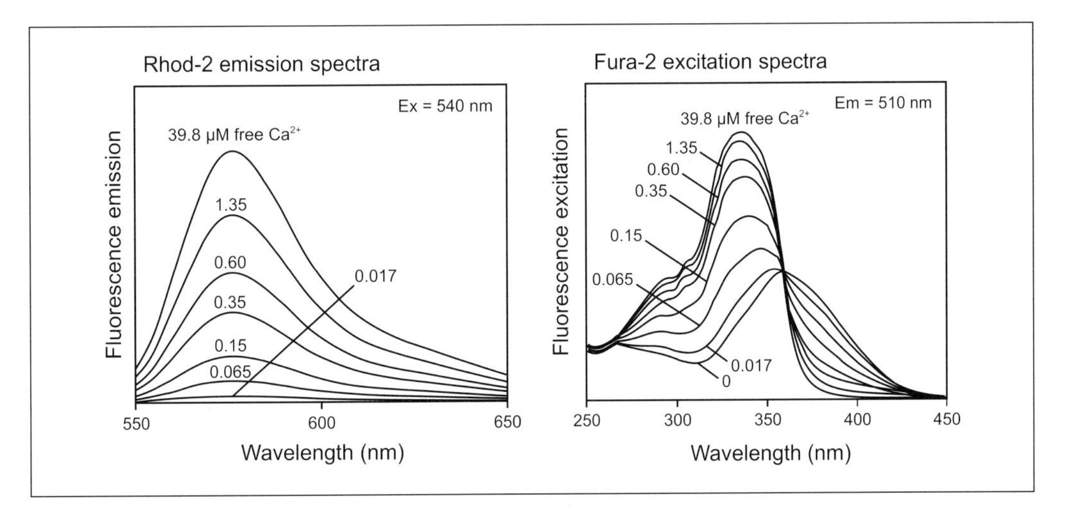

Fig. 9 Fluorescence spectra for rhod-2 and fura-2. The panel on the left shows emission spectra for rhod-2 at a series of Ca^{2+} concentrations using excitation light with a wavelength of 540 nm. The panel on the right shows excitation spectra for fura-2 at a series of Ca^{2+} concentrations as measured at a wavelength of 510 nm. Used with the permission of Molecular Probes (65).

Ca^{2+} dependence of the rhod-2 emission spectrum (Fig. 9) suggests that there would be substantial changes in the fluorescence intensity as cytoplasmic Ca^{2+} was increased from the usual resting level of 100 nM to levels in the low μM. Increases above that level, however, cause increasingly small changes in emission intensity. This reflects the affinity of rhod-2 for Ca^{2+}, which has an affinity constant (K_d) of about 570 nM. Ca^{2+}-sensitive indicator dyes are useful for the detection of $[Ca^{2+}]_i$ changes between about 0.1 K_d and 10 K_d, and the commonly used ones have K_ds in the hundreds of nanonolar. (These values are routinely determined *in vitro* in saline solutions, and it must be remembered that the affinity of a dye for Ca^{2+} in cytoplasm may be different.) Dyes have also been designed to have lower affinities for Ca^{2+} so that only very large increases in concentration will be detected. Examples are the low affinity indicator dyes Calcium Green-5N and Calcium Orange-5N (65), which have K_ds for Ca^{2+} of 14 μM and 20 μM, respectively. The low affinity of the protein indicator n-aequorin-J (a derivative of aequorin; see section 4.5) was exploited to image putative sites of neurotransmitter release at the squid giant synapse, which experience transient, very high levels of Ca^{2+} during depolarizations (66). Since only dye molecules in these 'microdomains' of Ca^{2+} influx were exposed to sufficiently high Ca^{2+} concentrations, the observed punctate patterns of fluorescence evoked during depolarization were thought to reflect sites of neurotransmitter release.

4.3.2 Dual wavelength dyes

A major disadvantage of the single wavelength dyes is that calibration is difficult and uncertain. The absolute amplitude of the fluorescence derived from such a dye in a cell depends on the cell geometry (due to out-of-focus light), the affinity of the interaction between Ca^{2+} and the dye in the cytoplasm, and the concentration of functional dye, none of which are known with certainty. Rigorous calibration of the signal therefore depends on post-experiment determination of the size of the signal in that cell in the absence of Ca^{2+} and in the presence of saturating Ca^{2+}. More commonly the signal is expressed as a ratio of a control image and an image obtained during or after a stimulus (such as an imposed depolarization). The size of the ratio in each pixel is a function of the change in Ca^{2+} at the corresponding position in the cell. The ratio image can therefore give an indication of where changes in Ca^{2+} levels occurred in the cell, and perhaps how quickly it occurred, but offers only an estimate of the magnitude of the changes.

To address this problem, dual wavelength ratiometric indicators were developed (67). An example is fura-2, which is the most widely used Ca^{2+} indicator dye. Ca^{2+} binding causes a change in the excitation spectrum of fura-2. This is illustrated in Fig. 9 (right panel), which shows the fluorescence emitted by fura-2 at 510 nm as a function of the excitation wavelength (between 250–450 nm) in the presence of different concentrations of Ca^{2+}. The observed shift in the excitation spectra with increased Ca^{2+}, from a peak excitation in 0 Ca^{2+} of 363 nm to one in saturating Ca^{2+} of 335 nm, provides a mechanism to estimate Ca^{2+} concentration that is independent of the absolute amplitude of fluorescence. This is typically accomplished by comparing the light emitted following excitation at 340 nm to that emitted following excitation at 380 nm. At low levels of Ca^{2+} this ratio will be small (see Fig. 9), but as the Ca^{2+} level increases the excitation at 340 nm will increase while that at 380 nm will decrease, causing an increase in the ratio. This ratio increases more than 50-fold from 0 Ca^{2+} to saturating Ca^{2+}, and can be correlated directly to the $[Ca^{2+}]_i$. Furapta, or mag-fura-2, is a chemical relative of fura-2 with a much lower affinity for Ca^{2+} (25 μM versus 145 nM), which makes it useful for imaging large changes in $[Ca^{2+}]_i$. Also note that when fura-2 is excited at a wavelength of 360 nm, the magnitude of evoked emission is independent of $[Ca^{2+}]_i$. This wavelength, called the isosbestic or isofluorescence point, is useful to monitor the changes in the concentration of functional dye molecules over time.

Using ratio measurements to calculate the $[Ca^{2+}]_i$ eliminates the problems of cell geometry and dye concentration that confound estimates of $[Ca^{2+}]_i$ made with single wavelength indicators. The use of fura-2 with ratiometric imaging therefore provides a means of accurately determining the magnitude as well as the location of changes in $[Ca^{2+}]_i$. The temporal resolution is not as great, however, because sequential measurements at two wavelengths are required to make the ratio. Another disadvantage is that the excitation energy required for fura-2 is in the ultraviolet range, which can activate autofluorescence in some preparations and which complicates the use of caged Ca^{2+} compounds (see Section 4.4).

Indo-1 is a ratiometric indicator for which Ca^{2+} causes a change in the emission rather than the excitation spectrum. Thus, instead of exciting the specimen at two wavelengths and detecting the emission at one, indo-1 is excited at a single wavelength and the emission measured at two. This could theoretically lead to a significant savings in time because the fluorescent emission could be divided and read by two separate detectors simultaneously. For Ca^{2+} imaging, however, problems associated with keeping the separate emission components properly aligned have made the use of indo-1 less convenient, and less utilized, than fura-2.

4.3.3 Loading and distribution of Ca^{2+} indicator dyes

Most frequently, Ca^{2+} indicator dyes are introduced into cells via a patch pipette. In experiments that do not require electrical control of voltage, however, this method may be impractical, or deleterious to cell function. A simpler and less invasive method is the use of esterified derivatives of Ca^{2+} indicator dyes (usually the acetoxymethyl or AM form) (68). These compounds are uncharged and are able to passively diffuse through cell membranes. Once inside the cell, the ester may be cleaved by the intracellular esterases found in most cell types to yield the functional indicator. This is particularly useful to study how externally applied agents influence intracellular Ca^{2+} dynamics. Ester loading of Ca^{2+} indicator dyes may also be combined with 'perforated patch-clamp' techniques to allow imaging of intracellular Ca^{2+} with voltage control without the disruption of cytoplasm that occurs during whole-cell patch-clamp recording. In conventional whole-cell patch-clamp, intracellular compounds involved in Ca^{2+} homeostasis may diffuse out of the cell. With the perforated patch technique, however, the electrical connection between the pipette and cell interior is made with pore-forming, polyene antibiotics such as nystatin. These compounds make holes in the cell membrane underneath the patch pipette allowing the passage of small monovalent ions, but excluding larger cytoplasmic components. Thus using a combination of ester loading of Ca^{2+} indicator dyes and perforated patch-clamp recording, it has been possible to image voltage gated Ca^{2+} influx at neurotransmitter release sites in cochlear hair cells with nearly intact cytoplasm (69).

Chemical modification of the indicator dyes has also been used to target their distribution within cells. An example is the conjugation of dyes with large dextran molecules, which decreases the likelihood of the dye being sequestered in intracellular organelles, or being bound to intracellular proteins. Either of these conditions can make the dye unresponsive to changes in cytoplasmic Ca^{2+} and thus bias the measurement. Lipophilic side chains have also been added to indicator dyes to target them to membranes to more accurately measure changes in Ca^{2+} close to Ca^{2+} voltage gated entry sites. A lipophilic derivative of fura-2 called FFP18 was used to measure the time course and amplitude of voltage gated submembranous Ca^{2+} changes in smooth muscle cells (70). A more specific strategy to target Ca^{2+} dyes was created by coupling a nuclear localization peptide with Calcium Green dextran, to measure Ca^{2+} changes within the nucleus (71). Ca^{2+} increases near the membrane may also be differentiated from increases in the bulk cytoplasm using excitation that is limited in its penetration of the cell. This has been done using total internal reflection fluorescence (which will be described in a following section). In this method the excitation light is limited to a space very close to the interface between the cell and the coverslip to which it is attached. This form of excitation was used to image submembranous Ca^{2+} increases in chemically stimulated neutrophils (72) and in cardiac muscle cells (73).

4.4 Caged Ca^{2+} compounds

A useful adjunct to the study of intracellular Ca^{2+} dynamics is the use of photosensitive 'caged compounds' (reviewed in refs 74 and 75). These are compounds whose activity may be abruptly altered by the absorption of a photon of UV light. Several types of caged compounds have been synthesized, including caged second messengers and caged transmitters, but the most relevant to studies of Ca^{2+} involvement in exocytosis are the caged Ca^{2+} molecules. These are again derivatives of Ca^{2+} chelator molecules, to which have been added photosensitive side groups that influence the affinity of the molecule for Ca^{2+}. The most popular example is called DM-nitrophen, which is a derivative of the Ca^{2+} chelator EDTA. This compound has a high affinity for Ca^{2+} ($K_d \approx 5$ nM), but following photolysis the affinity falls to about 3 mM. When DM-nitrophen bound to Ca^{2+} is introduced into a cell via a patch pipette, illumination with UV light (330–380 nm) will cause photolysis of a fraction of the DM-nitrophen and a release of bound Ca^{2+} into the cytoplasm. Other examples of caged Ca^{2+} compounds include the nitr family of compounds (which are derivatives of BAPTA) and nitrophenyl EGTA. Diazo-2, in contrast, is a photoactivatable Ca^{2+} chelator, which will rapidly decrease cytoplasmic Ca^{2+} following illumination with UV energy. In combination with Ca^{2+} imaging and a means to measure exocytosis, release of caged Ca^{2+} can be a useful tool in the elucidation of the Ca^{2+} dependence of the process of exocytosis. Photolysis of DM-nitrophen in combination with Ca^{2+} measurement with furapta and membrane capacitance measurements has been used to investigate the kinetics of secretion in pituitary melanotrophs (76), chromaffin cells (77), and mast cells (43), and to estimate the levels of Ca^{2+} required to evoke exocytosis in retinal bipolar terminals (78). The use of Ca^{2+} uncaging for these studies offers an important advantage, since it is possible to measure the homogeneous cytoplasmic increase evoked by uncaging Ca^{2+} from DM-nitrophen, whereas current Ca^{2+} imaging systems lack sufficient resolution to measure the very high $[Ca^{2+}]_i$ that occur in the narrow region close to voltage gated Ca^{2+} channels.

4.5 Proteinaceous Ca^{2+} indicators

A great deal of excitement has been generated recently over the possibilities of using recombinant engineered proteins to specifically target non-invasive Ca^{2+} indicators to specific sites in the cell (79–84). These proteins are derived from the two protein complex that is responsible for the light emitting properties of the jellyfish, *Aequorea victoria*. The first step in the generation of light is performed by the protein aequorin. Aequorin is bioluminescent, rather than fluorescent, meaning that in the presence of Ca^{2+} it will catalyse a chemical change in its covalently bound cofactor coelenterazine, leading to the emission of a photon of blue light (469 nm). Alternatively, the energy can be transferred to the second *A. victoria* protein, green fluorescent protein or GFP, which then becomes excited and releases a photon of green light (509 nm). GFP is itself also naturally fluorescent; it absorbs light with maxima at 395 nm and 470 nm and emits green light at a maxima of 509 nm. The fluorescence is inherent to the molecule and results from a post-translational cyclization reaction between serine, tyrosine, and glycine residues in positions 65–67 of the protein. Both aequorin (85) and GFP (86) have been cloned and have been transfected into a wide variety of cell types.

Aequorin has several advantages over fluorescent dyes as a Ca^{2+} indicator (81, 84). It appears to be less disruptive to cell function, contributes negligibly to Ca^{2+} buffering, has a very wide dynamic range (it may be used to measure Ca^{2+} from about 0.1 μM to greater than 100 μM), and has a very low background emission. On the other hand, the bioluminescent process is irreversible, meaning that each aequorin molecule can give off only one photon of light, and the amplitude of emitted light is low. Also, the necessity of introducing the aequorin via microinjection or by some form of cell permeabilization has until recently limited its utility. The cloning of the gene and the development of techniques to express the photoprotein in cell lines, however, has opened up a multitude of new possibilities. In particular, the addition of targeting sequences to the expressed protein has led to the specific localization of recombinant aequorin in various organelles (79, 87). Targeted locations include the mitochondrion (87), the nucleus (87), and, by fusing aequorin with SNAP-25, the plasma membrane (88).

Its intense fluorescence, small size, and benign nature have made GFP an ideal choice as a fluorescent tag to study the expression and localization of a large number of proteins (80–83). Furthermore, mutations of the GFP sequence have yielded proteins that are brighter and more stable, as well as variants with different excitation and emission spectra, including variants that emit blue rather than green light (89, 90) (reviewed in refs 81, 82, and 91). Although the fluorescence of GFP is not inherently sensitive to Ca^{2+}, linking the sequences of GFP variants with those of Ca^{2+} binding proteins have yielded hybrids that are capable of registering $[Ca^{2+}]_i$ (92–94). These strategies take advantage of a process known as fluorescence resonance energy transfer (FRET; see Fig. 10). This is a process in which the energy from one fluorescent molecule, without generating a photon, can excite a nearby molecule whose excitation spectrum overlaps with the emission spectrum of the first molecule. Therefore excitation of the first molecule causes the second molecule to emit a photon according to its emission spectrum (i.e. at a longer wavelength than the first molecule alone). The key is that FRET only occurs when the two molecules are very close together. This property has been exploited to produce Ca^{2+}-sensitive GFP hybrids by connecting two GFP variants using a linker whose conformation is somehow sensitive to the $[Ca^{2+}]_i$. The shift in conformation caused by Ca^{2+} binding either brings the variants closer together or moves them farther apart, thereby altering the amount of FRET. If the variants are distant from one another, excitation of the first

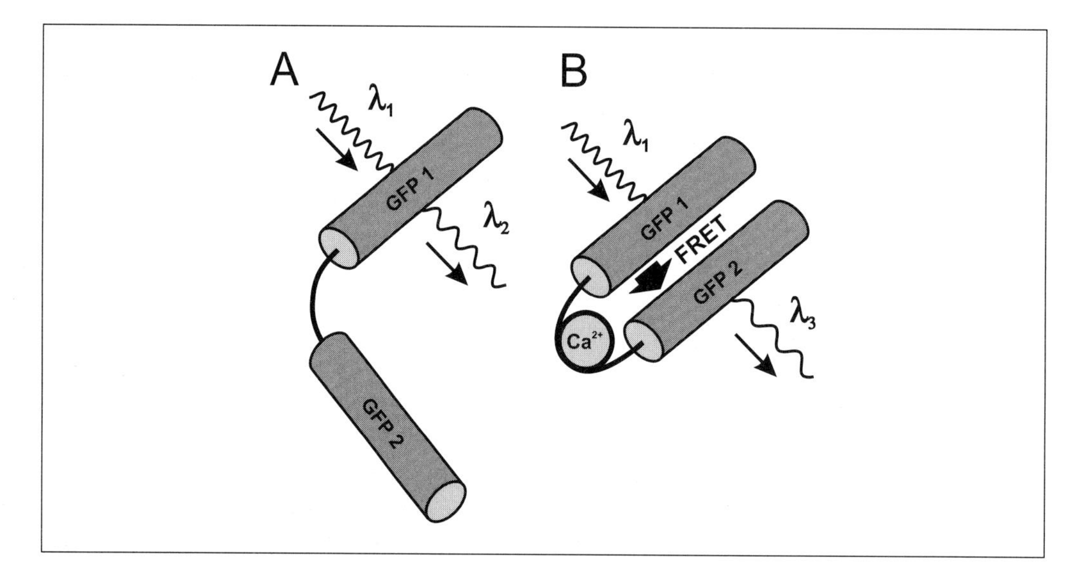

Fig. 10 Measurement of Ca^{2+} concentration using fluorescence resonance energy transfer (FRET). Two GFP variants with different fluorescent spectra (GFP 1 and GFP 2) are linked via a peptide (or combination of peptides) whose conformation may be altered by the binding of Ca^{2+}. (A) In the absence of Ca^{2+}, absorption of the excitation light (at a wavelength λ_1) by GFP 1 elicits the release of a photon from GFP 1 (at a wavelength λ_2). (B) The binding of Ca^{2+} to the linker peptide brings the two GFP variants closer together. At this shorter distance, the energy absorbed by GFP 1 may be passed to GFP 2 by FRET, resulting in the release of a photon from GFP 2 at a wavelength (λ_3) that is longer than λ_2. The ratio of light emitted at λ_2 and λ_3 is therefore proportional to the concentration of Ca^{2+}.

variant will result in emitted light according to its emission spectra. If the two variants are close together, however, the energy will be passed to the second variant by FRET and the emitted light will have a wavelength according to the emission spectrum of that variant. By monitoring the fluorescence emission at the expected frequencies for the two variants, one can assess the degree to which FRET is occurring, which reflects the conformation of the linker, and thus $[Ca^{2+}]_i$.

GFP hybrids are brightly fluorescent, innocuous, and can be made to detect Ca^{2+} over a very wide range by altering the sensitivity of the linker (92). Also, as with aequorin, they may be tagged with signal sequences to target them to different cellular organelles (92). The targeting of Ca^{2+}-sensitive photoproteins using molecular biological approaches could potentially yield very useful information about Ca^{2+} involvement in exocytosis by, for example, the targeting of aequorin to synaptic vesicles, or to dense core vesicles, or to the synaptic active zone.

4.6 New imaging techniques

In the following section we will describe some recently developed techniques for measuring intracellular Ca^{2+} levels using fluorescent indicator dyes. These techniques employ unique methods for the excitation and detection of fluorescence resulting in improved spatial and/or temporal resolution of Ca^{2+} changes.

4.6.1 Pulsed laser imaging

Pulsed laser imaging uses a novel strategy to approach the problem of imaging rapid changes in Ca^{2+} levels. Rather than increasing the rate of acquisition of images, with the attendant problems of decreased signal intensity and increased readout noise, pulsed laser imaging relies on a brief, intense stimulus to evoke a brief, intense response. A system for measuring Ca^{2+} influx using pulsed laser imaging was described originally by Monck *et al.* (95), and later expanded upon (96, 97). The system works by directing pulses of light from a flash lamp dye laser through a microscope objective. The timing of this excitation relative to a stimulus, such as a depolarization, is computer controlled. The resultant image can then be divided by a control image, taken one second before the stimulus, to determine the change in the fluorescent pattern. The brevity of the laser pulse (350 nsec) ensures the high temporal resolution of the measurement, since only dye molecules excited during the pulse will emit fluorescent light. Since only one image is taken for each pulse of laser light the readout noise is minimized and dark noise is minimized by the use of a cooled CCD camera as a detector. The intensity of the excitation pulse ensures that the emitted signal will have sufficient photon density to give a good SNR, while the shortness of the length of the pulse prevents the high intensity light from causing significant photobleaching of the indicator dye. In combination with a microscope with good optical capabilities, pulsed laser imaging is capable of offering sub-millisecond temporal resolution, as well as spatial resolution of less than a micrometre. The disadvantage is that since it gives only a single image per laser pulse, the time course of a response can only be reconstructed by repeating the stimulus and changing the time of the laser pulse. In combination with an ultraviolet pulsed laser used for uncaging, this system is capable of measuring the release of Ca^{2+} from DM-nitrophen with exquisite time resolution (97).

An example of an application of pulsed laser imaging is shown in Fig. 11 (96). The image on the left of the figure shows Ca^{2+} influx through voltage gated Ca^{2+} channels in a patch-clamped bovine chromaffin cell activated by a 50 msec depolarization to +20 mV. Influx is clearly highly localized to a small number of 'hotspots' (in this case three) that correspond to clusters of Ca^{2+} channels. This data has been combined with amperometric recordings (Fig. 11, right) to show that the locations of Ca^{2+} influx are associated with preferential sites of exocytotic release (96). This unexpected finding suggests that chromaffin cells have specialized release zones where Ca^{2+} channels and the exocytotic apparatus are co-localized.

4.6.2 Laser scanning confocal microscopy

When one considers the ubiquity and importance of confocal microscopy, it is startling to consider that the first commercial versions became available in the late 80s. Confocal microscopy offers a useful solution to the problem of image blurring

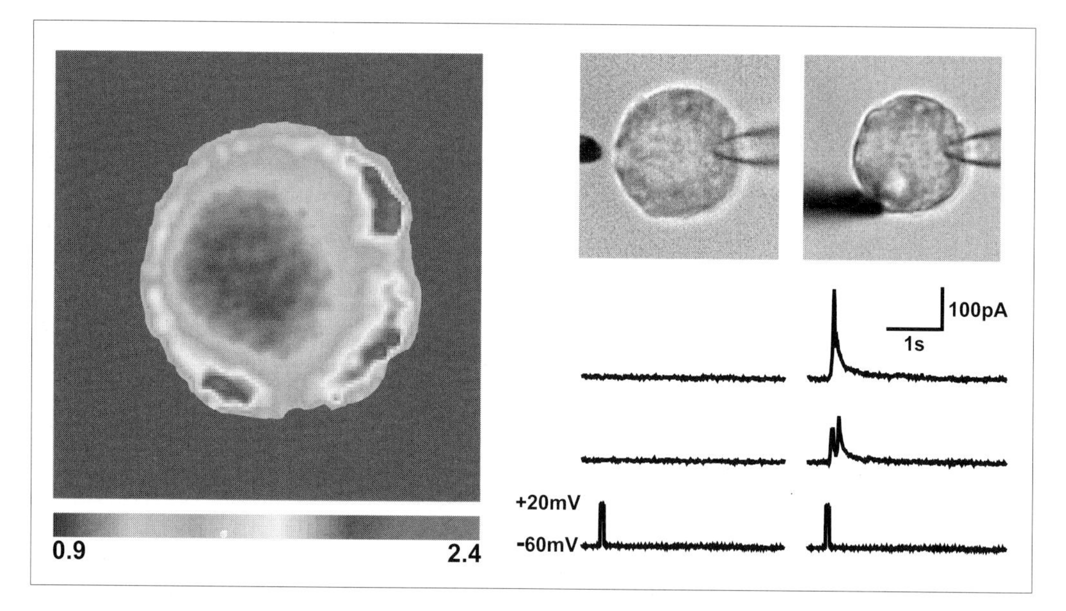

Fig. 11 Pulsed laser imaging of Ca^{2+} influx combined with amperometric detection of release. The image on the left shows the highly localized Ca^{2+} influx evoked in a patch-clamped chromaffin cell by a 50 msec depolarization from −60 mV to +20 mV. The false colours (see scale below image) indicate the ratio of Ca^{2+} levels compared to that in a control image. The upper two images on the right show the same cell with an amperometric carbon fibre in two different positions. When in the first position, which is not close to one of the hotspots of Ca^{2+} entry, the fibre does not detect release of catecholamines. When in the second position, however, which is close to one of the Ca^{2+} entry sites, depolarization of the cell evokes clear spikes in the amperometric trace. These data demonstrate that the sites of Ca^{2+} influx are associated with sites of preferential release of catecholamines. Reproduced from Robinson *et al.* (1995) *Proc. Natl. Acad. Sci. USA*, **92**, 2474–8, by copyright permission of The National Academy of Sciences, USA. See colour plate section between pages 46 and 47.

due to out-of-focus light. In conventional fluorescence microscopy, the quality of the measured image is degraded by the detection of light originating away from the focal plane. Confocal microscopy, rather than trying to resolve simultaneously the details in a full image of a specimen, measures the emitted light from tiny sections of the specimen individually. By scanning the image to measure sequentially the light emitted from each such section, the entire sample can be mapped. The principle behind the key technique used to exclude out-of-focus light from the detector is illustrated in Fig. 12A. When a small aperture or pin-hole (usually 0.5–10 mm) is placed in front of the emission detector, the light impinging on the detector is limited to that coming from a point source on the sample. Only the light from that spot will converge at the pin-hole to pass through to the detector. Light coming from points above or below the focal plane will be focused by the objective to points above or below the pin-hole such that little of the light will strike the detector. Likewise, light from points adjacent to the point of interest on the same plane will be focused by the objective away from the pin-hole and will thus be excluded. With the original confocal designs, this mechanism was matched by a similar pin-hole restricting illuminating light from the microscope condenser to a fine point. (The word confocal, in fact, refers to the coincidence of the focal planes for illumination and detection.) This restricted illumination further decreases the amount of out-of-focus light that reaches the detector, resulting in a high resolution image at the point of interest. The problem with the original designs employing a condenser is that it is difficult to scan the point of focus across the entire specimen to reconstruct a complete image. This could be achieved by moving the specimen relative to a stationary confocal spot; this method, however, is slow and inappropriate for the measurement of most Ca^{2+} signalling events. A much faster approach is to use laser illumination, which forms the basis of all commercially available confocal microscopes. In a typical laser scanning confocal microscope (see Fig. 12B), a laser beam source light is deflected via a dichroic mirror onto two sequential scanning mirrors and then through the microscope objective onto the specimen. The scanning mirrors rapidly oscillate to deflect the beam over the specimen in a raster pattern; one mirror moves the beam in the '*x*' dimension while the other moves it in the '*y*' dimension. They also serve to deflect the light emitted from the sample through the dichroic mirror (which blocks any reflected excitation light) and through the pin-hole aperture to the photomultiplier. The sequence of light intensities detected is then used to reconstruct the entire pattern of fluorescent light emitted by the sample on that particular focal plane. Note that in confocal microscopy the detector itself is therefore not required to differentiate admitted light spatially, since information is processed from a single defined point at a time. For this reason the detector used is a simple photomultiplier, which has a high sensitivity, dynamic range, and response rate, but captures no spatial information. Spatial information is collected by moving the point of interest throughout the specimen.

Laser illumination gives a well-defined, narrow excitation beam. In combination with the pin-hole aperture to block out-of-focus emitted light, confocal laser scanning Ca^{2+} imaging gives a fluorescent signal with excellent spatial resolution. The

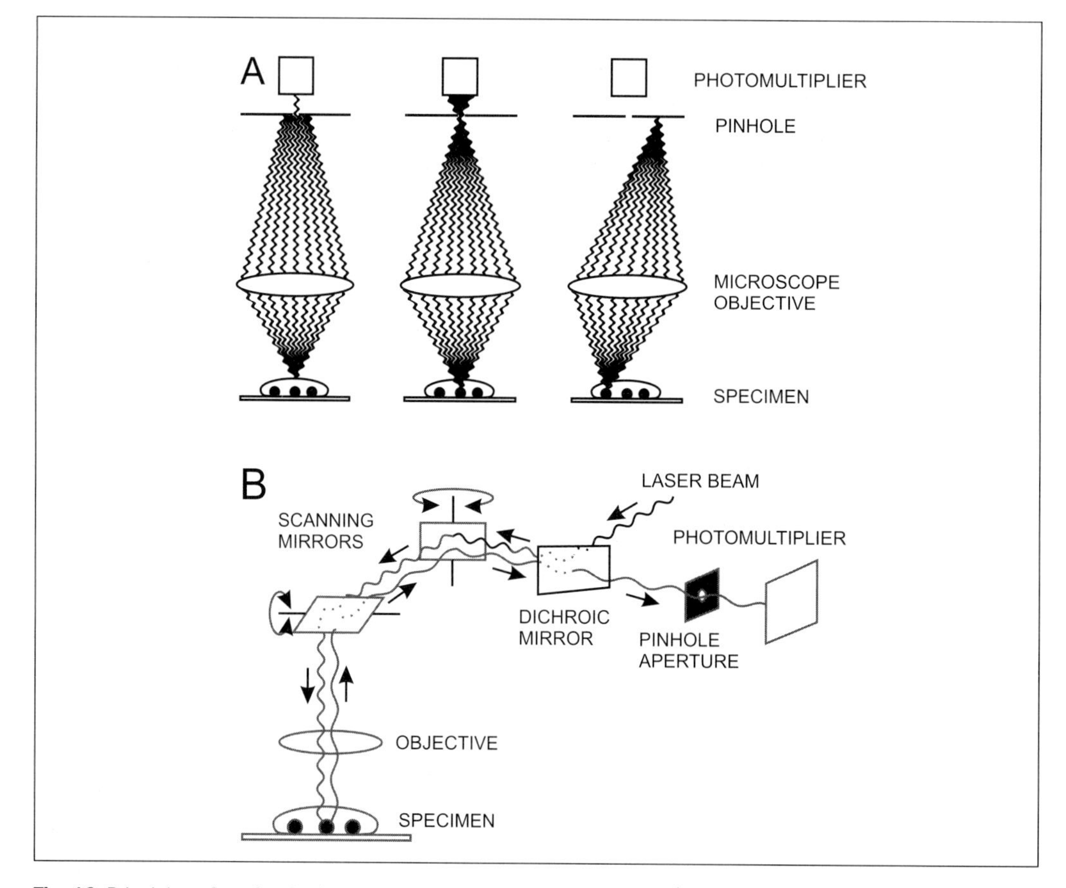

Fig. 12 Principles of confocal microscopy. (A) These illustrations demonstrate how a pin-hole aperture in front of the photomultiplier detection system of a confocal microscope can eliminate out-of-focus light. The focused light emanating from an object in the sample will pass through the pin-hole (centre) while most of the light from a different focal plane (left) or from an adjacent spot on the same focal plane (right) will be excluded. (B) This diagram illustrates a typical configuration for a laser scanning confocal microscope (see text for description).

laser also serves to limit photobleaching in the sample by minimizing the excitation field. A further, crucial advantage for the use of lasers in confocal Ca^{2+} imaging is the intensity of the illumination that it provides. To make a 512 × 512 pixel image of a specimen in one second it is necessary to capture each pixel emission in about four microseconds. In order to have sufficient photons to achieve a good SNR (particularly since most of the emitted light is blocked by the pin-hole aperture) it is therefore crucial to have an intense, and highly stable excitation source. The most commonly used laser sources are gas lasers made with either argon (which has principal emission wavelengths of 488 nm and 514 nm) or mixed argon–krypton (which has principal emission wavelengths of 488, 568, and 647 nm). It must also be remembered that when the source light is too intense the fluorophore might saturate (i.e. all of the molecules may become excited) thus degrading the image resolution.

Damage due to photobleaching also increases with laser intensity. These considerations therefore limit the temporal resolution of the method. Nevertheless, confocal laser scanning microscopy has been used at video rate (30/second) acquisition, for examples, to image hotspots of Ca^{2+} influx in cochlear hair cells (98) and to study Ca^{2+} dynamics in dendritic spines of cerebellar Purkinje neurons (99). An alternative way to increase temporal resolution, at the expense of spatial resolution, is to repetitively scan a single line. This is called line scan mode and since the scans can be acquired in a millisecond or less, this is a useful technique to resolve the time course of an event when its location is known.

Two excellent books covering all aspects of confocal microscopy are refs 100 and 101. Briefer descriptions are also available (102–104).

4.6.3 Two photon excitation

Two photon excitation is based on the ability of fluorophores to be excited by the simultaneous absorption of two photons, each of which has one-half the energy (i.e. calcium at twice the wavelength) of that usually required for excitation (105–107). For example Green, which has maximal absorption at 506 nm, can be excited with laser light at a wavelength of 850 nm (108). This phenomenon has been exploited as a novel means of excitation for laser scanning microscopy and has several exciting implications. Foremost is that the probability of excitation becomes a steep function of the distance from the focal plane. Since the simultaneous absorption of two photons is required, excitation is proportional to the square of the beam intensity. In a cone of illumination, the photon density is inversely proportional to the square of the distance from the focal plane. Therefore the probability of excitation with two photon excitation decreases in proportion to the fourth power of this distance. This enables the specific excitation of minute volumes of tissue ($\approx 1.0\ \mu m^3$) (109), with very little excitation or photobleaching occurring elsewhere in the light path. The area of excitation is so well defined that it is not generally necessary to reject out-of-focus light at the detector using a pin-hole aperture as with single photon confocal measurements. Therefore emitted light that is scattered during its path out of the tissue still contributes to the signal. This is an important advantage for two photon excitation since scattering of the fluorescence signal limits the depth to which confocal is useful. Furthermore, the longer wavelengths used for two photon excitation are less likely to be scattered by tissue than are the wavelengths for excitation used in confocal microscopy. The two photon excitation technique is therefore able to excite an area of interest at a greater depth than is regular confocal, and is also able to detect the fluorescence emitted from that depth with less loss due to light scattering.

Because the probability of two photons being absorbed simultaneously depends on the square of the local light intensity, the source beam must be intense. The time averaged intensity of the excitation, however, is limited by the photodamage and heating caused by absorption of the infrared excitation light along its path through the sample. This problem was solved by the use of a laser that gives repetitive bursts of high energy with a very small duty cycle. The optimum duty cycle was found to be laser pulses of 100 femtoseconds repeated at a rate of 100 MHz (i.e. once every 10 nanoseconds). This duty cycle (1:100 000) gives a level of excitation at the point of interest sufficient to generate a fluorescent image in a reasonable amount of time without excessive damage to the sample. The delay between pulses is ideal because it enables the fluorophores to return from the excited state, which has a lifetime of 1–10 nanoseconds (see Section 4.1), to the ground state. This rate therefore takes full advantage of the fluorescence cycle; faster rates would risk saturation, while slower rates leave the fluorophores idle.

Two photon excitation has been used to image Ca^{2+} increases in dendritic spines in slices of rat hippocampus (108) and to reconstruct the dendritic tree (to a depth of 500 μm!) of neocortical pyramidal neurons *in situ* (109). Two photon excitation has also been used for the highly localized release of caged neurotransmitters (110).

5. Imaging of vesicle dynamics

It is beyond the capability of light microscopy to visualize synaptic vesicles. Their size, which is on the order of 40–50 nm in diameter, is far below the theoretical limit to the resolution of light microscopy (about 0.2 μm). Our knowledge of the structure of synaptic vesicles therefore depends largely on electron microscopy of fixed cells (see for example, ref. 111). Large dense core vesicles, in contrast, can be seen in certain preparations under optimal conditions (see Fig. 13 for example), but not well enough to make practical measurements of the dynamics of exocytosis and endocytosis. To visualize vesicular dynamics is it therefore necessary to fluorescently tag the membranes to make them more visible. It is not sufficient to stain membranes non-specifically, however, since the resultant background would prevent the detection of the small changes due to exocytosis or endocytosis.

5.1 Activity-dependent labelling; the FM dyes

The key development that allowed experiments on vesicle dynamics to become viable was the invention of activity-dependent dyes. These are dyes that stain internal membranes only following secretory activity. These were first used at the snake neuromuscular junction (112). It was later discovered, however, that the dyes used were not useful in non-reptilian cells. Success in mammalian preparations was found with the styryl dye RH-414, originally designed as a potentiometric probe, and with a series of related compounds, known as FM dyes, that were synthesized for this purpose (113, 114). This series includes several compounds differing in chemical and spectral properties. The most widely used are FM1-43, which has excitation and emission spectra maxima at 480 nm and 600 nm, respectively, and FM4-64, which has spectra maxima at 505 nm and 670 nm (65). Their usefulness has been demonstrated at the neuromuscular junctions in several vertebrates (113–115), where intracellular staining was specific for nerve terminals, occurred only in response to activation, and, following removal of the dye from the media, could be destained by further activation. These dyes have since been used to examine vesicle dynamics at single synaptic boutons of hippocampal neurons in culture (116–120), in spinal neurons

(121), in chromaffin cells (122), in the synaptic terminals of goldfish retinal bipolar cells (123), and in the *Drosophila* neuromuscular junction (124).

The FM dyes have several properties that make them ideal activity-dependent dyes (125). Their amphiphilic structure (they have a charged head group and a lipophilic tail group) makes them water soluble, yet enables them to partition rapidly and reversibly into cell membranes. Thus, following addition of dye to the solution bathing a resting cell, only the plasma membrane will be stained. The charged head group prevents the dyes from crossing membranes, keeping them in the membrane outer leaflet and thereby maintaining the specificity of the labelling. Internal membranes will only be stained when there is direct contact between organellar membranes and the plasma membrane, such as during exocytosis. Finally, the dyes show little fluorescence when in aqueous solution, but show a dramatic increase in fluorescence when dissolved in cell membranes. It is thus possible to image membranes in the continued presence of dyes in the medium surrounding the imaged cell. In the absence of dye in the bathing solution, staining of the plasma membrane will disperse as the dye diffuses away. The only staining that remains in that situation will be the staining of internal membranes that resulted from endocytosis while the plasma membrane was stained. The subsequent loss of this internal staining will be activity-dependent and can therefore be used to monitor exocytotic activity. The staining of internal membranes using activity-dependent dyes is beautifully illustrated in Fig. 13 (126). Figure 13A (upper panel), shows selective staining of the plasma membrane of sea urchin eggs in the presence of RH-414. Upon fertilization of the eggs there was a rapid and massive internalization of membrane that is shown by the appearance of stained granules in the egg cytoplasm. Figure 13B depicts this process at the level of a single granule (in this case using FM1-43). By viewing the area just underneath the cell membrane with light microscopy (using differential interference contrast optics), and simultaneously monitoring fluorescence emission with a confocal microscope, Whalley *et al.* (126) were able to image the exocytosis and endocytosis of a single granule. While the left images show a granule that is not stained, the images on the right show that following exocytosis and endocytosis, the granule is stained with the fluorescent dye. The re-formed granule is invisible to DIC optics because its contents have been exchanged for the extracellular solution and thus have lost the refractive contrast which underlies DIC visualization.

FM1-43 has been used in combination with capacitance measurements in bovine chromaffin cells (122). Unlike at synapses, endocytosis was found to be delayed in chromaffin cells and the experimenters used the increase in staining at the membrane in the presence of FM1-43 as a measure of the insertion of new membrane due to exocytosis. These experiments showed that there is an excellent correlation between the measured increase in capacitance and the increase in cell surface staining caused by short depolarizations. This supports the hypothesis that the two methods are measuring the same phenomenon. After longer periods of depolarization, however, the capacitance changes levelled off while the fluorescence continued to increase. The simplest interpretation is that during prolonged stimulation the rate of

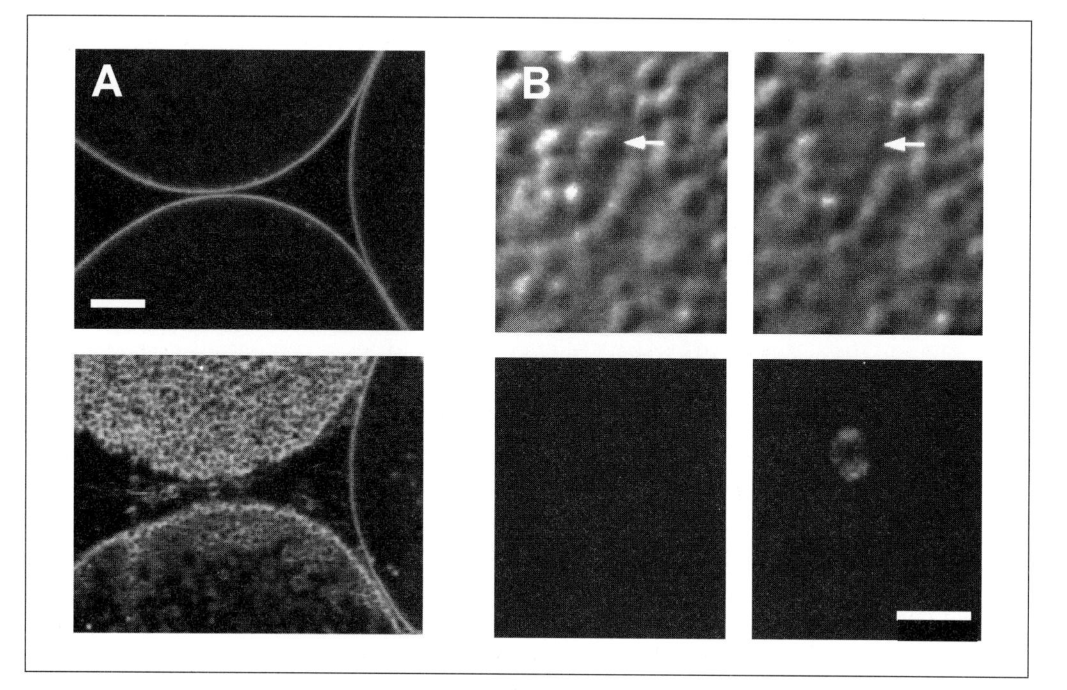

Fig. 13 Measurements of vesicle dynamics with activity-dependent styryl dyes. (A) The fluorescent labelling of sea urchin egg vesicles with RH-414. At rest (upper image), RH-414 labels only the plasma membranes. Fertilization causes a massive wave of exocytosis–endocytosis leading to the staining of internal vesicular membranes (lower image). The top egg was the first to be fertilized and the egg on the right has not yet been fertilized. Bar, 10 μm. (B) Correlation of granule fusion and the appearance of FM1-43 staining. The images show simultaneous DIC images and confocal laser images of FM1-43 fluorescence. Before fertilization the granules beneath the membrane are not stained with FM1-43 (lower left image). Fertilization causes the exocytosis and endocytosis of a granule causing the disappearance of the granule from the DIC image (arrows) and the appearance of FM1-43 staining in the re-formed granule (lower right image). Bar, 2.5 μm. Reproduced from *J. Cell Biol.* (1995) **131**, 1183–92, by copyright permission of The Rockefeller University Press.

endocytosis builds up to the point where there is no net change in capacitance; the rate of staining of the membrane, however, continued to increase reflecting only continued exocytosis. The difference between the two traces therefore provides a useful measure of the process of endocytosis. Results of uptake studies with FM1-43 have also been compared to results using a related strategy, uptake of a fluorescently labelled antibody to the lumenal domain of a synaptic vesicle protein (synaptotagmin) (118). These results suggested that the uptake of the labelled antibody may label a more specific vesicular compartment.

FM1-43 staining has also been used in combination with Ca^{2+} indicator dyes to obtain simultaneous measurements of Ca^{2+} influx and vesicle dynamics (120, 127, 128). In goldfish retinal bipolar terminals, FM1-43 and fura-2 measurements were used with the release of caged Ca^{2+} to estimate the vesicle turnover rate in response to Ca^{2+} increases and to demonstrate that this terminal is adapted for continuous secretion of neurotransmitter (123). FM1-43 has been used to assess vesicle turnover in cells from experimental animals with genetic alterations in the secretory process (124, 129, 130), and to study mechanisms of neurotransmitters and second messengers that modulate secretion (131–133).

5.1.1 Quantal measurements of release

By careful optimization of the conditions, it has been possible to refine the sensitivity of the FM1-43 technique to the visualization of the exocytosis and endocytosis of single synaptic vesicles in cultured hippocampal neurons (134–136). In these experiments the background fluorescence was minimized by the use of confocal laser imaging, sparse neuronal cultures, and minimal staining stimuli. Under these conditions, it was possible to see minute changes in fluorescence at putative release sites following small numbers (1–20) of evoked action potentials. The amplitude distributions of large numbers of these changes in many preparations (minus the background fluorescence remaining following sustained activity) displayed clear peaks with the amplitude of each peak being a multiple of the amplitude of the first. This quantized release is likely to represent the release of a specific number of stained vesicles. Since the quantal size of FM1-43 taken up by endocytosis in response to small stimuli was found to be the same size as that released by subsequent stimulation, Murthy and Stevens concluded that synaptic vesicles did not interact with internal membranes during the endocytic cycle (136).

5.2 Total internal reflection fluorescence microscopy

In contrast to the rapid turnover and refilling of synaptic vesicles, peptide-containing granules are released slowly and are replaced with granules manufactured in the cell soma. Peptide-secreting neuroendocrine cells may therefore not be good candidates for experiments with activity-dependent dyes. Secretory granules, however, may be labelled and monitored in other ways. Two groups recently used transfection of PC12 cells (a neuroendocrine cell line) with constructs made by fusing enhanced mutant GFPs with sequences for peptide prohormones (137, 138). Both groups showed that the expressed protein was properly packaged in dense core granules and was secreted in response to K^+-evoked depolarization in a Ca^{2+}-dependent fashion. Both groups were able to track the movement of individual, fluorescently tagged granules, for periods lasting tens of seconds, through the PC12 cell somata. Different methods were used for visualization of the granules, however. While Burke *et al.* (138) used conventional confocal imaging, Lang *et al.* (137) used a method that is new to the field of neurobiology: total internal reflection fluorescence microscopy (TIRFM).

This method constitutes a unique way of selectively illuminating the portion of the cell very near to the cover slide to which it is attached. When a coherent beam of light passing through one media strikes a second media with a lower refractive index, the light is bent, or refracted. If the angle of incidence of the incoming light exceeds a critical angle that depends on the ratio of the two refractive indices the beam is totally

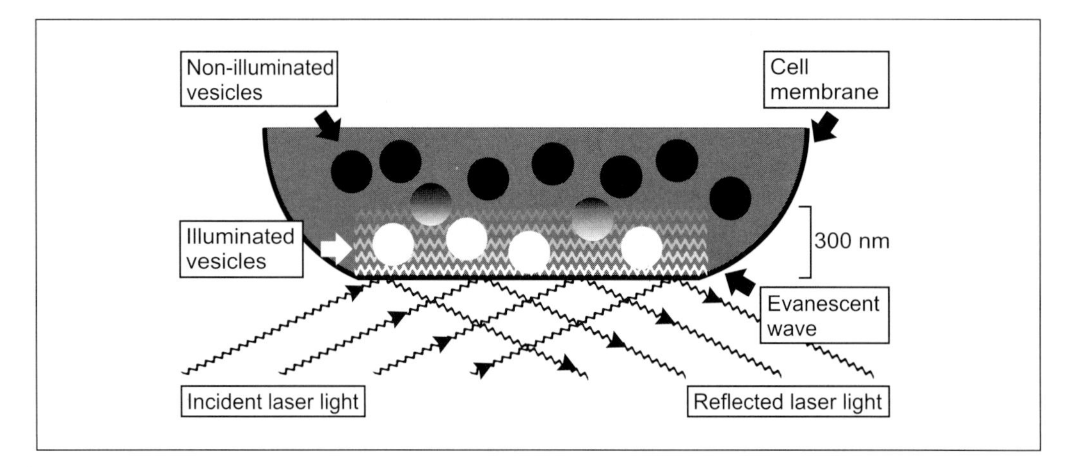

Fig. 14 Total internal reflection fluorescence microscopy. In TIRFM, the incident laser light is totally reflected at the interface between the glass coverslip and the aqueous bathing solution. However, interference between incident and reflected light creates an evanescent wave of light with an intensity that decays exponentially with distance from the interface. Fluorescently tagged granules within the evanescent wave (about 300 nm) are thus selectively excited.

reflected (see Fig. 14). Interference between the incident and reflected beams, however, results in a narrow electromagnetic field, 'the evanescent wave', that propagates parallel to the optical interface. The energy of this field falls off exponentially with distance from the interface, has a width on the order of hundreds of nanometres (depending on the wavelength and angle of the incident light, as well as the two refractive indices), and has a wavelength equal to that of the incident light.

This phenomenon has been exploited to achieve selective illumination of fluorescently stained large dense core granules near the cell membrane in PC12 cells (137) and in chromaffin cells (140, 141). In these examples granules were stained either by expression of GFP-tagged peptide hormones (see Section 4.5) (137) or with dyes that selectively stain acidic intracellular compartments (140, 141). Stained granules were excited by TIRFM such that only those within 300 nm of the coverslip/aqueous media interface were illuminated. Using this spatially resolved excitation, the authors were able to visualize a small subset of the total number of granules in the cell. This enabled them to identify and locate specific granules, to track their movement, to see them 'dock' to sites on the membrane, and to see them disappear in response to evoked secretion.

Like two photon excitation, therefore, TIRFM depends on a highly localized excitation light to derive a spatially resolved image, with little photobleaching due to unintended illumination. It also eliminates the need for the use of a pin-hole to remove out-of-focus light and the use of time-consuming scanning of an image, allowing a better temporal resolution. It lacks the ability of the confocal microscope, however, to scan different focal planes.

Acknowledgements

We express our gratitude to Professor Julio M. Fernandez for his support and encouragement, Steve Vogel for permission to use Fig. 13, and Julio Fernandez, Piotr Marszalek, Mariano Carrion-Vazquez, and Bruna Bonavia-Fisher for their helpful comments on the manuscript. This work was supported by NIH grants R01 NS 35866–3 and R01 GM46688–07 (to J. M. F.).

References

1. Cole, K. S. (1935) Electric impedance of Hipponoe eggs. *J Gen Physiol*, **18**, 877.
2. Hamill, O. P., Marty, A., Neher, E., Sakmann, B., and Sigworth, F. J. (1981) Improved patch-clamp techniques for high-resolution current recording from cells and cell-free membrane patches. *Pflugers Arch*, **391**, 85.
3. Neher, E. and Marty, A. (1982) Discrete changes of cell membrane capacitance observed under conditions of enhanced secretion in bovine adrenal chromaffin cells. *Proc Natl Acad Sci USA*, **79**, 6712.
4. Fernandez, J. M., Neher, E., and Gomperts, B. D. (1984) Capacitance measurements reveal stepwise fusion events in degranulating mast cells. *Nature*, **312**, 453.
5. Joshi, C. and Fernandez, J. M. (1988) Capacitance measurements. An analysis of the phase detector technique used to study exocytosis and endocytosis. *Biophys J*, **53**, 885.
6. Fidler, N. and Fernandez, J. M. (1989) Phase tracking: an improved phase detection technique for cell membrane capacitance measurements. *Biophys J*, **56**, 1153.
7. Lindau, M. and Fernandez, J. M. (1986) A patch-clamp study of histamine-secreting cells. *J Gen Physiol*, **88**, 349.
8. Penner, R. and Neher, E. (1989) The patch-clamp technique in the study of secretion. *Trends Neurosci*, **12**, 159.
9. Ashcroft, F. M., Proks, P., Smith, P. A., Ammala, C., Bokvist, K., and Rorsman, P. (1994) Stimulus-secretion coupling in pancreatic beta cells. *J Cell Biochem*, **55**, 54.
10. Matthews, G. (1996) Synaptic exocytosis and endocytosis: capacitance measurements. *Curr Opin Neurobiol*, **6**, 358.
11. Holevinsky, K. O. and Nelson, D. J. (1998) Membrane capacitance changes associated with particle uptake during phagocytosis in macrophages. *Biophys J*, **75**, 2577.
12. Gillis, K. D. (1995) Techniques for membrane capacitance measurements. In *Single-channel recording* (ed. B. Sakmann and E. Neher), p. 155. Plenum Press, New York.
13. Chandler, D. E. and Heuser, J. E. (1980) Arrest of membrane fusion events in mast cells by quick-freezing. *J Cell Biol*, **86**, 666.
14. Breckenridge, L. J. and Almers, W. (1987) Currents through the fusion pore that forms during exocytosis of a secretory vesicle. *Nature*, **328**, 814.
15. Alvarez de Toledo, G. and Fernandez, J. M. (1988) The events leading to secretory granule fusion. *Soc Gen Physiol Ser*, **43**, 333.
16. Lindau, M., Nusse, O., Bennett, J., and Cromwell, O. (1993) The membrane fusion events in degranulating guinea pig eosinophils. *J Cell Sci*, **104**, 203.
17. Valtorta, F., Fesce, R., Grohovaz, F., Haimann, C., Hurlbut, W. P., Iezzi, N., *et al.* (1990) Neurotransmitter release and synaptic vesicle recycling. *Neuroscience*, **35**, 477.
18. Monck, J. R. and Fernandez, J. M. (1994) The exocytotic fusion pore and neurotransmitter release. *Neuron*, **12**, 707.

19. Rahamimoff, R. and Fernandez, J. M. (1997) Pre- and postfusion regulation of transmitter release. *Neuron*, **18**, 17.
20. Spruce, A. E., Breckenridge, L. J., Lee, A. K., and Almers, W. (1990) Properties of the fusion pore that forms during exocytosis of a mast cell secretory vesicle. *Neuron*, **4**, 643.
21. Johnson, R. G. and Scarpa, A. (1984) Chemiosmotic coupling and its application to the accumulation of biological amines in secretory granules. *Soc Gen Physiol Ser*, **38**, 71.
22. Kelly, R. B. (1993) Storage and release of neurotransmitters. *Cell*, **72** Suppl, 43.
23. Unwin, P. N. and Zampighi, G. (1980) Structure of the junction between communicating cells. *Nature*, **283**, 545.
24. Monck, J. R., Alvarez de Toledo, G., and Fernandez, J. M. (1990) Tension in secretory granule membranes causes extensive membrane transfer through the exocytotic fusion pore. *Proc Natl Acad Sci USA*, **87**, 7804.
25. Oberhauser, A. F., Monck, J. R., and Fernandez, J. M. (1992) Events leading to the opening and closing of the exocytotic fusion pore have markedly different temperature dependencies. Kinetic analysis of single fusion events in patch-clamped mouse mast cells. *Biophys J*, **61**, 800.
26. Nanavati, C., Markin, V. S., Oberhauser, A. F., and Fernandez, J. M. (1992) The exocytotic fusion pore modeled as a lipidic pore. *Biophys J*, **63**, 1118.
27. Oberhauser, A. F. and Fernandez, J. M. (1993) Patch clamp studies of single intact secretory granules. *Biophys J*, **65**, 1844.
28. Monck, J. R. and Fernandez, J. M. (1992) The exocytotic fusion pore. *J Cell Biol*, **119**, 1395.
29. Nanavati, C. and Fernandez, J. M. (1993) The secretory granule matrix: a fast-acting smart polymer. *Science*, **259**, 963.
30. Monck, J. R., Oberhauser, A. F., and Fernandez, J. M. (1995) The exocytotic fusion pore interface: a model of the site of neurotransmitter release. *Mol Membr Biol*, **12**, 151.
31. Leszczyszyn, D. J., Jankowski, J. A., Viveros, O. H., Diliberto, E. J., Jr., Near, J. A., and Wightman, R. M. (1990) Nicotinic receptor-mediated catecholamine secretion from individual chromaffin cells. Chemical evidence for exocytosis. *J Biol Chem*, **265**, 14736.
32. Wightman, R. M., Jankowski, J. A., Kennedy, R. T., Kawagoe, K. T., Schroeder, T. J., Leszczyszyn, D. J., *et al.* (1991) Temporally resolved catecholamine spikes correspond to single vesicle release from individual chromaffin cells. *Proc Natl Acad Sci USA*, **88**, 10754.
33. Chow, R. H. and von Ruden, L. (1995) Electrochemical detection of secretion from single cells. In *Single-channel recording* (ed. B. Sakmann and E. Neher), p. 245. Plenum Press, New York.
34. Travis, E. R. and Wightman, R. M. (1998) Spatio-temporal resolution of exocytosis from individual cells. *Annu Rev Biophys Biomol Struct*, **27**, 77.
35. Bruns, D. and Jahn, R. (1995) Real-time measurement of transmitter release from single synaptic vesicles. *Nature*, **377**, 62.
36. Chow, R. H., von Ruden, L., and Neher, E. (1992) Delay in vesicle fusion revealed by electrochemical monitoring of single secretory events in adrenal chromaffin cells. *Nature*, **356**, 60.
37. Alvarez de Toledo, G., Fernandez-Chacon, R., and Fernandez, J. M. (1993) Release of secretory products during transient vesicle fusion. *Nature*, **363**, 554.
38. Urena, J., Fernandez-Chacon, R., Benot, A. R., Alvarez de Toledo, G. A., and Lopez-Barneo, J. (1994) Hypoxia induces voltage-dependent Ca^{2+} entry and quantal dopamine secretion in carotid body glomus cells. *Proc Natl Acad Sci USA*, **91**, 10208.
39. Huang, L., Shen, H., Atkinson, M. A., and Kennedy, R. T. (1995) Detection of exocytosis at

individual pancreatic beta cells by amperometry at a chemically modified microelectrode. *Proc Natl Acad Sci USA*, **92**, 9608.

40. Jaffe, E. H., Marty, A., Schulte, A., and Chow, R. H. (1998) Extrasynaptic vesicular transmitter release from the somata of substantia nigra neurons in rat midbrain slices. *J Neurosci*, **18**, 3548.
41. Wrona, M. Z. and Dryhurst, G. (1988) Further insights into the oxidation chemistry of 5-hydroxytryptamine. *J Pharm Sci*, **77**, 911.
42. Kawagoe, K. T., Zimmerman, J. B., and Wightman, R. M. (1993) Principles of voltammetry and microelectrode surface states. *J Neurosci Methods*, **48**, 225.
43. Oberhauser, A. F., Robinson, I. M., and Fernandez, J. M. (1996) Simultaneous capacitance and amperometric measurements of exocytosis: a comparison. *Biophys J*, **71**, 1131.
44. Jankowski, J. A., Schroeder, T. J., Ciolkowski, E. L., and Wightman, R. M. (1993) Temporal characteristics of quantal secretion of catecholamines from adrenal medullary cells. *J Biol Chem*, **268**, 14694.
45. Wightman, R. M., Schroeder, T. J., Finnegan, J. M., Ciolkowski, E. L., and Pihel, K. (1995) Time course of release of catecholamines from individual vesicles during exocytosis at adrenal medullary cells. *Biophys J*, **68**, 383.
46. Uvnas, B. and Aborg, C. H. (1983) Cation exchange–a common mechanism in the storage and release of biogenic amines stored in granules (vesicles)? *Acta Physiol Scand*, **119**, 225.
47. Volknandt, W. (1995) The synaptic vesicle and its targets. *Neuroscience*, **64**, 277.
48. Marszalek, P., Farrell, B., and Fernandez, J. M. (1996) Ion-exchange gel regulates neurotransmitter release through the exocytotic fusion pore. *Soc Gen Physiol Ser*, **51**, 211.
49. Marszalek, P. E., Farrell, B., Verdugo, P., and Fernandez, J. M. (1997) Kinetics of release of serotonin from isolated secretory granules. II. Ion exchange determines the diffusivity of serotonin. *Biophys J*, **73**, 1169.
50. Marszalek, P. E., Farrell, B., Verdugo, P., and Fernandez, J. M. (1997) Kinetics of release of serotonin from isolated secretory granules. I. Amperometric detection of serotonin from electroporated granules. *Biophys J*, **73**, 1160.
51. von Gersdorff, H. and Matthews, G. (1994) Inhibition of endocytosis by elevated internal calcium in a synaptic terminal. *Nature*, **370**, 652.
52. Burgoyne, R. D. (1995) Fast exocytosis and endocytosis triggered by depolarisation in single adrenal chromaffin cells before rapid Ca^{2+} current run-down. *Pflugers Arch*, **430**, 213.
53. Oberhauser, A. F., Robinson, I. M., and Fernandez, J. M. (1995) Do caged-Ca^{2+} compounds mimic the physiological stimulus for secretion? *J Physiol Paris*, **89**, 71.
54. McCormack, J. G. and Cobbold, P. H. (ed.) (1991) *Cellular calcium: a practical approach*. The Practical Approach Series, Oxford University Press, Oxford, England.
55. Nuccitelli, R. (ed.) (1994) *A practical guide to the study of calcium in living cells. Methods in cell biology*, Vol. 40. Academic Press, San Diego, CA.
56. Shotton, D. (ed.) (1993) *Electronic light microscopy: the principles and practice of video-enhanced contrast, digital intensified fluorescence, and confocal scanning light microscopy. Techniques in modern biomedical microscopy*. Wiley-Liss, New York.
57. Inoué, S. and Spring, K. R. (1997) *Video microscopy: the fundamentals* (2nd edn). Plenum Press, New York.
58. Spector, D. L., Goldman, R. D., and Leinwand, L. A. (ed.) (1998) *Cells: a laboratory manual*, Vol. 2. Cold Spring Harbor Laboratory Press, Cold Spring Harbor, NY.
59. Sluder, G. and Wolf, D. E. (ed.) (1998) *Video microscopy. Methods in cell biology*, Vol. 56. Academic Press, San Diego.

60. Tsien, R. Y. (1980) New calcium indicators and buffers with high selectivity against magnesium and protons: design, synthesis, and properties of prototype structures. *Biochemistry*, **19**, 2396.
61. Thomas, A. P. and Delaville, F. (1991) The use of fluorescent indicators for measurements of cytosolic-free calcium concentration in cell populations and single cells. In *Cellular calcium: a practical approach* (ed. J. G. McCormack and P. H. Cobbold), p. 1. Oxford University Press, Oxford.
62. Kao, J. P. Y. (1994) Practical aspects of measuring [Ca^{2+}] with fluorescent indicators. In *A practical guide to the study of calcium in living cells* (ed. R. Nuccitelli), p. 155. Academic Press, San Diego.
63. Thomas, A. P. (1994) Cell function studies using fluorescent Ca^{2+} indicators. *Methods Toxicol*, **1B**, 287.
64. Silver, R. B. (1998) Ratio imaging: practical considerations for measuring intracellular calcium and pH in living tissue. *Methods Cell Biol*, **56**, 237.
65. Haugland, R. P. (1996) *Handbook of fluorescent probes and research chemicals* (6th edn). Molecular Probes, Eugene, OR.
66. Llinas, R., Sugimori, M., and Silver, R. B. (1992) Microdomains of high calcium concentration in a presynaptic terminal. *Science*, **256**, 677.
67. Grynkiewicz, G., Poenie, M., and Tsien, R. Y. (1985) A new generation of Ca^{2+} indicators with greatly improved fluorescence properties. *J Biol Chem*, **260**, 3440.
68. Tsien, R. Y. (1981) A non-disruptive technique for loading calcium buffers and indicators into cells. *Nature*, **290**, 527.
69. Hall, J. D., Betarbet, S., and Jaramillo, F. (1997) Endogenous buffers limit the spread of free calcium in hair cells. *Biophys J*, **73**, 1243.
70. Etter, E. F., Minta, A., Poenie, M., and Fay, F. S. (1996) Near-membrane [Ca^{2+}] transients resolved using the Ca^{2+} indicator FFP18. *Proc Natl Acad Sci USA*, **93**, 5368.
71. Allbritton, N. L., Oancea, E., Kuhn, M. A., and Meyer, T. (1994) Source of nuclear calcium signals. *Proc Natl Acad Sci USA*, **91**, 12458.
72. Omann, G. M. and Axelrod, D. (1996) Membrane-proximal calcium transients in stimulated neutrophils detected by total internal reflection fluorescence. *Biophys J*, **71**, 2885.
73. Cleemann, L., DiMassa, G., and Morad, M. (1997) Ca^{2+} sparks within 200 nm of the sarcolemma of rat ventricular cells: evidence from total internal reflection fluorescence microscopy. *Adv Exp Med Biol*, **430**, 57.
74. Zucker, R. (1994) Photorelease techniques for raising or lowering intracellular Ca^{2+}. In *A practical guide to the study of calcium in living cells* (ed. R. Nuccitelli). *Methods in cell biology*, Vol. 40, p. 31. Academic Press, San Diego, CA.
75. Kao, J. P. Y. and Adams, S. R. (1993) Photosensitive caged compounds. In *Optical microscopy: emerging methods and applications* (ed. B. Herman and J. J. Lemasters), p. 27. Academic Press, San Diego.
76. Thomas, P., Wong, J. G., Lee, A. K., and Almers, W. (1993) A low affinity Ca^{2+} receptor controls the final steps in peptide secretion from pituitary melanotrophs. *Neuron*, **11**, 93.
77. Heinemann, C., Chow, R. H., Neher, E., and Zucker, R. S. (1994) Kinetics of the secretory response in bovine chromaffin cells following flash photolysis of caged Ca^{2+}. *Biophys J*, **67**, 2546.
78. Heidelberger, R., Heinemann, C., Neher, E., and Matthews, G. (1994) Calcium dependence of the rate of exocytosis in a synaptic terminal. *Nature*, **371**, 513.
79. De Giorgi, F., Brini, M., Bastianutto, C., Marsault, R., Montero, M., Pizzo, P., *et al.* (1996)

Targeting aequorin and green fluorescent protein to intracellular organelles. *Gene*, **173**, 113.

80. Stearns, T. (1995) Green fluorescent protein. The green revolution. *Curr Biol*, **5**, 262.
81. Kendall, J. M. and Badminton, M. N. (1998) Aequorea victoria bioluminescence moves into an exciting new era. *Trends Biotechnol*, **16**, 216.
82. Cubitt, A. B., Heim, R., Adams, S. R., Boyd, A. E., Gross, L. A., and Tsien, R. Y. (1995) Understanding, improving and using green fluorescent proteins. *Trends Biochem Sci*, **20**, 448.
83. Misteli, T. and Spector, D. L. (1997) Applications of the green fluorescent protein in cell biology and biotechnology. *Nat Biotechnol*, **15**, 961.
84. Miller, A. L., Karplus, E., and Jaffe, L. F. (1994) Imaging [Ca^{2+}] with aequorin using a photon imaging detector. In *A practical guide to the study of calcium in living cells* (ed. R. Nuccitelli), p. 305. Academic Press, San Diego.
85. Inouye, S., Noguchi, M., Sakaki, Y., Takagi, Y., Miyata, T., Iwanaga, S., *et al.* (1985) Cloning and sequence analysis of cDNA for the luminescent protein aequorin. *Proc Natl Acad Sci USA*, **82**, 3154.
86. Prasher, D. C., Eckenrode, V. K., Ward, W. W., Prendergast, F. G., and Cormier, M. J. (1992) Primary structure of the Aequorea victoria green-fluorescent protein. *Gene*, **111**, 229.
87. Rizzuto, R., Brini, M., and Pozzan, T. (1994) Targeting recombinant aequorin to specific intracellular organelles. In *A practical guide to the study of calcium in living cells* (ed. R. Nuccitelli), p. 339. Academic Press, San Diego.
88. Marsault, R., Murgia, M., Pozzan, T., and Rizzuto, R. (1997) Domains of high Ca^{2+} beneath the plasma membrane of living A7r5 cells. *EMBO J*, **16**, 1575.
89. Heim, R., Cubitt, A. B., and Tsien, R. Y. (1995) Improved green fluorescence. *Nature*, **373**, 663.
90. Cormack, B. P., Valdivia, R. H., and Falkow, S. (1996) FACS-optimized mutants of the green fluorescent protein (GFP). *Gene*, **173**, 33.
91. Heim, R. and Tsien, R. Y. (1996) Engineering green fluorescent protein for improved brightness, longer wavelengths and fluorescence resonance energy transfer. *Curr Biol*, **6**, 178.
92. Miyawaki, A., Llopis, J., Heim, R., McCaffery, J. M., Adams, J. A., Ikura, M., *et al.* (1997) Fluorescent indicators for Ca^{2+} based on green fluorescent proteins and calmodulin. *Nature*, **388**, 882.
93. Persechini, A., Lynch, J. A., and Romoser, V. A. (1997) Novel fluorescent indicator proteins for monitoring free intracellular Ca^{2+}. *Cell Calcium*, **22**, 209.
94. Romoser, V. A., Hinkle, P. M., and Persechini, A. (1997) Detection in living cells of Ca^{2+}-dependent changes in the fluorescence emission of an indicator composed of two green fluorescent protein variants linked by a calmodulin-binding sequence. A new class of fluorescent indicators. *J Biol Chem*, **272**, 13270.
95. Monck, J. R., Robinson, I. M., Escobar, A. L., Vergara, J. L., and Fernandez, J. M. (1994) Pulsed laser imaging of rapid Ca^{2+} gradients in excitable cells. *Biophys J*, **67**, 505.
96. Robinson, I. M., Finnegan, J. M., Monck, J. R., Wightman, R. M., and Fernandez, J. M. (1995) Colocalization of calcium entry and exocytotic release sites in adrenal chromaffin cells. *Proc Natl Acad Sci USA*, **92**, 2474.
97. Robinson, I. M., Yamada, M., Carrion-Vazquez, M., Lennon, V. A., and Fernandez, J. M. (1996) Specialized release zones in chromaffin cells examined with pulsed-laser imaging. *Cell Calcium*, **20**, 181.

98. Tucker, T. and Fettiplace, R. (1995) Confocal imaging of calcium microdomains and calcium extrusion in turtle hair cells. *Neuron*, **15**, 1323.
99. Eilers, J., Augustine, G. J., and Konnerth, A. (1995) Subthreshold synaptic Ca^{2+} signalling in fine dendrites and spines of cerebellar Purkinje neurons. *Nature*, **373**, 155.
100. Pawley, J. B. (ed.) (1995) *Handbook of biological confocal microscopy* (2nd edn). Plenum Press, New York.
101. Sheppard, C. and Shotton, D. (ed.) (1998) *Confocal laser scanning microscopy*. Springer, New York.
102. Diliberto, P. A., Wang, X. F., and Herman, B. (1994) Confocal imaging of Ca^{2+} in cells. In *A practical guide to the study of calcium in living cells* (ed. R. Nuccitelli), p. 243. Academic Press, San Diego.
103. Schild, D. (1996) Laser scanning microscopy and calcium imaging. *Cell Calcium*, **19**, 281.
104. Spector, D. L., Goldman, R. D., and Leinwand, L. A. (1998) Confocal microscopy and deconvolution techniques. In *Cells: a laboratory manual*, p. 96.1. Cold Spring Harbor Laboratory Press, Cold Spring Harbor, NY.
105. Denk, W. and Svoboda, K. (1997) Photon upmanship: why multiphoton imaging is more than a gimmick. *Neuron*, **18**, 351.
106. Denk, W., Delaney, K. R., Gelperin, A., Kleinfeld, D., Strowbridge, B. W., Tank, D. W., *et al.* (1994) Anatomical and functional imaging of neurons using 2-photon laser scanning microscopy. *J Neurosci Methods*, **54**, 151.
107. Denk, W., Piston, D. W., and Webb, W. W. (1995) Two photon molecular excitation in laser-scanning microscopy. In *Handbook of biological confocal microscopy* (ed. J. B. Pawley), p. 445. Plenum Press, New York.
108. Yuste, R. and Denk, W. (1995) Dendritic spines as basic functional units of neuronal integration. *Nature*, **375**, 682.
109. Svoboda, K., Denk, W., Kleinfeld, D., and Tank, D. W. (1997) *In vivo* dendritic calcium dynamics in neocortical pyramidal neurons. *Nature*, **385**, 161.
110. Denk, W. (1994) Two-photon scanning photochemical microscopy: mapping ligand-gated ion channel distributions. *Proc Natl Acad Sci USA*, **91**, 6629.
111. Peters, A., Palay, S. L., and Webster, H. D. (1991) *The fine structure of the nervous system: neurons and their supporting cells*, 3rd edn. New York: Oxford University Press.
112. Lichtman, J. W., Wilkinson, R. S., and Rich, M. M. (1985) Multiple innervation of tonic endplates revealed by activity-dependent uptake of fluorescent probes. *Nature*, **314**, 357.
113. Betz, W. J., Mao, F., and Bewick, G. S. (1992) Activity-dependent fluorescent staining and destaining of living vertebrate motor nerve terminals. *J Neurosci*, **12**, 363.
114. Betz, W. J. and Bewick, G. S. (1992) Optical analysis of synaptic vesicle recycling at the frog neuromuscular junction. *Science*, **255**, 200.
115. Betz, W. J. and Bewick, G. S. (1993) Optical monitoring of transmitter release and synaptic vesicle recycling at the frog neuromuscular junction. *J Physiol (Lond)*, **460**, 287.
116. Ryan, T. A., Reuter, H., Wendland, B., Schweizer, F. E., Tsien, R. W., and Smith, S. J. (1993) The kinetics of synaptic vesicle recycling measured at single presynaptic boutons. *Neuron*, **11**, 713.
117. Ryan, T. A. and Smith, S. J. (1995) Vesicle pool mobilization during action potential firing at hippocampal synapses. *Neuron*, **14**, 983.
118. Kraszewski, K., Mundigl, O., Daniell, L., Verderio, C., Matteoli, M., and De Camilli, P. (1995) Synaptic vesicle dynamics in living cultured hippocampal neurons visualized with CY3-conjugated antibodies directed against the lumenal domain of synaptotagmin. *J Neurosci*, **15**, 4328.

119. Liu, G. and Tsien, R. W. (1995) Properties of synaptic transmission at single hippocampal synaptic boutons. *Nature*, **375**, 404.
120. Reuter, H. (1995) Measurements of exocytosis from single presynaptic nerve terminals reveal heterogeneous inhibition by Ca($^{2+}$)-channel blockers. *Neuron*, **14**, 773.
121. Vogt, K., Luscher, H. R., and Streit, J. (1995) Analysis of synaptic transmission at single identified boutons on rat spinal neurons in culture. *Pflugers Arch*, **430**, 1022.
122. Smith, C. B. and Betz, W. J. (1996) Simultaneous independent measurement of endocytosis and exocytosis. *Nature*, **380**, 531.
123. Lagnado, L., Gomis, A., and Job, C. (1996) Continuous vesicle cycling in the synaptic terminal of retinal bipolar cells. *Neuron*, **17**, 957.
124. Ramaswami, M., Krishnan, K. S., and Kelly, R. B. (1994) Intermediates in synaptic vesicle recycling revealed by optical imaging of *Drosophila* neuromuscular junctions. *Neuron*, **13**, 363.
125. Betz, W. J., Mao, F., and Smith, C. B. (1996) Imaging exocytosis and endocytosis. *Curr Opin Neurobiol*, **6**, 365.
126. Whalley, T., Terasaki, M., Cho, M. S., and Vogel, S. S. (1995) Direct membrane retrieval into large vesicles after exocytosis in sea urchin eggs. *J Cell Biol*, **131**, 1183.
127. Cousin, M. A., Held, B., and Nicholls, D. G. (1995) Exocytosis and selective neurite calcium responses in rat cerebellar granule cells during field stimulation. *Eur J Neurosci*, **7**, 2379.
128. Reuter, H. and Porzig, H. (1995) Localization and functional significance of the Na^+/Ca^{2+} exchanger in presynaptic boutons of hippocampal cells in culture. *Neuron*, **15**, 1077.
129. Ryan, T. A., Li, L., Chin, L. S., Greengard, P., and Smith, S. J. (1996) Synaptic vesicle recycling in synapsin I knock-out mice. *J Cell Biol*, **134**, 1219.
130. Kuromi, H. and Kidokoro, Y. (1998) Two distinct pools of synaptic vesicles in single presynaptic boutons in a temperature-sensitive *Drosophila* mutant, *shibire*. *Neuron*, **20**, 917.
131. Isaacson, J. S. and Hille, B. (1997) GABA(B)-mediated presynaptic inhibition of excitatory transmission and synaptic vesicle dynamics in cultured hippocampal neurons. *Neuron*, **18**, 143.
132. Trudeau, L. E., Emery, D. G., and Haydon, P. G. (1996) Direct modulation of the secretory machinery underlies PKA-dependent synaptic facilitation in hippocampal neurons. *Neuron*, **17**, 789.
133. Wang, C. and Zucker, R. S. (1998) Regulation of synaptic vesicle recycling by calcium and serotonin. *Neuron*, **21**, 155.
134. Ryan, T. A., Reuter, H., and Smith, S. J. (1997) Optical detection of a quantal presynaptic membrane turnover. *Nature*, **388**, 478.
135. Murthy, V. N., Sejnowski, T. J., and Stevens, C. F. (1997) Heterogeneous release properties of visualized individual hippocampal synapses. *Neuron*, **18**, 599.
136. Murthy, V. N. and Stevens, C. F. (1998) Synaptic vesicles retain their identity through the endocytic cycle. *Nature*, **392**, 497.
137. Lang, T., Wacker, I., Steyer, J., Kaether, C., Wunderlich, I., Soldati, T., *et al.* (1997) Ca^{2+}-triggered peptide secretion in single cells imaged with green fluorescent protein and evanescent-wave microscopy. *Neuron*, **18**, 857.
138. Burke, N. V., Han, W., Li, D., Takimoto, K., Watkins, S. C., and Levitan, E. S. (1997) Neuronal peptide release is limited by secretory granule mobility. *Neuron*, **19**, 1095.
139. Axelrod, D. (1989) Total internal reflection fluorescence microscopy. *Methods Cell Biol*, **30**, 245.

140. Steyer, J. A., Horstmann, H., and Almers, W. (1997) Transport, docking and exocytosis of single secretory granules in live chromaffin cells. *Nature*, **388**, 474.
141. Oheim, M., Loerke, D., Stuhmer, W., and Chow, R. H. (1998) The last few milliseconds in the life of a secretory granule. Docking, dynamics and fusion visualized by total internal reflection fluorescence microscopy (TIRFM). *Eur Biophys J*, **27**, 83.

3 Nerve terminal membrane trafficking proteins: from discovery to function

CHRISTOPHER D. HAZUKA, DAVIDE L. FOLETTI, and
RICHARD H. SCHELLER

1. Introduction

Release of chemical messengers is the primary means by which neurons communicate with target cells. Chemical neurotransmitters are secreted from the presynaptic nerve terminal, travel across the synaptic cleft, and stimulate receptors on postsynaptic cells in a process termed neurotransmission (1). Initial physiological work demonstrated that neurotransmitter is released in packets or 'quanta', and that each quantum consists of a large number of neurotransmitter molecules (2, 3). Ultrastructural studies indicated that a major feature of the presynaptic nerve terminal is a local concentration of uniformly-sized lipid-bound vesicles, promoting the hypothesis that a quantum of neurotransmitter is stored in a vesicle (4). Subsequently, abundant research demonstrated the primacy of these vesicular structures in neurotransmitter release (5, 6). As each vesicle contains a quantum of neurotransmitter, exquisite control of nervous system communication can result from modulation of the behaviour of synaptic vesicles. Understanding how this pool of vesicles is established and maintained, and how vesicles regulate neurotransmitter release, has been a major topic of research for the last four decades.

A complex sequence of membrane trafficking events leads to the delivery of the vesicular contents into the plasma membrane and the synaptic cleft. First, vesicular components are synthesized in the cell body and are incorporated into membranous synaptic vesicle precursors in the Golgi complex (Fig. 1; Ia). These immature vesicles may begin a maturation process in the cell soma by fusing with the plasma membrane and endosomes prior to travelling down the axon via transport on the cytoskeleton (see Section 7). Fast transport to synaptic sites is likely mediated by microtubule-activated ATPases such as kinesin or myosinV (Fig. 1; Ib) (7–9). These vesicle precursors may further mature as they travel down the axon by fusing with the plasma membrane along the way (Fig. 1; Ic) (10). During this developmental

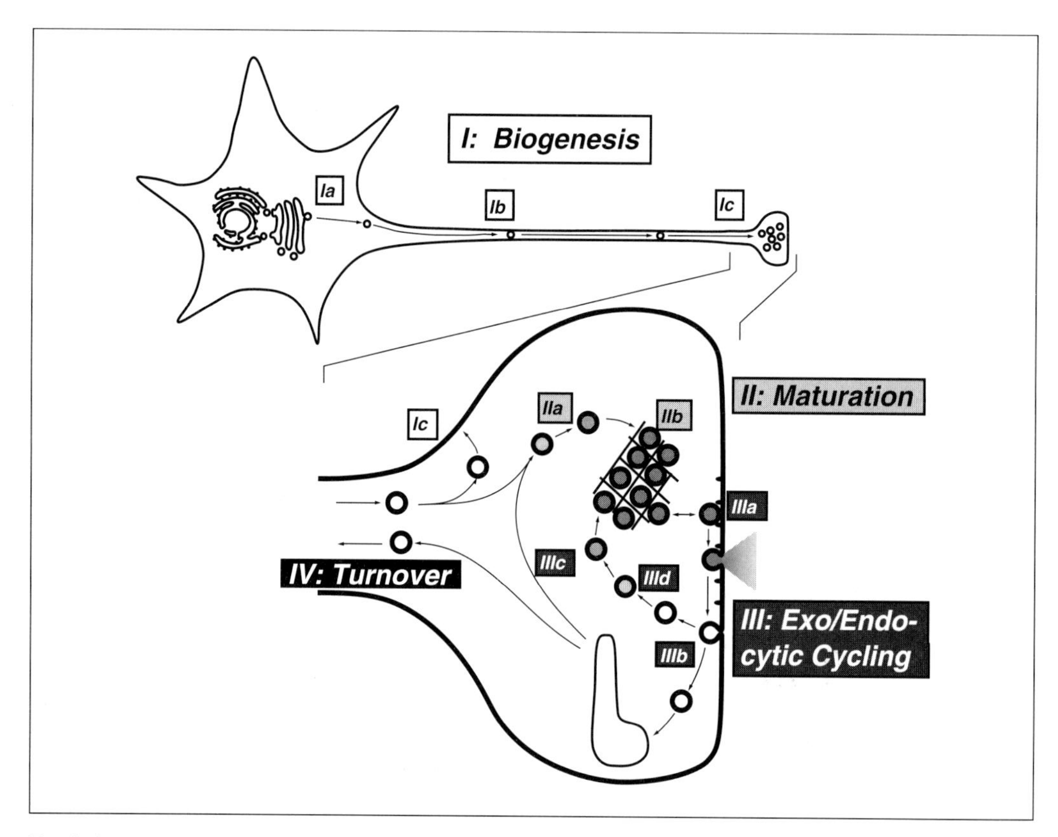

Fig. 1 Steps in the life of synaptic vesicles. See Section 1 in text for details.

process, synaptic vesicles accumulate at discrete plasma membrane domains, some of which give rise to mature presynaptic nerve terminals (see Section 7). Once the pool of vesicles is established in the mature presynaptic terminal, a local regenerating cycle of exo- and endocytosis begins and is maintained with minimal input of new material from the cell body. This property allows the synapse to function independently.

There are two major pools of vesicles in the nerve terminal: a reserve pool which is associated with the actin cytoskeleton (Fig. 1; IIb), and a readily-releasable or docked pool which is associated with the plasma membrane (Fig. 1; IIIa). Ca^{2+} entry through Ca^{2+} channels is required to trigger the fusion of docked vesicles with the plasma membrane and release of their contents into the synaptic cleft. After fusion, the vesicular components are retrieved via endocytosis (see Chapter 11) and the vesicles are refilled by specialized neurotransmitter transporters (Fig. 1; IIa) (see Chapter 5). The endocytic route is still the subject of investigation. Whereas previous models proposed recycling through endosomes (Fig. 1; IIIb and c), recent studies have suggested that endocytosed vesicles may bypass endosomal intermediates on their way to the reserve and docked pools (Fig. 1; IIId) (11). Finally, recycled vesicle components are either reused in future rounds of neurotransmission or returned to the cell body for turnover (Fig. 1; IV).

This chapter will present a review of the discovery of synaptic vesicle proteins (Section 2), the initial characterization of the proteins involved in vesicle fusion with the plasma membrane (Sections 2 and 3), attempts to develop models of the functions of proteins and protein complexes in membrane trafficking (Section 4), structural information about these complexes and their conformations (Section 5), the implications of recent data on models of vesicular protein function (Section 6), and molecules involved in the development of synaptic vesicle pools at synapses (Section 7). This chapter will not attempt to describe all the proteins involved in vesicle functions in detail but will instead focus on experiments which serve as examples of the techniques which can be utilized for characterizing membrane trafficking mechanisms.

2. Discovery of candidate proteins: methodology

2.1 Electric organ and rat brain synaptic vesicle purification

Original synaptic vesicle purification attempts relied on tissues containing high concentrations of one type of synaptic vesicle. The electric organ of the family of marine elasmobranchs was recognized as a source of a population of homogeneous cholinergic synaptic vesicles (12–14). A series of biochemical steps was utilized to obtain purified synaptic vesicles (13). First, the electric organ was homogenized and centrifuged in low and high speed steps to obtain a pellet of membranes and membrane-bound organelles (Fig. 2A). A population of membranes including synaptic vesicles, synaptosomes (membranes containing synaptic components including vesicles), and homogenization artefacts was obtained. Closed membranes such as synaptic vesicles are more dense than the open membrane artefacts. Thus, these membranes were resuspended in sucrose, applied to a tube containing a gradient of sucrose concentrations, and separated according to their densities by velocity centrifugation. These dense membranes were further purified by permeation chromatography through a column of glass beads (controlled-pore glass; cpg) containing 300 nm pores (Fig. 2B). This step was crucial because some membranes created during the original homogenization were likely fractured endoplasmic reticulum (ER) and Golgi complex membranes which had resealed into structures of comparable density to synaptic vesicles. These artefacts behaved like synaptic vesicles on the velocity sedimentation gradient, but were excluded in the cpg column step because of their larger size. The cpg-purified membranes were shown to have the same size, density, and electrophoretic mobility as was predicted for synaptic vesicles. Furthermore, electron microscopic analysis demonstrated that the membranes isolated in this way are electron lucent and are uniformly the size of the starting population of synaptic vesicles (Fig. 2C) (15).

This preparation of purified, homogeneous vesicles from the elasmobranch electric organ allowed the initial biochemical and biophysical characterization of

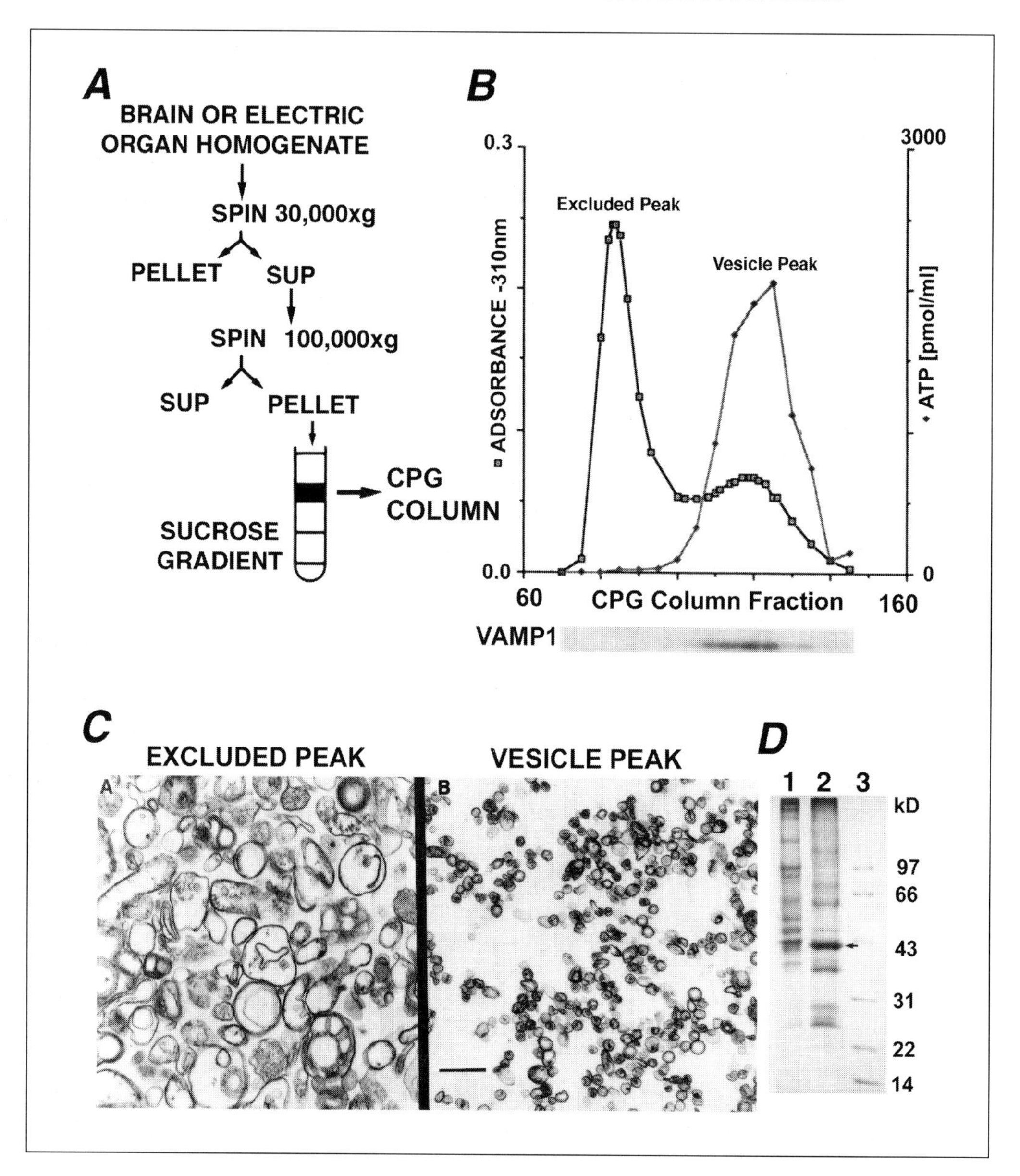

Fig. 2 Purification of synaptic vesicle proteins. (A) A source of synaptic vesicles such as brain or Torpedo electric organ is homogenized and subjected to steps of differential centrifugation and flotation equilibrium in sucrose density gradients. (B) After this initial crude purification, the pool of synaptic vesicles is further purified by permeation chromatography over a controlled-pore glass (cpg) bead column. Of the two peaks observed by monitoring the absorbance at 310 nm (which measures the total light scattering material) the vesicle peak is identified by its high content of ATP (13). A vesicle associated membrane protein (VAMP1) is shown by Western blot to be specific for the synaptic vesicle peak. (C) Electron microscopy further demonstrates the purification of synaptic vesicles after the cpg column. Scale bar, 400 nm. (D) An example of a gel separating proteins in (1) the starting crude electric organ material, (2) the purified fraction of synaptic vesicle proteins, and (3) molecular weight markers.

synaptic vesicles (14). The proteins and other molecules in the purified population of vesicles were precipitated, separated, identified, and quantified using either high-pressure liquid chromatography or sodium dodecyl sulfate polyacrylamide gel electrophoresis (SDS–PAGE). These analyses suggested that the cholinergic synaptic vesicles isolated in this study had a low protein-to-lipid ratio (1:5, w/w), were acidic (pH 6.4) and that their volume was $\sim$ 70% water. In addition, the vesicles contained high concentrations of phospholipid (16%; 3.7×10^4 molecules/vesicle), cholesterol (3.2%; 1.5×10^4 molecules/vesicle), protein (3%; 130 molecules/vesicle), acetylcholine (520 mM; 4.2×10^4 molecules/vesicle), and ATP (170 mM; 1.5×10^4 molecules/vesicle), and lower amounts of GTP (20 mM; 2.3×10^3 molecules/vesicle). Ignoring the contribution of water, salts, and non-phosphate-containing lipids, these vesicles were roughly 11% protein, 59% lipid, 16% ATP, and 14% acetylcholine. Mammalian synaptic vesicles likely contain different proportions of these components, but the point is that synaptic vesicle membranes from any species contain a population of important membrane proteins whose role in mediating neurotransmitter release can be experimentally tested.

These techniques were also used to purify synaptic vesicles from mammalian brain (16, 17). Experiments on these organelles opened the field of vesicular protein analysis (18–37). Originally, at least 20 proteins of different sizes were found in the purified pool of vesicles when separated by SDS–PAGE (14). An example of such a gel is shown in Fig. 2D. These proteins were likely integral or peripheral vesicular membrane proteins because they remained associated with the membrane even when vesicles were lysed by osmotic shock (14). Figures 3, 4, and 5 catalogue some of the proteins now known to be involved in vesicle trafficking, and depict their subcellular distribution.

2.2 Isolation of cDNAs encoding synaptic vesicle proteins

The genes and/or cDNAs encoding the proteins found on synaptic vesicle membranes have been isolated by several means. Originally, purified synaptic vesicles were injected into rabbits or mice to generate polyclonal and/or monoclonal antibodies (18–20). These antibodies were then characterized for their specific binding to different types of exocytic vesicles. Interestingly, antibodies raised against elasmobranch synaptic vesicles sometimes reacted with synaptic terminals in both the central and peripheral nervous system of rats (19). This was the first indication that the molecular mechanisms of neurotransmitter release are conserved across species.

These antibodies have been used to identify and characterize the proteins they recognize and to isolate the genes encoding synaptic vesicle proteins by several methods (examples will be discussed below). First, antibodies made to these proteins were used to screen expression libraries. Secondly, peptide sequences from the purified proteins were used to design degenerate oligonucleotides for DNA homology screens. Thirdly, peptide sequences were used to search protein and DNA sequence databases for previously reported proteins or protein fragments. Such gene fragments, frequently found in expressed sequence tag (EST) databases, are especially

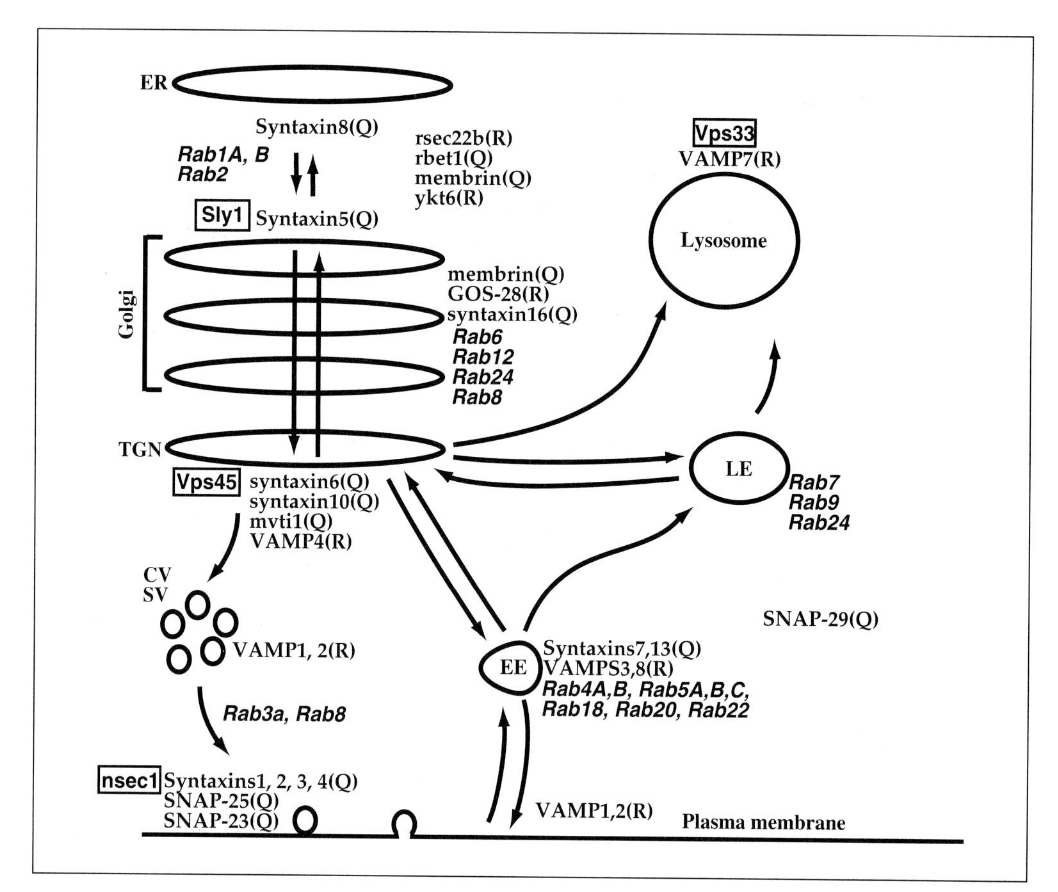

Fig. 3 Subcellular localization of proteins involved in membrane trafficking in mammalian cells. There are three families of proteins with members involved at different steps of the pathway: the SNARE (Q and R), Rab/Ypt, and sec1 families. Members of the Q-SNARE family: syntaxins1–5 (45, 107); syntaxin6 (173, 174); syntaxin7 (175, 176); syntaxin8 and SNAP-29 (177); syntaxin10 (178); syntaxin13 (179, 180); syntaxin16 (181); vti1 (179); SNAP-25 (68); SNAP-23 (182); membrin (183); rbet1 (184). R-SNARE family: R, VAMP1,2 (185); VAMP3 (186); VAMP4,7, and 8 (179); GOS-28 (187); rsec22b (183). nsec1 family (boxed): nsec1 (38); sly1 (188); vps33, vps45 (189). Rab family (italics) reviewed in ref. 190. The families are shown in their putative subcellular localization and/or at the trafficking step in which they are suggested to be involved. ER, endoplasmic reticulum; TGN, *trans*-Golgi network; CV, constitutive secretory vesicles; SV, synaptic vesicles; EE, early endosomes; LE, late endosomes.

valuable for generating probes useful in homology screens (38, 39). (EST databases have only more recently become available however, so many of the cDNAs encoding synaptic vesicle proteins were isolated without this advantage.) Using these techniques, the full-length open reading frames (ORFs) which encode the synaptic vesicle proteins were sequenced and characterized, allowing most of the experiments discussed in this chapter.

The synaptic vesicle proteins, *v*esicle-*a*ssociated *m*embrane *p*rotein (VAMP) (later also called synaptobrevin) and synaptotagmin, were isolated using these techniques.

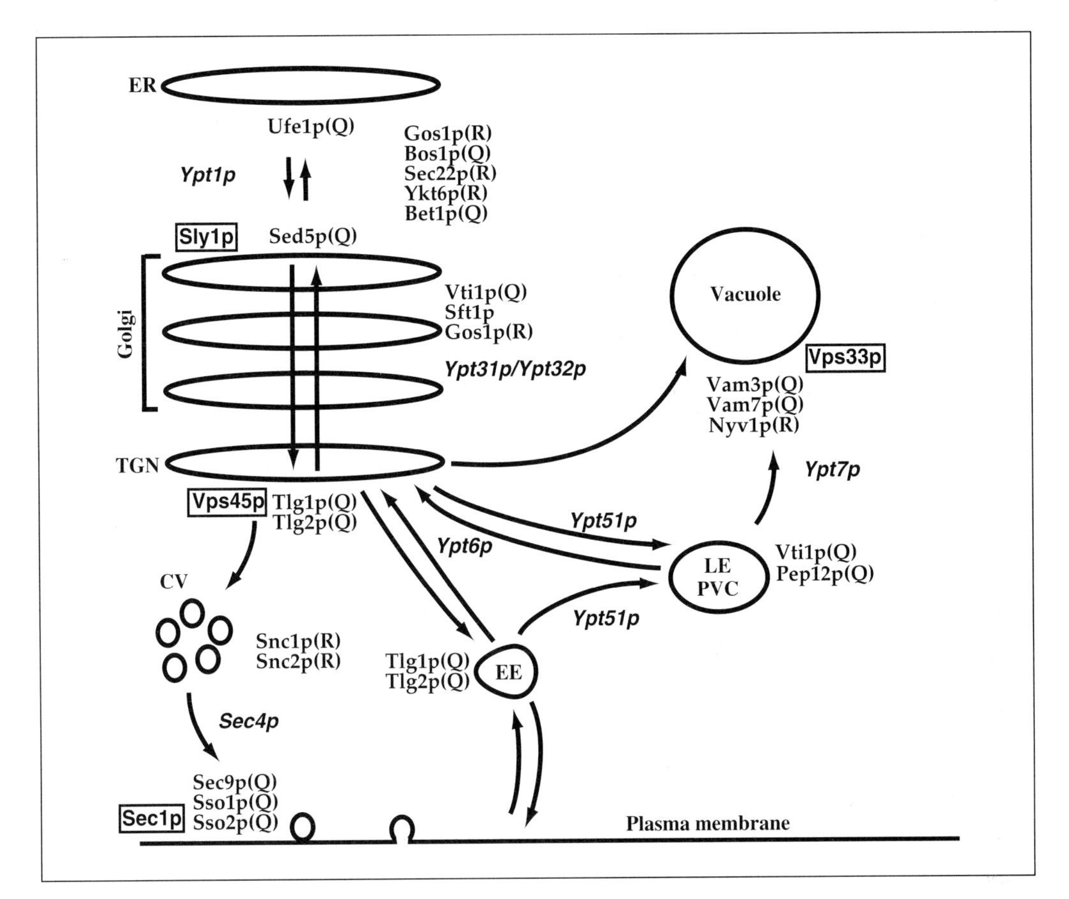

Fig. 4 Subcellular localization of proteins involved in membrane trafficking in *Saccharomyces cerevisiae*. There are three families of proteins with members involved at different steps of the pathway: the SNARE (Q and R), Rab/Ypt, and sec1 families. Members of the Q-SNARE family: Ufe1p (191); Bos1p and Bet1p (192); Vti1p (119); Tlg1p and Tlg2p (193); Vam3p (194); Vam7p (195); Pep12 (196); Sec9p (82); Sso1p and Sso2p (84); Sed5p (101). R-SNARE family: Sec22p (192); Ykt6p (197); Gos1p (198); Snc1p and Snc2p (199); Nyv1p (129); and Sft1p (unclear if R- or Q-SNARE) (206). nsec1 family (boxed): Sec1p (200); Sly1p (118, 201); Vps33p (202, 203); Vps45p (204, 205). Ypt/Rab family (italics) reviewed in ref. 139. The families are shown in their putative subcellular localization and/or at the trafficking step in which they are suggested to be involved. ER, endoplasmic reticulum; TGN, *trans*-Golgi network; CV, constitutive secretory vesicles; EE, early endosomes; PVC, prevacuolar compartment; LE, late endosomes.

VAMP was initially discovered by screening an expression library with an antibody raised against synaptic vesicles (23). The cDNA in the bacteria expressing the antigen which was recognized by the antibody was then sequenced and characterized. A cDNA encoding synaptotagmin was isolated using a different technique. Monoclonal antibodies were isolated from mice which were immunized with synaptic membranes. One of the antibodies recognized a protein band of 65 kDa which was present on synaptic vesicle membranes (20). This protein, called p65 (later renamed synaptotagmin), was purified and digested with proteases. Partial amino acid

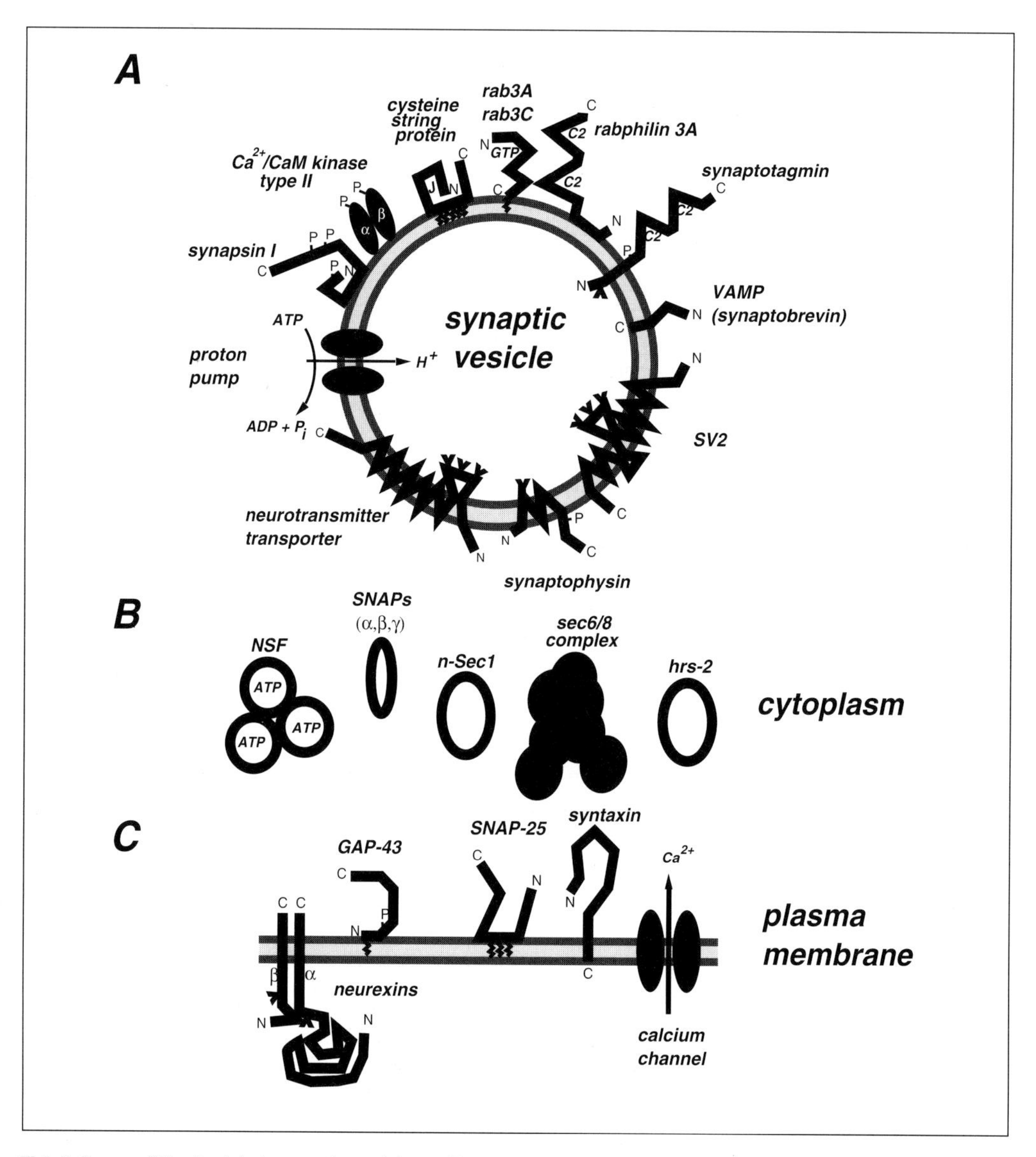

Fig. 5 Some of the isolated synaptic vesicle trafficking proteins and their distribution in the nerve terminal. (A) Integral and peripheral membrane proteins of the synaptic vesicle. (B) Cytosolic factors. (C) Membrane trafficking proteins located on the plasma membrane.

sequences informed the design of degenerate oligonucleotides for use in a polymerase chain reaction (PCR) to amplify a fragment of the DNA encoding p65 from a rat genomic DNA library. Using this fragment, a DNA library was screened providing several DNA clones containing the full-length ORF encoding p65 (40). Similar techniques also enabled the isolation of clones for the marine ray p65 (41) and synaptophysin (37, 42–44).

2.3 Protein interaction screens for protein discovery

Identification and characterization of the protein components of the synaptic vesicle membrane led to the discovery of proteins with which they interact. As vesicles change intracellular location during their life cycle, they interact with proteins of the axonal and presynaptic cytoskeleton and plasma membrane as well as with cytosolic factors. Some of the most exciting recent experiments have focused on isolating and characterizing non-vesicular proteins which interact with synaptic vesicles.

Immunoprecipitation is one technique which can be used to discover proteins that interact with vesicles. Immunoprecipitation experiments rely on the existence of a suitable antibody to a vesicle membrane protein. In one study, polyclonal antibodies raised against synaptotagmin were conjugated to an insoluble matrix such as Sepharose beads (45). Brain synaptosomes (membranes which include synaptic vesicles and synaptic plasma membranes) were purified, and their membrane proteins were solubilized using several different detergents. These membrane proteins were then incubated with beads coupled to anti-synaptotagmin antibodies. After non-specifically bound molecules were washed away, synaptotagmin, along with any associated molecules, was left behind. The proteins co-immunoprecipitated by antibodies to synaptotagmin were then eluted and separated by SDS–PAGE. In this way, a 35 kDa protein band was isolated and shown to be the integral plasma membrane protein, syntaxin1 (see Section 2.4). Syntaxin has since been shown to be critical for neurotransmission. Thus, using vesicle proteins as a starting point, non-vesicular proteins involved in neurotransmitter secretion were found.

In vitro binding assays have also been used to discover previously unknown interacting proteins. Generally, a known protein is tagged with a sequence that allows specific binding to insoluble beads. Examples of such tags are poly-histidine, myc, haemagglutinin (HA), and glutathione *S*-transferase (GST), all of which can be added to the protein sequence by standard molecular biology techniques. These tags specifically bind Ni^{2+}, anti-myc antibodies, anti-HA antibodies, or glutathione, respectively, all of which can be conjugated to insoluble beads. Once the tagged proteins are bound to the beads, they can be incubated with a source of potential interacting-proteins. If binding partners are present they will be purified through their specific interaction with proteins on the beads. One example of the utility of this technique is the discovery of a protein called tomosyn (46). Cytosolic proteins from rat cerebrum were poured over a column of beads to which GST–syntaxin1A was bound. After these beads were washed to remove non-specifically bound proteins, the remaining bound proteins were eluted and analyzed. A protein of 130 kDa specifically bound the GST–syntaxin1A column. Peptide fragments from this protein were obtained by proteolytic cleavage and sequenced. Degenerate oligonucleotides were designed based on these sequences and used to screen a rat brain cDNA library, thus obtaining full-length clones for the new protein. Characterization of tomosyn was then conducted using the protein expressed by the cDNA.

Although these *in vitro* binding techniques are useful for finding high-affinity binding proteins, weakly or transiently interacting proteins may not be detected during these experiments. The addition of chemical crosslinkers can assist in the isolation of such proteins. Chemical crosslinkers are molecules with two or more reactive groups spaced by an arm of defined length. Their reactive ends form a chemical bridge between proteins that physically interact. One protein isolated using the crosslinking technique is Rabphilin-3A (47). Rabphilin-3A was found in experiments designed to isolate Rab3a interacting proteins. Rab3a is a GTPase which is thought to associate with vesicles depending on whether it is bound to GTP or GDP (see Fig. 8, and Sections 3 and 6.2). Thus, it was particularly interesting to find proteins which bind Rab3a when it is in its GTP-bound, but not GDP-bound, state. Rab3a was incubated with a radioactive form of the non-hydrolysable GTP analogue, GTPγS. Rab3a–GTPγS was then incubated, in the presence of a crosslinker, with a crude fraction of proteins purified from brain. At the end of the incubation, the protein mixture containing the crosslinked products was separated by SDS–PAGE. A radioactive band of 110 kDa was observed and demonstrated to correspond to the 25 kDa Rab3a–GTPγS crosslinked to a novel 85 kDa protein which was named Rabphilin-3A.

A relatively new technique for protein interaction discovery and analysis is the yeast two-hybrid system (48). The basic principle behind this technique is that protein–protein interactions can be identified *in vivo* through reconstitution of the activity of a transcription factor. The method is based on the properties of the yeast GAL4 transcription factor, which consists of two separable domains, one which binds DNA and another which activates transcription. When separated, the two domains of the GAL4 transcription activator are not functional. However, when they are brought into close proximity, as when each is attached to one of two interacting proteins, they can promote the transcription of a reporter gene. Using this method, an ATPase, hrs-2, was discovered as a binding partner for SNAP-25, a synaptic plasma membrane protein (49). The ORF encoding SNAP-25 was spliced into a plasmid in-frame with the sequence encoding the activation domain of the GAL4 transcription factor. Plasmids containing fragments of a library of brain cDNAs fused to the sequence encoding the binding domain of the GAL4 protein were co-transformed with the plasmids containing the SNAP-25/GAL4 activation domain. Transformants were screened for GAL4-dependent reporter gene expression. The library plasmids from any positive yeast colonies were isolated and sequenced, yielding the identity of the interacting protein. In addition to uncovering weakly interacting proteins, this technique may also make possible the discovery of transiently interacting proteins which are not detected using bead-binding or immunoprecipitation techniques.

2.4 Clues from intracellular trafficking assays

Information about synaptic vesicle functioning can be obtained from studies of other intracellular membrane trafficking pathways. A pathway of intracellular constitutive vesicle trafficking was first described in pancreatic cells (50). Vesicles which bud from the ER travel to and fuse with the Golgi complex, thus delivering cargo for processing by Golgi complex enzymes (51) (see Figs 3 and 4). The *trans*-Golgi network (TGN) packages cargo into constitutive secretory vesicles (52) and, in neurons, synaptic vesicle precursors. Secretory vesicles fuse with the plasma membrane, releasing their contents. Components of the vesicle membrane are then retrieved and delivered to early endosomes. The components can then be targeted to lysosomes via late endosomes, or back to the *trans*-Golgi network for further use in exocytosis. The maintenance of intracellular compartment identity in the midst of dynamic membrane flow requires highly regulated transport of both retrograde and anterograde vesicles through the membrane compartments. Identification of proteins involved in these processes has indicated that homologous but distinct proteins are utilized at different compartments (Figs 3 and 4) (53).

A biochemical approach was undertaken to identify the molecules involved in vesicle trafficking through the Golgi complex. Membrane trafficking was monitored by following the transport of the vesicular stomatitis virus (VSV)-encoded glycoprotein (G protein) through different membrane-bound compartments *in vitro* (54–57). Donor membranes were prepared from Chinese hamster ovary (CHO) cells infected with VSV-G; these cells also lacked the enzyme *N*-acetylglucosaminyl (GlcNAc)-transferase I. Acceptor membranes were purified from wild-type CHO cells that possessed GlcNAc-transferase I. The two fractions were mixed and incubated at 37 °C for an hour in the presence of tritiated UDP–GlcNAc. If membrane trafficking occurred between donor and acceptor membranes, the VSV-G protein (from the donor compartment) would encounter the GlcNAc-transferase (in the acceptor compartment) resulting in incorporation of tritiated-GlcNAc into its polysaccharide chain. Thus, membrane trafficking between donor and acceptor membranes was monitored by the presence of radiolabelled VSV-G protein.

Using this assay, it was first shown that cytosol and ATP are required for VSV-G protein transport (56, 58). It was also discovered that inactivation of molecule(s) with the sulfydryl reagent *N*-ethylmaleimide (NEM) also inhibited membrane trafficking (55). The focus of research then became the identification of the NEM-inactivated molecule(s). By fractionating cytosol and applying the fractions to the assay described above, a single protein species was isolated which could restore membrane trafficking between NEM-treated membranes (59, 60). This protein was called NSF for NEM-sensitive factor and was present as a multimer of 76 kDa subunits. Interestingly, NSF was shown to be a membrane-associated ATPase, which is released into the cytosol upon ATP hydrolysis (61, 62). Furthermore, NSF was hypothesized to act catalytically rather than stoichiometrically because it was estimated that just one NSF multimer is sufficient to mediate the fusion of at least three vesicles.

Experiments were then undertaken to determine the molecules with which NSF interacts. Three related peripheral membrane proteins were purified which are required for membrane fusion at the same step as NSF (63, 64). These proteins, which were called α-, β-, and γSNAP (for *s*oluble *N*SF *a*ttachment *p*rotein) have molecular weights of 35, 36, and 39 kDa, respectively. The SNAPs can form mutually exclusive, high affinity, stoichiometric complexes with NSF. Independently, these complexes promote membrane fusion in varying degrees. The SNAPs do not act synergistically when combined. The three SNAPs are differentially expressed suggesting that different isoforms may be required in different tissues (65). Alternatively, each SNAP may have a unique function (66). The SNAPs were shown to promote the binding of NSF to a Golgi-complex integral membrane protein(s) (62–64). The concentrations of SNAP proteins necessary for membrane fusion were similar to those required for promoting NSF binding to membranes. These results confirmed the previous description of the role of NSF and αSNAP homologues in yeast (see Section 2.5).

A role for αSNAP and NSF in neurotransmission was suggested by the identification of the neuron-specific membrane protein, syntaxin1, as a membrane receptor for the NSF/αSNAP complex. Syntaxin, named from the Greek word, syntax, meaning to put together in an ordered way (45), had been previously implicated in neurotransmission. In experiments designed to characterize membrane receptors for the NSF/αSNAP complex or 'particle', the NSF/αSNAP particle was tagged with the myc sequence, attached to anti-myc beads, and incubated with Triton X-100 solubilized brain membranes (67). A non-hydrolysable form of ATP, ATPγS, was added to the NSF/αSNAP particle during the incubation with solubilized brain membranes to promote the interaction of NSF/αSNAP with the membrane receptors. After washing the non-specifically bound membrane proteins from the anti-myc beads, the specific interacting-proteins were eluted by disassembling the particle by exchanging ATPγS with ATP. The eluted proteins were microsequenced and their identity was found to be syntaxin1, VAMP2, and SNAP-25. While syntaxin and VAMP were formerly implicated in synaptic vesicle fusion, SNAP-25, a previously identified synaptosomal protein (68), had not been proposed to be an exocytic protein. These proteins were collectively termed SNAREs (*SNAP re*ceptors), and the complex they form with NSF and the SNAPs is commonly referred to as the 20S complex or particle, because it migrates at approximately 20S on glycerol gradients. It is likely that the NSF/SNAP particle acts at all stages of the secretory pathway, but acts through different SNARE homologues (see Figs 3 and 4). Thus, studies of intracellular membrane trafficking at the ER and Golgi complex steps have helped to identify proteins and shape hypotheses regarding the release of neurotransmitter at the nerve terminal.

2.5 Insights from yeast secretory mutants

The concept of homologous secretory proteins existing throughout the intracellular trafficking pathway in one cell type also extends to homologies across species. Cell types as divergent as yeast and neurons share similar exocytic protein machinery (69). Unlike mammalian brain cells, yeast cells are amenable to relatively quick screens for mutants in various cell biological processes. Screens have been conducted with the goal of identifying genes involved in vesicle trafficking throughout the secretory pathway. For example, several groups have identified genes in which mutations inhibit membrane trafficking to the yeast vacuole, a lysosome-like compartment (see Fig. 4) (70). Some of these genes are homologous to genes involved in neurotransmitter release (71–73). More directly related to neurotransmitter secretion, however, is a set of isolated mutants, called *sec* mutants, in which constitutive exocytosis at the plasma membrane step is blocked (74, 75). It was eventually appreciated that homologues of several of these late-acting 'Sec' genes were involved in multiple steps of the secretory pathway. Perhaps not surprisingly then, homologues of these yeast genes have been shown to play roles in exocytosis in the nerve terminal.

Specifically, 15 secretory mutants were identified which block fusion of Golgi complex-derived vesicles with the plasma membrane (75, 76). To isolate these mutants, yeast cells were treated with ethyl methane sulfonate (EMS) or nitrous acid to produce mutations in random genes. Cells containing temperature-sensitive mutations in genes specifically required for secretion were selected by the following criteria:

(a) An increase in cell density due to the production of protein and lipids in the absence of cell division or growth.
(b) A block in exocytosis of both invertase and acid phosphatase while protein synthesis continued at normal levels.
(c) Accumulation of vesicles as determined by electron microscopy.

A large number of cells were screened by density analysis and those that fit the above criteria were assigned to 23 complementation groups (of these, 15 are involved at the plasma membrane step); these were called the Sec genes. Identification and characterization of these genes and the functions of their products have continued since their discovery. Some of the proteins encoded by the Sec genes are involved in early or multiple steps of the secretory pathway. Other Sec proteins are involved in the final stages of exocytosis at the plasma membrane (see Figs 3 and 4). Some have provided insight into the molecular mechanisms of neurotransmitter release.

Several Sec genes are members of families which have been conserved throughout the secretory pathway and throughout evolution from yeast to man (69) (see Fig. 4). The first *sec* mutant described was *sec1* (74). Sec1p, the protein encoded by the wild-type allele of the gene mutated in the *sec1* strain, acts at the final stage of the secretory pathway and likely interacts with plasma membrane proteins of the syntaxin family. Sec4p is a member of a family of low molecular weight GTPases; a pool of Sec4p is present on exocytic vesicles (77, 78). Sec2p is a guanine nucleotide exchange factor (GEF) for Sec4p (79) (see Section 6.2 and Fig. 8). Sec3p, -5p, -6p, -8p, -10p, and -15p are members of a multisubunit complex which likely specifies domains of the plasma membrane for delivery of exocytic vesicles (80, 81) (see Section 7 and Fig. 9). Sec9p is a member of the SNAP-25 family of proteins (82). Sec17p is a member of the αSNAP family and Sec18p is a member of the NSF family (83). Both Sec17p and -18p are involved in multiple steps of the secretory pathway. The remaining Sec proteins (Sec7p, -11p, -12p, -13p, -14p, -16p, -19p, -20p, -21p, -22p, and -23p) are likely involved in earlier steps of the secretory pathway (76). Homologues of the syntaxin and VAMP families were probably not found in this original screen because they are present in two copies; the likelihood of randomly mutating both copies is extremely low. However, homologues of the syntaxin1 family (Sso1/2) (84) and the VAMP family (Snc1/2) (85) have since been described.

3. Ascribing possible function to synaptic vesicle-associated proteins

The techniques described in Section 2 have provided the identity of many membrane and cytosolic proteins that are likely important for synaptic vesicle exocytosis. Following the identification of the proteins, the challenge is to describe how the proteins function. Once the cDNAs which encode vesicle trafficking proteins are obtained, their predicted amino acid sequences can be analysed for homology to protein domains of known function and/or potential secondary and tertiary structural motifs. Several motifs have been described for many of the membrane trafficking proteins.

Based on these analyses and knowledge of important aspects of exocytosis, experimentally testable hypotheses of the functions of individual proteins have been established. For example, it is known that docked vesicles are thought to form a readily-releasable pool, awaiting the Ca^{2+} signal to begin fusion (see Fig. 1). It has also been proposed that, once docked, vesicles probably go through several biochemical transitions before fusing with the plasma membrane (see Fig. 6) (86). Furthermore, experiments indicate a requirement for GTP (87, 88) and ATP (89) in exocytosis. Capacitance measurements (see Chapter 2) of stimulated pituitary melanotrophs and chromaffin cells (these cells use regulated exocytic mechanisms like those observed at the nerve terminal) demonstrated that there are three kinetic phases of release after the ATP requiring step (see Fig. 6) (90). In the absence of MgATP, Ca^{2+} can trigger an exocytic burst of vesicles which are probably docked at the plasma membrane (86). While the initial burst is steeply dependent on Ca^{2+}, there are intermediate and longer phases which are sensitive to pH and temperature, respectively. Neither the biochemical composition nor mechanisms of action of these intermediates are known. Finally, vesicles rely on the cytoskeleton for their targeting (8, 91), organization in the nerve terminal (92, 93), and perhaps fusion itself (94). Protein motifs can suggest roles for proteins in many of these steps.

A hallmark of synaptic transmission is its sensitivity to Ca^{2+} concentration. Thus, any motif which suggests Ca^{2+} binding is a potentially interesting clue to the function of the protein. Motifs such as the C_2 domain, originally found in the C_2 region of protein kinase C (PKC) (95), are capable of binding both lipids and Ca^{2+}. Several synaptic vesicle trafficking proteins contain C_2 domains, including synaptotagmin (40, 41), Rabphilin-3A (96) and DOC2 (97). It was originally proposed that synaptotagmin (on the vesicle membrane), via its C_2 domain, binds the plasma membrane lipid bilayer in the presence of high concentrations of Ca^{2+}. In this way, lipid bilayer fusion between the vesicle and plasma membrane would be prompted by Ca^{2+} entry. However, there is no direct evidence of such a role for synaptotagmin

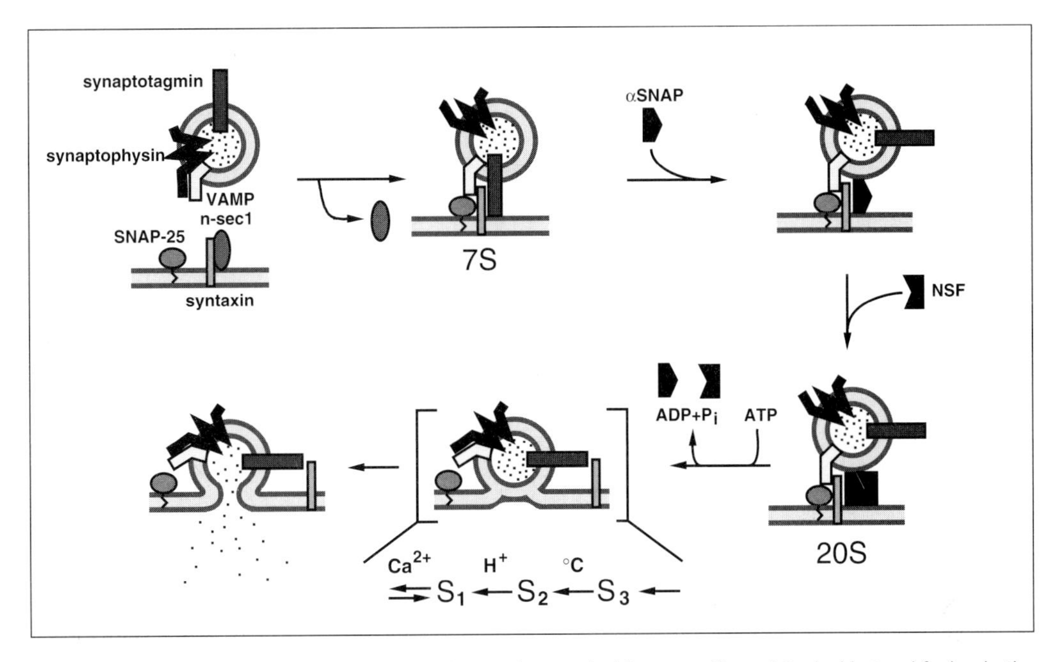

Fig. 6 A model of the sequence of molecular interactions underlying synaptic vesicle docking and fusion in the nerve terminal. Prior to vesicle docking at the plasma membrane, syntaxin and VAMP are bound to n-sec1 and synaptophysin, respectively. The vesicles are delivered to the plasma membrane and docked via the formation of the 7S complex composed of the two vesicle proteins, synaptotagmin and VAMP, and the two plasma membrane proteins, syntaxin and SNAP-25. αSNAP binds the 7S complex, removing synaptotagmin. NSF then binds to αSNAP thus forming the 20S complex. ATP hydrolysis by NSF promotes the dissociation of the 20S complex and leads to membrane fusion. There are at least three intermediates between the ATP-requiring step and fusion, which are sensitive to temperature, pH, and Ca^{2+} concentration.

in membrane fusion. More recent hypotheses suggest that synaptotagmin regulates SNARE complex formation and transmitter release through interactions, which are affected by Ca^{2+}, with SNAP-25 and syntaxin (see Section 6.3 and other chapters in this book). In addition, other aspects of the synaptic vesicle life cycle are likely sensitive to the concentration of intracellular presynaptic Ca^{2+} concentration; perhaps synaptotagmin regulates these events.

Vesicle trafficking is cyclical and thus requires mechanisms which regulate direction. Directionality might be conferred through trafficking proteins containing GTPase motifs (27, 77, 98). The GTPase domain likely acts as a switch changing the conformation of the protein and modulating interactions with effector proteins (see Section 6.2 and Fig. 8). Originally, GTPase activities of large G proteins which are involved in signalling cascades from the cell surface were described (99, 100). However, GTPase activity was also demonstrated in the Ras family of low molecular weight (small) G proteins which signal gene transcription. A set of low molecular weight GTPases homologous to Ras, called Rabs, were hypothesized to help in specifying trafficking of vesicles to the correct target membranes (see Fig. 6B). More than 40 mammalian (and 11 yeast) Rabs have been described (Figs 3 and 4) and distinct Rabs are found at different steps of the secretory pathway. Additionally, because some Rabs are expressed only in specific cell types, different cells may utilize unique Rabs for characteristic purposes.

In the fusion process, two energetically stable lipid bilayers must be broken, perhaps transiently exposing their hydrophobic cores to the aqueous environment. Energy is required to drive this lipid rearrangement. Thus any vesicle-associated protein with an ATPase domain might have a role in driving membrane fusion. NSF is a hexamer of ATP hydrolysing subunits. ATP hydrolysis by NSF has been shown to disassemble the complex consisting of syntaxin, VAMP, SNAP-25, NSF, and αSNAP *in vitro* (67), suggesting that the energy derived from ATP may be used to promote fusion. Another protein proposed to function in membrane trafficking which contains an ATPase domain is hrs-2 (49). As yet, it is unclear whether hrs-2 provides energy for membrane fusion or if the nucleotide binding domain is used as a switch in a similar fashion as the Rab GTPases.

Several vesicle trafficking proteins are predicted to contain an amphipathic α-helical motif with the potential to form a 'coiled-coil' structure with other proteins (101–103). This motif is composed of heptad repeats characterized by the presence of conserved hydrophobic residues in recurring positions along the helix. These hydrophobic residues form the interacting faces of the helices in the coiled-coil structure. For example, syntaxins, VAMPs, and SNAP-25 all contain these motifs and were originally proposed to interact via their coil domains (see Section 5.3). This hypothesis led to experiments in which site-directed mutagenesis was used to generate point mutants in these proteins (103). These experiments confirmed the importance of amino acid residues along the binding faces of the helices for both assembly and disassembly of the complexes containing VAMP, syntaxin, and SNAP-25.

4. A model of possible protein interactions underlying synaptic vesicle exocytosis based on protein binding properties

Based on experiments which revealed functional motifs, binding properties, and *in vitro* interactions of synaptic vesicle-associated proteins, a model of the sequence of molecular interactions underlying the final steps of membrane docking and fusion was proposed (104, 105) (Fig. 6). In this model, v-SNARES on the *v*esicle and t-SNAREs on the *t*arget plasma membrane specifically interact, ensuring that vesicles fuse only with appropriate membranes. Thus, at the nerve terminal, once the vesicle reaches the plasma membrane, v-SNAREs on the vesicle (synaptotagmin and VAMP) are thought to bind t-SNAREs on the plasma membrane (syntaxin and SNAP-25) forming a specific protein complex which might hold the vesicle docked at the active zone. This complex is known as the 7S complex because the four proteins, when detergent solubilized from brain extract, co-migrate on glycerol gradients at the 7S region.

In vitro binding experiments using bacterially expressed recombinant proteins focused on the mechanisms that led to 7S complex assembly and disassembly (103–106). These studies demonstrated that recombinant VAMP, syntaxin1a, and SNAP-25 all bind each other in paired reactions. However, VAMP binds the complex of syntaxin1a and SNAP-25 with higher affinity than it binds either syntaxin1a or SNAP-25 alone. These discoveries further supported the 'SNARE hypothesis' which stated that donor membranes (vesicles) contain unique v-SNAREs which specifically recognize t-SNAREs on acceptor membranes (45, 67, 104–107). This hypothesis further proposed that the various isoforms of the v-SNAREs and t-SNAREs ensure the fidelity of intracellular membrane trafficking, allowing vesicles to fuse only with the correct acceptor membrane.

Other molecules probably regulate the activity of vesicle and plasma membrane SNAREs for docking and fusion events. One possibility is that v-SNAREs and t-SNAREs are associated with other molecules prior to binding each other (see Fig. 6). Crosslinking experiments demonstrate that VAMP associates with another synaptic vesicle protein, synaptophysin (108–110). This association may regulate the introduction of VAMP into the 7S complex. Similarly, syntaxin1a binds, with high affinity, n-sec1, a soluble protein (38). n-sec1 inhibits the binding of syntaxin1a to VAMP and SNAP-25 (104) perhaps by holding syntaxin1a in a conformation in which the syntaxin N- and C-termini are bound to each other, a conformation different than that found in the 7S complex (103). Thus n-sec1 may constrain syntaxin1a to a fusion-inactive state. In addition, neither synaptophysin nor n-sec1 are found in the 7S complex, suggesting that they act at a step in the pathway preceding the formation of the 7S complex. However, it should be kept in mind that in yeast, Sec1 mutants accumulate late stage secretory vesicles, suggesting that Sec1 is not simply inhibitory (74). There are, of course, likely other, as yet unknown molecules which regulate the activity of these complexes.

An original tenet of the SNARE hypothesis was that a second complex forms after the 7S step, but before the actual fusion event. This complex was proposed to be the 20S complex containing VAMP, syntaxin, SNAP-25, NSF, and αSNAP (67, 104, 105) (see Fig. 6). The 20S complex has been shown to form *in vitro* in the presence of ATPγS and detergents. When ATP is added, the complex is disassembled, perhaps leading to membrane fusion. Although it was unclear exactly how or when these molecules are acting *in vivo*, the original SNARE hypothesis was attractive because it suggested molecular mechanisms of both how specific membrane interactions could occur (pairing of specific v- and t-SNARES) and how the energy required for fusion of two membranes (ATP hydrolysis by NSF) could be derived from a relatively small set of molecules.

Nevertheless, a number of questions remained unanswered by this hypothesis.

(a) How do the proteins of the 7S complex, which forms *in vitro*, interact *in vivo*?

(b) How are these proteins regulated to function only on the correct membranes? The localization of v- and t-SNAREs is not confined to specific intracellular sites during the vesicle fusion life cycle. For example, v-SNARE molecules must be present on the membranes of donor and acceptor compartments after fusion events because they are integral membrane proteins. What prevents v-SNAREs from acting when they are incorporated into the plasma membrane?

(c) In neurons, syntaxin and SNAP-25 are found on the plasma membrane outside of the active zone. According to the model then, vesicles should be able to dock at sites throughout the plasma membrane. However, vesicles accumulate only at specific plasma membrane domains (111).

(d) Why do vesicles appear morphologically docked in nerve terminals which have been exposed to the VAMP-cleaving neurotoxin, tetanus toxin (see Section 6.1)? The model predicts that in the absence of VAMP, vesicles would not be found apposed to discrete active zones.

(e) Is the order of interaction of these molecules correctly depicted by the model? For example, the role and timing of NSF activity is unclear; a number of biochemical transitions, downstream of the actions of NSF, likely occur before the fusion event (Fig. 6) (86).

These questions, and others, indicated that while the early models were a powerful force for suggesting experiments, much work remained.

5. Insights into the structures of proteins underlying exocytosis

Informative data have recently emerged from the use of powerful techniques for obtaining structural information: fluorescence resonance energy transfer experiments (FRET), electron microscopic rotary shadowing and immunological techniques, spin labelling electron paramagnetic resonance (EPR) spectroscopy, and X-ray crystallography. The first three techniques have given interesting information on general topographic structure and orientation of proteins in complexes. However, they are not able to provide detailed atomic structural data. To obtain such information, X-ray crystallography experiments are required. Obtaining crystals for these experiments can be very difficult, so the crystal structures of only a few of the synaptic vesicle trafficking proteins have been resolved. Furthermore, it is often the case that only fragments of the full-length proteins can be crystallized. Nevertheless, some attempts to crystallize membrane trafficking proteins have succeeded, providing data relevant to the molecular mechanisms of synaptic vesicle fusion.

5.1 Structures of NSF in the presence and absence of ATP

The quick-freeze/deep-etch rotary shadowing electron microscopy technique was used to visualize the structure of NSF. It is thought that NSF is composed of three functionally independent domains: two ATP-binding cassettes (an ATPase domain (called D1) and an ATP-dependent oligomerization domain (D2)) and a region for SNAP and SNARE binding (N). The full-length NSF protein was adsorbed onto mica chips in buffers either containing or lacking ATP. These chips were then frozen and 'deep-etched' to expose the adsorbed molecules. Finally, the chips were rotary shadowed with platinum and subjected to electron microscopic analysis. This study revealed that NSF is a hexagonal ring of about 15 nm in diameter with a 3–5 nm opening in the centre (112). This structure is about 10 nm in 'height'. Sometimes structures roughly twice as tall were observed; these were probably dimers. The D1 and D2 domains likely make up the basic cylindrical structure of NSF. Each may form a ring, and the two are likely stacked. The N domain possibly makes up the outside surface of the cylinder.

Because the function of NSF is critically dependent on ATP hydrolysis, the structure was examined in the absence and presence of ATP and ATPγS. Interestingly, in the absence of ATP, the structure dramatically changes. It becomes splayed, with 6–12 small (5–7 nm) globular structures emanating from a central ring of ~ 12 nm. The globular structures are sometimes observed arranged in groups which make up three 'lumpy arms'. The arms are likely composed of D1 and N domains. This structure was also observed in the presence of NEM and did not require ATP hydrolysis, but merely the lack of ATP. The compact, cylindrical structures re-formed after the addition of ATP. Furthermore, it appears that there are two distinct ATP-dependent conformational changes. One maintains the closed, cylindrical structure, while the other holds the oligomer together while the N domains are splayed. When the non-hydrolysable ATP analogue, ATPγS, was added to NSF, the rings of the NSF oligomers became more compact. However, six globular extensions at the base of the cylinder were observed. These are likely contributed by the N domain. These conformational changes may underlie the ability of NSF to dissociate complexes of SNARE proteins (113) (see Fig. 7).

The crystal structure of the D2 domain of NSF was recently solved (113, 114). Because this domain is involved in ATP-dependent hexamerization, knowledge of its structure helped to explain how complexes of NSF couple ATP hydrolysis to conformational changes. In addition, the D2 domain is similar to the D1 domain, allowing information obtained from crystals of the D2 domain to lead to a better understanding of how the 20S SNARE complex is dissociated by NSF ATP hydrolysis. ATP hydrolysis likely leads, by allosteric effects, to changes in the conformation at the hexamer interfaces in the D1 ring. Conformational changes in these domains may then lead to the dissociation of the SNARE subunits from the NSF hexamer.

The structure of the 20S complex was also analysed using electron microscopic techniques (112, 207). When the complex of NSF, αSNAP, syntaxin, SNAP-25, and VAMP was adsorbed to mica chips in the presence of ATPγS, a complex of NSF was observed with the core complex of SNAREs emerging from one end (see Fig. 7). This structure was not observed when the SNAREs, NSF, and αSNAP were incubated in the presence of ATP, which would promote the formation and disassembly of the 20S complex.

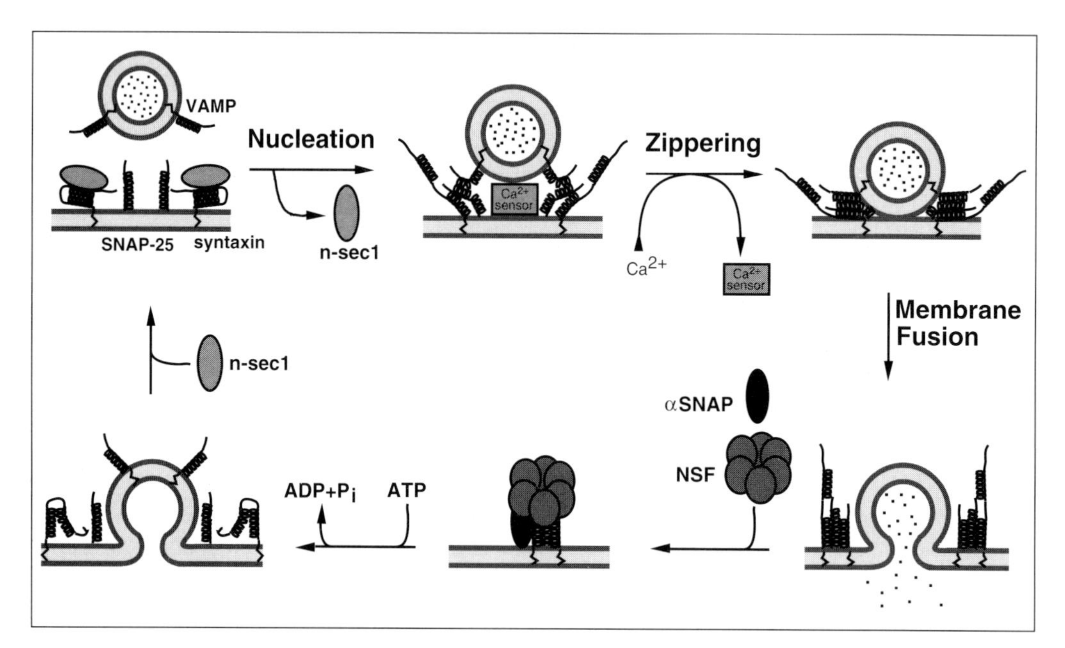

Fig. 7 A current model of the molecular mechanisms underlying synaptic vesicle exocytosis. Syntaxin is bound to n-sec1. Once free of n-sec1, a C-terminal helical domain from each of two syntaxin molecules binds two helical domains contributed by one SNAP-25 molecule (for simplicity, only one syntaxin molecule, and one SNAP-25 helix are depicted in the figure). In a process called nucleation, a helical domain of VAMP replaces one of the syntaxin helices, upon vesicle delivery to the plasma membrane. This core complex may be prevented from fully forming by a Ca^{2+} sensing molecule such as synaptotagmin. After the increase in intracellular Ca^{2+} concentration results in the displacement of the Ca^{2+} sensor, the core complex fully forms. This core complex consists of two helices contributed by SNAP-25, one helix contributed by syntaxin, and one helix provided by VAMP (only one of the two SNAP-25 helices is shown in the figure). This complex forces the two lipid bilayers into close apposition, leading to membrane fusion. After fusion, stable core complexes remain in the plasma membrane. The actions of NSF and αSNAP are required to separate the core complex. The SNAREs are then primed for another round of vesicle fusion.

5.2 Conformation of the VAMP/syntaxin/SNAP-25 complex

The complex comprised of VAMP, syntaxin, and SNAP-25 mediates binding and/or fusion of vesicles with the plasma membrane. It is therefore critical to understand the structure of this binding complex. Syntaxin and VAMP are type II membrane proteins: their membrane anchors are at the carboxyl-terminus. These two proteins probably interact before and after fusion, necessitating either a common binding structure in the two states, or a rearrangement of their binding structures during or after fusion. Furthermore, depending on how VAMP and syntaxin bind, the vesicle may or may not be in close proximity to the plasma membrane once it is docked. For example, if the helical binding domains of VAMP and synatxin1 are in an antiparallel orientation (the C- to N-terminal direction of both proteins being opposite when bound), the vesicle could be located at a distance the length of the complex of the two molecules apart from the plasma membrane. Alternatively, if the binding domains are oriented in a parallel fashion, the vesicle could be constrained to a location relatively close to the plasma membrane (see Fig. 7).

The binding orientation of the proteins was determined by two lines of experimentation. The basic idea for both was to label the carboxyl- or amino-termini of the two proteins with different markers and, after mixing the two proteins, assess the location of the label on one protein with respect to the label on the other protein. The relative locations were monitored using both FRET (115) and rotary shadowing electron microscopy (112) techniques. The FRET technique utilizes the phenomenon by which a donor fluorescent probe transfers energy to an acceptor probe in a distance-dependent manner. In the FRET experiments, the acceptor probe coupled to the amino-terminal end of VAMP received a greater extent of energy transfer from the donor probe if the donor probe was coupled on the amino-terminal end of syntaxin than when it was on the carboxy-terminal end, suggesting that the binding is parallel. In the rotary shadowing experiments, proteins were labelled with epitope tags, antibodies, and maltose binding protein. By imaging the location of these tags using electron microscopy, the parallel binding orientation of syntaxin and VAMP was also demonstrated. These results support a model whereby VAMP and syntaxin, by binding, bring the vesicle into very close apposition with the plasma membrane (see Section 6.1 and Fig. 7). Furthermore, the results suggest a hypothesis for how the two proteins may remain associated in the plasma membrane after fusion.

However, these experiments did not reveal the orientation of SNAP-25 in the VAMP2/syntaxin1A/SNAP-25 complex. In order to answer this question, this complex was subjected to gentle proteolysis to determine which regions contribute to a 'core complex'. This core complex was subjected to spin labelling EPR spectroscopy to determine its structure (116). These experiments were similar in concept to the FRET experiments. Cysteine residue 'spin labels' were incorporated, by site-directed mutagenesis, into the sequences of the core complex proteins at specific positions and used to monitor the orientation of binding of the core complex. These experiments indicated that the core complex is formed by four α-helices (two from SNAP-25 and one each from syntaxin and VAMP). Furthermore, the orientation of each helix was determined, corroborating previous data regarding syntaxin and VAMP. These experiments further revealed that the two SNAP-25 helices were oriented in a parallel fashion, necessitating a long linker to allow the second SNAP-25 helix to join the complex with the same orientation as the first. Spin labelling EPR can also provide information about the structure surrounding the incorporated spin labels. Using such information, a flexible site in the N-terminal SNAP-25 α-helix was discerned, suggesting a two-step hypothesis for a direct role of the core complex in membrane fusion. First, the complex could form partially, starting at the N-termini of the four helices, thus allowing the integral membrane proteins, syntaxin and VAMP, to remain in their respective membranes. The formation of the complex could be held in check at this point by a Ca^{2+} sensing molecule such as synaptotagmin (see Section 6.3 and Fig. 7). After initial binding, an increase in intracellular Ca^{2+} concentration would displace the Ca^{2+} sensor, allowing the helices to align completely and forcing the membranes into close proximity (see Fig. 7 and discussion below).

5.3 The structure of the core complex

A further step towards the understanding of the exocytosis process was taken by solving the crystal structure of the core complex. Crystals comprising the domains of syntaxin1A, VAMP2, and SNAP-25 that form the coiled-coil structure of the 7S complex were generated and analysed by X-ray diffraction (117). The basic topology consists of four α-helices arranged in a circular cylinder of about 120 angstroms in length. The four helices interact to form the coiled-coil structure. Syntaxin1A and VAMP2 each contributed one helix and SNAP-25 provided two helices to the core structure. All four helices were associated in the same direction thus giving the coil a parallel arrangement as previously shown by spin labelling EPR spectroscopy (116). To form such a structure, the two helices of SNAP-25 must be attached by a linker of at least 84 angstroms which allows one helix to fold around and associate with the other helices in a parallel fashion. This loop is also thought to contain the membrane anchor for SNAP-25 in the form of palmitoylated cysteine residues.

When bundled, the four associated α-helices provide conserved hydrophobic residues (leucine, isoleucine, and valine) that can be grouped into layers. The hydrophobic layers at the centre of the bundle most closely follow the packing features of parallel, tetrameric leucine zipper proteins as found in the general class of α-helix bundle proteins. Embedded within these hydrophobic layers, in the middle of the four-helix bundle, there is a highly conserved ionic layer (or 'bubble') that consists of an arginine (R) and three glutamine (Q) residues, each contributed by one of the four α-helices. The amino acid (either glutamine or arginine) at this position is the most highly conserved residue in all the SNARE proteins; mutation of this residue results in loss-of-function phenotypes in some SNAREs (75, 118, 119). v-SNAREs of the VAMP family contain a central arginine residue and t-SNAREs of the syntaxin and SNAP-25 families contain a central glutamine residue thus prompting a new nomenclature: R-SNAREs for v-SNAREs and Q-SNAREs for t-SNAREs. It was further hypothesized that NSF/αSNAP-mediated disassembly might occur by initially disrupting these ionic interactions. The surrounding leucine zipper is hypothesized to form a water-tight seal to shield and stabilize the ionic bubble from the surrounding aqueous environment. Mutations of these hydrophobic residues in some SNAREs have resulted in reduced stability of the complex and disruption of secretion and/or neurotransmission (82, 117, 120–124). Interactions between core complex helices could provide the structural determinants for the high stability of the 7S complex (125).

The structural information discussed above provides evidence for how these complexes might form and suggests explanations for previously observed biochemical properties. One hypothesis that emerges from these studies is that fusion is brought about by the binding of one VAMP helix to a previously existing complex comprised of two syntaxin helices (contributed by two syntaxin molecules) and two SNAP-25 helices (contributed by one SNAP-25 molecule) (117, 208). In this scenario, the bundle would consist of four helices even in the absence of VAMP. Upon vesicle docking, the VAMP helix would then displace one of the syntaxin molecules starting at the amino-terminus and ending at the carboxy-terminal end (a process called 'zippering'; see Fig. 7) thus forcing the vesicle and plasma membranes into close proximity. The helices might end their interactions very close to where they enter the membranes. Polar residues at the carboxy-terminus of the C-terminal coil domains of VAMP and syntaxin may interact with the plasma membrane, driving the fusion process. Alternatively, the polar residues could form an extended linker, thus holding the vesicle some distance from the membrane. Nevertheless, although these structural studies are of critical importance for our understanding of the molecular mechanisms potentially underlying neurotransmitter release, the data do not provide direct evidence that SNARE complexes are intimately involved in membrane fusion.

5.4 Structures of the C_2 domain of synaptotagmin when bound to Ca^{2+} and/or syntaxin

Synaptotagmin, a molecule that can bind up to four Ca^{2+} ions at a Ca^{2+} concentration of 100 μM (209), contains two C_2 domains within its cytoplasmically oriented C-terminus (40). The first of these C_2 domains binds phospholipids and Ca^{2+} in the range of Ca^{2+} concentrations which supports neurotransmitter release. Because the C_2 domains may be involved in Ca^{2+} regulation of synaptic vesicle fusion, the crystal structure of the first C_2 domain was determined with the intention of gaining insight into the protein–lipid interactions underlying membrane fusion (126). The structure is a 'β sandwich' comprised of two β sheets, each of which is made up of four antiparallel strands. One sheet is bent inward, forming a convex structure, while the other sheet is concave. Three loops extend from the top and the bottom of the sheets.

Ca^{2+} was diffused into crystals to study the Ca^{2+}-binding properties of the C_2 domain. A Ca^{2+} ion binds a cup-like cavity created by two loops that connect strands in a central, conserved domain of the sandwich structure, the 'C_2-key'. This structure is unlike that of previously described Ca^{2+}-binding proteins in that the two Ca^{2+} 'clamps' are provided by amino acid stretches distant from each other in the primary sequence. The Ca^{2+} ion binding site is surrounded by at least four aspartate side chains and the carbonyl oxygen of a phenylalanine. At least two of these residues are required for Ca^{2+}-dependent binding to lipids. Surprisingly, however, the conformation did not significantly change in the absence of Ca^{2+}. Perhaps, Ca^{2+}-binding causes a subtle change in the surface shape of the concave β sandwich which affects binding to other proteins. Alternatively, the *in vivo* Ca^{2+}-bound structure may differ significantly from the crystals to which Ca^{2+} was added after formation in the absence of Ca^{2+}.

This same C_2 domain also binds syntaxin in a Ca^{2+}-dependent manner. This interaction was analysed using nuclear magnetic resonance spectroscopy to monitor which residues of the synaptotagmin C_2 domain are involved in binding syntaxin (127). The spectra obtained from these experiments are very sensitive to the chemical environment. Thus, any change in binding properties can be monitored. It was found that 24 residues of the C_2 domain were likely involved in Ca^{2+}-dependent syntaxin binding. The amino acids that are most strongly perturbed by the binding of syntaxin were all located in the Ca^{2+}-binding loops of the C_2 domain. The more weakly perturbed residues were present at the boundaries of these loops. In general, structural changes caused by syntaxin/Ca^{2+} were small and were only found on the surface of the C_2 domain. Binding of syntaxin was hypothesized to result from changes in the distribution of charge (contributed by amino acid side chains) on the surface of synaptotagmin. Specifically, the Ca^{2+}-binding pocket is zwitterionic in the absence of Ca^{2+}, and its net charge becomes more positive upon Ca^{2+}-binding. This change in electrostatic potential may mediate binding to acidic residues on the surface of syntaxin. To test this hypothesis, site-directed mutagenesis methods were utilized to change basic residues of the Ca^{2+}-binding pocket with the goal of altering the net overall charge. While these mutants were shown to both fold correctly and bind Ca^{2+}, some of them did not bind syntaxin. Thus, Ca^{2+} ions may promote binding of syntaxin to synaptotagmin by altering the charge distribution of the C_2 domain.

6. Current models of protein interaction cascades underlying synaptic vesicle exocytosis

The model of vesicle trafficking and fusion in Fig. 6 was based on biochemical properties of the putative membrane trafficking proteins. However, recent data, including the structural results just presented, have suggested that the model is inaccurate. New data has suggested new working models (one is shown in Fig. 7) which differ with respect to the order of action of the various exocytic molecules.

6.1 Roles of NSF, αSNAP, and the SNAREs

An original tenet of the SNARE hypothesis was that the 20S complex composed of syntaxin, VAMP, SNAP-25, αSNAP, and NSF mediates the ultimate fusion event. The experiments mentioned above in pituitary melanotrophs suggested that ATPase activity does not act at the ultimate fusion steps. New data supports models which posit that, while ATP hydrolysis may provide energy to reorganize the pre- or post-fusion complex, it probably does not directly lead to lipid bilayer fusion. Instead, the energy derived from ATP hydrolysis may serve to dissociate the bound SNAREs after fusion, preparing them for future rounds of docking and fusion (Fig. 7). The molecules that are directly involved in fusing lipid bilayers might be the SNAREs themselves.

Experiments were conducted using yeast vacuoles to attempt to discover at which point during the fusion cycle sec17p (the αSNAP homologue) and sec18p (the NSF homologue) act. In yeast, vacuoles fuse with each other. Homotypic vacuolar fusion utilizes homologues of many of the proteins involved in neurotransmitter release. Taking advantage of these properties, an assay to monitor the effects of various proteins on membrane fusion was developed (128–130). By preparing vacuoles separately from independent strains, it was demonstrated that ATP- and NSF-dependent effects occur before the donor- and acceptor-membranes come into contact. In these experiments, vacuoles from a strain containing proPho8p, a proalkaline phosphatase, were separated and treated with NSF. After αSNAP was removed by washing, the vacuoles were mixed with vacuoles from a strain containing Pep4p, the protease that processes and thereby activates proPho8p. By monitoring phosphatase activity, fusion was demonstrated, indicating that αSNAP and NSF are not directly involved in docking or fusion.

Using mutant yeast strains deficient in various SNAREs, it was further demonstrated that Q-SNAREs are required on one population of vacuoles and R-SNAREs are required on the other, supporting the hypothesis that SNARE-pairing functions in membrane fusion. NSF was required before mixing of the vacuoles probably to separate SNARE complexes which were formed in previous fusion events. These studies have led to the hypothesis that ATP hydrolysis by NSF is important for priming the SNAREs for binding leading to fusion. Perhaps during this priming step energy is transferred and stored in the form of a specific SNARE conformation; by promoting the unpairing of SNARE complexes after fusion, the energy derived from ATP hydrolysis is transferred to the SNARE molecules (see Fig. 7). The energy from these individual molecules would then be harvested, via the formation of SNARE complexes, for membrane fusion. Although these studies give a powerful indication of the timing of individual molecular events, it remains to be shown if results from the vacuolar fusion system will extend to the synapse.

The squid giant synapse (SGS) has been used as a convenient preparation for physiological experimentation of properties of synaptic transmission because it is large enough to be easily manipulated. In one experiment, tetanus toxin, which cleaves VAMP, was injected into the presynaptic nerve terminal of the squid giant synapse (131). This toxin was shown to reduce neurotransmitter release as expected. Surprising results were obtained from analysis of electron micrographs of these toxin-treated nerve terminals. First, it was noted that there were more vesicles in the nerve terminal as compared to control synapses. Secondly, the number of vesicles 'tethered to' (immediately apposed to) the plasma membrane was twice the number found in control synapses. These data are consistent with the hypothesis that VAMP is required for fusion and therefore, without VAMP, the vesicles accumulate at the active zone. However, the results suggest that VAMP might not be necessary for vesicle docking at the plasma membrane.

These results from the SGS injection experiments were supported by experiments in *Drosophila*. In these flies, neurotransmission at the neuromuscular junction was analysed. VAMP was reduced by generating transgenic *Drosophila* strains expressing tetanus toxin (132). Syntaxin was perturbed by generating mutations in the syntaxin gene (133). Both of these manipulations blocked neurotransmission, while leaving the structure of the synapse similar to wild-type animals. Electron microscopic analysis was performed on these synapses (134). While there were similar numbers of vesicles in wild-type and syntaxin mutant *Drosophila*, there was a 50% increase in the number of vesicles in the terminals of VAMP-depleted synapses. Furthermore, more vesicles were 'morphologically docked' in both syntaxin- and VAMP-depleted animals as compared to wild-type *Drosophila*. When electrophysiological experiments were performed, neurotransmission was blocked in the VAMP-depleted flies (as it was in the SGS) and in the syntaxin mutants. However, when spontaneous, unstimulated neurotransmitter release was monitored, VAMP-depleted flies still exhibited neurotransmitter release. Interestingly, no spontaneous release was detected in syntaxin mutants. These data point to a role for both proteins in a post-docking step, and perhaps suggest a role for syntaxin in the fusion process itself.

Perhaps the SNAREs are involved in fusion in addition to a role in defining appropriate membrane interactions. If this is the case, do the SNAREs act directly and/or independently in the fusion process? One way of answering this question is to determine which proteins are necessary and sufficient for fusion of membranes *in vitro*. Using a system of artificial lipid vesicle (liposome) fusion, the effects of the presence of different combinations of vesicle proteins on liposome fusion was monitored (135). One population of liposomes was loaded with a mixture of fluorescent phospholipids and quenching lipids. Another population of liposomes contained no fluorescent phosholipids. Liposomes from the two groups were mixed. If lipid mixing occurred, the fluorescent phospholipids would then be diluted from the quenching molecules, leading to a rise in detectable fluorescence. These experiments demonstrated that the presence of VAMP in one population of liposomes, and syntaxin on the other were sufficient to support mixing of the two populations of lipids. Furthermore, VAMP in one membrane and syntaxin in another can form complexes before lipid mixing. Finally, when the formation of these SNARE complexes is blocked by toxin cleavage or by incubation with soluble SNAREs, lipid mixing is abolished. These results suggest that syntaxin and VAMP are both necessary and sufficient for membrane fusion. It is not clear if these results will extend directly to the events occurring *in vivo*. However, this data combined with the results from mutant flies and perturbed SGS suggests a central role for the SNAREs in the fusion event underlying neurotransmitter release.

A tenet of the SNARE hypothesis has been that the SNAREs define the specificity of membrane interactions as discussed above. However, the aforementioned evidence suggests that SNAREs are not involved in docking, but instead in fusion. While it would make sense that a vesicle chooses the appropriate membrane at the docking step, it could be that specificity is also defined or 'proof-read' at the fusion event; perhaps syntaxin both determines specificity and directly participates in lipid bilayer fusion. Another possibility is that the SNARE complex formation is involved in fusion and only partially, or not at all, involved in determining specificity. In this case, other molecules would impart specificity to the trafficking process either independently of, or in association with, SNAREs.

Recent data suggest that the SNAREs do not impart specificity on their own (125). SNARE complexes consisting of the core binding domains of one member of the syntaxin family, one of the SNAP-25 family, and one of the VAMP family were allowed to form. These are essentially the core regions of the 7S complex, the structure of which was discussed above. The structural stability of these complexes under varying temperatures was monitored using circular dichroism and SDS-sensitivity/resistance of complexes. Surprisingly, no significant differences were found in the stabilities of complexes formed using a variety of different SNARE combinations. For example, SNARE complexes are quite stable even when a plasma membrane SNARE is substituted for an endosomal SNARE. The patterns of stability do not correlate with known localizations and trafficking pathways for these proteins. It is possible that the 'non-core' domains might impart some degree of instability on some of the complexes but experiments using the full-length SNAP-25, syntaxin1, and VAMP showed very similar thermal stability to that found for the synaptic core complex (136). Perhaps the distinct subcellular localization of the SNAREs, possibly controlled by targeting determinants encoded in the divergent portions of their sequences, provides the basis for appropriate pairing and specificity. Additional proteins such as the Rabs and the sec6/8 complex could also be important in determining correct vesicle targeting (see Sections 6.2 and 7, and Figs 4, 8, and 9). Furthermore, the cytoskeletal direction of vesicle trafficking is certainly important for targeting and thus contributes to membrane fusion specificity.

6.2 The Rab3a cycle

Rab3a is a member of a large family of low molecular weight GTPases, which is thought to mediate directional vesicular trafficking (Fig. 8) (99, 100, 137–140). Rab3a, like all Rabs, cycles between an active, GTP-bound form and an inactive, GDP-bound form. Synaptic vesicles carrying Rab-GTP are targeted to and docked at the plasma membrane. Either during or after the fusion event, GTP hydrolysis stimulated by a GTPase activating protein (GAP) converts the Rab to its GDP-bound form (141). A cytosolic protein, GDP dissociation inhibitor (GDI) (142), retrieves Rab-GDP from the target membranes and delivers it back to the synaptic vesicles (143). Following GDI displacement, stimulated by a GDI displacement factor (GDF), Rab3a-GDP bound to the vesicle is converted to the active Rab3a-GTP form by the action of Rab3a GTP/GDP exchange factor (GEF), and the cycle can begin again.

Rab3a activity is controlled by other soluble proteins and is hypothesized to regulate the docking/fusion of the vesicle. However, the mechanisms by which Rab3a mediates this regulation are unknown. The yeast homologues of Rab3a, Sec4p and Ypt1p, interact genetically with yeast homologues of VAMP, n-sec1 and SNAP-25, indicating that Rab3a might play a role in 7S complex formation (82, 144, 145). However, direct binding of Rab3a to proteins of the 7S complex has not been observed. One reason for this may be that Rab3a may interact only transiently with the 7S proteins to perform a catalytic function making binding difficult to detect. Such a kinetic mechanism may make these other secretory proteins competent for docking and fusion (145, 146). Alternatively, because there may be a different Rab protein for each step of the pathway, each may provide a proof-reading mechanism ensuring that only the correct complement of R- and Q-SNARE proteins interact.

Rab3a may interact with the core complex indirectly through other proteins. For example, Rab3a binds Rabphilin-3a (96) and Rab interacting molecule (RIM) (147) in a GTP-dependent manner (Fig. 8). However, there is no evidence that Rabphilin-3a

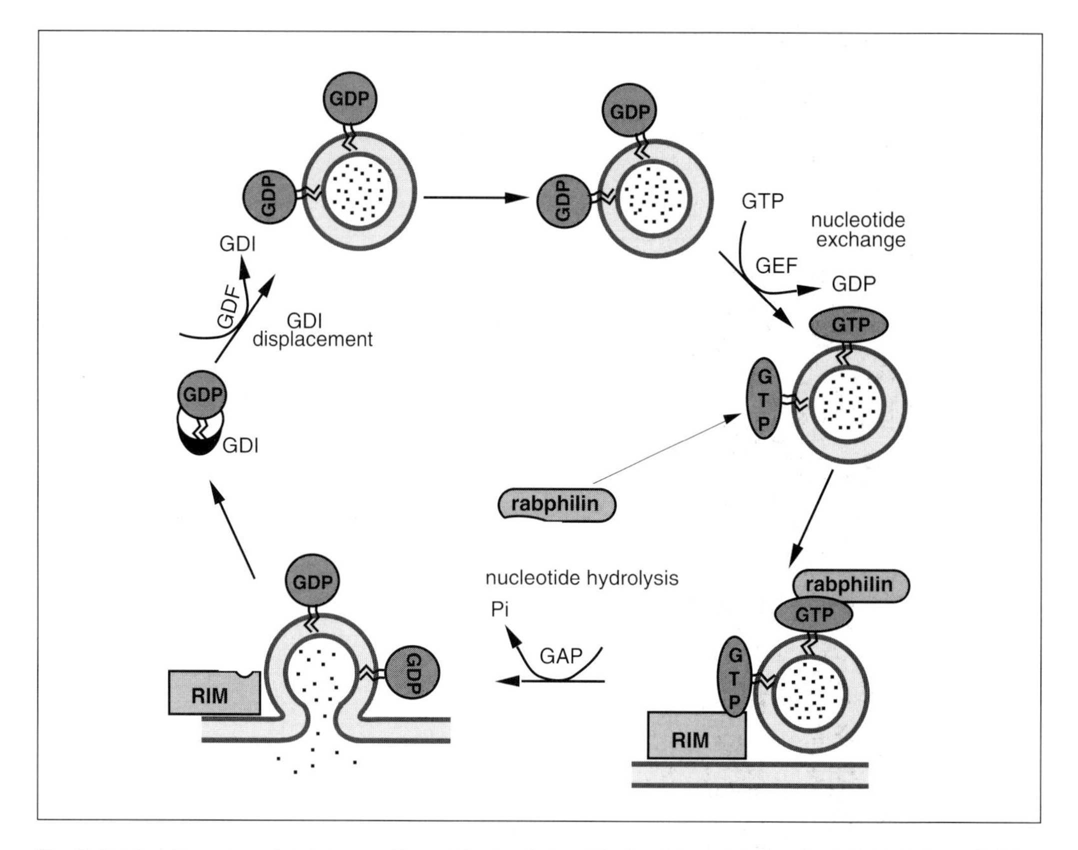

Fig. 8 The Rab3a cycle underlying synaptic vesicle docking and fusion. The low molecular weight GTPase, Rab3a, is associated with synaptic vesicles in both GTP- (shown as ovals labelled 'GTP') and GDP- (shown as circles labelled 'GDP') bound states. When bound to GTP, Rab3a can bind to the effector proteins, Rabphilin and RIM. In its GTP-bound state, Rab3a may promote the binding of the vesicle to the plasma membrane via interactions with RIM. GTPase activating protein (GAP) (141) promotes the hydrolysis of GTP to GDP and the subsequent dissociation of Rabphilin and RIM from Rab3a. Rab3a-GDP is removed from the membrane by guanine nucleotide dissociation inhibitor (GDI) (142). GDI displacement factor (GDF), displaces GDI, allowing Rab3a to bind to vesicles where Rab3a is converted to its GTP-bound form through the action of guanine nucleotide exchange factor (GEF), thus continuing the cycle.

or RIM interact with any of the 7S proteins. Rabphilin-3a may bind Rab3a simply in order to be directed to the vesicle because in Rab3a knock-out mice, levels of Rabphilin-3a in the synapse but not in the cell body are decreased (148). Interestingly RIM is located at the active zone, though it is likely also localized to other regions of the presynaptic terminal. RIM could play a role in the specific targeting of vesicles to active zones.

Expression of Rab3a in a strain of mice has been abolished by knock-out of the Rab3a gene (148). The phenotype is very subtle: the mice appear to be normal, however, upon repeated stimulation of hippocampal neurons, synaptic transmission fatigues more rapidly than in wild-type mice. This study suggested that Rab3a is involved in replenishing the readily-releasable pool of synaptic vesicles. Subsequent

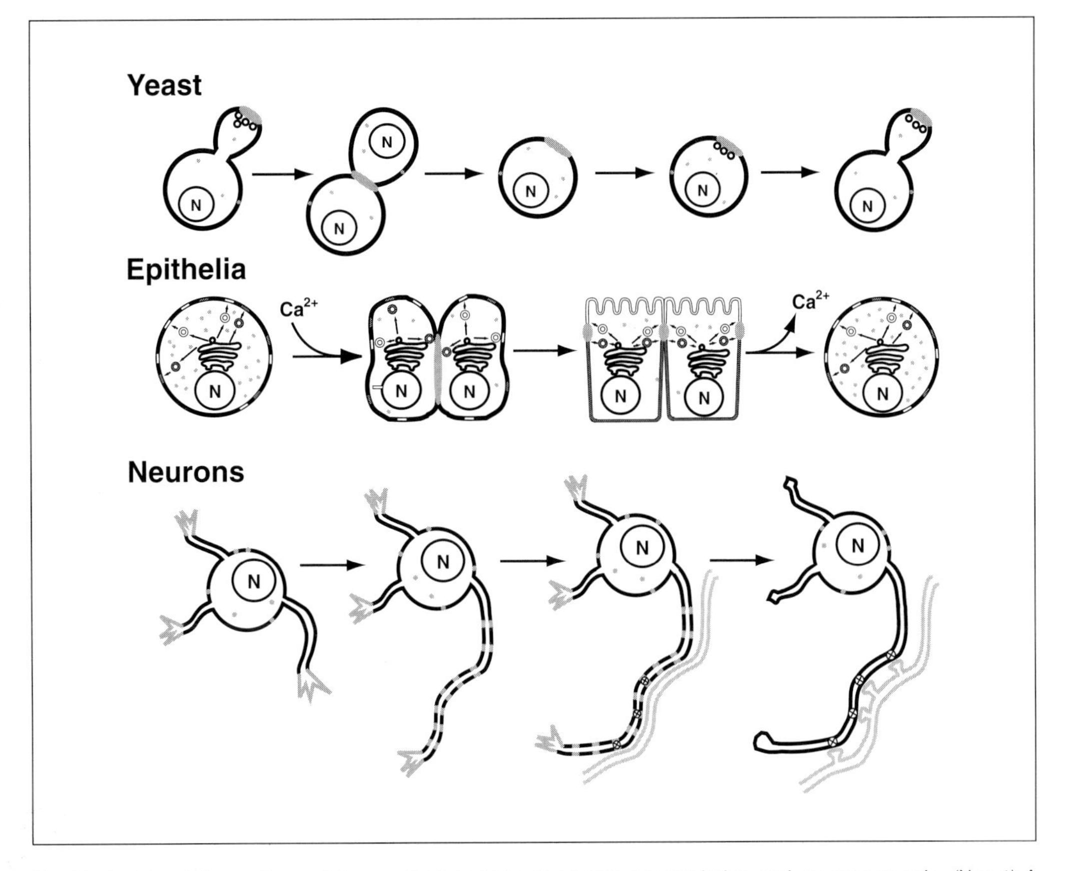

Fig. 9 Models of the actions of the sec6/8 complex in vesicle accumulation and synaptogenesis. (Yeast) A model of the actions of the sec6/8 complex during the life cycle of yeast cells. The sec6/8 complex (shown in light grey) is present at sites of secretion in the tip of the budding daughter cell, at sites of cytokinesis in dividing cells, and in a patch at the bud scar which designates the site of exocytosis underlying future bud formation (there is a cytosolic pool as well; light grey dots) (166). N, nucleus. (Epithelia) A model of the role of the sec6/8 complex in polarized secretion in MDCK epithelial cells. Unpolarized MDCK cells target vesicles (shown in dark grey and outlines) to random areas of the plasma membrane. The sec6/8 complex (light grey) is dispersed throughout the cytosol. Upon the addition of Ca^{2+}, two cells contact and cytoplasmic sec6/8 complex is reorganized to the contacting plasma membrane. Once polarized, the cells target vesicles either to the apical membrane (shown in outlines) or the basal-lateral membrane (shown in dark grey). The sec6/8 complex is located along the membrane near the tight junctions suggesting that the junction is a site to which vesicles are targeted. Upon depletion of Ca^{2+}, the cells become unpolarized and the sec6/8 complex is found dispersed in the cytosol (169). (Neurons) A model of the role of the sec6/8 complex in synaptogenesis. Neurons of increasing maturity are shown from left to right. In young neurons, the sec6/8 complex (light grey) is present in growth cones of neurites. Later, one neurite becomes the axon leaving the others as dendrites; the sec6/8 complex is organized into periodic domains along the axon. As the neuron matures, vesicle clusters are found in some of the sec6/8 domains (synapsin1-containing vesicle clusters are shown as circumscribed crosses). As synapses are formed between the axon and dendrites (a dendrite is shown in dark grey), local clusters of synaptic vesicles are stabilized and the sec6/8 complex is down-regulated (111).

work revealed that there was no effect on the size of the synaptic vesicle pools, or on the amount of neurotransmitter stored in vesicles, suggesting that Rab3a might act after docking (149). Interestingly, the amount of vesicle fusion per action potential appears to increase by 100% in the Rab3a knock-out mouse, suggesting that Rab3a may prevent multiple fusion events upon stimulation. At many synapses in the central nervous system, single vesicles are released in response to stimulation, a process possibly regulated by Rab3a.

Rab3a may act as a timer for vesicle docking. This hypothesis was suggested by experiments focusing on Rab5, a Rab3a homologue (150, 151). In this scenario, Rab3a would only allow binding of SNAREs while in its GTP-bound state. Perhaps the rate of binding of the SNAREs is critical because Rab3a may only be active long enough to allow certain SNARE complexes to form. There is also genetic data linking the function of the Rabs to the cytoskeleton (79, 152) and to the sec6/8 complex (see Section 7) (153). These experiments indicate that the role of Rab3a in neurotransmitter release is far from understood, and that it will be important to isolate new proteins which link the function of Rab3a with the SNAREs as well as other proteins.

6.3 Ca^{2+} regulation of neurotransmitter release

Another major research goal in the field is the elucidation of the mechanisms by which Ca^{2+} regulates release from neurons. While the fusion structures may be simply composed of the SNAREs, they must be uniquely regulated in the nerve terminal. A Ca^{2+}- binding protein such as synaptotagmin may hold the SNAREs in check, perhaps by preventing complete formation of the core complex (Fig. 7). Upon Ca^{2+} entry, the SNAREs would then be freed to drive membrane fusion. Such a model is supported by the structural data regarding the interaction between synaptotagmin and syntaxin (discussed in Section 6.4). Changes in the synaptotagmin-syntaxin binding properties upon Ca^{2+} binding might free the core complex from the inhibitory influence of synaptotagmin. Furthermore, when synaptotagmin is removed from mice by homologous recombination techniques, the fast component of Ca^{2+}-dependent neurotransmitter release is eliminated (154) (see other chapters in this book for a discussion of the physiological studies of synaptotagmin). However, a clear explanation of how synaptotagmin and Ca^{2+} act at the molecular level to regulate neurotransmission still requires further investigation.

Ca^{2+} regulation may be mediated by direct interaction between syntaxin/SNAP-25 and N-and/or Q-type Ca^{2+} channels (45). Syntaxin binds Ca^{2+} channels at an intracellular loop connecting domains II and III of the channel (155). Syntaxin and SNAP-25 were shown to interact with this loop with a maximal binding at 20 μM Ca^{2+}, a concentration in the physiological range capable of supporting exocytosis (156). Introduction of a peptide containing the region of the channel that interacts with syntaxin/SNAP-25 into neurons inhibited neurotransmitter release (157). A mechanism explaining how syntaxin and SNAP-25 affect regulated neurotransmitter release was suggested by experiments in *Xenopus* oocytes (158). When oocytes co-expressed syntaxin1A and the N-type Ca^{2+} channel, Ca^{2+} entry was inhibited because syntaxin1A held the Ca^{2+} channels in their inactive state. It is not yet clear if these results extend to neurons, but they argue for an important interaction between the two proteins. Perhaps components of the core complex can signal the Ca^{2+} channel when it is prepared for exocytosis. At the very least, such an interaction serves to localize the exocytic machinery very close to Ca^{2+} influx sites. Because Ca^{2+} concentration must be high to trigger exocytosis, vesicles must be maintained in close proximity to Ca^{2+} channels.

6.4 Current model of molecular mechanisms of membrane fusion

These recent experiments, along with structural data presented in Section 5, have prompted a revision of the model presented in Section 4. In the current model (Fig. 7), the SNAREs are first present in their high energy monomeric forms. Once the vesicle reaches the plasma membrane, the SNAREs begin to form the 7S complex. The core complex comprised of the helical domains of SNAP-25, syntaxin, and VAMP, form incompletely in a process termed 'nucleation'. This state may be stabilized by a Ca^{2+}-binding protein such as synaptotagmin which prevents full formation of the core complex. Upon Ca^{2+} influx, the inhibiting factor is removed and the core complex is allowed to completely form, or 'zipper'. The full binding of the four helices may constrain the vesicle and plasma membranes to conditions favourable for lipid bilayer fusion. Once the complex is formed it remains in a stable, low energy state in a single membrane (the plasma membrane). The actions of NSF/αSNAP and ATP hydrolysis are then required to dissociate this stable intermediate. Once dissociated, the SNAREs can be returned to the proper locations for future rounds of fusion.

7. Proteins involved in the organization of synaptic vesicle trafficking in the nerve terminal

The molecular mechanisms that regulate synaptic vesicle delivery to mature synapses and their fusion at active zones, are unclear. Furthermore, synaptic vesicle precursors and constitutive vesicles must also be specifically targeted during development of the neuron to support synapse formation and neurite outgrowth, respectively. These three processes are interrelated, making the mechanisms of specific vesicle delivery both interesting and difficult to study. Studies of vesicle proteins are already providing clues as to how the synaptic vesicle pool at the synapse develops.

Neurons are unique in their characteristic polarized morphologies. Their structures yield functionally distinct axons and dendrites, intercellular contacts, and synapses (159). The specific intracellular trafficking of lipid and protein components to correct locations of the cell is essential for this polarization. This trafficking supports the membrane addition required for development as well as dynamic morphological restructuring during neurite outgrowth. Trafficking of Golgi complex-derived vesicles to the plasma membrane is also crucial for establishing the synaptic vesicle pool at the synapse. While the proteins which mediate docking and fusion of vesicles with the plasma membrane certainly play important roles, it is not clear how the vesicles are targeted to specific locations at the plasma membrane.

Studies of synaptogenesis among dissociated hippocampal neurons in culture have provided a system in which to study the molecules involved in synaptic vesicle targeting and development. These experiments suggest that developing neurons probably contain, along their axons, the molecular components required for both the functioning of synaptic vesicles as well as the machinery needed for the constitutive exo- and endocytosis *prior* to contacting a postsynaptic cell (160–163). However, synapses are formed at distinct axonal regions and are of a defined size; they do not encompass the whole membrane. Thus, there must be a molecular signal at sites of future synapses which direct the delivery of synaptic vesicles and other synaptic components. These signals are probably not simply the SNARE proteins because Q-SNAREs of the syntaxin family are not specifically located at active sites of synaptic vesicle cycling (164).

Studies in yeast, in polarized epithelial cells, and in hippocampal cultures suggest that the mammalian sec6/8 complex may be a vesicle targeting candidate important in determining synaptic sites. The sec6/8 complex is found both in cytosolic and membrane-bound pools. The yeast sec6/8 complex is highly concentrated at the plasma membrane at sites of active vesicle exocytosis: both at the tip of a new growing bud (80, 165–167), and just prior to cytokinesis, at the necks of budded cells (168). Upon cell–cell contact and the development of asymmetric membrane domains in MDCK cells, the complex localizes to the tight junction region of the cell where it becomes membrane associated. Furthermore, antibodies directed against the rat brain sec8 protein block vesicle fusion at the basolateral membrane of polarized mammalian MDCK cells (169). These data are consistent with the hypothesis that the sec6/8 complex is essential for polarized vesicle targeting during the yeast cell cycle and in organizing trafficking pathways in mammalian cells.

Members of the mammalian brain sec6/8 complex associate to form a 17S complex (81). Analysis of the distribution of the mammalian sec6/8 complex in neurons during various stages of development of polarity and synaptogenesis was performed. Antibodies directed against sec6 label discrete, periodic domains of axons, the tips of immature, growing dendrites, growth cones, and both axonal and growth cone filopodia. All of these are areas of possible vesicle exocytosis. The observed periodic distribution of sec6 in axons was striking, suggesting that the mammalian sec6/8 complex might tag sites in the cell for delivery of vesicles. sec6 labelling in the tips of the processes—sites of constitutive exocytosis important for membrane addition—support this idea. Furthermore, the labelling seen along the axon shaft must represent sites of potential fusion along the axon because vesicle markers or functional vesicle cycling are not always found at sec6/8 sites. Immunolabelling of embryonic brain sections also reveals a role for sec6 in synaptogenesis; sec6 is highly expressed in the cortical subplate, a region of synapse formation in the developing cortex (111).

Much research has been devoted to understanding the location of vesicle exocytosis important for membrane addition in growing axons. Controversial results implicate either the cell body or the terminal growth cone as the important site (170, 171). Perhaps, however, at least in hippocampal neuronal cultures, vesicles fuse with the plasma membrane at sites along the axon shaft as well as at the cell body, dendrites, and the axon terminal (172). Additionally, other studies have suggested that *synaptic* vesicles are derived from a maturation of Golgi complex-derived vesicles involving presynaptic endosomes (10). These Golgi complex-derived, future synaptic vesicles may, as part of their maturation process, fuse with multiple sites along the axon as they travel toward their ultimate destination. For such synaptic vesicle maturation, plasma membrane machinery must be available for vesicle fusion. The molecules making up this machinery may be uniformly located along the entire membrane as is suggested by the distribution of the Q-SNARES, syntaxin and SNAP-25. However, the machinery important for exocytosis likely includes many soluble, membrane and cytoskeletal molecules. In a growing axon, which comprises a large membrane surface area, such exocytosis machinery may not be located at all points along the axonal membrane. Thus, it is possible that the sec6/8 complex is required along growing axons to mediate targeted exocytosis of constitutive and synaptic vesicles.

Another interesting observation was that sec6 immunoreactivity is down-regulated in mature synapses. Neurons contain intrinsic polarity and differentiation programs and do not divide. Perhaps the sec6/8 complex is important for directed exocytosis in immature neurons as in the other cell types. However, once the neuron reaches a mature state, a polarized secretory apparatus specific to synaptic vesicles may be in place, obviating the need of the sec6/8 complex for synaptic vesicle targeting. The sec6/8 complex might be important for targeting immature synaptic vesicles to presynaptic sites, as well as Golgi complex-derived vesicles for constitutive secretion in mature dendrites, axons and the cell body, and in newly forming axon terminals. Once the synaptic vesicles begin their local cycling process, the sec6/8 complex may be extraneous, at least for mature synaptic vesicles. This may be a further example of neuronal adaptation of the general mechanisms of secretion used by all cells to serve specialized neuronal-specific functions. The characterization of the sec6/8 complex illustrates the contributions of both yeast and non-neuronal mammalian cell studies to the elucidation of neurotransmitter release.

8. Conclusions

The multitude of steps in the life cycle of the synaptic vesicle requires a large and intricate set of interacting proteins. Although much progress has been made in discovering important proteins and in attempting to determine their function, there are still many unanswered questions. The future promises both the discovery of additional molecules important for neurotransmitter release as well as many years of experiments designed to determine their functions. Furthermore, the function of these proteins is likely to be important in the neuronal plasticity mechanisms that underlie learning and memory. Understanding these processes represents one of the ultimate challenges of modern biology.

Acknowledgements

The authors thank Drs Cindy Adams, Lino Gonzales, Peter Hyde, Greg Miller, and Martin Steegmaier for critical reading of the manuscript.

References

1. Katz, B. (1969) *The release of neural transmitter substances.* Liverpool University Press, Liverpool.
2. Del Castillo, J. B. and Katz, B. (1954) Quantal components of the end plate potential. *J. Physiol.*, **124**, 560.
3. Del Castillo, J. B. and Katz, B. (1956) Biophysical aspects of neuromuscular transmission. *Prog. Biophys. Biophys. Chem.*, **6**, 121.
4. Heuser, J. E. and Reese, T. S. (1973) Evidence for recycling of synaptic vesicle membrane during transmitter release at the frog neuromuscular junction. *J. Cell Biol.*, **57**, 315.
5. DeCamilli, P. and Jahn, R. (1990) Pathways to regulated exocytosis in neurons. *Annu. Rev. Physiol.*, **52**, 625.
6. Trimble, W. S., Linial, M., and Scheller, R. H. (1991) Cellular and molecular biology of the presynaptic nerve terminal. *Annu. Rev. Neurosci.*, **14**, 93.
7. Brady, S. T. (1991) Molecular motors in the nervous system. *Neuron*, **7**, 521.
8. Hirokawa, N. (1993) Axonal transport and the cytoskeleton. *Curr. Opin. Neurobiol.*, **3**, 724.
9. Gindhart, J. G., Desai, C. J., Beushausen, S., Zinn, K., and Goldstein, L. S. B. (1998) Kinesin light chains are essential for axonal transport in *Drosophila*. *J. Cell Biol.*, **141**, 443.
10. Nakata, T., Terada, S., and Hirokawa, N. (1998) Visualization of the dynamics of synaptic vesicle and plasma membrane proteins in living axons. *J. Cell Biol.*, **140**, 659.
11. Murthy, V. N. and Stevens, C. F. (1998) Synaptic vesicles retain their identity through the endocytic cycle. *Nature*, **392**, 497.
12. Whittaker, V. P., Essman, W. B., and Dowe, G. H. (1972) The isolation of pure cholinergic synaptic vesicles from the electric organs of elasmobranch fish of the family Torpedinidae. *Biochem. J.*, **128**, 833.
13. Carlson, S. S. and Kelly, R. B. (1983) A highly antigenic proteoglycan-like component of cholinergic synaptic vesicles. *J. Biol. Chem.*, **258**, 11082.
14. Wagner, J. A., Carlson, S. S., and Kelly, R. B. (1978) Chemical and physical characterization of cholinergic synaptic vesicles. *Biochemistry*, **17**, 1199.
15. Ngsee, J. K., Miller, K., Wendland, B., and Scheller, R. H. (1990) Multiple GTP-binding proteins from cholinergic synaptic vesicles. *J. Neurosci.*, **10**, 317.
16. Huttner, W. B., Schiebler, W., Greengard, P., and De Camilli, P. (1983) Synapsin1 (protein 1), a nerve terminal-specific phosphoprotein. III. Its association with synaptic vesicles studied in a highly purified synaptic vesicle preparation. *J. Cell Biol.*, **96**, 1374.
17. Bennett, M. K., Calakos, N., Kreiner, T., and Scheller, R. H. (1992) Synaptic vesicle membrane proteins interact to form a multimeric complex. *J. Cell Biol.*, **116**, 761.
18. Buckley, K. and Kelly, R. B. (1985) Identification of a transmembrane glycoprotein specific for secretory vesicles of neural and endocrine cells. *J. Cell Biol.*, **100**, 1284.
19. Hooper, J. E., Carlson, S. S., and Kelly, R. B. (1980) Antibodies to synaptic vesicles purified from *Narcine* electric organ bind a subclass of mammalian nerve terminals. *J. Cell Biol.*, **87**, 104.
20. Matthew, W. D., Tsavaler, L., and Reichardt, L. F. (1981) Identification of a synaptic vesicle-specific membrane protein with a wide distribution in neuronal and neurosecretory tissue. *J. Cell Biol.*, **91**, 257.

21. Wiedenmann, B. and Franke, W. W. (1985) Identification and localization of synaptophysin, an integral membrane glycoprotein of Mr 38 000 characteristic of presynaptic vesicles. *Cell*, **41**, 1017.
22. Jahn, R., Schiebler, W., Ouimet, C., and Greengard, P. (1985) A 38 000-dalton membrane protein (p38) present in synaptic vesicles. *Proc. Natl. Acad. Sci. USA*, **91**, 12487.
23. Trimble, W. S., Cowan, D. M., and Scheller, R. H. (1988) VAMP-1: a synaptic vesicle-associated integral membrane protein. *Proc. Natl. Acad. Sci. USA*, **85**, 4538.
24. Baumert, M., Maycox, P. R., Navone, F., DeCamilli, P., and Jahn, R. (1989) Synaptobrevin: and integral membrane protein of 18#000 daltons present in small synaptic vesicles of rat brain. *EMBO J.*, **8**, 379.
25. Baumert, M., Takei, K., Hartinger, J., Burger, P. M., Fischer von Mollard, G., Maycox, P. R., *et al.* (1990) P29: a novel tyrosine-phosphorylated membrane protein present in small clear vesicles of neurons and endocrine cells. *J. Cell Biol.*, **110**, 1285.
26. Matsui, Y., Kikuchi, A., Kondo, J., Hishida, T., Teranishi, Y., and Takai, Y. (1988) Nucleotide and deduced amino acid sequences of a GTP-binding protein family with molecular weights of 25 000 from bovine brain. *J. Biol. Chem.*, **15**, 11071.
27. Zahraoui, A., Touchot, N., Chardin, P., and Tavitian, A. (1989) The human *Rab* genes encode a family of GTP-binding proteins related to yeast YPT1 and SEC4 products involved in secretion. *J. Biol. Chem.*, **264**, 12394.
28. Chidon, S. and Shira, T. S. (1989) Characterization of a H^1-ATPase in rat brain synaptic vesicles. Coupling to L-glutamate transport. *J. Biol. Chem.*, **264**, 8288.
29. Fischer von Mollard, G., Mignery, G. A., Baumert, M., Perin, M. S., Hanson, T. J., Burger, P. M., *et al.* (1990) rab3 is a small GTP-binding protein exclusively localized to synaptic vesicles. *Proc. Natl. Acad. Sci. USA*, **87**, 1988.
30. Südhof, T. C., Czernik, A. J., Kao, H.-T., Takei, K., Johnston, P. A., Horiuchi, A., *et al.* (1989) Synapsins: mosaics of shared and individual domains in a family of synaptic vesicle phosphoproteins. *Science*, **245**, 1474.
31. Yamagata, S. K., Noremberg, K., and Parsons, S. M. (1989) Purification and subunit composition of a cholinergic synaptic vesicle glycoprotein, phosphointermediate-forming ATPase. *J. Neurochem.*, **53**, 1345.
32. Carlson, S. S. and Kelly, R. B. (1983) A highly antigenic proteoglycan-like component of cholinergic synaptic vesicles. *J. Biol. Chem.*, **258**, 11082.
33. Floor, E., Leventhal, P. S., and Schaeffer, S. F. (1990) Partial purification and characterization of the vacuolar H(1)-ATPase of mammalian synaptic vesicles. *J. Neurochem.*, **55**, 1663.
34. DeCamilli, P., Harris, S. M., Huttner, W. B., and Greengard, P. (1983) Synapsin I (protein I), a nerve terminal-specific phosphoprotein. II. Its specific association with synaptic vesicles demonstrated by immunocytochemistry in agarose-embedded synaptosomes. *J. Cell Biol.*, **96**, 1355.
35. Stadler, H. and Dowe, G. H. (1982) Identification of a heparan sulphate-containing proteoglycan as a specific core component of cholinergic synaptic vesicles from Torpedo marmorata. *EMBO J.*, **1**, 1381.
36. Südhof, T. C., Baumert, M., Perin, M. S., and Jahn, R. (1989) A synaptic vesicle membrane protein is conserved from mammals to *Drosophila*. *Neuron*, **2**, 1475.
37. Buckley, K. M., Floor, E., and Kelly, R. B. (1987) Cloning and sequence analysis of cDNA encoding p38, a major synaptic vesicle protein. *J. Cell Biol.*, **105**, 2447.
38. Pevsner, J., Hsu, S.-C., and Scheller, R. H. (1994) n-Sec1: a neural-specific syntaxin-binding protein. *Proc. Natl. Acad. Sci. USA*, **91**, 1445.

39. Hazuka, C. D., Hsu, S.-C., and Scheller, R. H. (1997) Characterization of a cDNA encoding a subunit of the rat brain rsec6/8 complex. *Gene*, **187**, 67.
40. Perin, M. S., Fried, V. A., Mignery, G. A., Jahn, R., and Südhof, T. C. (1990) Phospholipid binding by a synaptic vesicle protein homologous to the regulatory region of protein kinase C. *Nature*, **345**, 260.
41. Wendland, B., Miller, K. G., Schilling, J., and Scheller, R. H. (1991) Differential expression of the p65 gene family. *Neuron*, **6**, 993.
42. Leube, R. E., Kaiser, P., Seiter, A., Zimbelmann, R., Franke, W. W., Rehm, H., *et al.* (1987) Synaptophysin: molecular organization and mRNA expression as determined from cloned cDNA. *EMBO J.*, **2**, 1265.
43. Südhof, T. C., Lottspeich, F., Greengard, P., Mehl, E., and Jahn, R. (1987) A synaptic vesicle protein with a novel cytoplasmic domain and four transmembrane regions. *Science*, **238**, 1142.
44. Cowan, D. M., Linial, M., and Scheller, R. H. (1990) *Torpedo* synaptophysin: evolution of a synaptic vesicle protein. *Brain Res.*, **509**, 1.
45. Bennett, M. K., Calakos, N., and Scheller, R. H. (1992) Syntaxin: a synaptic protein implicated in docking of synaptic vesicles at presynaptic active zones. *Science*, **257**, 255.
46. Fujita, U., Shirataki, H., Sakisaka, T., Asakura, T., Ohya, T., Kotani, H., *et al.* (1998) Tomosyn: a syntaxin-1-binding protein that forms a novel complex in the neurotransmitter release process. *Neuron*, **20**, 905.
47. Shirataki, H., Kaibuchi, K., Yamaguchi, T., Wada, K., Horiuchi, H., and Takai, Y. (1992) A possible target protein for *smg*-25A/*rab*3A small GTP-binding protein. *J. Biol. Chem.*, **267**, 10946.
48. Fields, S. and Song, O. (1989) A novel genetic system to detect protein-protein interactions. *Nature*, **340**, 245.
49. Bean, A. J., Seifert, R., Chen, Y. A., Sacks, R., and Scheller, R. H. (1997) Hrs-2 is an ATPase implicated in calcium-regulated secretion. *Nature*, **385**, 826.
50. Palade, G. (1975) Intracellular aspects of the process of protein synthesis. *Science*, **189**, 347.
51. Kaiser, C. and Ferro-Novick, S. (1998) Transport from the endoplasmic reticulum to the Golgi. *Curr. Opin. Cell Biol.*, **10**, 477.
52. Jung, L. J. and Scheller, R. H. (1991) Peptide processing and targeting in the neuronal secretory pathway. *Science*, **251**, 1330.
53. Hay, J. C. and Scheller, R. H. (1997) SNAREs and NSF in targeted memmbrane fusion. *Curr. Opin. Cell Biol.*, **9**, 505.
54. Fries, E. and Rothman, J. E. (1980) Transport of vesicular stomatitis virus glycoprotein in a cell-free extract. *Proc. Natl. Acad. Sci. USA*, **77**, 3870.
55. Balch, W. E., Dunphy, W. G., Braell, W. A., and Rothman, J. E. (1984) Reconstitution of the transport of protein between successive compartments of the Golgi complex measured by the coupled incorporation of N-acetylglucosamine. *Cell*, **39**, 405.
56. Balch, W. E., Glick, B. S., and Rothman, J. E. (1984) Sequential intermediates in the pathway of intercompartmental transport in a cell-free system. *Cell*, **39**, 525.
57. Orci, L., Glick, B. S., and Rothman, J. E. (1986) A new type of coated vesicular carrier that appears not to contain clathrin: its possible role in protein transport within the Golgi complex stack. *Cell*, **46**, 171.
58. Braell, W. A., Balch, W. E., Dobbertin, D. C., and Rothman, J. E. (1984) The glycoprotein that is transported between successive compartments of the Golgi complex in a cell-free system resides in stacks of cisternae. *Cell*, **39**, 511.

59. Glick, B. S. and Rothman, J. E. (1987) A possible role for fatty acyl-coenzyme A in intracllular protein transport. *Nature*, **326**, 309.
60. Block, M. R., Glick, B. S., Wilcox, C. A., Wieland, F. T., and Rothman, J. E. (1988) Purification of an *N*-ethylmaleimide-sensitive fusion protein (NSF) catalyzing vesicular transport. *Proc. Natl. Acad. Sci. USA*, **85**, 7852.
61. Malhotra, V., Orci, L., Glick, B. S., Block, M. R., and Rothman, J. E. (1988) Role of an *N*-ethylmaleimide-sensitive transport component in promoting fusion of transport vesicles with cisternae of the Golgi complex stack. *Cell*, **54**, 221.
62. Weidman, P. J., Melancon, P., Block, M. R., and Rothman, J. E. (1989) Binding of an *N*-ethylmaleimide-sensitive fusion protein to Golgi complex membranes requires both a soluble protein(s) and an integral membrane receptor. *J. Cell Biol.*, **108**, 1589.
63. Clary, D. O., Griff, E. C., and Rothman, J. E. (1990) SNAPS, a family of NSF attachment proteins involved in intracellular membrane fusion in animals and yeast. *Cell*, **61**, 709.
64. Clary, D. O. and Rothman, J. E. (1990) Purification of three related peripheral membrane proteins needed for vesicular transport. *J. Biol. Chem.*, **265**, 10109.
65. Whiteheart, S. W., Griff, I. C., Brunner, M., Clary, D. O., Mayer, T., Buhrow, S. A., *et al.* (1993) SNAP family of NSF attachment proteins includes a brain-specific isoform. *Nature*, **362**, 353.
66. Schiavo, G., Gmachl, M. J. S., Stenbeck, G., Söllner, T. H., and Rothman, J. E. (1995) A possible docking and fusion particle for synaptic transmission. *Nature*, **378**, 733.
67. Söllner, T., Whiteheart, S. W., Brunner, M., Erdjument-Bromage, H., Geromanos, S., Tempst, P., *et al.* (1993) SNAP receptors implicated in vesicle targeting and fusion. *Nature*, **362**, 318.
68. Oyler, G. A., Higgins, G. A., Hart, R. A., Battenberg, E., Billingsley, M., Bloom, F. E., *et al.* (1989) The identification of a novel synaptosomal-associated protein, SNAP-25, differentially expressed by neuronal subpopulations. *J. Cell Biol.*, **109**, 3039.
69. Bennett, M. K. and Scheller, R. H. (1993) The molecular machinery for secretion is conserved from yeast to neurons. *Proc. Natl. Acad. Sci. USA*, **90**, 2559.
70. Bankiatis, V. A., Johnson, L. M., and Emr, S. D. (1986) Isolation of yeast mutants defective in protein targeting to the vacuole. *Proc. Natl. Acad. Sci. USA*, **83**, 9075.
71. Kornfeld, S. and Mellman, I. (1989) The biogenesis of lysosomes. *Annu. Rev. Cell Biol.*, **5**, 483.
72. Horazdovsky, B. F., DeWald, D. B., and Emr, S. D. (1995) Protein transport to the yeast vacuole. *Curr. Opin. Cell Biol.*, **7**, 544.
73. Darsow, T., Rieder, S. E., and Emr, S. D. (1997) A multispecificity syntaxin homologue, Vam3p, essential for autophagic and biosynthetic protein transport to the vacuole. *J. Cell Biol.*, **138**, 517.
74. Novick, P. and Schekman, R. (1979) Secretion and cell-surface growth are blocked in a temperature-sensitive mutant of *Saccharomyces cerevisiae*. *Proc. Natl. Acad. Sci.USA*, **76**, 1858.
75. Novick, P., Field, C., and Schekman, R. (1980) Identification of 23 complementation groups required for post-translational events in the yeast secretory pathway. *Cell*, **251**, 205.
76. Novick, P., Ferro, S., and Schekman, R. (1981) Order of events in the yeast secretory pathway. *Cell*, **25**, 461.
77. Salminen, A. and Novick, P. (1987) A *ras*-like protein is required for a post-Golgi complex event in yeast secretion. *Cell*, **49**, 527.
78. Goud, B., Salminen, A., Walworth, N. C., and Novick, P. (1988) A GTP-binding protein required for secretion rapidly associates with secretory vesicles and the plasma membrane in yeast. *Cell*, **53**, 753.

79. Walch-Solimena, C., Collins, R. N., and Novick, P. J. (1997) Sec2p mediates nucleotide exchange on Sec4p and is involved in polarized delivery of post-Golgi complex vesicles. *J. Cell Biol.*, **137**, 1495.
80. TerBush, D. R., Maurice, T., Roth, D., and Novick, P. (1996) The exocyst is a multiprotein complex required for exocytosis in *Saccharomyces cerevisiae*. *EMBO J.*, **15**, 6483.
81. Hsu, S. C., Ting, A. E., Hazuka, C. D., Davanger, S., Kenny, J. W., Kee, Y., *et al.* (1996) The mammalian brain rsec6/8 complex. *Neuron*, **17**, 1209.
82. Brennwald, P., Kearns, B., Champion, K., Keranen, S., Bankaitis, V., and Novick, P. (1994) Sec9 is a SNAP-25-like component of a yeast SNARE complex that may be the effector of Sec4 function in exocytosis. *Cell*, **79**, 245.
83. Dunphy, W. G., Pfeffer, S. R., Clary, D. O., Wattenberg, B. W., Glick, B. S., and Rothman, J. E. (1986) Yeast and mammals utilize similar cytosolic components to drive protein transport through the Golgi complex. *Proc. Natl. Acad. Sci. USA*, **83**, 1622.
84. Aalto, M. K., Ronne, H., and Keranen, S. (1993) Yeast syntaxins Sso1p and Sso2p belong to a family of related membrane proteins that function in vesicular transport. *EMBO J.*, **12**, 4095.
85. Gerst, J. E., Rodgers, L., Riggs, M., and Wigler, M. (1992) *SNC1*, a yeast homolog of the synaptic vesicle-associated membrane protein/synaptobrevin gene family: genetic interactions with the *RAS* and *CAP* genes. *Proc. Natl. Acad. Sci. USA*, **89**, 4338.
86. Parsons, T. D., Coorssen, J. R., Horstmann, H., and Almers, W. (1995) Docked granules, the exocytic burst, and the need for ATP hydrolysis in endocrine cells. *Neuron*, **15**, 1085.
87. Hess, S. D., Doroshenko, P. A., and Augustine, G. J. (1993) A functional role for GTP-binding proteins in synaptic vesicle cycling. *Science*, **259**, 1169.
88. Okano, K., Monck, J. R., and Fernandez, J. M. (1993) GTPγS stimulates exocytosis in patch-clamped rat melanotrophs. *Neuron*, **11**, 165.
89. Dunn, L. A. and Holz, R. W. (1983) Catacholamine secretion from digitonin-treated adrenal medullary chromaffin cells. *J. Biol. Chem.*, **258**, 4989.
90. Thomas, P., Wong, J. G., Lee, A. K., and Almers, W. (1993) A low affinity Ca^{2+} receptor controls the final steps in peptide secretion from pituitary melanotrophs. *Neuron*, **11**, 93.
91. Bloom, G. S. and Goldstein, L. S. B. (1998) Cruising along microtubule highways: how membranes move through the secretory pathway. *J. Cell Biol.*, **140**, 1277.
92. Landis, D. M. D., Hall, A. K., Weinstein, L. A., and Reese, T. S. (1988) The organization of cytoplasm at the presynaptic active zone of a central nervous system synapse. *Neuron*, **1**, 201.
93. Hirokawa, N., Sobue, K., Kanda, K., Harada, A., and Yorifuji, H. (1989) The cytoskeletal architecture of the presynaptic terminal and molecular structure of synapsin 1. *J. Cell Biol.*, **108**, 111.
94. Vitale, M. L., Seward, E. P., and Trifaro, J.-M. (1995) Chromaffin cell cortical actin network dynamics control the size of the release-ready vesicle pool and the inital rate of exocytosis. *Neuron*, **14**, 353.
95. Nishizuka, Y. (1988) The molecular heterogeneity of protein kinase C and its implications for cellular regulation. *Nature*, **334**, 661.
96. Shirataki, H., Kaibuchi, K., Sakoda, T., Kishida, S., Yamaguchi, T., Wada, K., *et al.* (1993) Rabphilin-3A, a putative target protein for smg p25A/rab3A p25 small GTP-binding protein related to synaptotagmin. *Mol. Cell. Biol.*, **13**, 2061.
97. Orita, S., Sasaki, T., Naito, A., Komuro, R., Ohtsuka, T., Maeda, M., *et al.* (1995) Doc2: a novel brain protein having two repeated C_2-like domains. *Biochem. Biophys. Res. Commun.*, **206**, 439.

98. Segev, N., Mulholland, J., and Botstein, D. (1988) The yeast GTP-binding YPT1 protein and a mammalian counterpart are associated with the secretion machinery. *Cell*, **52**, 915.
99. Bourne, H. R., Sanders, D. A., and McCormick, F. (1990) The GTPase superfamily: a conserved switch for diverse cell functions. *Nature*, **348**, 125.
100. Bourne, H. R., Sanders, D. A., and McCormick, F. (1991) The GTPase superfamily: conserved structure and molecular mechanism. *Nature*, **349**, 117.
101. Hardwick, K. G. and Pelham, H. R. (1992) SED5 encodes a 39-kD integral membrane protein required for vesicular transport between the ER and the Golgi complex. *J. Cell Biol.*, **119**, 513.
102. Chapman, E. R., An, S., Barton, N., and Jahn, R. (1994) SNAP-25, a t-SNARE which binds to both syntaxin and synaptobrevin via domains that may form coiled coils. *J. Biol. Chem.*, **269**, 27427.
103. Kee, Y., Lin, R., Hsu, S.-C., and Scheller, R. H. (1995) Distinct domains of syntaxin are required for synaptic vesicle fusion complex formation and dissociation. *Neuron*, **14**, 991.
104. Pevsner, J., Hsu, S.-C., Braun, J. E. A., Calakos, N., Ting, A. E., Bennett, M. K., *et al.* (1994) Specificity and regulation of a synaptic vesicle docking complex. *Neuron*, **13**, 353.
105. Söllner, T., Bennett, M. K., Whiteheart, S. W., Scheller, R. H., and Rothman, J. E. (1993) A protein assembly-disassembly pathway *in vitro* that may correspond to sequential steps of synaptic vesicle docking, activation, and fusion. *Cell*, **75**, 409.
106. Calakos, N., Bennett, M. K., Peterson, K. E., and Scheller, R. H. (1994) Protein-protein interactions contributing to the specificity of intracellular vesicular trafficking. *Science*, **263**, 1146.
107. Bennett, M. K., Garcia-Arraras, J. E., Elferink, L., Peterson, K., Fleming, A. M., Hazuka, C. D., *et al.* (1993) The syntaxin family of vesicular transport receptors. *Cell*, **74**, 863.
108. Calakos, N. and Scheller, R. H. (1994) VAMP and synaptophysin are associated on the synaptic vesicle. *J. Biol. Chem.*, **269**, 24534.
109. Edelmann, L., Hanson, P. I., Chapman, E. R., and Jahn, R. (1995) Synaptobrevin binding to synaptophysin: a potential mechanism for controlling the exocytotic fusion machine. *EMBO J.*, **14**, 224.
110. Washbourne, P., Schiavo, G., and Montecucco, C. (1995) Vesicle-associated membrane protein-2 (synaptobrevin-2) forms a complex with synaptophysin. *Biochem. J.*, **305**, 721.
111. Hazuka, C. D., Foletti, D. L., Hsu, S.-C., Kee, Y., Hopf, F. W., and Scheller, R. H. (1999) The sec6/8 complex is located at neurite outgrowth and axonal synapse-assembly domains. *J. Neurosci.*, **19**, 1324.
112. Hanson, P. I., Roth, R., Morisaki, H., Jahn, R., and Heuser, J. E. (1997) Structure and conformational changes in NSF and its membrane receptor complexes visualized by quick-freeze/deep-etch electron microscopy. *Cell*, **90**, 523.
113. Lenzen, C. U., Steinmann, D., Whiteheart, S. W., and Weis, W. (1998) Crystal structure of the hexamerization domain of *N*-ethylmaleimede-sensitive fusion protein. *Cell*, **94**, 525.
114. Yu, R. C., Hanson, P. I., Jahn, R., and Brunger, A. T. (1998) Structure of the ATP-dependent oligomerization domain of *N*-ethylmaleimide sensitive factor complexed with ATP. *Nature Struct. Biol.*, **5**, 803.
115. Lin, R. C. and Scheller, R. H. (1997) Structural organization of the synaptic exocytosis core complex. *Neuron*, **19**, 1087.
116. Poirier, M. A., Xiao, W., Macosko, J. C., Chan, C., Shin, Y.-K., and Bennett, M. K. (1998) The synaptic SNARE complex is a parallel four-stranded helical bundle. *Nature Struct. Biol.*, **5**, 765.

117. Sutton, R. B., Fasshauer, D., Jahn, R., and Brunger, A. T. (1998) Crystal structure of a SNARE complex involved in synaptic exocytosis at 2.4 Å resolution. *Nature*, **395**, 347.
118. Ossig, R., Dascher, C., Trepte, H. H., Schmitt, H. D., and Gallwitz, D. (1991) The yeast SLY gene products, suppressors of defects in the essential GTP-binding Ypt1 protein, may act in endoplasmic reticulum-to-Golgi transport. *Mol. Cell. Biol.*, **11**, 2980.
119. Fischer von Mollard, G., Nothwehr, S. F., and Stevens, T. H. (1997) The yeast v-SNARE Vti1p mediates two vesicle transport pathways through interactions with the t-SNAREs Sed5p and Pep12p. *J. Cell Biol.*, **137**, 1511.
120. Rossi, G., Salminen, A., Rice, L. M., Brunger, A. T., and Brennwald, P. (1997) Analysis of a yeast SNARE complex reveals remarkable similarity to the neuronal SNARE complex and a novel function for the C terminus of the SNAP-25 homolog, Sec9. *J. Biol. Chem.*, **272**, 16610.
121. Stone, S., Sacher, M., Mao, Y., Carr, C., Lyons, P., Quinn, A. M., *et al.* (1997) Bet1p activates the v-SNARE Bos1p. *Cell*, **8**, 1175.
122. Wuestehube, L. J., Duden, R., Eun, A., Hamamoto, S., Korn, P., Ram, R., *et al.* (1996) New mutants of *Saccharomyces cerevisiae* affected in the transport of proteins from the endoplasmic reticulum to the Golgi complex. *Genetics*, **142**, 393.
123. Saifee, O., Wei, L., and Nonet, M. L. (1998) The Caenorhabditis elegans unc-64 locus encodes a syntaxin that interacts genetically with synaptobrevin. *Mol. Biol. Cell*, **9**, 1235.
124. Hao, J. C., Salem, N., Peng, X. R., Kelly, R. B., and Bennett, M. K. (1997) Effect of mutations in vesicle-associated membrane protein (VAMP) on the assembly of multimeric protein complexes. *J. Neurosci.*, **17**, 1596.
125. Yang, B., Gonzalez, L., Prekeris, R., Steegmaier, M., Advani, R. J., and Scheller, R. H. (1999) Membrane fusion specificity is not determined by the interactions between SNAREs. *J. Biol. Chem.*, **274**, 5649.
126. Sutton, R. B., Daveltov, B. A., Berghuis, A. M., Südhof, T. C., and Sprang, S. R. (1995) Structure of the first C_2 domain of synaptotagmin1: a novel Ca^{2+}/phospholipid-binding fold. *Cell*, **80**, 929.
127. Shao, X., Li, C., Fernandez, I., Zhang, X., Südhof, T. C., and Rizo, J. (1997) Synaptotagmin-syntaxin interaction: the C_2 domain as a Ca^{2+}-dependent electrostatic switch. *Neuron*, **18**, 133.
128. Mayer, A., Wichner, W., and Haas, A. (1996) Sec18p (NSF)-driven release of Sec17p (αSNAP) can precede docking and fusion of yeast vacuoles. *Cell*, **85**, 83.
129. Nichols, B. J., Ungermann, C., Pelham, H. R., Wickner, W. T., and Haas, A. (1997) Homotypic vacuolar fusion mediated by t- and v-SNAREs. *Nature*, **387**, 199.
130. Ungermann, C., Nichols, B. J., Pelham, H. R. B., and Wickner, W. (1998) A vacuolar v-t-SNARE complex, the predominant form *in vivo* and on isolated vacuoles, is disassembled and activated for docking and fusion. *J. Cell Biol.*, **140**, 61.
131. Hunt, J. M., Bommert, K., Charlton, M. P., Kistner, A., Habermann, E., Augustine, G. J., *et al.* (1994) A post-docking role for synaptobrevin in synaptic vesicle fusion. *Neuron*, **12**, 1269.
132. Sweeney, S. T., Broadie, K., Keane, J., Niemann, H., and O'Kane, C. J. (1995) Targeted expression of tetanus toxin light chain in *Drosophila* specifically eliminates synaptic transmission and causes behavioral defects. *Neuron*, **14**, 341.
133. Schulze, K. L., Broadie, K., Perin, M. S., and Bellen, H. J. (1995) Genetic and electrophysiological studies of *Drosophila* syntaxin-1A demonstrate its role in nonneuronal secretion and its essential role in neurotransmitter release. *Cell*, **80**, 311.

134. Broadie, K., Prokop, A., Bellen, H. J., O'Kane, C. J., Schulze, K. L., and Sweeney, S. T. (1995) Syntaxin and synaptobrevin function downstream of vesicle docking in *Drosophila*. *Neuron*, **15**, 663.
135. Weber, T., Zemelman, B. V., McNew, J. A., Westermann, B., Gmachl, M., Parlati, F., *et al.* (1998) SNAREpins: minimal machinery for membrane fusion. *Cell*, **92**, 759.
136. Nicholson, K. L., Munson, M., Miller, R. B., Filip, T. J., Fairman, R., and Hughson, F. M. (1998) Regulation of SNARE complex assembly by an N-terminal domain of the t-SNARE Sso1p. *Nature Struct. Biol.*, **5**, 793.
137. Südhof, T. C. (1997) Function of Rab3 GDP-GTP exchange. *Neuron*, **18**, 519.
138. Bean, A. J. and Scheller, R. H. (1997) Better late than never: a role for rabs late in exocytosis. *Neuron*, **19**, 751.
139. Lazar, T., Götte, M., and Gallwitz, D. (1997) Vesicular transport: how many Ypt/Rab GTPases make a eukaryotic cell? *Trends. Biochem. Sci.*, **22**, 468.
140. Schimmöller, F., Simon, I., and Pfeffer, S. R. (1998) Rab GTPases, directors of vesicle docking. *J. Biol. Chem.*, **273**, 22161.
141. Burnstein, E. S. and Macara, I. G. (1992) Characterization of a guanine nucleotide-releasing factor and a GTPase-activating protein that are specific for the *ras*-related protein $p25^{rab3A}$. *Proc. Natl. Acad. Sci. USA*, **89**, 1154.
142. Matsui, Y., Kikuchi, A., Araki, S., Hata, Y., Kondo, J., Teranishi, Y., *et al.* (1990) Molecular cloning and characterization of a novel type of regulatory protein (GDI) for smg p25A, a ras p21-like GTP-binding protein. *Mol. Cell. Biol.*, **10**, 4116.
143. Soldati, T., Shapiro, A. D., Dirac Svejstrup, A. B., and Pfeffer, S. R. (1994) Membrane targeting of the small GTPase Rab9 is accompanied by nucleotide exchange. *Nature*, **369**, 76.
144. Sogaard, M., Tani, K., Ye, R. R., Geromanos, S., Tempst, P., Kirchhausen, T., *et al.* (1994) A rab protein is required for the assembly of SNARE complexes in the docking of transport vesicles. *Cell*, **78**, 937.
145. Lupashin, V. V. and Waters, M. G. (1997) t-SNARE activation through transient interaction with a rab-like guanosine triphosphatase. *Science*, **276**, 1255.
146. Lian, J. P., Stone, S., Jiang, Y., Lyons, P., and Ferro-Novick, S. (1994) Ypt1p implicated in v-SNARE activation. *Nature*, **372**, 698.
147. Wang, Y., Okamoto, M., Schmitz, F., Hofmann, K., and Südhof, T. C. (1997) Rim is a putative Rab3 effector in regulating synaptic-vesicle fusion. *Nature*, **388**, 593.
148. Geppert, M., Bolshakov, V. Y., Siegelbaum, S. A., Takei, K., De Camilli, P., Hammer, R. E., *et al.* (1994) The role of rab3A in neurotransmitter release. *Nature*, **369**, 493.
149. Geppert, M., Goda, Y., Stevens, C. F., and Südhof, T. C. (1997) The small GTP-binding protein Rab3A regulates a late step in synaptic vesicle fusion. *Nature*, **387**, 810.
150. Stenmark, H., Parton, R. G., Steele-Mortimer, O., Lutcke, A., Gruenberg, J., and Zerial, M. (1994) Inhibition of Rab5 GTPase activity stimulates membrane fusion in endocytosis. *EMBO J.*, **13**, 1287.
151. Rybin, V., Ullrich, O., Rubino, M., Alexandrov, K., Simon, I., Seabra, M. G., *et al.* (1997) GTPase activity of Rab5 acts as a timer for endocytic membrane fusion. *Nature*, **383**, 266.
152. Peranen, J., Auvinen, P., Virta, H., Wepf, R., and Simons, K. (1996) Rab8 promotes polarized membrane transport through reorganization of actin and microtubules in fibroblasts. *J. Cell Biol.*, **135**, 153.
153. Bowser, R., Muller, H., Govindan, B., and Novick, P. (1992) Sec8p and Sec15p are components of a plasma membrane-associated 19.5S particle that may function downstream of Sec4p to control exocytosis. *J. Cell Biol.*, **118**, 1041.

154. Geppert, M., Goda, Y., Hammer, R. E., Li, C., Rosahl, T. W., Stevens, C. F., *et al.* (1994) Synaptotagmin I: A major Ca^{2+} sensor for transmitter release at a central synapse. *Cell*, **79**, 717.
155. Sheng, Z.-H., Rettig, J., Cook, T., and Catterall, W. A. (1996) Calcium-dependent interaction of N-type calcium channels with the synaptic core-complex. *Nature*, **379**, 451.
156. Sheng, Z.-H., Rettig, J., Takahashi, M., and Catterall, W. A. (1994) Identification of a syntaxin-binding site on N-type calcium channels. *Neuron*, **13**, 1303.
157. Mochida, S., Sheng, Z.-H., Baker, C., Kobayashi, H., and Caterall, W. A. (1996) Inhibition of neurotransmission by peptides containing the synaptic protein interaction site on N-type Ca^{2+} channels. *Neuron*, **17**, 781.
158. Bezprozvanny, I., Scheller, R. H., and Tsien, R. W. (1995) Functional impact of syntaxin on gating of N-type and Q-type calcium channels. *Nature*, **378**, 623.
159. Craig, A. M. and Banker, G. (1994) Neuronal polarity. *Annu. Rev. Neurosci.*, **17**, 267.
160. Matteoli, M., Takei, K., Perin, M. S., Südhof, T. C., and De Camilli, P. (1992) Exo-endocytotic recycling of synaptic vesicles in developing processes of cultured hippocampal neurons. *J. Cell Biol.*, **117**, 849.
161. Fletcher, T. L., De Camilli, P., and Banker, G. (1994) Synaptogenesis in hippocampal cultures: evidence indicating that axons and dendrites become competent to form synapses at different stages of neuronal development. *J. Neurosci.*, **14**, 6695.
162. Kraszewski, K., Mundigl, O., Daniell, L., Verderio, C., Matteoli, M., and De Camilli, P. (1995) Synaptic vesicle dynamics in living cultured hippocampal neurons visualized with CY3-conjugated antibodies directed against the lumenal domain of synaptotagmin. *J. Neurosci.*, **15**, 4328.
163. Dai, Z. and Peng, H. B. (1996) Dynamics of synaptic vesicles in cultured spinal cord neurons in relationship to synaptogenesis. *Mol. Cell Neurosci.*, **7**, 443.
164. Galli, T., Garcia, E. P., Mudigl, O., Chilcote, T. J., and De Camilli, P. (1995) v- and t-SNAREs in neuronal exocytosis: a need for additional components to define sites of release. *Neuropharmacology*, **34**, 1351.
165. TerBush, D. R. and Novick, P. (1995) Sec6, Sec8, and Sec15 are components of a multisubunit complex which localizes to small bud tips in *Saccharomyces cerevisiae*. *J. Cell Biol.*, **130**, 299.
166. Finger, F. P., Hughes, T. E., and Novick, P. (1998) Sec3p is a spatial landmark for polarized secretion in budding yeast. *Cell*, **92**, 559.
167. Finger, F. P. and Novick, P. (1997) Sec3p is involved in secretion and morphogenesis in *Saccharomyces cerevisiae*. *Mol. Biol. Cell*, **8**, 647.
168. Mondesert, G., Clarke, D. J., and Reed, S. I. (1997) Identification of genes controlling growth polarity in the budding yeast *Saccharomyces cerevisiae*: a possible role of N-glycosylation and involvement of the exocyst complex. *Genetics*, **147**, 421.
169. Grindstaff, K. K., Yeaman, C., Anandasabapathy, N., Hsu, S. C., Rodriguez-Boulan, E., Scheller, R. H., *et al.* (1998) Sec6/8 complex is recruited to cell-cell contacts and specifies transport vesicle delivery to the basal-lateral membrane in polarized epithelial cells. *Cell*, **93**, 731.
170. Popov, S., Brown, A., and Poo, M. (1993) Forward plasma membrane flow in growing nerve processes. *Science*, **259**, 244.
171. Craig, A. M., Wyborski, R. J., and Banker, G. (1995) Preferential addition of newly synthesized membrane protein at axonal growth cones. *Nature*, **375**, 592.
172. Futerman, A. and Banker, G. (1996) The economics of neurite outgrowth-the addition of new membrane to growing axons. *Trends Neurosci.*, **19**, 144.

173. Bock, J. B., Klumperman, J., Davanger, S., and Scheller, R. H. (1997) Syntaxin 6 functions in trans-Golgi network vesicle trafficking. *Mol. Biol. Cell*, **8**, 1261.
174. Bock, J. B., Lin, R. C., and Scheller, R. H. (1996) A new syntaxin family member implicated in targeting of intracellular transport vesicles. *J. Biol. Chem.*, **271**, 17961.
175. Wang, H., Frelin, L., and Pevsner, J. (1997) Human syntaxin 7: a Pep12p/Vps6p homologue implicated in vesicle trafficking to lysosomes. *Gene*, **199**, 39.
176. Wong, S. H., Xu, Y., Zhang, T., and Hong, W. (1998) Syntaxin 7, a novel syntaxin member associated with the early endosomal compartment. *J. Biol. Chem.*, **273**, 375.
177. Steegmaier, M., Yang, B., Yoo, J.-S., Huang, B., Shen, M., Yu, S., *et al.* (1998) Three novel proteins of the syntaxin/SNAP-25 family. *J. Biol. Chem.*, **273**, 34171.
178. Tang, B. L., Low, D. Y., Tan, A. E., and Hong, W. (1998) Syntaxin 10: a member of the syntaxin family localized to the trans-Golgi network. *Biochem. Biophys. Res. Commun.*, **242**, 345.
179. Advani, R. J., Bae, H.-R., Bock, J. B., Chao, D. S., Doung, Y.-C., Prekeris, R., *et al.* (1998) Seven novel mammalian SNARE proteins localize to distinct membrane compartments. *J. Biol. Chem.*, **273**, 10317.
180. Prekeris, R., Klumperman, J., Chen, Y. A., and Scheller, R. H. (1998) Syntaxin13 mediates cycling of plasma membrane proteins via tubulovesicular recycling endosomes. *J. Cell Biol.*, **143**, 957.
181. Simonsen, A., Bremnes, B., Ronning, E., Aasland, R., and Stenmark, H. (1998) Syntaxin-16, a putative Golgi t-SNARE. *Eur. J. Cell Biol.*, **75**, 223.
182. Ravichandran, V., Chawla, A., and Roche, P. A. (1996) Identification of a novel syntaxin- and synaptobrevin/VAMP-binding protein, SNAP-23, expressed in non-neuronal tissues. *J. Biol. Chem.*, **271**, 13300.
183. Hay, J. C., Chao, D. S., Kuo, C. S., and Scheller, R. H. (1997) Protein interactions regulating vesicle transport between the endoplasmic reticulum and Golgi apparatus in mammalian cells. *Cell*, **89**, 149.
184. Hay, J. C., Hirling, H., and Scheller, R. H. (1996) Mammalian vesicle trafficking proteins of the endoplasmic reticulum and Golgi apparatus. *J. Biol. Chem.*, **271**, 5671.
185. Elferink, L. A., Trimble, W. S., and Scheller, R. H. (1989) Two vesicle-associated membrane protein genes are differentially expressed in the rat central nervous system. *J. Biol. Chem.*, **264**, 11061.
186. McMahon, H. T., Ushkaryov, Y. A., Edelmann, L., Link, E., Binz, T., Niemann, H., *et al.* (1993) Cellubrevin is a ubiquitous tetanus-toxin substrate homologous to a putative synaptic vesicle fusion protein. *Nature*, **364**, 346.
187. Nagahama, M., Orci, L., Ravazzola, M., Amherdt, M., Lacomis, L., Tempst, P., *et al.* (1996) A v-SNARE implicated in intra-Golgi transport. *J. Cell Biol.*, **133**, 507.
188. Peterson, M. R., Hsu, S.-C., and Scheller, R. H. (1996) A mammalian homologue of SLY1, a yeast gene required for transport from endoplasmic reticulum to Golgi. *Gene*, **169**, 293.
189. Pevsner, J., Hsu, S.-C., Hyde, P. S., and Scheller, R. H. (1996) Mammalian homologues of yeast vacuolar protein sorting (vps) genes implicated in Golgi-to-lysosome trafficking. *Gene*, **183**, 7.
190. Takai, Y., Sasaki, T., Shirataki, H., and Nakanishi, H. (1996) Rab3A small GTP-binding protein in Ca^{2+}-dependent exocytosis. *Genes to Cells*, **1**, 615.
191. Lewis, M. J. and Pelham, H. R. (1996) SNARE-mediated retrograde traffic from the Golgi complex to the endoplasmic reticulum. *Cell*, **85**, 205.
192. Newman, A. P., Shim, J., and Ferro-Novick, S. (1990) *BET1, BOS1,* and *SEC22* are

members of a group of interacting yeast genes required for transport from the endoplasmic reticulum to the Golgi complex. *Mol. Cell Biol.*, **10**, 3405.

193. Holthius, J. C., Nichols, B. J., Dhruvakumar, S., and Pelham, H. R. (1998) Two syntaxin homologues in the TGN/endosomal system of yeast. *EMBO J.*, **17**, 113.
194. Wada, Y., Nakamura, N., Ohsumi, Y., and Hirata, A. (1997) Vam3p, a new member of syntaxin related protein, is required for vacuolar assembly in the yeast *Saccharomyces cerevisiae*. *J. Cell Sci.*, **110**, 1299.
195. Sato, T. K., Darsow, T., and Emr, S. D. (1998) Vam7p, a SNAP-25-like molecule, and Vam3p, a syntaxin homolog, function together in yeast vacuolar protein trafficking. *Mol. Cell Biol.*, **18**, 5308.
196. Becherer, K. A., Rieder, S. E., Emr, S. D., and Jones, E. W. (1996) Novel syntaxin homolog, Pep12p, required for the sorting of lumenal hydrolases to the lysosome-like vacuole in yeast. *Mol. Biol. Cell*, **7**, 579.
197. McNew, J. A., Sogaard, M., Lampen, N. M., Machida, S., Ye, R. R., Lacomis, L., *et al.* (1997) Ykt6p, a prenylated SNARE essential for endoplasmic reticulum-Golgi complex transport. *J. Biol. Chem.*, **272**, 17776.
198. McNew, J. A., Coe, J. G., Sogaard, M., Zemelman, B. V., Wimmer, C., Hong, W., *et al.* (1998) Gos1p, a *Saccharomyces cerevisiae* SNARE protein involved in Golgi complex transport. *FEBS Lett.*, **435**, 89.
199. Protopopov, V., Govindan, B., Novick, P., and Gerst, J. E. (1993) Homologs of the synaptobrevin/VAMP family of synaptic vesicle proteins function on the late secretory pathway in *S. cerevisiae*. *Cell*, **74**, 855.
200. Aalto, M. K., Ruohnen, L., Hosono, K., and Keranen, S. (1991) Cloning and sequencing of the yeast *Saccharomyces cerevisiae* SEC1 gene localized on chromosome IV. *Yeast*, **7**, 643.
201. Dascher, C., Ossig, R., Gallwitz, D., and Schmitt, H. D. (1991) Identification and structure of four yeast genes (SLY) that are able to suppress the functional loss of YPT1, a member of the RAS superfamily. *Mol. Cell Biol.*, **11**, 872.
202. Banta, L. M., Vida, T. A., Herman, P. K., and Emr, S. D. (1990) Characterization of yeast vps33p, a protein required for vacuolar protein sorting and vacuole biogenesis. *Mol. Cell Biol.*, **10**, 4638.
203. Wada, Y., Kitamoto, K., Knabe, T., Tanaka, K., and Anruku, Y. (1990) The SLP1 gene of *Saccharomyces cerevisiae* is essential for vacuolar morphogenesis and function. *Mol. Cell Biol.*, **10**, 2214.
204. Cowles, C. R., Emr, S. D., and Horazdevsky, B. F. (1994) Mutations in the VPS45 gene, a SEC1 homologue, result in vacuolar protein sorting defects and accumulation of membranes vesicles. *J. Cell Sci.*, **107**, 3449.
205. Piper, R. C., Whitters, E. A., and Stevens, T. H. (1994) Yeast Vps45p is a Sec1p-like protein required for the consumption of vacuole-targeted, post-Golgi transport vesicles. *Eur. J. Cell Biol.*, **65**, 305.
206. Banfield, D. K., Lewis, M. J., and Pelham, H. R. (1995) A SNARE-like protein required for traffic through the Golgi complex. *Nature*, **375**, 806.
207. Hohl, T. M., Parlati, F., Wimmer, C., Rothman, J. E., Sollner, T. H., and Engelhardt, H. (1998) Arrangement of subunits in 20S particles consisting of NSF, SNAPs, and SNARE complexes. *Mol. Cell*, **2**, 539.
208. Fasshauer, D., Otto, H., Eliason, W. K., Jahn, R., and Brunger, A. T. (1997) Structural changes are associated with soluble *N*-ethylmaleimide-sensitive fusion protein attachment protein receptor complex formation. *J. Biol. Chem.*, **272**, 28036.

209. Brose, N., Petrenko, A. G., Sudhof, T. C., and Jahn, R. (1992) Synaptotagmin: a calcium sensor on the synaptic vesicle surface. *Science*, **256**, 1021.

4 The role of phospholipids in neurosecretion

THOMAS F. J. MARTIN

1. Introduction

As emphasized elsewhere in this volume, the function of the nervous system depends upon chemical communication mediated by extracellular signals such as neurotransmitters, neuropeptides, and neurotrophins. The regulated release of these signals into the extracellular space as well as the responses elicited in cells by these signals depends upon integrated molecular events that are membrane based. Neurotransmitter release requires the fusion of secretory vesicles with the plasma membrane whereas regulation by transmitters is mediated by cell surface trans-membrane receptors and associated signal transduction complexes. Research on understanding the molecular mechanisms that underlie these processes has understandably focused on identifying key protein participants that operate within or on membrane bilayers. In contrast, the membrane itself has frequently been regarded as a relatively inert platform upon which protein–protein interactions occur. Research over the last 20 years has, however, progressively revealed a dynamic set of regulatory events that involves membrane phospholipids, in particular the inositol phospholipids. These highly phosphorylated lipids exhibit high rates of turnover, and function both as precursors for intracellular messengers as well as key regulators of protein recruitment to and protein function within the membrane. This chapter provides an overview of the multiple roles of inositol phospholipids in neural cell function.

2. Phospholipid composition of biological membranes

Membranes in neurons have a phospholipid composition similar to that of other vertebrate cells. Phospholipids such as phosphatidylcholine (PC), phosphatidyl-ethanolamine (PE), phosphatidylserine (PS), and phosphatidylinositol (PI) and the sphingolipid sphingomyelin (SM) constitute the principal bilayer-forming lipid components of a membrane (Table 1) (1, 2). The other major lipid constituent, cholesterol, intercalates with phospholipids and sphingolipids to increase the rigidity and impermeability of the bilayer. Limited studies of model membranes such the erythrocyte plasma membrane suggest that the two leaflets of a bilayer exhibit compositional asymmetry with choline-containing lipids (PC and SM) principally in the non-cytoplasmic leaflet of the bilayer and with PE, PS, and PI enriched in the cytoplasmic leaflet. Phosphorylated lipids derived from PI are also preferentially distributed in the cytoplasmic leaflet. In cell types where purification of organelles can be achieved, phospholipid analyses reveal differences in the phospholipid composition of cellular organelles (Table 1). Cholesterol is enriched in the plasma membrane, reflective of the increasing concentration of cholesterol between the endoplasmic reticulum, where it is synthesized, and the late Golgi cisternae. Golgi-derived secretory vesicles and granules also contain a higher cholesterol content (Table 1). Otherwise the lipid composition of neuronal secretory vesicles (synaptic vesicles) is quite similar to that of other cellular membranes. In contrast, dense-core secretory granules (e.g. from adrenal chromaffin and pancreatic acinar tissue) exhibit an unusually high content of lysoPC although the functional significance of this is not understood (3). In general, the functional role for the distinct but overlapping phospholipid composition of organellar membranes is poorly understood. There is evidence that compositionally distinct domains of Golgi membrane enriched in cholesterol and sphingomyelin (rafts) function in the sorting of proteins into transport vesicles targeted for apical/axonal destinations, and that the membrane composition within the Golgi plays a role in the retention of resident proteins (4, 5).

Table 1 Phospholipid composition of cell membranes[a]

Membrane	Lipid[b] (% of total lipid)								
	Chol	PC	LPC	PE	PS	PI	PG	CL	SM
Plasma	30	18	–	11	9	4	0	0	14
Golgi	8	40	–	15	4	6	0	0	10
SER	10	50	–	21	0	7	0	2	12
RER	6	55	–	16	3	8	0	0	3
Nuclear	10	55	–	20	3	7	0	0	3
Lysosome	14	25	–	13	0	7	0	5	24
Mitochondrial									
inner	3	45	–	24	1	6	2	18	3
outer	5	45	–	23	2	13	3	4	5
Synaptosomal	17	41	1	34	14	4	0	0	5
Synaptic vesicle	22	42	1.5	36	12	3	0	0	5
Chromaffin granule	21	26	17	35	8	3	0	0	12
Zymogen granule	21	25	11	34	2	15	0	0	9

[a] Data compiled from refs 1–3.

[b] Abbreviations are: Chol, cholesterol; PC, phosphatidylcholine; LPC, lysophosphatidylcholine; PE, phosphatidylethanolamine; PS, phosphatidylserine; PI, phosphatidylinositol; PG, phosphatidylglycerol; CL, cardiolipin; SM, sphingomyelin.

3. Unique diversity of inositol phospholipids achieved by phosphorylation

PI is a unique phospholipid in eukaryotic cells since it plays an important role as a precursor for signalling molecules in addition to its structural role in cell membranes. PI is a relatively major component of membranes (Table 1) that is utilized for the synthesis of many less abundant phospholipids that are derived from it by ATP-dependent phosphorylations catalysed by lipid kinases (6–9). PI and its derivatives are generally termed phosphoinositides with the phosphorylated phospholipids termed polyphosphoinositides. The inositol headgroup of PI contains five hydroxyls and the presently known diversity of the polyphosphoinositides derive from phosphorylation at the 3-, 4-, and 5-position hydroxyls, D-3, D-4, and D-5 (Fig. 1). Identified polyphosphoinositides in mammalian cells consist of PI(4)P, PI(4,5)P_2, PI(3)P, PI(3,5)P_2, PI(3,4)P_2, and PI(3,4,5)P_3. Some of these are present at very low concentrations (e.g. PI(3,5)P_2) or only in agonist-activated cells (e.g. PI(3,4)P_2 and PI(3,4,5)P_3). PI(4)P and PI(4,5)P_2 are more abundant consisting of $\sim$ 10% of the cellular phosphoinositides. The extraordinary diversity of phosphoinositides situated in the cytoplasmic leaflet of membranes provides a basis for their multiple roles in intracellular signalling and regulatory events.

4. Lipid kinases and transport proteins

Figure 1 also summarizes the defined classes of lipid kinases that have been identified that phosphorylate the D-3, D-4, and D-5 positions of PI (6–9). Whereas certain characterized lipid kinases (PI 4-kinase, type III PI 3-kinase) exhibit a relatively narrow substrate specificity, other lipid kinases catalyse a broader range of reactions. Type I and II PI 3-kinases catalyse D-3 phosphorylation of PI, PI(4)P, or PI(4,5)P_2. PIP

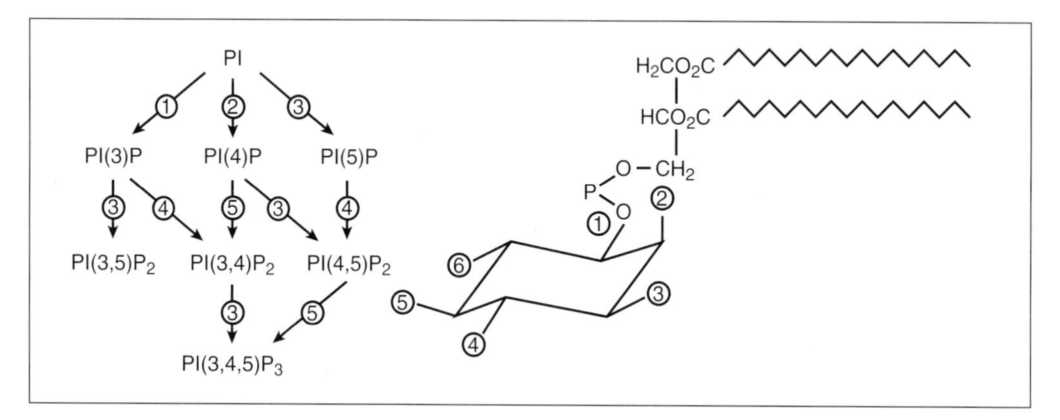

Fig. 1 Diversity of phosphoinositides and their lipid kinases. Phosphorylation reactions occur at the D-3, D-4, and D-5 positions. The known lipid kinases that catalyse these reactions are: 1. PI 3-kinase (type III); 2. PI 4-kinase (type II and III); 3. PI(P) 5-kinase (aka type I PIP kinase); 4. PIP 4-kinase (aka type II PIP kinase); 5. PI 3-kinase (types I and II).

5-kinases catalyse D-5 phosphorylation of PI(3)P, PI(4)P, and PI(3,4)P whereas PIP 4-kinases catalyse D-4 phosphorylation of PI(5)P and PI(3)P (10–12). The reactions shown in Fig. 1 for these kinases represent their major activities but a larger number of interconversions are possible at least *in vitro*.

Phosphoinositides are both precursors for second messenger formation as well as membrane-associated signals. For either role, the localization of phosphoinositides in specific organellar membranes or in domains within a membrane is important for the spatial localization of signalling events (Figs 2 and 6). Factors that localize and activate lipid kinases dictate where and when polyphosphoinositide synthesis occurs within a membrane domain, which will initiate phosphoinositide-dependent reactions such as protein recruitment to a membrane. Although a molecular characterization of many of the major phosphoinositide kinases has been achieved (7–9), the basis for their cellular localization and the regulation of their activity remains in most cases unclear. Surprisingly, given their membrane-localized substrates, the characterized PI and PIP kinases are hydrophilic proteins that lack transmembrane domains (6–8). In some cases, lipid kinases constitutively associate with membranes as peripherally-bound proteins. PI 4-kinases reside on several organellar membranes including secretory granules, although the basis for this membrane association is unknown. None the less, this enzyme catalyses the first step in the synthesis of $PI(4,5)P_2$ on these membranes, which is essential for exocytosis (see below). PIP 4-kinase, which may catalyse the synthesis of some pools of $PI(4,5)P_2$ for signal transduction, associates with the cytoplasmic domain of several transmembrane receptor proteins, which may localize this enzyme to sites of receptor signalling (13). The small G proteins Rho and Rac regulate PIP 5-kinase activity, which may be an important element for the phosphoinositide regulation of localized cytoskeletal rearrangements, but the precise

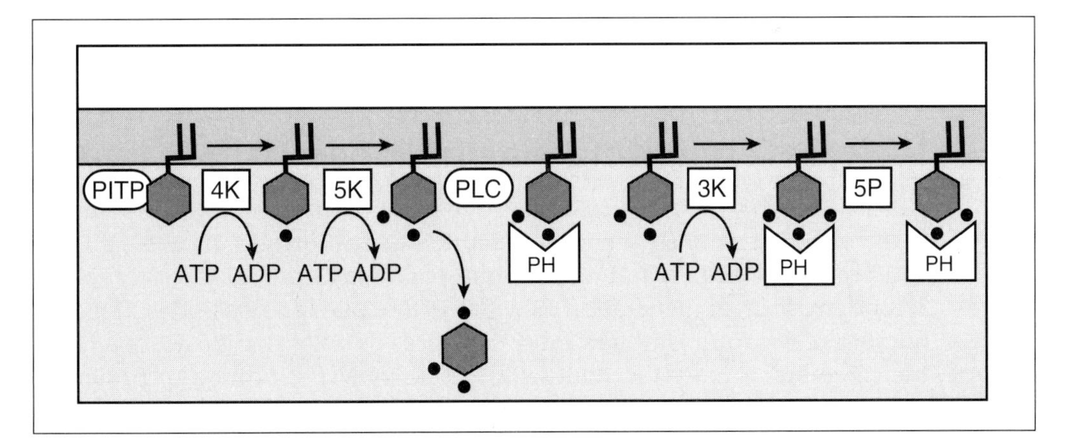

Fig. 2 Phosphoinositide phosphorylation in membrane domains. PI, bound to PITP, is shown to undergo sequential phosphorylation by PI 4-kinase (4K), PIP 5-kinase (5K), and PI 3-kinase (3K) reactions, and dephosphorylation by a phosphoinositide 5-phosphatase (5P). PH domain-containing proteins can interact with $PI(4,5)P_2$, $PI(3,4)P_2$, or $PI(3,4,5)P_3$. $PI(4,5)P_2$ can be hydrolysed to $Ins(1,4,5)P_3$ by phospholipase C (PLC) or phosphorylated to $PI(3,4,5)P_3$ by a PI 3-kinase.

nature of the regulation by Rho and Rac has not been established (14). The basis for the membrane association and activation of PI 3-kinases, which consist of types I–III, is better understood (15, 16). For a type III PI 3-kinase (yeast Vps34p, VPS designating vacuolar protein sorting genes), membrane association and activity depend upon a protein kinase (Vps15p), which presumably localizes the synthesis of PI(3)P to regions of the Golgi that generate transport vesicles that mediate protein trafficking to the vacuole (16). Type I PI 3-kinases are recruited to activated, tyrosine phosphorylated membrane growth factor receptors via SH2 (src homology 2) domains present in a regulatory (p85) subunit that activates the catalytic (p110) subunit for the synthesis of PI(3,4,5)P_3 from PI(4,5)P_2 at the plasma membrane for protein recruitment events (see below).

Phospholipid synthesis, including that for PI, occurs in the endoplasmic reticulum. An important but poorly understood aspect of membrane biogenesis concerns the mechanisms by which phospholipids are transported to other cellular membranes. While phospholipid transport is mediated in part through vesicular trafficking in the endoplasmic reticulum–Golgi system, additional mechanisms are present to provide phospholipids to organelles such as mitochondria that lie outside of this pathway. Phospholipid transport is maintained in yeast cells in which vesicular transport has been genetically impaired. Cytosolic proteins that transport phospholipids between donor and acceptor membranes *in vitro* have been suggested to mediate phospholipid trafficking *in vivo* (17). PI transfer protein (PITP) mediates the *in vitro* transfer of PI to a membrane in exchange for PC or vice versa, but there is as yet little evidence that PITP serves a transport role *in vivo* (18). Instead it has been established that PITP serves as a cofactor in PI phosphorylation reactions (Fig. 2) since PI bound to PITP is utilized preferentially as a substrate for phosphorylation by PI 3- and 4-kinases (19). PITP is essential for several membrane trafficking reactions including secretory vesicle formation in the late Golgi and neurotransmitter secretion. In these membrane trafficking steps, PITP functions as a cofactor for PI phosphorylation (see below).

5. Receptor signalling via phosphoinositide hydrolysis

As early as 1955, it was noticed that PI in brain tissue exhibits an unusually high rate of turnover in response to neurotransmitter activation (20). By 1983, this observation had been clarified into an understanding of a major pathway for receptor-regulated second messenger generation involving the phosphoinositides (Fig. 3). A large number of neurotransmitters exert their pre- or postsynaptic actions through seven transmembrane receptors that are coupled to GTP-binding (G) proteins. Receptor subtypes for these neurotransmitters (Table 2) regulate the activity of a phospholipase C (PLCβ) that selectively utilizes PI(4,5)P_2 as substrate. Receptor–PLCβ coupling is achieved through the activation of G proteins of the $G_{q/11}$ or G_i family. The PLCβ isoforms involved in signal transduction are directly activated by the α subunits of $G_{q/11}$ or alternatively by βγ subunits that result from activation of the more abundant G_i proteins (21).

A second mechanism for PLC activation is associated with growth factor receptors that are protein tyrosine kinases or that associate with non-receptor protein tyrosine kinases (Fig. 3). Phosphorylation of receptor cytoplasmic domain tyrosine residues results in the recruitment of PLCγ isoforms containing SH2 domains. Receptor recruitment is accompanied by the tyrosine phosphorylation of PLCγ that results in its activation for hydrolysis of $PI(4,5)P_2$ (21).

Table 2 Neurotransmitter receptors coupled to $PI(4,5)P_2$ hydrolysis

Neurotransmitter	Receptor subtype
Serotonin	$5\text{-}HT_{2A}$, $5\text{-}HT_{2B}$, $5\text{-}HT_{2C}$
Norepinephrine, adrenergic	α_{1A}, α_{1B}, α_{1D}
Acetylcholine, muscarinic	M_1, M_3, M_5
Glutamate, metabotropic	mGlu1a, mGlu5a
Bombesin	BB1, BB2
Bradykinin	B_1, B_2
Histamine	H_1
Purinergic	$P2Y_1$, $P2Y_2$, $P2Y_4$, $P2Y_6$
Substance P, K	NK_1, NK_2, NK_3
Vasopressin	V_{1A}, V_{1B}

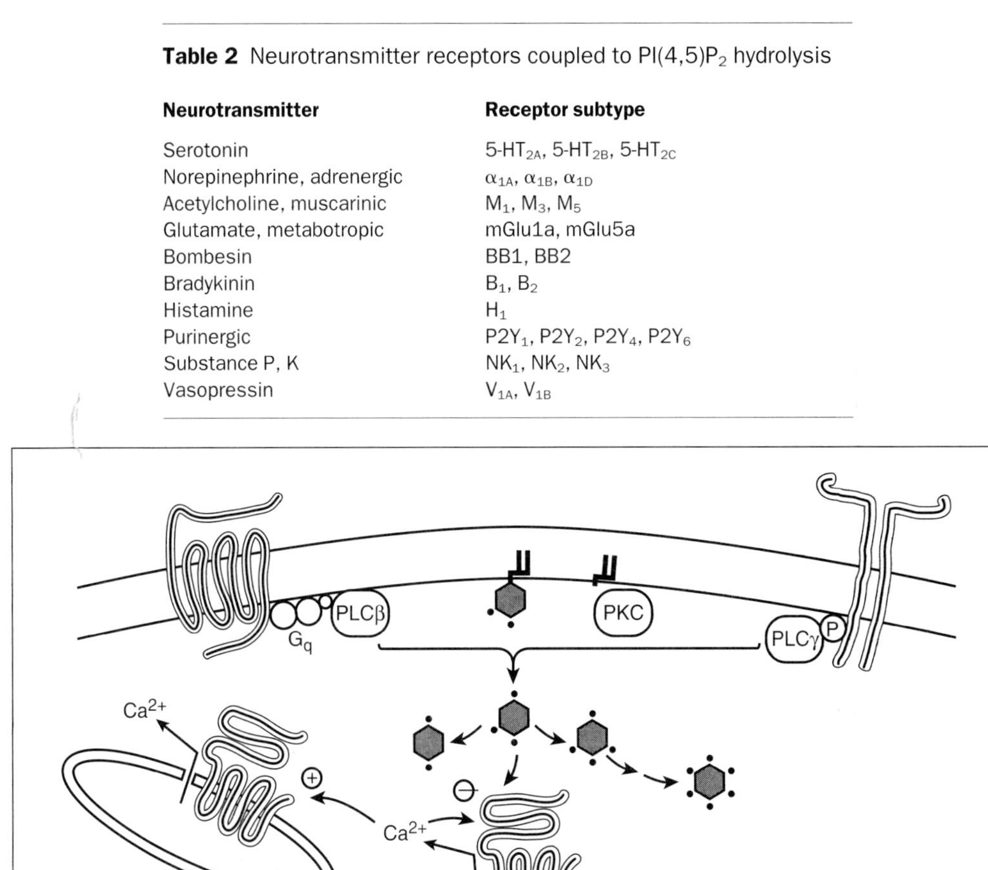

Fig. 3 Signal transduction pathways that employ PLC catalysed hydrolysis of $PI(4,5)P_2$. G protein-coupled receptors activate a PLCβ isoform by via α_q or βγ subunits of G proteins. Protein tyrosine kinase receptors activate a PLCγ isoform by recruitment to receptor phosphotyrosines via SH2 domain interactions followed by phosphorylation of PLCγ on tyrosine residues. DAG generated from $PI(4,5)P_2$ hydrolysis activates protein kinase C (PKC) isoforms. $Ins(1,4,5)P_3$ generated causes calcium mobilization from stored pools by activating IP_3 receptors. IP_3 receptors as well as ryanodine receptors can mediate calcium-induced calcium release (indicated by +), which propagates a calcium wave across the cytoplasm. $Ins(1,4,5)P_3$ is inactivated as a calcium-mobilizing second messenger either by dephosphorylation to $Ins(1,4)P_2$ or phosphorylation to $Ins(1,3,4,5)P_4$. Highly phosphorylated inositol polyphosphates such as $Ins(1,2,3,4,5,6)P_6$ can be formed through a series of conversions.

Activated PLC rapidly hydrolyses PI(4,5)P_2 to generate the membrane-associated diacylglycerol (DAG) backbone and the water soluble headgroup Ins(1,4,5)P_3, which both serve second messenger roles. DAG activates isoforms of protein kinase C (PKC), which requires PS on the cytoplasmic leaflet of the membrane, by reducing the calcium requirement for activation (22). The generation of DAG is ordinarily accompanied by a cytoplasmic calcium elevation, which serves to synergistically activate calcium-dependent PKCs. PKC phosphorylates a broad range of protein substrates (22) that are involved in neurotransmitter synthesis and secretion as well as activated gene transcription in neurons.

Ins(1,4,5)P_3 is an activating ligand for the IP_3 receptor (IP_3R), an endoplasmic reticulum-associated ion channel (Fig. 3). Gating of the IP_3R opens a cation conductance that allows lumenal calcium to flow into the cytoplasm. Positive regulation of the IP_3R at low calcium renders calcium mobilization 'all or none', which can generate highly localized increases in calcium (23). Negative regulation of the IP_3R at high calcium confers a self-limiting property to calcium mobilization and renders the calcium rise transient. Self-propagation of a calcium signal from the point of initial mobilization across the cytoplasm can occur through calcium-induced calcium release (Fig. 3). Calcium-induced calcium release can be mediated by IP_3R receptor channels in the endoplasmic reticulum or by ryanodine receptor channels. These mechanisms are responsible for calcium oscillations and waves that have been observed in agonist-activated cells (24). In secretory cells such as pituitary gonadotropes, oscillations in cytoplasmic calcium have been shown to drive oscillations in hormone secretion (25). IP_3R receptors may be present on some secretory granules, which could enable calcium mobilization from the granules and a localized calcium rise that would promote exocytosis of the granules (26). In highly polarized neurons, cytoplasmic calcium increases can oscillate in spatially delimited cellular compartments such as the dendrite or soma (27).

The D-4 and D-5 phosphates of Ins(1,4,5)P_3 are crucial for high affinity IP_3R binding and calcium mobilization (23). Inositol polyphosphate 5-phosphatases that remove the D-5 phosphate result in inactivation of Ins(1,4,5)P_3. Ins(1,4,5)P_3 can also be inactivated for calcium mobilization by conversion of Ins(1,4,5)P_3 to Ins(1,3,4,5)P_4 (Fig. 3). This polyphosphate can undergo extensive additional modification to generate a large variety of inositol polyphosphates including Ins(1,2,3,4,5,6)P_6 as well as pyrophosphorylated derivatives (28). InsP_6 has been found to inhibit neurotransmitter secretion (29) and regulate ion channel activity (28) but a bona fide cellular role for inositol polyphosphates other than Ins(1,4,5)P_3 remains to be established.

In addition to generating Ins(1,4,5)P_3, the hydrolysis of PI(4,5)P_2 can also directly affect membrane enzymes or ion channels that are regulated by this phosphoinositide. Membrane K^+ channels are highly sensitive to levels of PI(4,5)P_2 in the plasma membrane (30, 31). ATP-sensitive K^+ channels (K_{ATP}) and inward rectifier K channels (K_{ir}) interact with PI(4,5)P_2 via a C-terminal domain. This interaction antagonizes ATP inhibition of K_{ATP} channels and stabilizes the open channel configuration activated by G protein $\beta\gamma$ subunits for K_{ir} channels. Neurotransmitter receptors that activate PLC lead to decreased levels $PI(4,5)P_2$, which results in increased sensitivity of the K_{ATP} channel to ATP inhibition or to a deactivation of K_{ir} channels.

6. Receptor signalling via PI 3-kinase

A second major signal transduction pathway was discovered in 1988 with the finding that $PI(4,5)P_2$ undergoes further conversion to $PI(3,4,5)P_3$ in growth factor-activated cells (32). Specific isoforms of PI 3-kinase that mediate this conversion are capable of utilizing PI, PI(4)P, or $PI(4,5)P_2$ as substrates *in vitro* but $PI(4,5)P_2$ is probably the major substrate *in vivo*. Occupancy of a large variety of receptors results in the activation of the type I PI 3-kinase family of enzymes. For protein tyrosine kinase receptors, this is achieved by receptor tyrosine phosphorylation and the recruitment of PI 3-kinase to the receptor and membrane via SH2 (src homology 2) domains in the p85 regulatory subunit, which activates the p110 catalytic subunit of PI 3-kinase (Fig. 4). PI 3-kinase recruitment and activation is essential for the growth promoting and maintenance effects of a variety of growth factors and trophic peptides including the neurotrophins (e.g. NGF and IGF-1). G protein-coupled receptors are also capable of activating $PI(4,5)P_2$ phosphorylation to $PI(3,4,5)P_2$ mediated by a type I PI 3-kinase that is activated by the $\beta\gamma$ subunits of heterotrimeric G proteins (15) (Fig. 4).

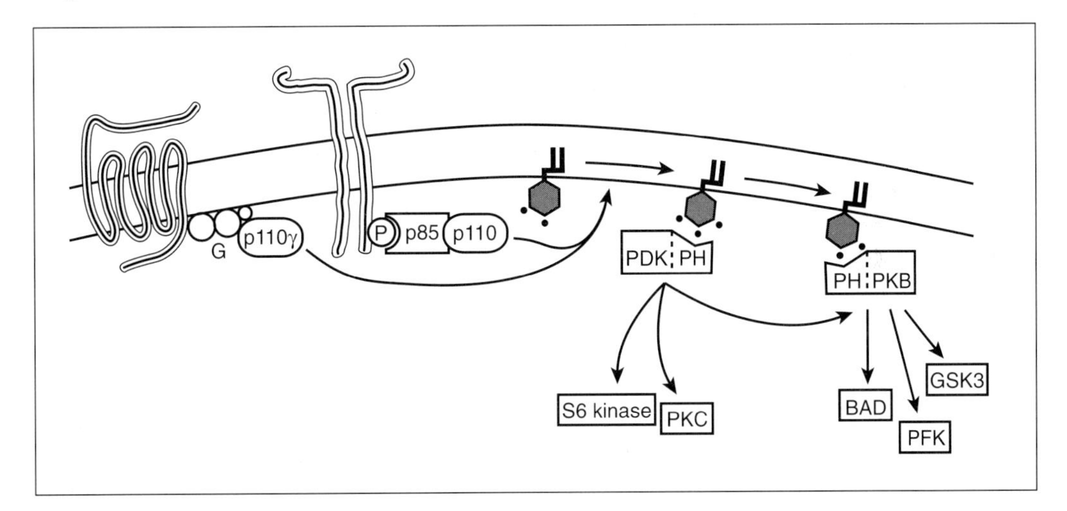

Fig. 4 Signal transduction through PI 3-kinase activation. $PI(3,4,5)P_3$ formation from $PI(4,5)P_2$ is catalysed by type I PI 3-kinases that undergo activation either by the recruitment of SH2 domain-containing p85 regulatory subunits to tyrosine-phosphorylated receptors or by stimulation by $\beta\gamma$ subunits of G proteins. $PI(3,4)P_2$ formation also occurs either by dephosphorylation of $PI(3,4,5)P_3$ by a 5-phosphatase or by phosphorylation of PI(4)P by PI 3-kinase. The formation of these D-3 phosphoinositides in the plasma membrane recruits PH domain-containing proteins such as PKB (by binding $PI(3,4)P_2$) or activates protein kinases such as PDK (activated by $PI(3,4,5)P_3$). PDK is a phosphoinositide-dependent protein kinase that phosphorylates and activates several protein kinases including PKC, S6 kinase, and PKB. PKB in turn phosphorylates a variety of substrates such as Bad (Bcl–xL/Bcl-2-associated death promoter), GSK3 (glycogen synthase kinase 3) and PFK (6-phosphofructose 2-kinase).

PLC enzymes are incapable of hydrolysing $PI(3,4,5)P_3$ for second messenger generation. The $Ins(1,3,4,5)P_4$ present in cells results from the 3-phosphorylation of $Ins(1,4,5)P_3$ rather than from the hydrolysis of $PI(3,4,5)P_3$. The lack of phospholipases that metabolize D-3 phosphoinositides to inositol polyphosphates suggested that phosphoinositide lipids themselves serve as membrane-localized signals in cells (32). This concept has been strongly supported by the identification of downstream components in the PI 3-kinase signal transduction pathway consisting of proteins with $PI(3,4,5)P_3$-binding domains.

7. Phosphoinositide-binding proteins as effectors for membrane-based signalling

Long sought effector proteins that function downstream in PI 3-kinase signalling pathways have recently been characterized (33). Two of these, Akt/PKB (Akt proto-oncogene/protein kinase B) and PDK (phosphoinositide-dependent protein kinase), were identified as phosphoinositide-regulated protein kinases that undergo activation in response PI 3-kinase stimulation and increased $PI(3,4,5)P_3$ synthesis (Fig. 4). Akt/PKB undergoes membrane recruitment in response to increased PI 3-kinase activity (34). PDK is activated by $PI(3,4,5)P_3$ and catalyses the phosphorylation and activation of Akt/PKB. Both kinases are regulated by D-3 phosphoinositides via interactions with the PH (pleckstrin homology) domains present in these proteins. Several protein substrates for activated Akt/PKB have been identified including glycogen synthase kinase 3 (GSK3), 6-phosphofructo-2-kinase (PFK), and BAD (Bcl-xL/Bcl–2-associated death promoter), which mediate the effects of growth factors and trophic peptides on metabolism, protein synthesis, and cell survival (33).

Although $PI(4,5)P_2$ is employed in signal transduction mechanisms as a precursor for $Ins(1,4,5)P_3$ and DAG generation as well as a precursor for $PI(3,4,5)P_3$ synthesis, the intact phospholipid also serves as a membrane signal (35, 36). Receptor activation and PLC mechanisms only have access to a compartmentalized pool of $PI(4,5)P_2$ corresponding to the phospholipid localized within proximity to hormone receptors. PI(4)P and $PI(4,5)P_2$ are present in other pools at the plasma membrane as well as on intracellular membranes where they function to recruit and activate $PI(4,5)P_2$-binding proteins. A large number of proteins involved in regulation of the cytoskeleton and membrane trafficking events have been found to bind phosphoinositides including $PI(4,5)P_2$ (Table 3).

Several motifs in proteins mediate interactions with phosphoinositides. The best characterized of these, present in a large number ($\sim$ 100) of proteins, is the PH domain (Table 3). While highly divergent in primary amino acid sequence, PH domains exhibit a highly conserved tertiary structure (37). Individual PH domains in different proteins are distinct and exhibit specificity for particular phosphoinositides. Akt/PKB and PDK bind $PI(3,4)P_2$ and $PI(3,4,5)P_3$ with high affinity in preference to $PI(4,5)P_2$. Hence, these proteins are capable of functioning as effectors for PI 3-kinase signalling since they interact with D-3 phosphoinositides in preference to the more

Table 3 Phosphoinositide-binding proteins

Protein[a]	**Binding motif**	**Cellular role**
PKB/Akt	PH[b]	Phosphoinositide-regulated protein kinase
βARK	PH	G protein-coupled receptor kinase
α-Actinin	PH	Actin crosslinking protein
CAPS	PH	Required for regulated secretion
Dynamin	PH	GTPase required for endocytosis
PLCδ_1	PH	Hydrolysis of PI(4,5)P_2
Grp1	PH	ARF–GTP exchange factor
Gelsolin	KR[c]	Capping and severing of F actin
Profilin	KR	Binding of G actin
AP2, AP180	KR	Clathrin adaptors for endocytosis
Synaptotagmin	C2[d]	Required for regulated secretion
Rabphilin	C2	Required for regulated secretion
EEA1	FYVE[e]	Required for endosome fusion

[a] See refs 37–45.
[b] PH, pleckstrin homology domain.
[c] KR, lysine/arginine-rich domain.
[d] C2 domain.
[e] FYVE finger domain.

abundant PI(4,5)P_2. In contrast, other PH domains (e.g. those of dynamin, CAPS, PLCδ_1) either preferentially bind PI(4,5)P_2 or exhibit little specificity for particular phosphoinositides. Such proteins are potential effectors for PI(4,5)P_2 signalling (38). PH domains on several proteins have been demonstrated to be essential for recruitment of the proteins to the membrane in response to phosphoinositide phosphorylation (39). Targeting to specific membranes such as the Golgi or secretory granules may involve additional interactions with membrane constituents other than phosphoinositides (40). For example, the β adrenergic receptor kinase (βARK) is recruited to the membrane in proximity to G protein coupled receptors via PH domain interactions with phosphoinositides as well as flanking sequence interactions with the βγ subunits of G proteins (39).

Another type of phosphoinositide-binding domain is present on many other proteins that are cytoskeletal elements (41). Although PI(4,5)P_2-binding domains in gelsolin are homologous to those in several other actin-binding proteins such as villin, binding sites in other proteins such as profilin and α-actinin are distinct (42). Binding sites in these proteins are rich in lysine and arginine residues that presumably contact inositol ring phosphates as well as in hydrophobic residues that could interact with the diacylglycerol portion. The C2 domain of several proteins such as synaptotagmin and Rabphilin has also been shown to bind phosphoinositides at a region that is similarly rich in basic and hydrophobic residues (43). Phosphoinositide interactions with clathrin adaptor proteins (e.g. AP2 and AP180) or other coat proteins may be mediated through similar regions rich in basic residues (42).

A distinct binding domain is present on proteins that bind PI(3)P. This phosphoinositide is constitutively present on intracellular membranes where it plays a role in membrane trafficking (see below). The domain that interacts with PI(3)P, termed a FYVE finger domain for the proteins identified to possess it (*F*ab1, *Y*OTB/ZK632.12, *V*ac1, and *E*arly endosomal antigen *E*EA1), mediates the membrane association of proteins via PI(3)P binding (44, 45).

A general conclusion from the preceding is that many proteins are capable of binding phosphoinositides either through highly stereoselective interactions or through general charge interactions. If PI phosphorylation occurs in specific spatial domains on membranes, or on the membranes of specific organelles, a unique localized environment will exist within the membrane that serves as a site for protein recruitment or activation. Protein interactions with phosphoinositides generate segregated membrane domains in which the phosphoinositides and their binding proteins are concentrated while excluding non-phosphoinositide lipids and other proteins (46). Such domains create a specialized local interface between the cytosol and membrane surface at which recruited proteins can initiate biochemical events crucial for cellular processes (Fig. 6). In addition to signal transduction complexes assembled in response to PI(3,4,5)P_3 formation discussed previously, current evidence suggests that many reactions required for membrane trafficking involve similar membrane-protein interactions (35, 36).

8. Inositol phospholipids in membrane traffic

Secretory and membrane proteins are synthesized in the rough endoplasmic reticulum and further processed and transported through the Golgi prior to their delivery to the extracellular space and cell surface via Golgi-derived transport vesicles (Fig. 5). Whereas all cell types traffic proteins to the cell surface and extracellular space by a constitutive transport pathway, neural and endocrine cells possess additional transport pathways where delivery to the cell surface is dependent upon cellular activation mechanisms that elevate cytoplasmic calcium concentrations (Fig. 5). One of these pathways utilizes dense-core vesicles that are assembled in the *trans*-Golgi cisternae. The second pathway utilizes secretory vesicles (in neurons known as synaptic vesicles) that are derived from constitutive vesicles that recycle by endocytosis. In addition to these anterograde export pathways, endocytic pathways are utilized to retrieve membrane and proteins from the cell surface and extracellular space. Phosphoinositides serve as important signals for directing membrane traffic in the anterograde and retrograde directions.

The first indication for a role for phospholipids in general and inositol phospholipids specifically in membrane trafficking emerged from genetic studies in the yeast *S. cerevisae* where it was shown that mutants in the gene encoding the yeast PITP (SEC14) exhibited a transport block in the Golgi (47). Subsequent studies in mammalian cells discovered that PITP was required for the formation of regulated and constitutive transport vesicles in the Golgi (48). The structurally unrelated yeast Sec14p PITP was capable of substituting for the mammalian PITP. As noted previously, PITPs enhance PI phosphorylation reactions for formation of PI(3)P or

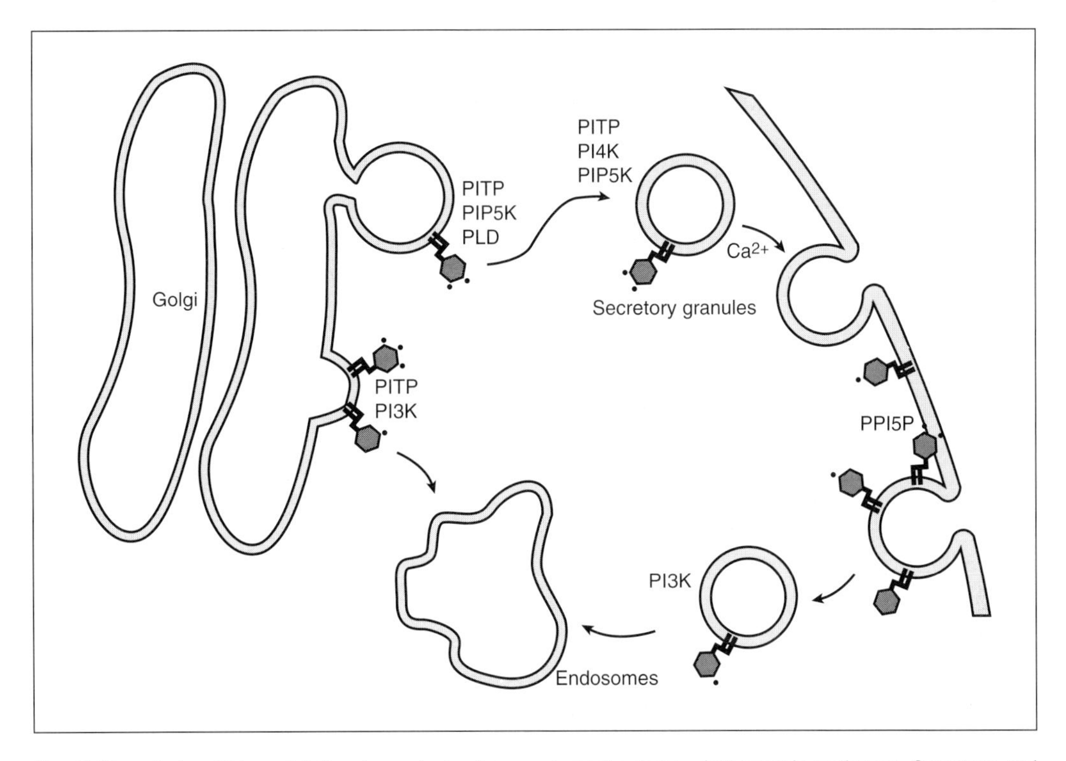

Fig. 5 Phosphoinositide regulation in vesicular transport mechanisms. *Anterograde pathways.* Secretory and membrane proteins are conveyed to the cell surface via the endoplasmic reticulum–Golgi pathway of vesicular transport. In neural and endocrine cells, secretory granules derived from the Golgi undergo exocytosis with the plasma membrane in a calcium-dependent reaction. In neural cells, synaptic vesicles (not shown) are derived from endosomes and also undergo calcium-dependent exocytosis. PI(4,5)P_2 functions at secretory granule formation in the Golgi and at an ATP-dependent step in secretory granule fusion. During exocytosis, PI(4,5)P_2 is converted to PI(4)P. Vesicular traffic from the Golgi can also be directed to the cell surface via constitutive vesicles or to the endosomal–lysosomal pathway. PI(3)P or PI(3,4,5)P_3 function at constitutive vesicle formation from the Golgi and in the formation of vesicles destined for the endosomal–lysosomal pathway. *Retrograde pathways.* Membrane is retrieved from the cell surface by endocytosis, commonly through clathrin-coated vesicle intermediates. Endosomes undergo homotypic fusion to assemble and join with late endosomes from which vesicular trafficking to lysosomes or Golgi can occur. PI(3)P is required for homotypic fusion reactions between endosomes. Phosphoinositides (PI(4,5)P_2 or PI(3,4,5)P_3) may function at an earlier stage in endocytosis to recruit clathrin adapter proteins or to regulate dynamin activity, which is required for pinching off clathrin-coated vesicles. A polyphosphoinositide 5-phosphatase synaptojanin (PPI5P) associates with dynamin.

PI(4)P suggesting that PI phosphorylation plays a role in Golgi vesicle budding reactions. A PI 3-kinase has been found to be involved in the formation of a specific class of constitutive vesicles (49) whereas PI(4,5)P_2 synthesis has been implicated in the formation of regulated secretory granules (50, 51). In the latter, PI(4,5)P_2 may serve as an activator of phospholipase (PLD) (50) and/or as a regulator of cytoskeletal assembly reactions associated with Golgi budding (51).

The activity of PLD is dramatically influenced by the presence of phosphoinositides in the membrane (52). PLD1 is a Golgi-localized enzyme that catalyses the cleavage of PC to phosphatidic acid (PA) in a reaction dependent upon ARF (ADP ribosylation factor), a small G protein. PLD1 activity is entirely dependent upon inclusion of PI(4,5)P_2 or PI(3,4,5)P_3 in the bilayer. The activity of ARF and its activation and deactivation by GTP exchange and hydrolysis are also dependent upon phosphoinositides (53). This dependence of PLD activity, as well as that of its ARF activators, on phosphoinositides has suggested that PLD may be an important effector for the regulation of membrane trafficking reactions by PI(4,5)P_2 (54). Since PI(4)P 5-kinase is a PA-activated enzyme, high local concentrations of PI(4,5)P_2 may be formed as a result of PLD activity via an autocatalytic cycle (Fig. 6) that could facilitate vesicle budding. The ability of ARF to stimulate Golgi PI(4,5)P_2 levels independent of PLD presumably by activating phosphoinositide kinases has also been reported (71). High local concentrations of PI(4,5)P_2 and PA within this domain may promote curvature required for budding and membrane fission, promote recruitment of proteins such as clathrin adaptors or other coat proteins to the membrane for coating reactions, or nucleate actin cytoskeletal rearrangements at the bud site (53, 54).

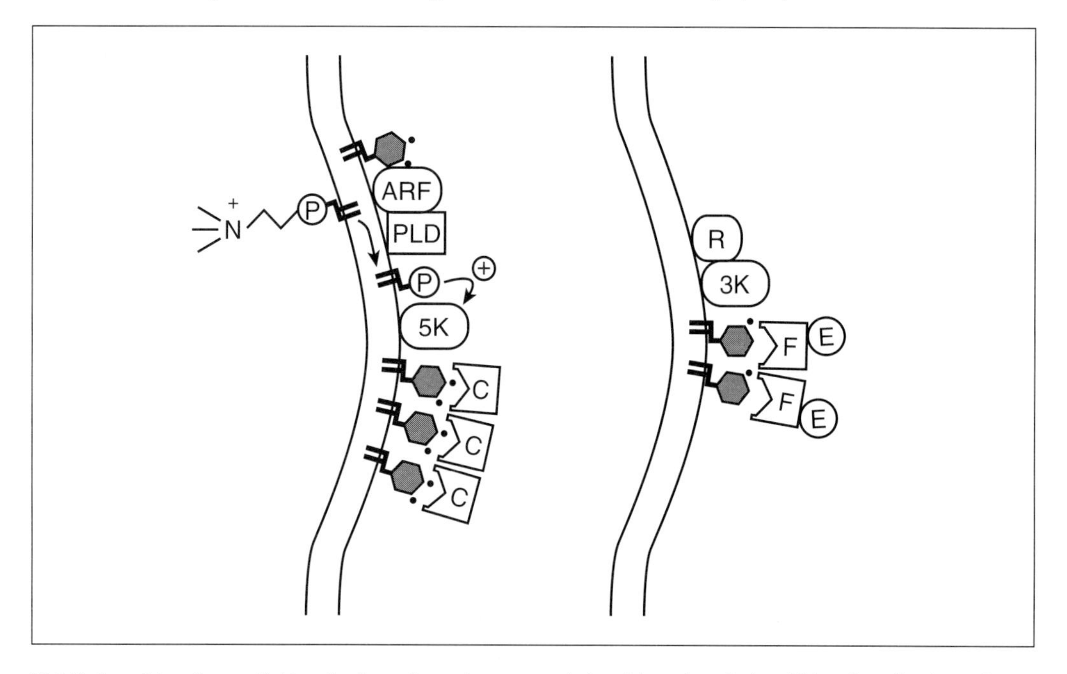

Fig. 6 Paradigms for spatial localization of membrane events involving phosphoinositides. Localized membrane reactions involving activation of phosphoinositide kinases result in the formation of membrane domains rich in phosphoinositides. These sites serve to recruit proteins to the membrane and activate phosphoinositide-dependent enzymes in the membrane. *Left.* PI(4,5)P_2 is a cofactor that allows activation of PLD, an enzyme that catalyses PC to PA conversion. ARF, with bound GTP, is a positive regulator of PLD, and PI(4,5)P_2 functions as a cofactor in the GTP exchange process that activates ARF. The formation of PA could result in increased PI(4,5)P_2 formation since PA can enhance PI(4)P 5-kinase activity. Increased concentrations of PI(4,5)P_2 may recruit coat proteins (C) to the membrane for vesicle budding. *Right.* Regulated production of PI(3)P in the membrane (shown as regulatory protein R association with PI 3-kinase) can serve to recruit PI(3)P-binding proteins (F for FYVE domain-containing protein) that can mediate the membrane association of effector proteins (E). See text for discussion.

The first indication for a role of lipid kinases in membrane trafficking reactions was provided from studies in the yeast *S. cerevisiae* where a gene essential for trafficking of proteins to the yeast vacuole (lysosome) named VPS34 was found to encode a PI-specific 3-kinase, which is the primary lipid kinase in yeast responsible for PI(3)P formation (55). Other gene products acting downstream of Vps34p (Vac1p and Vps27p) that possess PI(3)P-binding FYVE finger domains are the likely effectors for phosphoinositides in this membrane trafficking reaction (56). PI(3)P formation is also essential for endosomal–lysosomal targeting of proteins in mammalian cells. Inhibitors of PI 3-kinases such as wortmannin inhibit a restricted repertoire of transport events that are associated with traffic between the late Golgi, endosomes, and lysosomes (57). *In vitro* studies also indicate that D-3 phosphorylation of phosphoinositides is required for the formation of constitutive transport vesicles in the Golgi (49) and the fusion of early endosomes (58). EEA1, an endosome-localized protein that is redistributed from endosomes to the cytosol in wortmannin-treated cells, is a FYVE finger domain-containing effector of PI(3)P (59). EEA1 is involved in mediating the endosomal membrane localization of Rab5 (Rab = *ras* gene from *ra*t *b*rain), a small G protein, for participation in membrane fusion events.

PI(4,5)P_2 plays a critical role in the calcium-regulated fusion reactions that lead to exocytosis of dense-core secretory vesicles (Fig. 7). Reconstitution studies in neuroendocrine PC12 cells found that PITP and PI(4)P 5-kinase are required factors that act synergistically in an ATP-dependent priming step in the dense-core vesicle exocytotic pathway (60, 61). A PI 4-kinase that is resident on dense-core vesicles is also required for the ATP-dependent priming reaction (62). That PI(4,5)P_2 was the essential phosphoinositide involved in regulated secretion was indicated by the fact that a PLC and PI(4,5)P_2 antibodies inhibit calcium triggered neurotransmitter release (61). The precise role of PI(4,5)P_2 in regulated dense-core vesicle exocytosis has not been established but it is likely that specific protein effectors that bind PI(4,5)P_2 function in this pathway. A primary candidate effector is CAPS (calcium-dependent activator protein for secretion), a protein that is required for the final calcium triggered fusion step. CAPS contains a PH domain and exhibits specific PI(4,5)P_2-binding that promotes a conformational change in the protein (63). This conformational change could allow CAPS, which is found on both dense-core vesicles and plasma membrane (64), to promote a closer association of the secretory granule and plasma membranes as a prelude to fusion. Other potential effector proteins that are required for regulated secretion that bind PI(4,5)P_2 include synaptotagmin, a proposed calcium sensor for regulated exocytosis, and Rabphilin, a Rab3-interacting protein (65, 66). These proteins exhibit calcium-dependent interactions with PI(4,5)P_2 through their C2 domains. Although PI(4,5)P_2 synthesis is required for regulated dense-core vesicle exocytosis, it may not be essential for regulated synaptic vesicle exocytosis (67, but see ref. 68).

The membrane and proteins that reach the cell surface by exocytosis are retrieved by the process of endocytosis. PI(4,5)P_2 also appears to regulate early steps in the endocytic pathway. Adaptor proteins (AP2) involved in the assembly of clathrin-coated vesicles bind phosphoinositides, which regulate clathrin assembly reactions

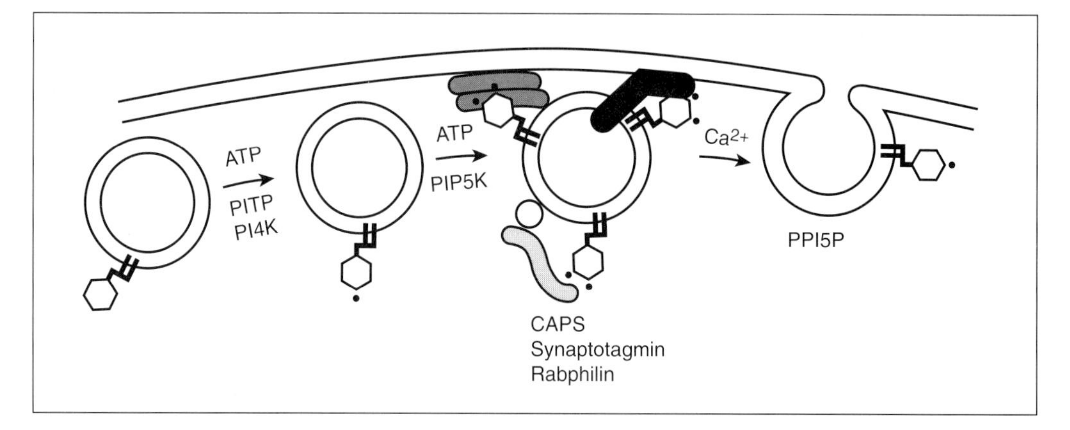

Fig. 7 Role of $PI(4,5)P_2$ in the regulated exocytosis of dense-core vesicles. Dense-core secretory vesicles, formed in the Golgi, translocate to and dock at the plasma membrane. ATP-dependent priming reactions are required prior to the calcium triggered fusion of granules with the plasma membrane. ATP-dependent priming requires PITP, PI 4-kinase, and PIP 5-kinase to catalyse the synthesis of $PI(4,5)P_2$. The role(s) of $PI(4,5)P_2$ in calcium triggered exocytosis may be to regulate the activity of phosphoinositide-binding proteins such as CAPS (which contains a PH domain) or synaptotagmin or Rabphilin (which contain C2 domains). Dephosphorylation of $PI(4,5)P_2$ at the D-5 phosphate occurs upon exocytosis.

(69). Dynamin, which is required for the fission reactions that pinch off endocytic vesicles, contains a PH domain that binds $PI(4,5)P_2$, which regulates dynamin activity on membranes (70). Another protein that associates with dynamin, synaptojanin, is a phosphoinositide 5-phosphatase that may regulate some aspect of dynamin or adaptor protein recruitment by hydrolysing $PI(4,5)P_2$ (35, 36, 42, 68). While the mechanisms remain to be fully clarified, $PI(4,5)P_2$ may serve as a signal that functions to direct the secretory vesicle membrane toward exocytic and endocytic destinations (Figs 5 and 7).

9. Perspectives

Phosphoinositides such as $PI(4,5)P_2$ serve as critical precursors for generating the second messengers for receptors that activate phosphoinositide hydrolysis by PLC. Research over the last ten years has revealed additional major regulatory pathways that utilize intact phosphoinositides as membrane-bound signals. D-3, D-4, or D-5 phosphorylated inositides function as cellular signals in the two-dimensional surface of the membrane where they act to promote spatially-localized biochemical events by recruiting phosphoinositide-binding proteins to the membrane or activating phosphoinositide-dependent enzymes within the membrane. Site-specific signalling at the interface of a membrane with cytosol is of particular importance for the events involved in signal transduction, cytoskeletal assembly, and membrane trafficking. Overall phosphoinositide signalling represents a unique example where membrane constituents play a dynamic role in regulating events on the membrane.

References

1. Lehninger, A. L., Nelson, D. L., and Cox, M. M. (1993) *Principles of biochemistry*. Worth Publishers, New York.
2. Graham, J. M. (1992) The composition and structure of membranes. In *Cell biology LabFax* (ed. G. B. Dealtry and D. Rickwood). Academic Press, New York.
3. Westhead, E. W. (1987) Lipid composition and orientation in secretory vesicles. *Ann. N. Y. Acad. Sci.*, **493**, 92.
4. Munro, S. (1998) Localization of proteins to the Golgi apparatus. *Trends Cell Biol.*, **8**, 11.
5. Simons, K. and Ikonen, E. (1997) Functional rafts in cell membranes. *Nature*, **387**, 569.
6. Carpenter, C. L. and Cantley, L. C. (1990) Phosphoinositide kinases. *Biochemistry*, **29**, 11147.
7. Carpenter, C. L. and Cantley, L. C. (1996) Phosphoinositide kinases. *Curr. Opin. Cell Biol.*, **8**, 153.
8. Loijens, J. C., Boronenkov, I. V., Parker, G. J., and Anderson, R. A. (1996) The phosphatidylinositol 4-phosphate 5-kinase family. *Adv. Enzyme Regul.*, **36**, 115.
9. Domin, J. and Waterfield, M. D. (1997) Using structure to define the function of phosphoinositide 3-kinase family members. *FEBS Lett.*, **410**, 91.
10. Zhang, X., Loijens, J. C., Boronenkov, I. V., Parker, G. J., Norris, F. A., Chen, J., *et al.* (1997) Phosphatidylinositol-4-phosphate 5-kinase isozymes catalyze the synthesis of 3-phosphate-containing phosphatidylinositol signaling molecules. *J. Biol. Chem.*, **272**, 17756.
11. Rameh, L. E., Tolias, K. F., Duckworth, B. C., and Cantley, L. C. (1997) A new pathway for synthesis of phosphatidylinositol-4,5-bisphosphate. *Nature*, **390**, 192.
12. Tolias, K. F., Rameh, L. E., Ishihara, H., Shibasaki, Y., Chen, J., Prestwich, G. D., *et al.* (1998) Type I phosphatidylinositol-4-phosphate 5-kinases synthesize the novel lipids phosphatidylinositol 3,5-bisphosphate and phosphatidylinositol 5-phosphate. *J. Biol. Chem.*, **273**, 18040.
13. Castellino, A. M., Parker, G. J., Boronenkov, I. V., Anderson, R. A., and Chao, M. V. (1997) A novel interaction between the juxtamembrane region of the p55 tumor necrosis factor receptor and phosphatidylinositol-4-phosphate 5-kinase. *J. Biol. Chem.*, **272**, 5861.
14. Tolias, K. F., Couvillon, A. D., Cantley, L. C., and Carpenter, C. L. (1997) Characterization of a rac-1 and RhoGDI-associated lipid kinase signaling complex. *Mol. Cell Biol.*, **18**, 762.
15. Vanhaesebroeck, B., Leevers, S. J., Panayotou, G., and Waterfield, M. D. (1997) Phosphoinositide 3-kinases: a conserved family of signal transducers. *Trends Biochem. Sci.*, **22**, 267.
16. Horazdovsky, B. F., DeWald, D. B., and Emr, S. D. (1995) Protein transport to the yeast vacuole. *Curr. Opin. Cell Biol.*, **7**, 544.
17. Wirtz, K. W. A. (1997) Phospholipid transfer proteins revisited. *Biochem. J.*, **324**, 353.
18. Kearns, B. G., Alb, J. G., and Bankaitis, V. A. (1998) Phosphatidylinositol transfer proteins: the long and winding road to physiological function. *Trends Cell Biol.*, **8**, 276.
19. Cockcroft, S. (1997) Phosphatidylinositol transfer proteins: requirements in phospholipase C signaling and in regulated exocytosis. *FEBS Lett.*, **410**, 44.
20. Hokin, L. E. and Hokin, M. R. (1955) Effects of acetylcholine on phosphate turnover in phospholipids of brain cortex *in vitro*. *Biochim. Biophys. Acta*, **16**, 229.
21. Rhee, S. G. and Bae, Y. S. (1997) Regulation of phosphoinositide-specific phospholipase C isozymes. *J. Biol. Chem.*, **272**, 15045.
22. Jaken, S. (1996) Protein kinase C isozymes and substrates. *Curr. Opin. Cell Biol.*, **8**, 168.
23. Mikoshiba, K. (1997) The $InsP_3$ receptor and intracellular Ca^{2+} signaling. *Curr. Opin. Neurobiol.*, **7**, 339.

24. Berridge, M. J. (1997) The AM and FM of calcium signalling. *Nature*, **386**, 759.
25. Tse, A., Tse, F. W., Almers, W., and Hille, B. (1993) Rhythmic exocytosis stimulated by GnRH-induced calcium oscillations in rat gonadotropes. *Science*, **260**, 82.
26. Gerasimenko, O. V., Gerasimenko, J. V., Belan, P. V., and Petersen, O. H. (1996) Inositol trisphosphate and cyclic ADP-ribose-mediated release of Ca^{2+} from single isolated pancreatic zymogen granules. *Cell*, **84**, 473.
27. Ghosh, A. and Greenberg, M. E. (1995) Calcium signaling in neurons: molecular mechanisms and cellular consequences. *Science*, **268**, 239.
28. Shears, S. B. (1996) Inositol pentakis- and hexakisphosphate metabolism adds versatility to the actions of inositol polyphosphates. Novel effects on ion channels and protein traffic. *Sub-Cell. Biochem.*, **26**, 187.
29. Fukuda, M., Moreira, J. E., Lewis, F. M. T., Sugimori, M., Niinobe, M., Mikoshiba, K., *et al.* (1995) Role of the C2B domain of synaptotagmin in vesicular release and recycling as determined by specific antibody injection into the squid giant synapse preterminal. *Proc. Natl. Acad. Sci. USA*, **92**, 10708.
30. Huang, C. L., Feng, S., and Hilgemann, D. W. (1998) Direct activation of inward rectifier potassium channels by PIP_2 and its stabilization by $G\beta\gamma$. *Nature*, **391**, 803.
31. Shyng, S.-L. and Nichols, C. G. (1998) Membrane phospholipid control of nucleotide sensitivity of K_{ATP} channels. *Science*, **282**, 1138.
32. Toker, A. and Cantley, L. C. (1997) Signalling through the lipid products of phosphoinositide-3-kinase. *Nature*, **387**, 673.
33. Downward, J. (1998) Mechanisms and consequences of activation of protein kinase B/Akt. *Curr. Opin. Cell Biol.*, **10**, 262.
34. Andjelkovic, M., Alessi, D. R., Meier, R., Fernandez, A., Lamb, N. J. C., Frech, M., *et al.* (1997) Role of translocation in the activation and function of protein kinase B. *J. Biol. Chem.*, **272**, 31515.
35. Martin, T. F. J. (1997) Phosphoinositides as spatial regulators of membrane traffic. *Curr. Opin. Neurobiol.*, **7**, 331.
36. DeCamilli, P. D., Emr, S. D., McPherson, P. S., and Novick, P. (1996) Phosphoinositides as regulators in membrane traffic. *Science*, **271**, 1533.
37. Rebecchi, M. J. and Scarlata, S. (1998) Pleckstrin homology domains: a common fold with diverse functions. *Annu. Rev. Biophys. Biomol. Struct.*, **27**, 503.
38. Kavran, J. M., Klein, D. E., Lee, A., Falasca, M., Isakoff, S. J., Skolnik, E. Y., *et al.* (1998) Specificity and promiscuity in phosphoinositide binding by pleckstrin homology domains. *J. Biol. Chem.*, **273**, 30497.
39. Lemmon, M. A., Falasco, M., Ferguson, K. M., and Schlessinger, J. (1997) Regulatory recruitment of signalling molecules to the cell membrane by pleckstrin homology domains. *Trends Cell Biol.*, **7**, 237.
40. Levine, T. P. and Munro, S. (1998) The pleckstrin homology domain of oxysterol-binding protein recognizes a determinant specific to Golgi membranes. *Curr. Biol.*, **8**, 729.
41. Janmey, P. A. (1994) Phosphoinositides and calcium as regulators of cellular actin assembly and disassembly. *Annu. Rev. Physiol.*, **56**, 169.
42. Martin, T. F. J. (1998) Phosphoinositide lipids as signaling molecules: common themes for signal transduction, cytoskeletal regulation and membrane trafficking. *Annu. Rev. Cell Dev. Biol.*, **14**, 231.
43. Fukuda, M., Aruga, J., Niinobe, M., Aimoto, S., and Mikoshiba, K. (1994) Inositol-1,3,4,5-tetrakisphosphate binding to C2B domain of IP4BP/Synaptotagmin II. *J. Biol. Chem.*, **269**, 29206.

44. Gaullier, J.-M., Simonsen, A., D'Arrigo, A., Bremnes, B., and Stenmark, H. (1998) FYVE fingers bind PtdIns(3)P. *Nature*, **394**, 432.
45. Patki, V., Lawe, D. C., Corvera, S., Virbasius, J. V., and Chawla, A. (1998) A functional PtdIns(3)P-binding motif. *Nature*, **394**, 433.
46. Glaser, M., Wanaski, S., Buser, C. A., Boguslavsky, V., Rashidzada, W., Morris, A., *et al.* (1996) Myristoylated alanine-rich C kinase substrate (MARCKS) produces reversible inhibition of phospholipase C by sequestering phosphatidylinositol 4,5-bisphosphate in lateral domains. *J. Biol. Chem.*, **271**, 26187.
47. Bankaitis, V. A., Aitken, J. R., Cleves, A. E., and Dowhan, W. (1990) An essential role for a phospholipid transfer protein in yeast Golgi function. *Nature*, **347**, 561.
48. Ohashi, M., deVries, K. J., Frank, R., Snoek, G., Bankaitis, V., Wirtz, K., *et al.* (1995) A role for phosphatidylinositol transfer protein in secretory vesicle formation. *Nature*, **377**, 544.
49. Jones, S. M., Alb, J. G., Phillips, S. E., Bankaitis, V. A., and Howell, K. E. (1998) A phosphatidylinositol 3-kinase and phosphatidylinositol transfer protein act synergistically in formation of constitutive transport vesicles from the trans-Golgi network. *J. Biol. Chem.*, **273**, 10349.
50. Chen, Y.-G., Siddhanta, A., Austin, C. D., Hammond, S. M., Sung. T.-C., Frohman, M. A., *et al.* (1997) Phospholipase D stimulates release of nascent secretory vesicles from the trans-Golgi network. *J. Cell Biol.*, **138**, 495.
51. Tuscher, O., Lorra, C., Bouma, B., Wirtz, K. W., and Huttner, W. B. (1997) Cooperativity of phosphatidylinositol transfer protein and phospholipase D in secretory vesicle formation from the TGN- phosphoinositides as a common denominator? *FEBS Lett.*, **419**, 271.
52. Liscovitch, M., Chalifa, V., Pertile, P., Chen, C.-S., and Cantley, L. C. (1994) Novel function of phosphatidylinositol 4,5-bisphosphate as a cofactor for brain membrane phospholipase D. *J. Biol. Chem.*, **269**, 1994.
53. Roth, M. G. and Sternweis, P. C. (1997) The role of lipid signaling in constitutive membrane traffic. *Curr. Opin. Cell Biol.*, **9**, 519.
54. Liscovitch, M. and Cantley, L. C. (1995) Signal transduction and membrane traffic: the PITP/phosphoinositide connection. *Cell*, **81**, 659.
55. Herman, P. K., Stack, J. H., and Emr, S. D. (1992) An essential role for a protein and lipid kinase complex in secretory protein sorting. *Trends Cell Biol.*, **2**, 363.
56. Burd, C. G. and Emr, S. D. (1998) Phosphatidylinositol(3)-phosphate signaling mediated by specific binding to RING FYVE domains. *Mol. Cell*, **2**, 157.
57. Shepherd, P. R., Reaves, B. J., and Davidson, H. W. (1996) Phosphoinositide 3-kinases and membrane traffic. *Trends Cell Biol.*, **6**, 92.
58. Li, G., D'Souza-Schorey, C., Barbieri, M. A., Roberts, R. L., Klippel, A., Williams, L. T., *et al.* (1995) Evidence for phosphatidylinositol 3-kinase as a regulator of endocytosis via activation of Rab5. *Proc. Natl. Acad. Sci. USA*, **92**, 10207.
59. Simonsen, A., Lippe, R., Christoforidis, S., Gaullier, J.-M., Brech, A., Callaghan, J., *et al.* (1998) EEA1 links PI(3)K function to Rab5 regulation of endosome fusion. *Nature*, **394**, 494.
60. Hay, J. C. and Martin, T. F. J. (1993) Phosphatidylinositol transfer protein required for ATP-dependent priming of Ca^{2+}-activated secretion. *Nature*, **366**, 572.
61. Hay, J. C., Fisette, P. L., Jenkin, G. H., Fukami, K., Takenawa, T., Anderson, R. A., *et al.* (1995) ATP-dependent inositide phosphorylation is required for Ca^{2+}-activated exocytosis. *Nature*, **374**, 173.
62. Wiedemann, C., Schafer, T., and Burger, M. M. (1996) Chromaffin granule-associated phosphatidylinositol 4-kinase activity is required for stimulated secretion. *EMBO J.*, **15**, 2094.

63. Loyet, K. M., Kowalchyk, J. A., Chaudhary, A., Chen, J., Prestwich, G. D., and Martin, T. F. J. (1998) Specific binding of phosphatidylinositol 4,5-bisphosphate to CAPS, a potential phosphoinositide effector protein for regulated exocytosis. *J. Biol. Chem.*, **273**, 8337.
64. Berwin, B., Floor, E., and Martin, T. F. J. (1998) CAPS (Mammalian UNC-31) protein localizes to membranes involved in dense-core vesicle exocytosis. *Neuron*, **21**, 137.
65. Schiavo, G., Gu, Q.-M., Prestwich, G. D., Sollner, T. H., and Rothman, J. E. (1996) Calcium-dependent switching of the specificity of phosphoinositide binding to synaptotagmin. *Proc. Natl. Acad. Sci. USA*, **93**, 13327.
66. Chung, S. H., Song, W. J., Kim, K., Bednarski, J. J., Chen, J., Prestwich, G. D., *et al.* (1998) The C2 domains of Rabphilin3A specifically bind phosphatidylinositol 4,5-bisphosphate containing vesicles in a Ca^{2+}-dependent manner. *J. Biol. Chem.*, **273**, 10240.
67. Khvotchev, M. and Sudhof, T. C. (1998) Newly synthesized phosphatidylinositol phosphates are required for synaptic norepinephrine but not glutamate or γ-aminobutyric acid release. *J. Biol. Chem.*, **273**, 21451.
68. Wiedemann, C., Schafer, T., Burger, M. M., and Sihra, T. S. (1998) An essential role for a small synaptic vesicle-associated phosphatidylinositol 4-kinase in neurotransmitter release. *J. Neurosci.*, **18**, 5594.
69. Gardarov, I., Chen, Q., Falck, J. R., Reddy, K. K., and Keen, J. H. (1996) A functional phosphatidyl 3,4,5-trisphosphate/phosphoinositide binding domain in the clathrin adaptor AP-2 a subunit. *J. Biol. Chem.*, **271**, 20922.
70. Schmid, S. L., McNiven, M. A., and DeCamilli, P. D. (1998) Dynamin and its partners: a progress report. *Curr. Opin. Cell Biol.*, **10**, 504.
71. Godi, A., Santone, I., Pertile, P., Devarajan, P., Stabach, P. R., Morrow, J. S., *et al.* (1998) ADP ribosylation factor regulates spectin binding to the Golgi complex. *Proc. Natl. Acad. Sci. USA*, **95**, 8607.

5 | Neurotransmitter transporters

DAVID E. KRANTZ, FARRUKH A. CHAUDHRY, and
ROBERT H. EDWARDS

1. Introduction

Synaptic transmission involves two distinct types of neurotransmitter transport activity. Plasma membrane reuptake is generally considered to terminate the action of neurotransmitter in the extracellular space. Transport from the cytoplasm into secretory vesicles packages transmitter for regulated release by exocytosis. Although the two activities thus have important if not essential roles in synaptic transmission, the mechanisms of active transport, their regulation and roles in signalling remain poorly understood. Molecular analysis has also suggested previously unanticipated functions for plasma membrane and vesicular neurotransmitter transporters.

Neurotransmitter transport was originally discovered using whole animal and tissue preparations. In the case of plasma membrane transport, peripheral tissues were found to accumulate systemically injected catecholamine, suggesting the presence of a transport activity at the cell surface (1) that was subsequently confirmed by direct flux assays. Antidepressants and psychostimulants such as cocaine potently inhibit plasma membrane transport of monoamines, further demonstrating the biological significance of reuptake for behaviour (2, 3). In the case of vesicle transport, drugs such as the antihypertensive reserpine inhibit packaging into secretory vesicles and induce a syndrome resembling depression (4), indicating the importance of this activity as well for transmitter release and behaviour.

The identification of neurotransmitter transport activities served to focus attention on the biochemical mechanisms responsible. Both plasma membrane and vesicle transport activities can generate concentration gradients of up to 10^6 and so perform active transport rather than facilitated diffusion. Similar to many other secondary active transport systems driven by electrochemical gradients, the translocation of neurotransmitter is coupled to the movement of ions. Both plasma membrane and vesicular transport also depend on ATP. Ouabain inhibits plasma membrane transport, suggesting that ATP hydrolysis by the Na^+/K^+-ATPase generates ionic gradients which directly drive transmitter uptake (5). Indeed, manipulation of the ionic gradients across membranes prepared from the different tissues demonstrated that plasma membrane transport depends on the inwardly directed (out > in) Na^+ gradient across the plasma membrane (3, 6). In contrast, vesicular transport depends on the H^+ electrochemical gradient directed out of the vesicle (7). In plasma membrane transport, the movement of Na^+ down its electrochemical gradient drives the co-transport of transmitter (Fig. 1). Vesicle transport involves the exchange of lumenal protons for cytoplasmic transmitter. The charge of the substrate and the stoichiometry of neurotransmitter to ions provide the rules of the reaction that together with the size of the driving force, determine the concentration gradient of transmitter reached at equilibrium. The relatively slow pace of the coupling mechanism requires considerable time for active transport to reach equilibrium, but can eventually achieve concentration gradients of 10^6. Active transport thus differs markedly from the function of channels which mediate the extremely rapid and apparently non-stoichiometric movement of ions down their electrochemical gradient.

The characterization of transport activity in membrane preparations rather than intact cells further suggested the possibility of purifying the transport proteins by functional reconstitution in artificial membranes. Functional reconstitution involves solubilization of the transport activities in detergent followed by their incorporation into purified brain lipid or other lipids, generation of an appropriate driving force, and the measurement of transmitter flux. Heroic purification using functional reconstitution as the assay for activity (8) resulted in isolation of the first cDNA encoding a neurotransmitter transporter, the GABA transporter GAT-1 (9). Ingenious expression cloning of a plasma membrane norepinephrine transporter using the clinical imaging ligand meta-iodobenzylguanidine then identified a related protein (10) and together with GAT-1, defined a novel gene family now known to include many

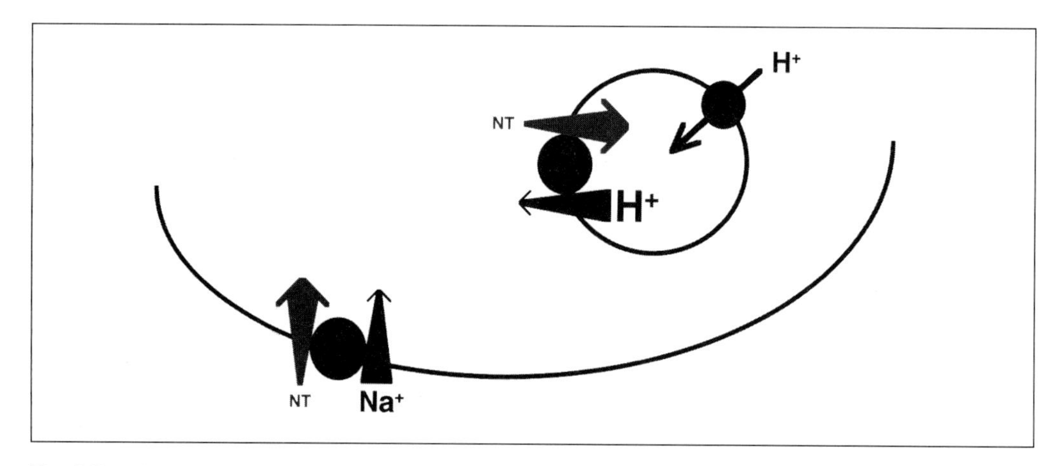

Fig. 1 Two distinct activities transport neurotransmitters across membranes. Plasma membrane transporters (blue) use the inwardly directed Na^+ gradient to co-transport neurotransmitter from the extracellular space. Vesicular transporters (red) use a proton electrochemical gradient generated by the vacuolar H^+-ATPase to drive the packaging of transmitter by an exchange mechanism.

neurotransmitter transporters. Similar purification and expression cloning strategies resulted in isolation of the first cDNAs encoding the plasma membrane transport of excitatory amino acid transmitters (11–13). Expression cloning also identified the first vesicular neurotransmitter transporters (14–16) and genetic studies using the nematode *C. elegans* have helped to identify additional members of both this family (17) and another distinct family that mediates the packaging of amino acid transmitters into neurosecretory vesicles (18). Thus, two families of proteins transport neurotransmitters at the plasma membrane and two distinct families catalyse transport into vesicles.

Molecular cloning of the neurotransmitter transporters has enabled study of the structural basis for their transport activity, their regulation, and their role in synaptic transmission (19, 20). Site-directed mutagenesis in combination with heterologous expression has begun to identify residues responsible for different features of the transport cycle such as substrate recognition and bioenergetic coupling. However, the elimination of active transport by a particular mutation does not indicate its role in the transport cycle. This requires assays for the partial reactions that make up the transport cycle, and these can be difficult to assess with standard flux assays. On the other hand, electrophysiological analysis has provided a powerful tool to dissect transporter function. In the process, it has also uncovered previously unsuspected features of transporter activity. In addition to the small currents that result from the coupled, stoichiometric movement of transmitter and ions, neurotransmitter transporters exhibit uncoupled charge movement and leak currents that occur even in the absence of transmitter. These observations have indicated novel ways to measure transmitter release and appear to have physiological significance of their own. Knock-out mice have also begun to reveal the biological role of neurotransmitter transporters. However, we still know very little about their regulation. Many psychoactive drugs interfere directly with neurotransmitter transport, indicating the potential for regulation of transport activity to affect behaviour, but the implications of these striking observations remain largely unexplored.

2. Na^{+}- and Cl^{-}-dependent plasma membrane transporters

2.1 GABA transport

Isolation of the GAT-1 cDNA by protein purification (9) and the related norepinephrine transporter cDNA by expression cloning (10) defined a novel family of proteins with 12 predicted transmembrane domains (TMDs), N- and C-termini in the cytoplasm, and a large extracellular loop between TMDs 3 and 4 (21). We now know that this family includes four distinct plasma membrane GABA transporters. Before cloning, pharmacological studies had suggested two distinct classes of GABA transporter, a high affinity neuronal form sensitive to inhibition by *cis*-1,3-aminocyclohexane carboxylic acid (ACHC) and nipecotic acid, and a lower affinity glial form sensitive to β-alanine. The original GAT-1, a high affinity transporter sensitive to nipecotic acid and insensitive to β-alanine, occurs as anticipated predominantly in the presynaptic terminals of GABAergic neurons, but also appears in astrocytic processes in the cortex (22, 23) and presumably the other brain regions where it is expressed. The anticonvulsant tiagabine also appears to act by inhibiting the function of GAT-1 (24).

Identified as a cDNA related to GAT-1, GAT-2 shows strong sequence similarity to the betaine transporter BGT1 and both transporters occur in the kidney (25). However, GAT-2 but not BGT1 occurs in the brain. GAT-2 also shows a significantly lower affinity for GABA (in the middle micromolar range) than GAT-1, and β-alanine but not nipecotic acid or ACHC inhibit its activity (26). In the brain, arachnoid and ependymal cells express GAT-2, consistent with the expression by non-neuronal cells suggested by its functional characteristics (27). In the retina, GAT-2 similarly occurs in the retinal pigment epithelium and non-pigmented ciliary epithelium in the rat retina (28).

GAT-3 shows a high affinity for GABA in the low micromolar range but both nipecotic acid and β-alanine inhibit its activity, indicating classically defined features of both neuronal and glial GABA transport (29, 30). Indeed, both neurons and glia appear to express GAT-3, which also occurs at high levels in the kidney (31). In the brain, GAT-3 appears abundant in the spinal cord, brainstem, and thalamus, with considerably lower levels in cerebral cortex, hippocampus, striatum, and cerebellum. Its localization to neurons or glia also appears to depend on the region, with expression by GABAergic neurons in the basal forebrain but by glia in cortex and hippocampus (32, 33).

GAT-4 mediates GABA transport with very low to sub-micromolar affinity and nipecotic acid inhibits its activity. Consistent with the pharmacology suggesting expression by neurons, GAT-4 occurs predominantly in the brain (34). Expressed diffusely for the first few weeks after birth, it subsequently becomes restricted to olfactory bulbs, the midbrain, deep cerebellar nuclei, medulla, and spinal cord (35). A related taurine transporter also occurs in the brain as well as the kidney (34, 36). In the kidney, its expression is induced by hypertonicity (37), suggesting novel modes of regulation in the brain. However, in brain it appears restricted to white matter (34). Another protein in this subfamily was originally considered to transport choline (38) but has since been shown to transport creatine (39). The creatine transporter occurs in heart and skeletal muscle as well as brain. Like the taurine transporter, the function of the creatine transporter in brain is unclear. Indeed, the individual contributions of the four GABA transporters to synaptic transmission also remain uncertain.

The four GABA transporters presumably all function to clear GABA from the extracellular space but their location on neurons or glia suggests important differences. Neuronal transport has the additional potential to recycle GABA for subsequent rounds of release whereas glial transport cannot do so, at least directly. Further, the electrogenic nature of GABA transport raises the possibility that depolarization may suffice to release GABA from cytoplasmic stores by flux reversal through the GATs (40, 41). This appears responsible for the non-vesicular, Ca^{2+}-

independent release of GABA reported in a retinal neuron (42) but could in principle also occur from glia. Finally, neuronal transport may serve to limit the concentration of GABA in the synaptic cleft in which it was released, whereas glial transport may influence diffusion to more remote synapses. In light of the limited pharmacology, knock-out mice will probably be required to address the physiological roles of GABA transport.

2.2 Monoamine transport

Classical observations had suggested distinct transporters for the different monoamine transmitters (3). Consistent with this expectation, expression cloning of the plasma membrane norepinephrine transporter (10) enabled the identification of closely related dopamine (43–45) and serotonin transporters (46, 47) that all show strong sequence similarity to the GABA transporters and have 12 predicted TMDs with a large loop between TMD3 and 4. Cocaine inhibits all three of these proteins but mazindol shows particular selectivity for the dopamine transporter (DAT), the tricyclic antidepressant desipramine for the norepinephrine transporter (NET), and newer antidepressants such as paroxetine and fluoxetine for the serotonin transporter (SERT) (48). Further, DAT shows restricted expression only in dopaminergic neurons, NET in noradrenergic neurons (and the adrenal medulla), and SERT in serotonergic neurons (and also in the adrenal medulla). However, functional characterization of the isolated sequences demonstrates less selectivity. In particular, both DAT and NET have a higher apparent affinity for dopamine than norepinephrine (48). In addition, DAT appears to transport both substrates at a much higher rate than NET. Drug binding can be used to estimate the number of transport proteins in a preparation and so determine turnover numbers for DAT of $\sim$ 115/min, for NET of $\sim$ 6.5/min, and for SERT of $\sim$ 110/min (48). Thus, the different monoamine transporters exhibit distinct functional characteristics but may have overlapping roles *in vivo*. More recently, an epinephrine transporter has also been identified in the bullfrog *R. catesbiana* (49). It will be interesting to determine whether a homologue exists in mammals.

The monoamine transporters appear to have several physiological and pharmacological roles in addition to transmitter clearance. First, their presynaptic location allows them to recycle transmitter for subsequent rounds of release. DAT knock-out mice show dramatically reduced levels of dopamine and fail to maintain release with repeated stimulation (50). Although biosynthesis has generally been considered the major source of dopamine for exocytotic release, reuptake thus has a surprisingly important role in maintaining vesicular stores. Despite the depleted storage pool, DAT knock-out mice are very hyperactive due to the extremely slow clearance and very high levels of extracellular dopamine (50).

Secondly, the monoamine transporters all mediate the selective accumulation of *N*-methyl-4-phenylpyridinium (MPP$^+$), the active metabolite of the potent neurotoxin *N*-methyl-1,2,3,6-tetrahydropyridine (MPTP) (51, 52) by monoamine neurons. Although this accounts for some of the selective toxicity observed in this model of

Parkinson's disease (53–55), the functional characteristics of the different monoamine transporters and their levels of expression do not appear to account for the selective degeneration of dopamine rather than noradrenergic or serotonergic neurons after exposure to MPTP (48, 56, 57).

Thirdly, amphetamines induce flux reversal by the plasma membrane monoamine transporters (58–60). The precise mechanism for this non-vesicular release remains uncertain and could involve either exchange of extracellular amphetamine for cytoplasmic monoamine followed by diffusion of amphetamine back out of the cell for another round of exchange (exchange-diffusion) or simple uptake of amphetamine followed by diffusion out of the cell, dissipating the Na^+ gradient that normally drives uptake and prevents efflux. None the less, this action together with an effect on vesicular stores presumably accounts for the psychostimulant effect of amphetamines and may also contribute to the associated neurotoxicity (61). Indeed, DAT knock-out mice are resistant to the effects of amphetamines (62) as well as cocaine, including the amphetamine-induced toxicity (63). Despite this resistance to psychostimulants, DAT knock-outs still self-administer cocaine, indicating additional sites of action possibly including SERT (64). The action of many antidepressants such as fluoxetine at SERT and tricyclics at NET further suggests an important role for these transporters in behaviour.

Surprisingly, SERT knock-out mice exhibit relatively normal behaviour. Although SERT inhibitors severely disrupt craniofacial development, the knock-outs also show no such abnormalities (65). Despite a drastic reduction in brain serotonin, the main abnormality observed in these animals thus far is a defect in the locomotor response to methylenedioxymethamphetamine (ecstasy) but not amphetamine, which presumably acts primarily on DAT.

The identification of human sequences encoding the monoamine transporters has also enabled assessment of their role in human neuropsychiatric disease. The action of antidepressant drugs on these proteins has suggested that an alteration in their function may contribute to normal variations in behaviour. Interestingly, a polymorphism in the regulatory region of the human SERT gene shows association with anxiety-related personality traits and makes a small but statistically significant contribution to the variance (66). However, the promoter variant associated with anxiety shows reduced transcriptional activation and cells from individuals show reduced SERT activity, an effect opposite to that predicted from the action of antidepressants which also inhibit SERT but are used to alleviate anxiety.

2.3 Glycine and proline transport

Molecular cloning has identified two distinct plasma membrane transporters for glycine (67–69). Similar to other members of this family, glycine transporters 1 and 2 depend on extracellular Na^+ and Cl^- and their sequences predict 12 TMDs. The GLYT1 cDNA encodes glycine transport with a $K_m \sim 100\ \mu M$ and inhibition by sarcosine but not alanine or methyl-aminoisobutyric acid (67, 68). It may correspond to the previously described system Gly and occurs in peripheral tissues as well as the brain (70). GLYT1 also appears to undergo alternative RNA splicing that affects the length of the cytoplasmically disposed N-terminus (69). *In situ* hybridization has indicated strong expression by the brainstem and spinal cord, where glycine has a prominent role in inhibitory neurotransmission. However, GLYT1 mRNA also occurs in other brain regions where it has been proposed to modulate excitatory transmission through glutamatergic NMDA receptors (67). In contrast, GLYT2 occurs only in the brainstem, spinal cord, and cerebellum where it presumably has a more specific role in glycinergic transmission (69). Indeed, GLYT2 occurs in neurons at presynaptic sites and in some glial elements, consistent with a role in glycine recycling as well as clearance (71). GLYT1, on the other hand, occurs essentially only in glia, around both glycinergic and non-glycinergic neurons. GLYT2 also shows somewhat higher affinity for glycine ($K_m \sim 20\ \mu M$) and resists inhibition by sarcosine (69).

A plasma membrane proline transporter also belongs to this family of proteins (72). It remains unclear how proline transport contributes to synaptic transmission but its expression by subpopulations of glutamatergic neurons suggests a role in excitatory transmission. Interestingly, the proline transporter appears to localize to synaptic vesicles (73). Since it depends on a Na^+ gradient rather than a H^+ electrochemical gradient, the proline transporter presumably only functions as a transporter at the plasma membrane rather than in synaptic vesicles. The activity that induces synaptic vesicle fusion may therefore also increase proline transport and possibly contribute to synaptic plasticity.

2.4 Orphan transporters

The family of proteins defined by GAT-1 and NET includes several members without defined substrates. Although similar in sequence to the other members, the orphan transporter rat-XT1 (rXT1, also known as NTT4) (74, 75) contains an exceptionally large extracellular loop between TMDs 7 and 8 and an unusually large cytoplasmic C-terminus (76). *In situ* hybridization shows expression in the granule and Purkinje cell layers of the cerebellum, in the pyramidal and dentate gyrus granule cells of the hippocampus, the cortex, and the thalamus. Post-embedding electron microscopy shows the protein to be enriched on synaptic vesicles in excitatory terminals, as well as to occur at lower levels in inhibitory terminals (F. A. Chaudhry, personal communication). Lesion studies and immunoelectron microscopy further suggest localization of the protein to presynaptic structures in neurons (77). Thus, the distribution of rXT1 generally suggests a role in excitatory transmission but the expression in Purkinje cells raises other possibilities as well. Another orphan, v7–3, appears widely expressed in the cortex, hippocampus, cerebellum as well as in the ventral horn of the spinal cord, motor nuclei of the brainstem, and multiple monoamine cell groups (78, 79). However, the orphan rB21a occurs in the meninges rather than grey or white matter (80), suggesting a completely different biological role. Additional orphan transporters continue to be described.

2.5 Transport mechanism

2.5.1 Ionic requirements and stoichiometry

Secondary active transporters such as the plasma membrane neurotransmitter transporters use the movement of ions down their electrochemical gradient to drive the concentration of substrate against its concentration gradient. In particular, the plasma membrane neurotransmitter transporters described here rely on the movement of both Na^+ and Cl^- (Fig. 2). The translocation (and not simply the binding) of these ions somehow provides the energy to drive active transport. The stoichiometry in turn determines the theoretical magnitude of the concentration gradient reached at equilibrium. Coupling to multiple ions increases the free energy available, producing larger gradients than those possible with fewer coupling ions. Interestingly, the Na^+-/Cl^--dependent neurotransmitter transporters differ in their stoichiometry. Manipulation of the ionic gradients in artificial membrane vesicles reconstituted with solubilized brain GABA transporter and in membrane vesicles prepared from heterologous cells expressing the GAT-1 cDNA shows a sigmoidal dependence of transport on Na^+ but a hyperbolic dependence on GABA and Cl^-, consistent with a stoichiometry of 2 Na^+:1 Cl^-:1 GABA (81). Since GABA is a zwitterion at physiological pH, this stoichiometry predicts a net charge movement of +1 into the cell. GAT-3 shows a similar stoichiometry (30), as does the plasma membrane dopamine transporter DAT (48). However, the closely related biogenic amine transporters NET and SERT exhibit a different stoichiometry of 1 Na^+:1 Cl^-:1 monoamine (48, 82). Since these transporters appear to recognize the protonated form of monoamines, they translocate +1 net charge into the cell and hence appear electrogenic. SERT function, on the other hand, appears to involve the counter-transport of 1 K^+ ion, resulting in

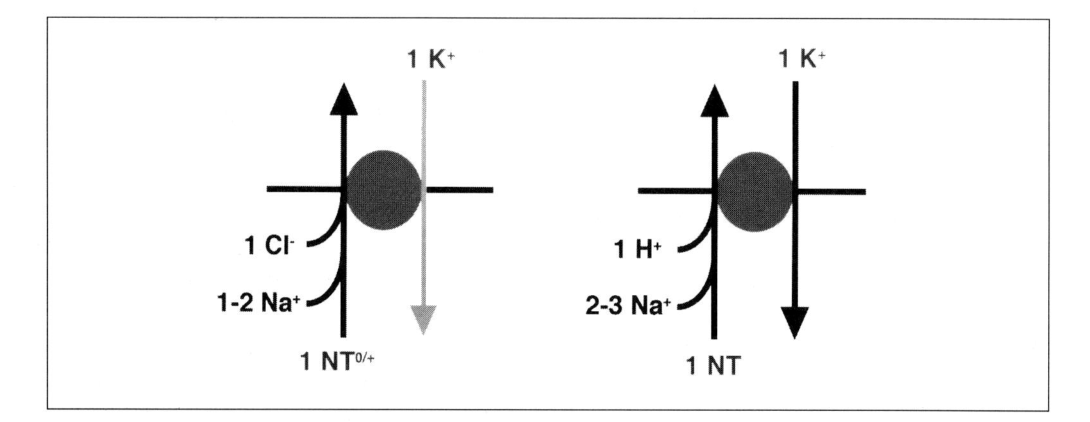

Fig. 2 Two mechanistically distinct classes of transport activity mediate neurotransmitter uptake at the plasma membrane. The Na^+- and Cl^--dependent family of transporters (left) co-transports either positively charged (monoamines) or neutral (GABA) substrates (NT) with 1 Cl^- and either 1 Na^+ (in the case of the norepinephrine and serotonin transporters) or 2 Na^+ (in the case of the GABA transporters and the dopamine transporter). Within this family, only the serotonin transporter appears to require counter-transport of K^+. In the excitatory amino acid transporter family (right), all members mediate K^+ counter-transport as well as the co-transport with anionic transmitter of 2–3 Na^+ and 1 H^+.

electroneutrality (83, 84) (Fig. 2). Unlike the co-transported Na^+, however, counter-transported K^+ can be replaced by protons, accounting for the transport activity detected in the absence of K^+ (85).

2.5.2 Coupled and uncoupled charge movement

The stoichiometry of substrate and ion flux determines whether transport is electroneutral, as in the case of SERT (84), or electrogenic, as has thus far been shown for other members of this family including GAT-1, -3, and the monoamine transporters. Electrophysiological measurements of steady state currents under different ionic conditions confirm these predictions in the case of GAT-1 (86–89). However, the magnitude of the currents observed often vastly exceeds that predicted by the stoichiometry, the turnover rate of the transporter, and the level of transporter expression.

Several transporters show a discrepancy between charge movement and flux (90). Despite the apparent electroneutrality of flux, SERT expressed in *Xenopus* oocytes translocates 5–12 net positive charges with each molecule of serotonin (91). GAT-1 also shows variable stoichiometry (92). In addition, human DAT expressed in oocytes exhibits charge movement that generally exceeds the 2 Na^+ predicted by the stoichiometry for flux and that varies at different membrane potentials (93). Similarly, human NET and rat GAT-1 expressed in mammalian cell lines displayed the same dependence of charge movement on ionic gradients as flux, but the magnitude of the currents vastly exceeded (by more than two orders of magnitude) those predicted by the stoichiometry (89, 94). Patch-clamp analysis of the charge movement mediated by NET shows that substrate induces rare, brief events reminiscent of channel opening (95). Analysis of *Drosophila* SERT further demonstrates that the channel mode occurs approximately once for every 500 serotonin molecules transported, but moves 10000 charges for each opening (96). High substrate concentrations increase the likelihood of channel opening and inhibitors block these currents as well as flux. However, another class of conductances associated with the transporters occurs only in the absence of substrate or inhibitor.

Reaction cycles that couple the active transport of substrate to movement of an ion down its electrochemical gradient have the potential to slip occasionally (97). Ion translocation may thus occur in the absence of substrate movement. Indeed, substrate or inhibitor should reduce the likelihood of slippage and events with these properties clearly occur with SERT (91), DAT (93), and GAT-1 (98). Patch-clamp analysis of GAT-1 and rat SERT have also detected large, infrequent current bursts in the absence of substrate (99, 100). Interestingly, this constitutive leak current may carry either the same or different ions from those that drive transport. In the case of SERT, GAT-1, and DAT, the leak currents carry Li^+ and K^+ (91, 93, 99), ions that do not drive transport. This observation suggests that the ion permeation pathway involved in the leak differs from that involved in coupled ion translocation.

The observation of uncoupled as well as coupled charge movements mediated by neurotransmitter transporters has important implications (101). First, the uncoupled charge movements raise questions about the mechanistic relationship between transporters and ion channels. Transporters are generally considered to function according to an alternating access model where a gate on one side of the membrane opens to admit one molecule of transmitter, then closes before the opening of a second gate on the other side of the membrane (Fig. 3). The rules for coupling to ion translocation enable these gates to function appropriately despite the development of large concentration gradients of substrate. Simultaneous opening of the gates on both sides of the membrane would presumably destroy these gradients. However, simultaneous opening of both gates might occur rarely and briefly, accounting for the uncoupled charge movement observed for several neurotransmitter transporters. Alternatively, transporters may not differ significantly from ion channels, with a permeation pathway that requires binding and unbinding of particular ions for appropriate gating behaviour. In principle, this does not differ from the alternating access model and might even allow for such phenomena as counter-flow and exchange which have generally been associated only with transporters. Counter-flow and exchange involve exchange of substrate on one side of the membrane for substrate on the other, resulting in no net flux, and channels are generally considered incapable of this coupling phenomenon, but with the appropriate gating behaviour, they remain possible. However, the fundamental question remains whether the uncoupled charge movement observed with transporters involves the same pathway used for active transport. If not, then the same protein presumably has two conduction pathways, one for transport and the other for channel-like behaviour.

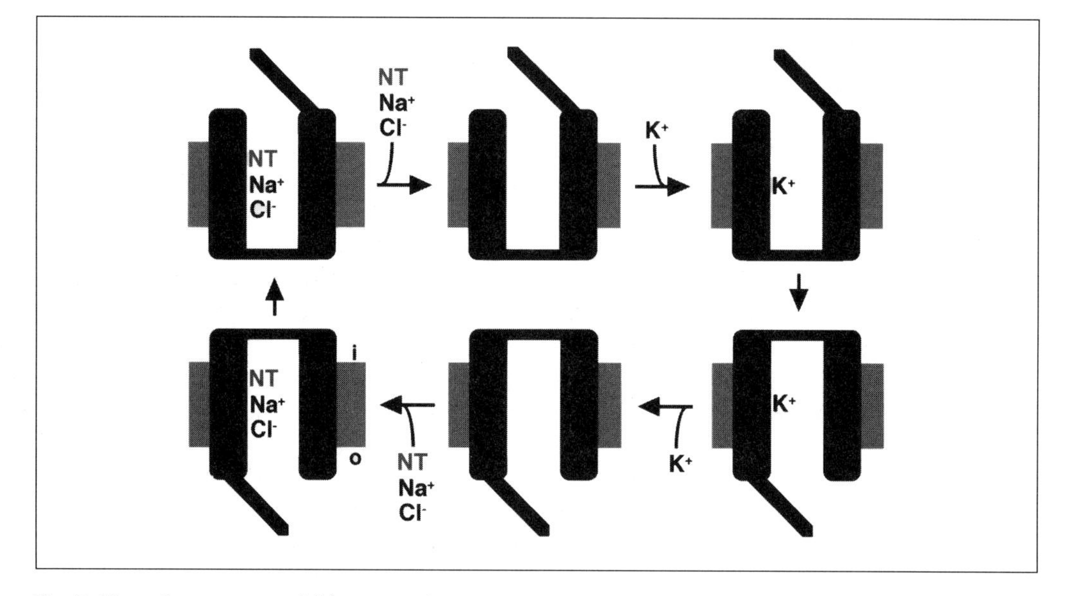

Fig. 3 Alternating access model for serotonin transport. An external gate opens to admit neurotransmitter (NT), Na^+, and Cl^- from the extracellular space (o) (lower left-hand corner). This gate then closes and the internal gate opens to the cell interior (i), allowing the translocation of neurotransmitter and ions (upper left-hand corner). After dissociation (middle, top), K^+ binds to the transporter (upper right-hand corner), the internal gate closes, the external gate opens (lower right-hand corner), and K^+ translocates out of the cell (middle, bottom).

The charge movements associated with neurotransmitter transporters may also have physiological significance. The depolarization associated with either coupled or uncoupled currents has the potential to depolarize the cell. In the skate retina, GABA induces depolarization and activates Ca^{2+} flux (102). The pharmacology strongly implicates GABA transporters which presumably depolarize the cell and so activate voltage-sensitive Ca^{2+} channels. In the case of the monoamine transporters, the conductances observed also enable a distinction to made between substrates such as amphetamines, which induce uncoupled as well as coupled charge movement but inhibit leak conductances, and inhibitors such as cocaine, which block leakage (93). The family of excitatory amino acid transporters exhibits still more examples of physiologically relevant charge movement (see Sections 3.3.1 and 3.3.2).

2.5.3 Structure and mechanism

A variety of observations support the originally predicted topology of 12 TMDs with the N- and C-termini in the cytoplasm. Anti-peptide antibodies show the expected dependence on permeabilization to recognize different domains in human NET by immunofluorescence (103). The large loop between TMDs 3 and 4 predicted to reside outside the cell also contains glycosylated residues in GLYT1 and SERT (104, 105). Insertion of *N*-glycosylation sites into various domains of both GAT-1 and GLYT1 further supports the general topology (106, 107). However, the analysis has also suggested that TMD1 of both transporters may not fully span the membrane. A long re-entrant loop at this position similar to the pore loops found in ion channels might even contribute to the observed channel-like behaviour.

To determine how Na^+ and Cl^- drive the active transport of neurotransmitters, investigators have focused on substrate recognition and ion translocation. The transport cycle presumably involves the binding of Na^+, Cl^-, and neurotransmitter on the extracellular face of the plasma membrane, followed by translocation, dissociation of ions and substrate in the cytoplasm, and return of the unloaded carrier to face the outside of the cell. Thus, changes in substrate recognition and ion binding on the two faces of the membrane presumably drive the transport cycle. The structural domains responsible for these events apparently reside within the 12 TMD core of the proteins since protease digestion of the N- and C-termini as well as their deletion by site-directed mutagenesis do not affect the activity of rat GAT-1 (108). The C-terminus appears responsible only for the proper trafficking and processing of the transport protein (109). In contrast, several of the external loops appear important for transport activity (110, 111).

Charged residues within TMDs presumably interact directly with ions and neurotransmitter. Surprisingly, the only charged residue in a predicted TMD essential for the activity of GAT-1 is Arg-69, in the re-entrant loop of TMD1 (112). Its positive charge suggests a possible interaction with chloride, but it could have many other roles. Although there are no acidic residues in predicted TMDs that are essential for activity, aromatic residues can use their π electrons to bind cations and mutagenesis of Trp-68 and -222 in GAT-1 impair activity (113). Interestingly, the Trp68Leu mutant appears to lock Na^+ onto the transporter and prevent its dissociation in the cytoplasm (114). The electrophysiological analysis also supports previous work suggesting that Na^+ and Cl^- bind to GAT-1 first and so enable GABA to bind with high affinity. In addition, Glu-101 in GAT-1 appears essential for activity (115). It also eliminates Na^+ transients detected electrophysiologically, suggesting an interaction with Na^+. In the serotonin transporter, replacement of Ser-545 in TMD11 with alanine enables Li^+ as well as Na^+ to drive transport, suggesting that this residue also influences the interaction with cations (116).

A variety of experimental approaches have been used to understand substrate recognition by this family of transporters. Site-directed mutagenesis of GAT-1 has shown a selective role for Tyr-140 in substrate and not ion recognition as measured electrophysiologically (117). DAT also has an aspartate and serines that contribute to the recognition of neurotransmitter and cocaine (118). These residues resemble those of G protein-coupled receptors for monoamines which have been shown to interact with the primary amino group and hydroxyl groups of dopamine, respectively. Comparison of closely related transporters through the production and analysis of chimeric and subsequently site-directed mutants also has the potential to identify residues critical for substrate selectivity (111, 119, 120), although the results have often been difficult to interpret. Indeed, a major problem with all of these studies is the difficultly in distinguishing direct interactions with ion or substrate from indirect effects through changes in protein structure. In the absence of high-resolution structural information, which will presumably be difficult to obtain for these polytopic membrane proteins, cysteine-scanning mutagenesis provides an alternative way to study both interactions and structure. Cysteine-reactive reagents such as the methane thiosulfonates modify membrane proteins and thus have the potential to disrupt their activity. Although the neurotransmitter transporters contain multiple cysteines, many of them can be replaced by serine with little effect in activity (121), enabling the introduction of cysteines at other sites in the protein. In addition, differently charged methane thiosulfonate derivatives vary in their ability to cross membranes. The ability of membrane permeant but not impermeant reagents to influence transporter behaviour thus helps to determine transmembrane topology (122). Further, substrate and ions may protect against the modification at particular sites, providing more direct support for the role of these domains in recognition (123). Interestingly, cocaine actually increases the reactivity of a cysteine in human DAT (121). These systematic approaches combined with the functional analysis of the transport cycle by flux or electrophysiological assays will eventually help us to understand the mechanism of active transport.

Recent work has also begun to elucidate the mechanistic basis for uncoupled conductances associated with this family of transporters. The rat but not human serotonin transporter shows a larger associated conductance at acidic than at neutral pH (124). Site-directed mutagenesis has identified a single residue in the extracellular loop between TMD9 and 10 (a threonine in the rat sequence and a lysine in the human) that is responsible for this difference and that does not appear to affect other aspects of transporter function. It will be very important to determine when the conductance occurs during the transport cycle and these investigators specifically propose that the conductance occurs as a side-reaction of the inwardly oriented transporter, when the normally closed external gate inadvertently opens (124).

2.6 Regulation

The profound behavioural effects of drugs that act on neurotransmitter transporters have suggested that endogenous regulation of transport activity may contribute in very important ways to normal behaviour. However, we still know very little about the regulation of transport activity. In addition, there is very little evidence for regulation of intrinsic transport activity. An early study reported up-regulation of SERT activity by stimulation of adenosine A3 receptors, coupling through nitric oxide and cGMP (125), but the effect predominantly involved an increase in V_{max}. Indeed, the most clearly documented forms of transporter regulation involve changes in subcellular location.

Stimulation of protein kinase C (PKC) with phorbol esters increases the activity of GAT-1 in the plasma membrane of injected *Xenopus* oocytes (126). The effect involves an increase in V_{max} rather than an effect on K_m and is accompanied by an increase in GAT-1 protein at the cell surface. Additional experiments have further indicated a role for SNARE proteins such as the t-SNARE syntaxin in the increased plasma membrane expression (127), raising the possibility that GAT-1 may, like the proline transporter, occur on synaptic vesicles *in vivo* and undergo regulated translocation to the plasma membrane upon stimulation. Conversely, phorbol esters induce a loss of human SERT and rat DAT from the plasma membrane of mammalian cells (128–130). Staurosporine inhibits the reduction in V_{max}, suggesting a role for kinases, and both proteins undergo phosphorylation in response to phorbol esters (131, 132). However, elimination of the phosphorylation sites does not appear to affect the down-regulation, suggesting that phorbol esters promote internalization by activating different protein(s) that may contribute directly to the endocytotic event. Interestingly, transport inhibitors also appear to promote the internalization of human NET, raising the possibility that the presence of the neurotransmitter itself may regulate plasma membrane expression (133).

Neurotransmitter transporters also vary in the site of cell surface expression. GAT-1 sorts to the axonal rather than somatodendritic membrane of primary hippocampal neurons (134). Consistent with the analogy between axonal and apical (and somatodendritic and basal) surfaces of epithelia, GAT-1 sorts to the apical surface of the epithelial Madin-Darby canine kidney (MDCK) cell line. GAT-3 also localizes to the apical surface of transfected MDCK cells but GAT-2 resembles the endogenous renal betaine transporter BGT-1 and sorts to the basolateral surface (135). However, BGT-1 appears on both axonal and somatodendritic processes when expressed in primary hippocampal neurons. Further, all the monoamine transporters sort to the basolateral surface of the polarized LLC-PK1 cells but in MDCK cells, only NET and SERT sort to the basolateral surface whereas DAT trafficks to the apical surface (136). Thus, different cell lines sort the same transport proteins in different ways, raising questions about their trafficking *in vivo*. Interestingly, SERT shows a polarized distribution on adrenal chromaffin cells, a cell type generally not considered to have significant polarity (137). In the central nervous system, the dopamine transporter localizes to both the striatal terminals and the cell bodies of midbrain dopamine neurons (138–140). Within the midbrain, DAT localizes to intracellular membranes in perikarya and proximal dendrites and only on the plasma membrane in distal dendrites. Preliminary observations thus suggest that neurotransmitter transporters may differ in steady state localization as well as regulation but the physiological consequences remain to be explored.

3. Excitatory amino acid transporters

3.1 Identification and isolation

Plasma membrane transporters for glutamate differ in several important respects from the other plasma membrane neurotransmitter transporters (6, 141). Membrane vesicles as well as synaptosomes from rat brain show Na^+-dependent glutamate (and aspartate) uptake (142). However, the activity does not have a requirement for extracellular Cl^- as well as Na^+ (Fig. 2). In addition, glutamate uptake requires the counter-transport of K^+ and glutamate efflux from pre-loaded membranes requires external K^+ (143). Classical studies have also demonstrated that the process of active uptake is electrogenic. On the other hand, electroneutral exchange does not require K^+, suggesting that translocation of this ion serves to reorient the unloaded carrier (144), similar to the role of counter-transported K^+ in the serotonin transporter (Fig. 3). Interestingly, exchange also shows a dependence on internal but not external Na^+, suggesting that Na^+ binds before glutamate on the external face of the membrane but dissociates from the transporter before glutamate on the internal face. Further, glutamate transport involves either the co-transport of a H^+ or the counter-transport of a OH^- (145).

Purification of a glutamate transporter by functional reconstitution of the activity described above led to identification of the first excitatory amino acid transporter in 1992 (11, 146). Surprisingly, alternative approaches yielded two cDNAs encoding distinct plasma membrane glutamate transport at essentially the same time. GLAST (also known as excitatory amino acid transporter-1 or EAAT1) was identified in the course of purifying a galactosyl transferase from brain (12). The sequence of the protein showed similarity to bacterial glutamate transporters and heterologous expression of the cDNA conferred glutamate transport. Purification of the glutamate transport activity from rat brain led to the identification of GLT-1, also known as EAAT2 (11). Importantly, both cDNAs predicted proteins with 8–10 TMDs and no sequence similarity to the Na^+-/Cl^--dependent transporters. Both EAAT1 and EAAT2 also occur specifically in the brain but EAAT1 appears particularly abundant in the cerebellum whereas EAAT2 occurs diffusely throughout the brain and spinal cord (147–149). In addition, both EAAT1 and 2 occur within glia by *in situ* hybridization and immunocytochemistry. Ultrastructural analysis further indicates that the highest levels of EAAT1 and 2 occur in glial processes ensheathing excitatory synapses (148), supporting a role in transmitter clearance. Despite these observations *in vivo*, neurons express EAAT2 in primary culture (150). However, the localization of EAAT2 in neurons appears post- rather than presynaptic.

The third isoform initially identified, EAAT3 (originally known as excitatory amino acid carrier-1 or EAAC1), was isolated by expression cloning for glutamate transport in *Xenopus* oocytes (13). It shows sequence similarity to EAAT1 and 2 but occurs in the kidney as well as the brain. Within the brain, EAAT3 is expressed at high levels in the hippocampus but also occurs in other regions. In addition, EAAT3 appears neuronal rather than glial (147). In contrast to the neuronal Na^+-/Cl^--coupled neurotransmitter transporters which generally occur presynaptically where they mediate transmitter recycling as well as clearance, however, EAAT3 usually occurs postsynaptically. In GABAergic Purkinje cells of the cerebellum, EAAT3 occurs on axons as well as cell bodies (147).

Consistent with previous work suggesting heterogeneity in glutamate transport (151), the sequence similarity among EAAT1–3 subsequently enabled the identification of two related carriers, EAAT4 and 5. EAAT4 was isolated from the cerebellum and appears relatively restricted to this brain region (152). Within the cerebellum, parasagittal stripes of EAAT4 immunoreactivity occur in Purkinje cell bodies and dendrites (153–155). Thus, EAAT4 has a postsynaptic localization similar to EAAT3. Ultrastructural analysis indicates expression at excitatory synapses, but with higher levels apposing glial rather than neuronal membranes (155). EAAT5 appears selectively expressed by the retina and also encodes Na^+-dependent glutamate transport by heterologous expression (156). Retina expresses the other glutamate transporters as well and in a variety of cell types (157), supporting the diverse roles suggested by physiological studies to be discussed below. Interestingly, EAAT5 contains a C-terminal sequence that occurs in other membrane proteins known to interact with the PDZ domains on synaptic proteins such as PSD-95 (158). Presumably, a similar interaction influences the distribution of EAAT5 at the synapse.

Additional proteins related to the EAATs include several that mediate the transport activity previously described as system ASC. System ASC mediates the Na^+-dependent uptake of zwitterionic amino acids such as alanine, serine, and cysteine rather than acidic amino acids such as aspartate or glutamate. ASCT1 occurs in the brain as well as many peripheral tissues but ASCT2 occurs primarily outside the nervous system (159, 160). The ATB^o protein appears identical to ASCT2 and may correspond to the previously described Na^+-dependent, broad specificity amino acid transport activity observed in the apical surface of epithelia (161). In contrast to the EAATs, however, these transporters do not appear to require counter-transport of K^+ and operate in an electroneutral manner. Indeed, the ASCTs may simply function as exchangers (160, 162).

3.2 Pharmacology

The plasma membrane glutamate transporters exhibit relatively subtle differences in their affinity for substrates. All transport glutamate with an apparent affinity in the low to mid-micromolar range (21) but EAAT4 has the lowest K_m (~ 2–3 μM), with the others 10–100 μM. In addition, the transporters all recognize specifically the L-enantiomer of glutamate. They transport aspartate with similar affinity but, surprisingly, do not discriminate between the two enantiomers of this amino acid. In contrast to the ASCTs, they also have a low (high micromolar) affinity for zwitterionic amino acids.

In terms of pharmacology, the EAATs exhibit more substantial differences (21, 163). With regard to non-transported inhibitors, human EAAT1 and 3 show sensitivity to L-serine-O-sulfate (SOS) with a K_i ~ 100 μM, but less sensitivity to the dihydrokainate (DHK), kainic acid (KA), and L-α-aminoadipate (LαAA). In contrast, EAAT2 is sensitive to inhibition by DHK and KA but not LαAA or SOS, and EAAT4 is sensitive to LαAA but not DHK or KA. However, the rat isoforms differ from the human in several respects: rodent EAAT2 and 3 are both relatively sensitive to LαAA, with micromolar K_is (13). The compounds threo-3-hydroxy-DL-aspartate (THA) and L-*trans*-pyrrolidine-2,4-dicarboxylic acid (PDC) act as competitive substrates for most of the glutamate transporters (with an apparent affinity in the low micromolar range), but more specifically block the function of EAAT5 at lower concentrations (156). Several new compounds have been developed over the last few years (164, 165) but the pharmacology of the EAATs remains relatively primitive.

3.3 Mechanism

3.3.1 Stoichiometry

To assess the stoichiometry of glutamate transport, classical studies used the activation of flux by different concentrations of substrate and co- or counter-transported ions to determine the Hill coefficient. This analysis indicated a first-order dependence of glutamate transport on [glutamate] and internal [K^+] but a sigmoidal dependence on external [Na^+], suggesting the translocation of 2–3 Na^+ for each glutamate transported and each K^+ counter-transported (143, 166). Electrophysiological analysis of substrate-evoked current has yielded similar results, with 2 Na^+ (in the case of EAAT3) (167) or 3 Na^+ (in the case of EAAT1, 2, and 3) (168–170) translocated for each glutamate and H^+ co-transported and each K^+ counter-transported. The analysis of capacitative, pre-steady state currents induced by changes in the membrane potential in cells expressing EAAT2 and presumably due to the binding and unbinding of Na^+ from the transporter further suggests that Na^+ may bind before substrate (171). Interestingly, EAAT1–3 all transport L-cysteine in a manner independent of external pH (172), suggesting that the transport of cysteine resembles the co-transport of 1 H^+ with each glutamate and hence has the same net charge stoichiometry. The stoichiometry also predicts that transport is electrogenic, moving 1–2 charges per glutamate. However, the preponderance of currents associated with glutamate transport are not coupled to the flux of transmitter.

3.3.2 Uncoupled chloride flux

Previous work on cone photoreceptors in the retina of the tiger salamander had suggested the association of unusual currents with glutamate transport (173–175). Glutamate activates a Cl^- conductance in these cells that depends entirely on the presence of extracellular Na^+. Typical glutamate receptor blockers fail to inhibit this current and its dependence on Na^+ implicates Na^+-dependent transporters. Indeed, expression of EAAT4 in *Xenopus* oocytes produces a glutamate-activated chloride conductance (152) and EAAT1–3 exhibit similar properties (176). The charge movement does not simply represent electrogenic transport because the reversal potential varies with the equilibrium potential for chloride. Oocytes express a calcium-activated chloride channel but calcium chelators and chloride channel blockers do not block the glutamate-activated conductance, supporting a permeation pathway specific to the transporter. Although the currents due to chloride can be quite substantial (in excess of 20-fold over electrogenic transport alone for EAAT4) (152), the selectivity of the conductance is $SCN^- > NO_3^- > I^- > Br^- > Cl^- > F^-$. These observations have raised important questions about the function of the EAATs as both glutamate transporters and anion channels.

The glutamate transport activity and anion conductance exhibited by the EAATs appear quite distinct from each other. Removal of chloride from the external medium eliminates outward currents without affecting glutamate transport (176). Similarly, membrane potentials above or below the equilibrium potential for Cl^- do not affect the rate of glutamate flux. However, the anion conductance requires activation by glutamate and Na^+, similar to transport. The pharmacology also suggests that the binding of glutamate to a single site on the transport protein activates both processes. In addition, the reversal potential observed for the different transporters appears independent of glutamate concentration (176). Since the reversal potential for electrogenic transport and the equilibrium potential for Cl^- differ markedly, this indicates that both transport and anion conductance are activated in similar proportions by different concentrations of glutamate. Glutamate must therefore activate both processes with identical potency. However, L-glutamate and D-aspartate produce different reversal potentials in the case of EAAT1 (176). In addition, EAAT1 has a reversal potential closer to 0 mV whereas EAAT3 and especially EAAT2 have reversal potentials ~ +40–80 mV, depending on the substrate. Thus, the proportion of electrogenic transport to anion conductance can vary between different substrates and among different transporters, with EAAT1 showing a higher anion conductance than EAAT2 or 3.

Further analysis of the anion conductance of EAAT1 has provided additional important information about transporter function. Expression of EAAT1 confers a small but detectable leak current in the absence of substrate but importantly, the anion conductance exhibits no significant permeability to glutamate (177). The anion conductance will therefore not dissipate the concentrations of glutamate accumulated by coupled transport. Analysis of EAAT1 in inside-out patches also indicates an apparent affinity for efflux only tenfold lower than for uptake from the external face of the plasma membrane (177), suggesting that additional steps in the transport cycle account for the extremely large concentration gradients predicted by the stoichiometry of coupled transport. In addition, coupled transport demonstrates a much greater temperature dependence than the anion conductance (177), consistent with the increased number of conformational changes presumably involved in transport. However, the small size of the currents precludes direct analysis of unitary events. Noise analysis combined with estimates of transporter number suggests a subfemtosiemen conductance with an extremely low probability (177). The anion conductance also activates very rapidly, consistent with a physiologically significant role in the modulation of signalling, but currents induced by L-glutamate decay to a much greater extent and much more quickly than currents induced by D-aspartate (177), raising the possibility of different responses to different physiological ligands.

3.3.3 Structural analysis

The structure of the glutamate transporters has assumed particular importance with the identification of channel-like as well as transport function. However, the topology of the glutamate transporters has appeared more difficult to determine than that for the Na^+-/Cl^--dependent neurotransmitter transporters. The sequence of the cDNAs originally suggested six α-helical TMDs at the N-terminus with both N- and C-termini in the cytoplasm (11–13), but the sequence of the C-terminus has alternatively suggested two or four TMDs or a β sheet structure, despite a higher level of conservation than the N-terminus. To resolve these discrepancies, a variety of approaches have been used to determine the topology directly. The insertion of potential *N*-linked glycosylation sites at the C-terminus of the EAAT1 protein truncated at various sites suggests a topology with six α-helical TMDs followed by four β sheet structures (178). On the other hand, analysis of the related bacterial transporter GltT using phoA fusions where alkaline phosphatase activity depends on expression in the periplasm has suggested ten α-helical TMDs (179). However, the accessibility of cysteine-scanning mutants in EAAT2 (GLT-1) to sulfhydryl-modifying agents has most recently suggested eight TMDs with a structure reminiscent of a pore loop between TMD7 and 8 that could potentially account for the associated anion conductance (180). This experimental approach may provide the most reliable information about topology.

The analysis of chimeras has suggested domains important for EAAT function. Replacement of a 76 residue segment from the C-terminal TMDs of EAAT1 with a homologous region from EAAT2 conferred the sensitivity to kainate charcteristic of EAAT2 but did not change other transport properties (181). Additional chimeric analysis of the C-terminal TMDs in EAAT1 and 2 confirm the importance of this region (182). However, the analysis of site-directed mutants has provided more information about the function of the glutamate transporters.

To identify residues involved in the recognition of glutamate, one group has focused on positively charged residues present in the EAATs but not the zwitterionic amino acid transporter ASCT1 (183). Replacement of Arg-122 and Arg-280 in EAAT1 by the corresponding non-polar residues in ASCT1 alter only the apparent affinity for aspartate but not for glutamate. However, mutagenesis of Arg-479 and Tyr-405, which is also conserved in the EAATs but not ASCT1, eliminates transport function without affecting protein stability or trafficking to the plasma membrane. Another group has focused on conserved histidines and found a requirement for His-326 in the function of EAAT2 (184).

To identify residues that mediate the coupling to Na^+, K^+, or H^+, site-directed mutagenesis has also been used to alter the highly conserved, negatively charged residues predicted to reside within the TMDs of EAAT2. Mutagenesis of either Asp-398, Glu-404, or Asp-470 each eliminates the transport of glutamate (185). Surprisingly, these mutations do not affect aspartate transport and their effects are not simply due to the loss of negative charge. However, subsequent analysis has now shown that the flux observed in intact cells expressing the E404D mutant is not electrogenic (186). Indeed, it requires internal excitatory amino acids and thus reflects electroneutral exchange rather than net uptake. Sodium dependence and anion conductance appear unaffected by the mutation but net flux induced by the counter-transport of K^+ is abolished. Thus, Glu-404 contributes specifically and perhaps directly to K^+ translocation. Interestingly, replacement of each of residues 396–400 by cysteine eliminates transport function but replacement of Tyr-403 by cysteine makes EAAT2 sensitive to inhibition by positively charged methane thiosulfonate reagents (187). Further, Na^+ can protect against this inactivation, the non-transported inhibitor dihydrokainate accelerates the inactivation and transported substrates slow the inactivation. Supporting the importance of this region, replacement of Tyr-403 by phenylalanine also results in obligate electroneutral exchange, but unlike E404D (188), Y403F shows an increased affinity for Na^+. Unlike wild-type EAAT2 that requires specifically Na^+, the Y403F mutant tolerates replacement with lithium or cesium. The immediately adjacent Tyr-403 and Glu-404 thus influence the ionic selectivity for Na^+ and K^+, providing the first insight into the mechanism of coupling by these transporters. However, none of the positively or negatively charged residues implicated in glutamate transport function has thus far been shown to form an electrostatic pair that could participate in the charge relay system invoked to account for active transport catalysed by the lac permease of *E. coli* (189).

Glutamate transporters may also form multimers. The original purification of EAAT2 suggested function as a monomer (146). However, crosslinking in native brain membranes and after solubilization suggests dimerization and trimerization of EAAT1–3 (190). The three glutamate transporters do not multimerize with each other and radiation inactivation supports an oligomeric structure. However, it remains unclear whether multimerization is required for transport activity or has a distinct role in, for example, membrane trafficking and subcellular localization.

3.4 Physiological role

3.4.1 Synaptic transmission

Glutamate transporters presumably serve to terminate the action of excitatory amino acid transmitters in the synaptic cleft, but their precise physiological role remains uncertain. Indeed, initial studies had suggested no role for glutamate transporters in shaping the time course of synaptic currents (191–193). Rather, diffusion was presumed to have the predominant role, with transport acting outside the synaptic cleft to lower extracellular glutamate, prevent extrasynaptic spread of transmitter, and maintain a large concentration gradient. However, recent studies have shown that at climbing and parallel fibre synapses onto Purkinje cells in the cerebellum, inhibiting glutamate uptake prolongs the decay of the evoked postsynaptic currents (EPSCs) (194). Postsynaptic neuronal transporters as well as glial glutamate transporters have a role because D-aspartate in the Purkinje cell reduces uptake and slows the decay of the climbing fibre EPSC (195). In addition, inhibition of glutamate transport in micro-island cultures of hippocampal neurons reduces the amplitude of both NMDA and AMPA receptor EPSCs and increases their duration (196).

The slow turnover rate of glutamate transporters raises important questions about their ability to clear synaptic transmitter within a time scale relevant for the postsynaptic response. In contrast to postsynaptic currents which generally last < 5 msec, glutamate transporters have turnover numbers ~ 15/second and thus cycle every ~ 70 msec (171, 177). Glutamate transport thus cannot account for the time course of postsynaptic potentials. However, binding of glutamate to the transporters occurs very rapidly (171). Together with the high density of transporters, binding alone could therefore contribute to shaping the postsynaptic response. Indeed, inhibition of glutamate transport with THA increases the amplitude but not the duration of both evoked and spontaneous AMPA receptor currents in micro-island cultures, consistent with an extremely rapid action for the transporter in buffering synaptic glutamate (197). Estimates of glutamate concentration using the AMPA receptor antagonist kynurenate support a role for glutamate transporters in the submillisecond time scale (198), presumably through glutamate binding but not transport (199). Analysis of the anion conductance associated with glutamate transport further indicates that postsynaptic neuronal uptake alone, without including glial uptake, removes at least 22% of the glutamate released at a climbing fibre–Purkinje cell synapse, apparently due to the high density of transporters relative to glutamate receptors (200). At Schaffer collateral–commissural synapses in the hippocampus, however, the analysis of anion conductances shows no significant activation of neuronal glutamate transporters (201). Rather, only astrocytes at this synapse show the glutamate-activated anion conductance characteristic of glutamate transporters.

The anion conductance associated with glutamate transport has also provided a novel and powerful way to estimate the synaptic concentration of glutamate (202–204). Recording from astrocytes in stratum radiatum of CA1 in the hippocampus as well as dissociated neuronal culture reveals evoked responses with the characteristics of glutamate transporter currents: inhibition by the transported THA as well as non-transported DHK, and by the replacement of internal K^+ with Na^+ (203). As anticipated from the analysis of glutamate transporters in oocytes, replacement of external Cl^- by SCN^- also increases the magnitude of the astrocyte currents, facilitating their use as a tool to monitor synaptic glutamate. The glial currents appear with the same time delay after stimulation as postsynaptic currents, but peak and decay more slowly, suggesting that the transporters occur outside the synaptic cleft, in a position where they can regulate diffusion to neighbouring synapses. A similar delay in the evoked currents occurs in recordings from glial cells at climbing fibre synapses in the cerebellum (205). Further, the astrocyte currents sense presynaptic changes in neurotransmission such as paired-pulse facilitation and post-tetanic potentiation (204). However, this method fails to detect an increase in synaptic glutamate in long-term potentiation, supporting a primarily postsynaptic locus for this model of neural plasticity (204, 206).

In addition to its use as a tool to monitor synaptic glutamate, the anion conductance associated with glutamate transporters may have an important physiological role. A chloride conductance with the pharmacological properties characteristic of glutamate transporters was previously described in the retina of the tiger salamander (157, 174, 175), as noted above. More recent work has characterized the biophysical properties of this conductance. Both cone and rod photoreceptors have a conductance ~ 0.7 pS with an open time of 2–3 msec (207, 208). Glutamate and Na^+ affect the probability of opening but not the conductance or mean open time. Since the reversal potential for chloride appears more negative than the resting membrane potential, activation of the glutamate transporters on photoreceptors has the potential to serve an auto-inhibitory role. Indeed, glutamate can suppress in these cells the increase in intracellular Ca^{2+} induced by depolarization (209).

A chloride conductance with the properties of a glutamate transporter also occurs in the bipolar cells of perch retina (210). It requires activation by extracellular Na^+ as well as glutamate and shows partial inhibition by THA and PDC. Localization to the dendrites of bipolar cells suggests a crucial role in the response to glutamatergic input from photoreceptors. In this case, the conductance would hyperpolarize the bipolar cell and so contribute to the light-dark response.

Unlike the Na^+-/Cl^--dependent neurotransmitter transporters, glutamate transporters do not appear to be expressed at nerve terminals that use glutamate as a transmitter. Rather, they appear only in glia or if expressed in neurons, occur postsynaptically. Thus, the glutamate transporters cannot contribute directly to the recycling of neurotransmitter at the nerve terminal in the same way that DAT contributes to the recycling of dopamine (50). However, glial uptake does facilitate the conversion to glutamine by astrocytes. Release from glia and subsequent reuptake of glutamine by excitatory neurons then enables the conversion to glutamate by glutaminase (211–213). Thus, glial transport may contribute to the recycling of glutamate needed to replenish the large vesicular pools required to maintain the high rates of transmitter release observed at central synapses. None the less, additional mechanisms may help to maintain glutamate stores in the face of massive release. Interestingly, an inorganic phosphate transporter (EAT-4) appears important specifically for the release of glutamate in *C. elegans* (214). A vertebrate homologue known as the brain-specific Na^+-dependent phosphate transporter (BNPI) has also been isolated and its expression appears restricted to a subset of glutamatergic neurons (215, 216). Since phosphate activates the glutaminase that produces glutamate for release as a transmitter, accounting for the designation of this enzyme as the phosphate-activated glutaminase (PAG), the transporter may elevate cytosolic phosphate and hence increase glutamate production. Supporting this possibility, the nematode *eat-4* mutant shows protection against excitotoxicity (217). Further, the BNPI protein localizes to excitatory synapses and specifically to synaptic vesicles, suggesting regulation of its function at the plasma membrane by neural activity (218). However, the physiological role of this mechanism in the maintenance of vesicular glutamate requires further investigation and additional mechanisms may also contribute.

3.4.2 Excitotoxicity

The clearance of extracellular glutamate by transporters has the potential to influence excitotoxicity as well as synaptic transmission. In the case of motor neuron disease or amyotrophic lateral sclerosis (ALS), the cerebrospinal fluid from patients shows an increase in glutamate and aspartate (219, 220). Spinal cord and affected brain regions (e.g. motor cortex) show a reduction in Na^+-dependent glutamate transport (221). Importantly, unaffected brain regions (e.g. hippocampus) show no significant reduction and the transport of GABA appears normal. To test the hypothesis that a defect in glutamate transport normally protects against this form of excitotoxicity to motor neurons, glutamate transport was inhibited in organotypic spinal cord explants (222). Inhibition of glutamate transport increased the levels of glutamate in the medium and produced excitotoxicity in the motor neurons that could be prevented with glutamate receptor antagonists, supporting the hypothesis. In addition, GLT1 (EAAT2) appears selectively reduced in the tissue from ALS patients (223). Further, antisense inhibition of glutamate transporters GLAST and GLT-1 (EAAT1 and 2) also produces toxicity *in vivo* and results in progressive paralysis (224).

The ability of glutamate transports to protect against excitotoxicity raises questions about the nature of the defect in ALS. Mutations in the EAAT2 gene have not been found in families with ALS (225). Transcription of the EAAT2 also appears unimpaired (226). However, defects in the splicing of RNA transcripts for EAAT2 have recently been described in a substantial subset of affected individuals (227). These defects include intron retention and exon skipping, and dramatically affect the protein encoded by these transcripts. The precise nature of the defect in RNA splicing remains unknown but appears remarkably specific for EAAT2. Expression of the mutant proteins also demonstrates either marked instability or a dominant negative effect on the expression and activity of co-expressed wild-type EAAT2. Knock-out mice and inhibition with antisense sequences have helped to explore further the physiological and pathological roles of glutamate transporters. Targeted disruption of the EAAT2 gene in mice reduces synaptosomal glutamate transport to ~ 5% of wild-type, indicating that EAAT2 is indeed the predominant glutamate transporter in the brain (228). EAAT2-deficient mice die within three to four months after birth, apparently as a result of seizures. The time course of the EPSC in CA1 of the hippocampus does not appear changed, but use of the rapidly dissociating NMDA antagonist L-2-amino-5-phosphonopentanoic acid (L-AP5) suggests that the peak concentration of synaptically released glutamate is elevated in the knock-out mice, consistent with a role for EAAT2 in the synaptic clearance of transmitter (228). In addition, the knock-out mice show selective neuronal degeneration in the hippocampus that varies considerably from animal to animal, possibly reflecting the extent of seizure-related damage. The animals are also more susceptible to oedema after cold-induced injury, supporting a role for EAAT2 in preventing excitotoxicity related to trauma.

Importantly, EAAT2-deficient mice do not apparently show degeneration of motor neurons (228). They may not survive long enough to develop degeneration of these cells, but antisense inhibition produced pathology after only several weeks (224). Perhaps a toxic gain of function similar to that suggested by the defects in RNA splicing, rather than the simple loss of function produced in knock-out mice, contributes to the neural degeneration. Supporting this possibility, glutamate transporters have the potential to release massive amounts of glutamate from intracellular stores under the appropriate conditions (41, 229). The stoichiometry of transport suggests that the transporters can generate concentration gradients of 10^6:1 and may not reverse under even pathological conditions (169). However, an increase in extracellular K^+ with hypoxia or ischaemia may enable the counter-transported ion to drive reverse transport (230) and thus dramatically increase extracellular glutamate. Since H^+ are co-transported with glutamate, acidosis might tend to oppose the non-vesicular release of glutamate and low pH does inhibit both forward and reverse flux mediated by the transporters (231), potentially mitigating the extent of flux reversal under pathological conditions.

The gene encoding neuronal EAAT3 has also been disrupted by homologous recombination and inhibited using antisense. Antisense inhibition did not elevate extracellular glutamate, but did produce mild neural toxicity and epilepsy (224). EAAT3 knock-out mice develop dicarboxylic amino aciduria and show reduced spontaneous locomotor activity but do not exhibit any neural degeneration (232). The nature of the defect in synaptic transmission, if any, remains to be elucidated.

3.4.3 Regulation

Neurons appear to regulate the expression of glutamate transporters by glia. Alone in primary culture, astrocytes express predominantly GLAST (EAAT1) (233). However, co-culture with neurons or treatment with dibutyryl-cAMP (db-cAMP) induces the expression of GLT-1 (EAAT2). The increased expression appears to occur at the level of mRNA and addition of db-cAMP does not substantially increase the expression of GLT-1 beyond that observed with co-cultured neurons, suggesting a shared step in the signalling pathway (234). On the other hand, inhibition of cAMP-dependent protein kinase (PKA) does not block the effect of co-culture with neurons, indicating distinct mechanisms. The signal from neurons appears diffusible but remains otherwise uncharacterized (234). The glutamate transporters also undergo developmental changes in expression, with GLAST increasing postnatally from low prenatal levels and appearing in the cerebellum before other brain regions, and GLT-1 present at substantial levels in the embryo and similarly increasing after birth (235). EAAT4 remains restricted to the cerebellum but increases in expression after birth whereas EAAT3 occurs at higher levels before birth than after.

In terms of post-translational mechanisms, membrane trafficking appears to influence the expression of glutamate transporters as well as the Na^+-/Cl^--dependent neurotransmitter transporters. Activation of protein kinase C (PKC) by phorbol esters stimulates the phosphorylation of GLT-1 (EAAT2) in C6 glioma cells and increases plasma membrane glutamate uptake (236). Mutagenesis of a consensus site for phosphorylation by PKC (Ser-113 between TMD2 and 3) (180) eliminates the increase, suggesting an increase in intrinsic transport activity. However, more recent analysis of endogenous EAAT3 in C6 glioma cells shows that phorbol esters cause an increase in both cell surface activity and expression, and inhibition of protein kinase C (PKC) prevents this increase (237). Wortmannin, an inhibitor of PI 3-kinase, reduces baseline activity and cell surface expression but surprisingly, blocks only the increase the cell surface expression by phorbol esters, not the increase in activity. Thus, stimulation of PKC appears to activate the glutamate transporters through two distinct mechanisms, one involving translocation to the plasma membrane and the other an increase in intrinsic activity, possibly due to the phosphorylation previously described for EAAT2 (236).

Among known second messengers, arachidonic acid has been extensively implicated in the regulation of glutamate transport (238–240). Specifically, arachidonic acid has been reported to inhibit glutamate transport in a variety of systems, including purified transporter reconstituted into artificial membranes (241). However, a postsynaptic anion conductance in cerebellar Purkinje cells (presumably EAAT3 or 4) undergoes facilitation for more than 20 minutes after transient depolarization (and the associated Ca^{2+} influx) and arachidonic appears to mediate this form of regulation (242). Calcium influx presumably activates the phospholipase A2 present in Purkinje cells to release arachidonate.

Analysis of the specific glutamate transporter subtypes also indicates divergent effects of arachidonic acid. Arachidonate reduces the rate of transport by EAAT1 but increases the apparent affinity of EAAT2 for substrate (243). Further, arachidonic acid has recently been found to alter the currents associated with cerebellar EAAT4 (244, 245). Importantly, arachidonic acid does not alter glutamate flux or the anion conductance associated with EAAT4. Rather, it activates a previously unrecognized H^+ conductance that appears intrinsic to the transporter. The reversal potential of the activated current varies with the pH gradient and drugs that block ion exchangers intrinsic to the oocyte do not affect the current. Thus, arachidonic acid has dramatic and distinct effects on different glutamate transporters. Although the physiological role of this regulation remains unknown, it may contribute to neural plasticity.

Zinc can also influence different glutamate transporter subtypes in distinct ways. Concentrations well below those estimated to occur synaptically during transmitter release inhibit both forward and reverse flux mediated by a transporter in retinal Muller cells but increase the associated anion conductance and reduce the anion conductance associated with glutamate transport in cone photoreceptors (EAAT5) (246).

The ability of glutamate transporters to protect against excitotoxicity has raised the possibility that they may themselves undergo damage that accelerates the degenerative process. Free radicals generated by hydrogen peroxide impair the function of glutamate transporters in cortical astrocytes and reducing agents can reverse this inhibition, suggesting the formation of disulfides (247). Further, the transporters may possess redox-sensing mechanisms that link their function to the redox state of the cell (248). Peroxynitrite, a biological oxidant, inhibits glutamate transport and specifically reduces the V_{max} of EAAT1–3 (249). It will be particularly interesting to identify those modifications that affect uptake but not flux reversal and thus might have a dominant negative phenotype.

4. Vesicular neurotransmitter transporters

4.1 Identification and characterization of transport activity

Previous work has identified four distinct activities that transport classical neurotransmitters into secretory vesicles. Monoamine transport into chromaffin granules has received the most attention. This activity depends on a proton electrochemical gradient ($\Delta\tilde{\mu}_{H+}$) (7) generated by the vacuolar H^+-ATPase, a multisubunit protein with separable integral membrane and cytosolic domains that most strongly resembles the F_1F_0 H^+-ATPase of mitochondria (250). $\Delta\tilde{\mu}_{H+}$ consists of both a chemical component (ΔpH) and an electrical component ($\Delta\Psi$). Dense core vesicles and synaptic vesicles generally exhibit a ΔpH ~ 1.5–2 units and a $\Delta\Psi$ ~ 50 mV (7), but variations in the vesicular chloride conductance have the potential to alter these values. Increased chloride conductance would enable cytoplasmic chloride to neutralize the positive charge associated with the H^+ pump and hence generate a larger ΔpH gradient at the expense of $\Delta\Psi$. Conversely, a reduced vesicular anion conductance would enable the pump to generate a greater $\Delta\Psi$ but relatively little ΔpH.

The monoamine transport activity in chromaffin granules uses the energy stored in the form of $\Delta\tilde{\mu}_{H+}$ by exchanging two lumenal protons for one cytoplasmic amine (251) (Fig. 4). The transporter appears to recognize the charged form of the substrate since it transports permanently charged compounds such as the neurotoxin *N*-methyl-4-phenylpyridinium (MPP^+) (252, 253). Vesicular transport thus involves the net efflux of two protons but only +1 charge and hence depends to a greater extent on ΔpH than on $\Delta\Psi$. None the less, the observations indicate that vesicular monoamine transport is electrogenic. The stoichiometry of monoamine transport further predicts that for a pH gradient across the vesicle membrane of 1.5–2 units and an internal

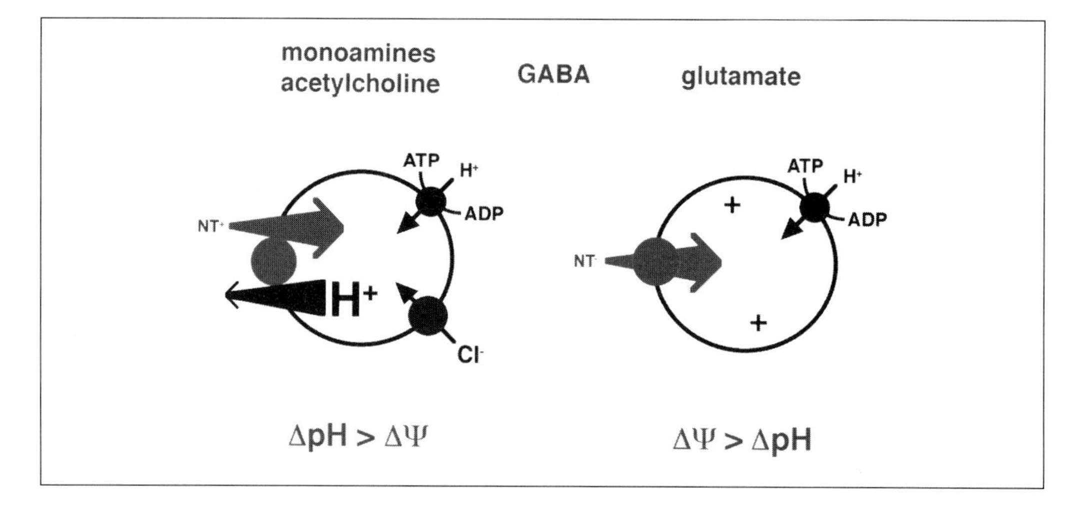

Fig. 4 Dependence of the different vesicular neurotransmitter transport activities on the chemical (ΔpH) and electrical ($\Delta\Psi$) components of the proton electrochemical gradient $\Delta\bar{\mu}_{H+}$. The entry of chloride through a specific vesicular conductance serves to neutralize the H^+ pumped in by the vacuolar H^+-ATPase and so contributes to the development of ΔpH at the expense of $\Delta\Psi$. Vesicular monoamine and acetylcholine transport involve the exchange of two lumenal H^+ for one cytoplasmic molecule of protonated transmitter, resulting in the net efflux of only +1 charge, and hence depend predominantly on ΔpH. In contrast, vesicular glutamate transport apparently involves the exchange of one to two lumenal H^+ for one anionic transmitter, resulting in the net efflux of +2–3 charges, and hence depends predominantly on $\Delta\Psi$. Vesicular GABA transport depends more equally on both ΔpH and $\Delta\Psi$.

positive potential of ~ 50 mV, vesicles should accumulate transmitter at concentrations 10^4 times greater than those in the cytoplasm (254). This prediction is consistent with the analysis of chromaffin granules, which can contain molar concentrations of monoamine, some of which forms insoluble complexes with other granule components (7).

Synaptic vesicles from the electric organ of *Torpedo californica* have been used to study the vesicular transport of acetylcholine (ACh). Similar to vesicular monoamine transport, ACh transport into vesicles depends principally on ΔpH, but the K^+ ionophore valinomycin that dissipates $\Delta\Psi$ reduces transport activity, indicating an electrogenic component (255). Since ACh is permanently positively charged, the electrogenic nature of transport dictates the exchange of two lumenal protons for one cytoplasmic molecule of transmitter (256). However, unlike vesicular monoamine transport, the predicted stoichiometry for ACh does not generate more than two orders of magnitude higher ACh levels inside synaptic vesicles than outside, due to the leakage of transmitter or protons nonspecifically through the membrane or perhaps even through the transporter itself (257).

The vesicular transport of amino acid transmitters also depends on $\Delta\bar{\mu}_{H+}$, but relies to a greater extent than monoamine and ACh transport on the electrical component of the gradient, $\Delta\Psi$. Synaptic vesicles prepared from rat brain accumulate the excitatory amino acid transmitter glutamate and show maximal transport activity under conditions that promote $\Delta\Psi$ at the expense of ΔpH (258, 259) (Fig. 4). Conversely, increasing concentrations of chloride or permeant anions that dissipate $\Delta\Psi$ by neutralizing the charge associated with proton translocation into the vesicle, inhibit glutamate transport activity. Further, glutamate transport acidifies synaptic vesicles, presumably by neutralizing the charge associated with proton translocation and thus enabling the development of larger pH gradients (259). However, low concentrations of chloride (2–5 mM) appear necessary for glutamate transport activity (260). Chloride may simply act at an allosteric site to promote transport and the inhibition of both glutamate uptake and efflux by the anion blocker 4,4′-diisothiocyanatostilbene-2,2′-disulfonic acid (DIDS) supports this possibility (261). On the other hand, chloride will also support the development of a larger pH gradient than possible in the absence of anions, and the Na^+/H^+ ionophore nigericin that dissipates ΔpH also reduces glutamate transport (262), suggesting the exchange of at least one lumenal proton for one cytoplasmic molecule of glutamate. Indeed, this stoichiometry dictates the outward movement of +2 charges along with one proton for each transport cycle, and is consistent with the observed preferential dependence of glutamate transport on $\Delta\Psi$ relative to ΔpH. A mechanism involving proton exchange is also consistent with larger vesicular concentration gradients than those predicted by a strictly membrane potential-driven process under the usual conditions, although differences in chloride conductance have the potential to vary both ΔpH and $\Delta\Psi$ and so provide driving forces particularly suited for either monoamine/ACh or glutamate transport.

Vesicular transport of the inhibitory transmitters GABA and glycine differs from glutamate transport in a greater dependence on ΔpH. Varying the concentrations of chloride has little effect on GABA transport into brain synaptic vesicles and dissipation of $\Delta\Psi$ has less effect on GABA than glutamate transport (263). None the less, vesicular GABA transport still depends on $\Delta\Psi$ to a greater extent than the VMATs. This more equal dependence on ΔpH and $\Delta\Psi$ predicts a stoichiometry with equal outward translocation of protons and charge. Exchange of one lumenal proton for the presumably neutral, zwitterionic transmitter dictates the outward translocation of +1 charge, consistent with these observations.

4.2 Vesicular monoamine transport

The availability of chromaffin granules expressing robust transport activity and of specific inhibitors has greatly facilitated the analysis of vesicular monoamine transport. In contrast to the vesicular transport activities for ACh, GABA and glutamate which have K_ms in the millimolar range, monoamine transport has a K_m in the low micromolar range (7). However, the activity in chromaffin granules recognizes all monoamine transmitters as substrates, with the highest apparent affinity for serotonin followed by dopamine, norepinephrine, and epinephrine. Although catecholamine and indolamine substrates require an NH_2- group and OH- groups on the aromatic ring for recognition, the neurotoxin MPP^+ and histamine lack hydroxyl groups and still act as substrates (252, 264), suggesting multiple modes of translocation.

In terms of pharmacology, the potent antihypertensive drug reserpine inhibits vesicular monoamine transport, supporting the biological importance of this transport activity. However, reserpine also induces a syndrome resembling depression (4), limiting its utility as an antihypertensive but providing some of the first evidence implicating monoamines in general and vesicular transport in particular in affective disorders. Monoamine substrates compete with reserpine for the inhibition of transport and also inhibit binding of [^{3}H]reserpine to the transporter with potencies very similar to their apparent affinities as substrates (265), suggesting that reserpine acts at the site of substrate recognition. Further, [^{3}H]reserpine binds to the transporter very slowly in the absence of $\Delta\bar{\mu}_{H+}$, and imposition of $\Delta\bar{\mu}_{H+}$ greatly accelerates binding (266–268), suggesting that the outward translocation of at least one of the two protons reorients the substrate recognition site towards the cytoplasmic face of the membrane where it can bind to another molecule of transmitter (Fig. 5). The inhibitor tetrabenazine appears to act in a different way from reserpine: only very high concentrations of monoamines inhibit the binding of [^{3}H]tetrabenazine; and imposition of $\Delta\bar{\mu}_{H+}$ does not accelerate [^{3}H]tetrabenazine binding (268, 269). Thus, tetrabenazine does not appear to interact at the site of substrate recognition, but it does inhibit the binding of [^{3}H]reserpine, suggesting that the two sites interact. Ketanserin, generally considered a serotonin receptor antagonist, also appears to

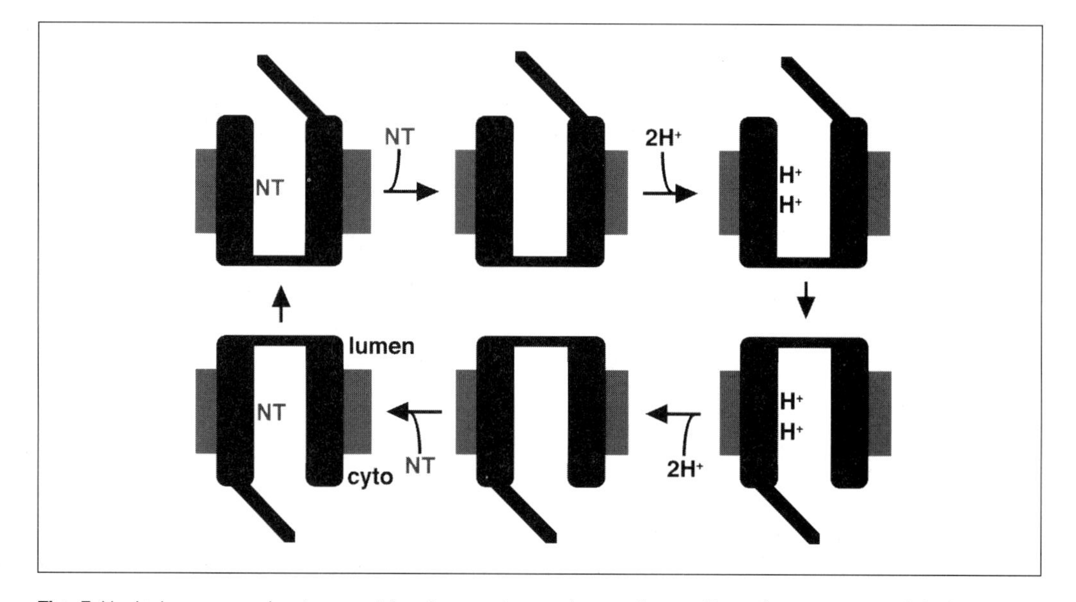

Fig. 5 Vesicular monoamine transport involves proton exchange. In an alternating access model of transport similar to that shown for serotonin transport in Fig. 3, two H^{+} rather than one K^{+} serve to reorient the substrate recognition site back toward the cytoplasmic face of the membrane. Reserpine appears to compete with substrates for binding to the cytoplasmically disposed carrier (middle, bottom). In addition, imposition of accelerates reserpine binding. Reserpine binding together with flux assays have helped to identify residues in the VMATs specifically involved in substrate recognition and proton translocation. Outward proton translocation appears to serve as a ratchet that prevents reversal of the transport cycle and hence enables active transport against large concentration gradients. Amphetamines and other weak bases promote flux reversal by dissipating $\Delta\bar{\mu}_{H+}$.

inhibit vesicular monoamine transport in the same way as tetrabenazine. In addition, binding of the transporter to reserpine, tetrabenazine, and ketanserin has aided initial attempts to isolate the transport protein.

Reconstitution of detergent solubilized vesicular monoamine transporter in artificial membrane vesicles yields activity with the same properties as native biological membranes (270). However, it has been difficult to purify the protein by this means alone, and different groups have relied on the binding of the protein to radiolabelled inhibitors. [^{3}H]Reserpine binds almost irreversibly to vesicular monoamine transporters and the inclusion of trace amounts in detergent extracts has enabled the development of a purification scheme and the isolation of functional transport protein (270). Radiolabelled tetrabenazine and ketanserin have similarly been used to identify and purify the transporter (271, 272). None the less, expression cloning led to the first identification of sequences encoding vesicular monoamine transport.

The potent toxin 1-methyl-1,2,3,6-tetrahydropyrine (MPTP) produces a clinical Parkinsonian syndrome and results in the death of dopamine neurons in the substantia nigra (56), similar to the pathology of idiopathic Parkinson's disease (PD). As a lipophilic compound, MPTP readily penetrates the blood–brain barrier, and then undergoes conversion by monoamine oxidase-B (MAO-B) to the active metabolite MPP^+ (273, 274). Plasma membrane monoamine transporters then recognize MPP^+ as a substrate and, since they occur only on monoamine neurons, result in the selective accumulation of MPP^+ within these cells (275). MPP^+ then enters mitochondria and kills the cell by inhibiting respiration (276, 277). Surprisingly, cells such as chromaffin cells in the adrenal medulla accumulate large amounts of MPP^+ but show little toxicity, whereas other cells that lack plasma membrane monoamine transport may show more sensitivity to the toxin (14). To understand this differential sensitivity, sequences were transferred from MPP^+-resistant PC12 cells to a more sensitive fibroblast line and selected in the toxin. One cell clone showed considerable resistance to MPP^+ and reserpine reversed this resistance, suggesting that expression of a vesicular monoamine transporter protected against MPP^+. Indeed, the sequences that confer resistance to MPP^+ also confer vesicular amine transport by heterologous expression (15).

The sequence of the cDNA responsible for resistance to MPP^+ as well as for the uptake of [^{3}H]serotonin in an autoradiographic screen (16) shows no similarity to plasma membrane neurotransmitter transporters (15). However, the sequence does predict 12 TMDs and remote but definite similarity to a class of bacterial antibiotic resistance proteins. Interestingly, these bacterial proteins confer resistance to antibiotics by transporting them out of the cell, a direction topologically equivalent to transport into intracellular vesicles (278–280). In addition, the bacterial proteins act as proton exchangers (281) and one even shows inhibition by reserpine. Together with the isolation by selection in MPP^+, the sequence similarity suggests that vesicular monoamine transporters (VMATs) have evolved from ancient detoxification systems. Their high apparent affinity relative to other vesicular transporters further supports a role in neural protection as well as signalling. Indeed, the normal neurotransmitter dopamine oxidizes readily and exhibits considerable toxicity, suggesting that the VMATs may protect against their normal neurotransmitter substrates by sequestering them inside secretory vesicles. The evidence of oxidative stress in idiopathic PD also raises the possibility that abnormal VMAT function may contribute to the human disorder.

The mechanism of amphetamine action further implicates cytosolic monoamine levels and VMAT function in neural degeneration. As described in Section 2.2, amphetamines promote flux reversal by the plasma membrane amine transporters. However, the cytosolic concentrations of transmitter are presumably very low, and amphetamines also appear to act by promoting efflux from vesicular storage pools into the cytoplasm. In particular, amphetamines may act by exchange–diffusion at the vesicle as well as at the plasma membrane, but considerable evidence indicates that they can also act as weak bases to dissipate the driving force for VMAT activity, $\Delta\bar{\mu}_{H^+}$ (282–285). Indeed, previous work has shown that dissipation of $\Delta\bar{\mu}_{H^+}$ promotes efflux from chromaffin granules pre-loaded with [^{3}H]serotonin, although reserpine and tetrabenazine fail to block this efflux, suggesting non-specific efflux across the membrane rather than a specific role for the VMATs (286). Consistent with the importance of cytosolic monoamine levels for degeneration, amphetamines produce substantial neural toxicity, with methamphetamine injuring both dopaminergic and serotonergic axon terminals, and 3,4-methylenedioxymethamphetamine (MDMA, ecstasy) more selective for serotonergic projections (287, 288). In addition, free radical injury produced by methamphetamine in primary cultures of dopamine neurons appears focal (289), suggesting that efflux from vesicular stores causes local injury to surrounding structures. None the less, the evidence that dopamine toxicity can produce neural degeneration and contribute to idiopathic PD, although compelling, remains almost entirely circumstantial. Further, differences in the expression of plasma membrane and vesicular monoamine transporters fail to account for the selective vulnerability of dopamine neurons in the substantia nigra to MPTP toxicity or PD (254).

Molecular cloning has identified two VMAT isoforms (15). VMAT1 occurs in chromaffin cells of the adrenal medulla and other non-neuronal cell populations such as small intensely fluorescent (SIF) cells in sympathetic ganglia and enterochromaffin cells in the gut (290, 291). In contrast, VMAT2 occurs almost exclusively in neurons, including central dopaminergic, noradrenergic, and serotonergic populations, but also in histamine-containing cells in the gut and presumably mast cells (290–292).

In terms of functional characteristics, transport assays using either membrane vesicles prepared from transfected cells or transfected cells semi-permeabilized with digitonin show that both VMAT1 and VMAT2 recognize dopamine, norepinephrine, epinephrine, and serotonin as substrates (293, 294), consistent with the expression of VMAT2 in all central monoamine neurons. Both VMATs protect against MPP^+ toxicity, but VMAT2 has a generally two- to fourfold higher apparent affinity for these substrates relative to VMAT1. Interestingly, both VMATs exhibit a three- to fivefold higher turnover rate for dopamine than serotonin (295). In addition, VMAT2 but not VMAT1 recognizes histamine as a high affinity substrate (264, 296). Further, although reserpine inhibits both VMAT1 and 2, tetrabenazine selectively inhibits VMAT2 (15, 293). Using binding to these radiolabelled drugs to determine the number of transporters, both VMAT1 and 2 have a turnover number at saturating substrate concentrations of ~ 400/minute at 29 °C (293). Although similar to the turnover numbers calculated for other neurotransmitter transporters, this slow rate predicts that the filling of synaptic vesicles would, even with the higher rates predicted for 37 °C, take minutes if one assumes one to two transporters per vesicle and 5–20 000 molecules of amine released per vesicle.

4.3 Vesicular ACh transport

Together with identification of the VMATs, genetic studies in *C. elegans* led to identification of the protein responsible for ACh transport into synaptic vesicles. *Unc-17* belongs to a class of *C. elegans* mutants that show resistance to the organophosphorus toxin aldicarb (17). Aldicarb inhibits acetylcholinesterase, the enzyme that degrades ACh, and produces paralysis by massively increasing synaptic levels of ACh, hence inactivating the postsynaptic ACh receptors. Worms resistant to aldicarb presumably have a defect in the release of ACh and the sequence of *unc-17* showed strong similarity to the previously identified VMATs, suggesting a specific role in the packaging of ACh into synaptic vesicles (17). Indeed, UNC-17 localizes to synaptic vesicles in the worm and vertebrate homologues isolated by PCR and low stringency hybridization reside specifically in the synaptic vesicles of cholinergic neurons (297–299).

Despite initial difficulty demonstrating that these sequences mediate the active transport of ACh into vesicles, recent studies have conclusively shown transport activity with the expected characteristics (300). Further, this activity does not require expression in neural cells such as PC12 cells, but has also been detected by heterologous expression in a fibroblast line (301). In contrast to the VMATs which show a high apparent affinity for monoamines, the VAChT cDNA confers a K_m for ACh ~ 0.5 mM. In addition, the drug vesamicol potently inhibits VAChT function, consistent with results using *Torpedo* synaptic vesicles. However, recent data differ from previous studies in *Torpedo* vesicles that had suggested vesamicol might bind to a protein distinct from the transporter (302). Rather, vesamicol binds directly to VAChT, but still differs from reserpine with regard to the large concentrations of substrate (~ 10 mM ACh) required to inhibit the binding of [^{3}H]vesamicol (A. Roghani, R. H. E., in preparation; L. B. Hersh *et al.*, in press).

The VAChT gene has a particularly interesting chromosomal organization. It resides at the same locus as the biosynthetic enzyme for ACh, choline acetyltransferase (ChAT) (298, 303). Indeed, VAChT and ChAT share promoters in *C. elegans* and *D. melanogaster* as well as mammals, with the entire VAChT protein coding sequence located within a single intron of the ChAT gene (304, 305). This arrangement presumably accounts for the observed co-ordinate regulation of both genes (306). The contiguous protein coding sequence of VAChT also differs markedly from the genes for other neurotransmitter transporters, in which distinct exons encode each TMD.

4.4 Vesicular amino acid transport

Genetic studies in *C. elegans* have led to identification of the first vesicular transporters for amino acid neurotransmitters. Laser ablation of GABAergic neurons revealed the behaviour that results from a defect in GABAergic neurotransmission, and a genetic screen based on this phenotype resulted in the isolation of multiple mutants (307, 308). The ability of exogenous GABA to rescue the phenotype then indicated whether the defects occurred pre- or postsynaptically. Immunostaining of the presynaptic mutants for GABA further indicated two distinct subtypes, including those without GABA (e.g. due to a defect in the biosynthetic enzyme glutamic acid decarboxylase, or GAD) and those with normal or higher than normal levels of GABA. *Unc-46* and *-47* showed increased immunoreactivity for GABA, suggesting a defect in release. The sequence subsequently found to complement the defect in *unc-47* mutants predicts multiple transmembrane domains, raising the specific possibility of a defect in the transport of GABA into synaptic vesicles (18). Supporting a role in transmitter packaging, UNC-47 localizes to synaptic vesicles.

Characterization of a vertebrate homologue for *unc-47* confirms a role in vesicular GABA transport. First, *in situ* hybridization indicates expression of the sequences restricted to GABAergic populations in the rat brain, in a pattern essentially identical to that of the biosynthetic enzyme GAD (18). Secondly, an antibody to the rat protein labels synapses in primary hippocampal culture. Thirdly, heterologous expression of the cDNA in PC12 cells confers GABA transport activity on synaptic-like microvesicles isolated from transfected cells. Further, the activity shows a relatively low apparent affinity ($K_m \sim 5$ mM) and a greater dependence on $\Delta\Psi$ than the VMAT isoform (VMAT1) expressed on the same population of secretory vesicles, consistent with the previous observations using native synaptic vesicles. γ-Vinyl-GABA inhibits the expressed transport with a potency similar to GABA (18), the plasma membrane GABA transport blocker nipecotic acid inhibits it poorly, and glutamate does not inhibit it at all, consistent with the existence of a distinct vesicular glutamate transporter. However, glycine inhibits vesicular GABA transport with moderate potency (18) and VGAT immunoreactivity occurs in glycinergic as well as GABAergic neurons (309, 310), supporting previous evidence that the two transmitters compete for uptake into synaptic vesicles (311). Indeed, electrophysiological studies have detected the release of both inhibitory transmitters from the same synaptic vesicles (312) and VGAT recognizes as substrates both glycine and GABA (R. J. R., E. G., R. H. E., unpublished observations). Since glycine occurs in all cells, the decision to release GABA versus glycine in an individual neuron presumably reflects expression of the biosynthetic enzyme GAD, of the plasma membrane transporters for GABA and glycine that increase cytosolic concentrations, and the relative affinity of VGAT for the two transmitters (313).

The sequence of VGAT and *unc-47* predicts a protein with multiple TMDs, possibly ten, and with the N- and C-termini in the cytoplasm, similar to the other neurotransmitter transporters. However, VGAT and *unc-47* show no sequence similarity to the VMATs, VAChT, or the plasma membrane transporters. On the other hand, they do belong to a large family of proteins, with multiple members in nematodes, yeast, plants, and vertebrates. Relatively little is known about most members of this family, but several sequences in *Arabidopsis thaliana* and *Nicotiana sylvestris* encode amino acid transport driven by a proton electrochemical gradient, although they appear to act as co-transporters at the plasma membrane rather than as proton exchangers in secretory vesicles (314). The family thus appears to confer primarily amino acid transport driven by $\Delta\bar{\mu}_{H+}$, but members vary in their site and mode of function. This family presumably also includes the synaptic vesicle transporter for glutamate.

4.5 Additional vesicular transport activities

Considerable evidence including the identification of both ionotropic and G protein-coupled receptors for nucleotides indicates that ATP can act as a transmitter in multiple neural systems (315). Consistent with this role, ATP occurs at substantial levels in both chromaffin granules and cholinergic synaptic vesicles from the *Torpedo* electric organ (316, 317). Inside these secretory vesicles, ATP appears to form a complex with the cationic neurotransmitters (7). The formation of an insoluble complex may then facilitate accumulation of extremely high transmitter concentrations (318). Atractyloside, an inhibitor of the ATP/ADP exchanger in mitochondria, was originally thought to inhibit nucleotide uptake into secretory vesicles (319). However, the formation of an insoluble complex inside chromaffin granules complicates the analysis of uptake. In one report, chromaffin granule 'ghosts' produced by repeated osmotic lysis showed no concentrative transport of ATP, entering simply by non-saturable diffusion (320). None the less, it seems very likely that the accumulation of this highly charged molecule inside secretory vesicles depends on a specific carrier. More recent studies with chromaffin granules and ghosts indicate specific uptake dependent on $\Delta\Psi$ that differs from the mitochondrial nucleotide carrier in the recognition of GTP and UTP as well as ATP, and by weak inhibition with atractyloside (318, 321). In addition, the anion transport inhibitor DIDS inhibits vesicular nucleotide transport.

Secretory vesicles also contain large amounts of ascorbate. Ascorbate acts as a cofactor for several enzymes located in the vesicle lumen, including dopamine-β-hydroxylase which converts dopamine to norepinephrine, and peptide α-amidase, which modifies neural peptides. Since these enzymatic reactions result in the oxidation of ascorbate, secretory vesicle express a cytochrome (cytochrome *b561*) which transfers electrons specifically from reduced ascorbate in the cytosol to oxidized ascorbate in vesicles (322, 323). However, the mechanism by which ascorbate itself enters secretory vesicles remains controversial, with some studies providing no evidence for energy-dependent uptake (324). More recent work has further suggested storage of ascorbate in multiple secretory compartments and an extremely slow rate of uptake into chromaffin granules (325, 326). Ascorbate may indeed differ from ATP and classical transmitters in its action as a renewable cofactor, but the large amounts of ascorbate released during synaptic transmission do implicate a mechanism to replenish these vesicular stores.

In addition to the activities described above, molecular cloning has identified a family of vesicular transporters unrelated to the known neurotransmitter transporters. SV2 was originally identified as an antigen on synaptic vesicles from *Torpedo* that also occurred on all synaptic vesicles and dense core vesicles in many other vertebrate species including mammals (327). Indeed, the expression of SV2 specifically on neurosecretory vesicles has made it an excellent anatomic and biochemical marker for nerve terminals. Molecular cloning has further shown that SV2 contains 12 TMDs with a large N-terminal cytoplasmic domain recognized by the monoclonal antibody, exceptionally large cytosoplasmic and lumenal loops between TMD6–7 and 7–8, respectively, and no sequence similarity to other neurotransmitter transporters (328–330). However, unlike many of the neurotransmitter transporters, SV2 shows strong sequence similarity to a large family of carbohydrate and nutrient transporters from bacteria to humans. Importantly, these related transporters use a variety of mechanisms for transport, from passive, facilitated diffusion to Na^+- or H^+-coupled processes. Thus, this family of proteins appears to share substrate specificity but not mechanism. Both ATP and ascorbate have carbohydrate moieties, suggesting that SV2 may recognize them as substrates, but repeated efforts in multiple systems have failed to demonstrate the biochemical function of SV2. Alternatively, SV2 may contribute to the development of a secretory vesicle matrix (331). Additional studies have identified two isoforms for SV2, both recognized by the original monoclonal antibody, which differ only slightly in sequence (332). SV2A appears more ubiquitous, expressed in essentially all neuronal and endocrine cells, whereas SV2B occurs only in certain neuronal populations (333). The N-terminus of SV2A may also interact with an isoform of synaptotagmin (334), implicating the protein in other aspects of the synaptic vesicle cycle, but both the cellular and biochemical roles of SV2 remain unknown.

4.6 Mechanisms of vesicular neurotransmitter transport

The availability of ligands that interact in distinct ways with the vesicular transporters for monoamines and ACh has helped to identify domains and residues involved in substrate recognition. Replacement with asparagine of a conserved aspartate in TMD1 of VMAT2 abolishes transport activity (335). Importantly, this mutant remains coupled to the driving force $\Delta\bar{\mu}_{H+}$ because imposition of $\Delta\bar{\mu}_{H+}$ accelerates its binding to [^{3}H]reserpine. However, monoamine transmitters no longer potently inhibit the binding of [^{3}H]reserpine to this mutant, indicating a specific defect in substrate recognition. Mutation of a set of the serines located near TMD3 of VMAT2 produces a similar biochemical phenotype, suggesting that the vesicular transporters may, like G protein-coupled receptors and plasma membrane transporters for monoamines (118), use the acidic residue to interact with the amino group of the ligand and the serines to interact with the hydroxyl groups on the aromatic ring (336, 337). Surprisingly, the highly conserved aspartate in TMD1 of VAChT does not appear required for activity (L. B. Hersh *et al.*, in press).

The physiological and pharmacological differences between VMAT1 and 2 have also provided a tool to identify the residues important for ligand recognition. The analysis of chimeric transporters has identified domains required for the high affinity interactions of VMAT2 with monoamine transmitters, including histamine, and with the inhibitor tetrabenazine (338). Site-directed mutagenesis within these domains has further implicated specific residues in the differential ligand recognition (295, 339). In particular, Tyr-434 in TMD11, Lys-446 near TMD12, and Asp-461 in TMD12 contribute to the interaction with multiple ligands. Comparison of serotonin with related compounds such as tryptamine further suggests that the hydroxyl group on the tryptophan ring of serotonin directly contacts Tyr-434. However, in the absence of structural information, many of the residues identified may have indirect rather than direct effects on ligand recognition.

Crosslinking can also be used to identify the sites of ligand interaction on VMAT2. VMAT2 has been expressed at relatively high levels in a baculovirus system and although activity may be compromised, it retains binding to [^{3}H]dihydrotetrabenazine (340). Photolabelling of the purified transporter with [^{125}I]7-azido-8-iodoketanserin followed by peptide mapping and radiosequencing has shown crosslinking to Lys-20 at the N-terminus (341). In contrast, photolabelling with [^{125}I]2-*N*-[(3′-iodo-4′-azidophenyl)proprionyl]tetrabenazine showed derivatization to a peptide between TMD10 and 11, consistent with the results of mutagenesis in this region. Since tetrabenazine and ketanserin both protect against the labelling with both compounds, the results suggest the juxtaposition of TMD1 and 10/11. Supporting this possibility, the conserved lysine in TMD2 of VMAT2 and a similarly conserved aspartate in TMD11 form a charge pair: neutralization of either residue alone abolishes transport activity but neutralization of both restores function, with only a slightly reduced apparent affinity (342). These results provide some of the first information about the structure of the VMAT/VAChT family of transporters.

To understand how $\Delta\bar{\mu}_{H+}$ drives neurotransmitter accumulation, several groups have focused on acidic residues predicted to reside in TMDs that might participate directly in proton translocation. After excluding the residues described above, neutralization of the conserved aspartates in TMD6 and 12 does not substantially impair transport (342). This leaves the highly conserved aspartate in TMD10 available for proton translocation. Indeed, neutralization of this residue in VMAT1 abolishes transport activity (343). Further, this mutant does not bind to [^{3}H]reserpine, indicating either a defect in folding or failure to couple to $\Delta\bar{\mu}_{H+}$. Interestingly, replacement of this aspartate with glutamate still permits function, but with an altered pH profile, further supporting a role for this residue in proton translocation (343) confirmed by additional studies of synaptic transmission.

4.7 Regulation of vesicular neurotransmitter transport

4.7.1 Regulation of quantal size

Although electrophysiological analysis of synaptic transmission has generally interpreted changes in quantal size as due to changes in the sensitivity of the postsynaptic neuron, changes in quantal size may also occur at the level of the presynaptic neuron as a result of changes in vesicle filling. Repetitive stimulation at the neuromuscular junction transiently reduces quantal size, but hypertonic solution and other treatments increase quantal size over hours (344). In particular, cAMP, monoamines, insulin, and calcitonin gene-related peptide (CGRP) all increase quantal size at the neuromuscular junction up to twofold (345), whereas nicotinic agonists antagonize this increase (346) and stimulation of protein kinase C with phorbol esters reduces quantal size (347). Importantly, these changes in quantal size occur in the absence of any changes in postsynaptic receptor sensitivity, indicating a presynaptic mechanism. In addition, many of the changes observed are blocked by hemicholinium, the plasma membrane choline transport blocker that presumably reduces the supply of choline available for ACh biosynthesis, by dissipation of the $\Delta\bar{\mu}_{H^+}$ that drives ACh packaging, and by the vesicular uptake inhibitor vesamicol (348, 349). Interestingly, the observed increases in quantal size appear due to the increased filling of vesicles already available for release: addition of either hemicholinium or the false transmitter monoethylcholine, which acts less effectively at ACh receptors, have no effect on quantal size until hours after addition to unstimulated nerve terminals; but both agents reduce quantal size within minutes when added to a stimulated preparation (350–352). Indeed, this phenomenon may account for the observed preferential release of newly synthesized transmitter.

Central synapses also exhibit a considerable variation in quantal size. Originally interpreted as due to the activation of multiple synapses with varying distances from the recording electrode, more recent studies activating individual boutons suggests considerable variation at a single synapse (353, 354). Presynaptic regulation of quantal size presumably accounts for this variation, and variation in the amplitude of GABAergic miniature endplate potential amplitude indeed appears presynaptic, with little evidence for saturation of postsynaptic receptors (355, 356). Further, the extent of variation in quantal size at GABAergic synapses correlates with the extent of variation in size of synaptic vesicles. In contrast, the changes in quantal size observed at the neuromuscular junction do not involve concomitant changes in synaptic vesicle size (344), arguing in favour of a specific change in vesicle filling.

Multiple presynaptic mechanisms have the potential to regulate quantal size. Changes in the cytoplasmic concentration of transmitter influence the concentration achieved inside vesicles. Indeed, exogenous L-dopa increases quantal size, presumably by increasing the cytosolic concentration of dopamine (357). Conversely, D2 dopamine receptor activation reduces quantal size, apparently by inhibiting the activity of the biosynthetic enzyme tyrosine hydroxylase (358). Reuptake of released transmitter at the plasma membrane also influences transmitter storage, as demonstrated by the failure of dopamine transporter knock-out mice to sustain quantal release (50). Increased cytoplasmic glycine mediated by GLYT2 may also contribute to the packaging of glycine by VGAT as noted above. In addition, the size of the vesicle might be regulated. *Drosophila* mutants in the clathrin adaptor protein AP180 show both larger synaptic vesicles by electron microscopy and larger miniature endplate currents electrophysiologically (359). The magnitude of the driving force $\Delta\bar{\mu}_{H+}$ across the vesicle membrane also influences the amount of transmitter stored. In parafollicular cells of the thyroid gland, thyroid-stimulating hormone and extracellular calcium activate a chloride conductance in secretory vesicles that increases ΔpH and the release of serotonin (360, 361). Thus, both cytoplasmic transmitter concentration and vesicle driving force affect the concentration of vesicular transmitter achieved at equilibrium. In contrast, regulation of vesicular transport itself might be expected to affect specifically the rate of filling.

4.7.2 Vesicular transport limits quantal size

A series of recent studies have indicated that vesicular transport activity limits quantal size (362). Knock-out mice lacking both functional alleles of VMAT2 feed poorly and do not survive for more than a few days after birth, presumably because central monoamine neurons cannot release dopamine, norepinephrine, or serotonin by regulated exocytosis (363–365). The brains of these mice show no major anatomic difference from wild-type in terms of monoamine cell groups or projections, but they do contain extremely low (1–5% wild-type) levels of dopamine, norepinephrine, and serotonin. Interestingly, VMAT2 +/– heterozygotes show substantial reductions in brain dopamine and serotonin relative to wild-type littermates (364), suggesting that VMAT2 limits the storage and release of these transmitters. Midbrain cultures prepared from these mice have helped to characterize the effect of reduced VMAT2 expression. VMAT2 –/– homozygous cultures show no depolarization-evoked release of dopamine, but amphetamines still induce release (albeit reduced from wild-type), indicating that VMAT2 promotes but is not required for the action of these drugs (364). Indeed, amphetamines can substantially promote the survival of VMAT2 –/– homozygotes. VMAT2 +/– cultures show depolarization-evoked dopamine release but at levels ~ 50% wild-type, supporting the observations *in vivo*. Further, the heterozygous animals show increased sensitivity to MPTP relative to wild-type and a reduction in conditioned place preference, a measure of the reward pathway implicated in drug abuse (363). The animals also show a markedly increased sensitivity to the motor effects of amphetamine, apomorphine, and cocaine (365). Additional studies using PC12 cells show that increased VMAT expression also increases monoamine release *in vitro* (E. Pothos, D. E. Krantz, R. H. Edwards, and D. Sulzer, manuscript submitted).

Vesicular transport also limits transmitter release at a cholinergic synapse, the neuromuscular junction. Heterologous expression of rat VAChT in *Xenopus* motor neurons increases quantal size when measured postsynaptically in co-cultured muscle cells (366). In addition, heterologous expression of one VAChT mutant abolishes the increase in quantal size over the background due to the endogenous *Xenopus* transporter, whereas another mutant actually reduces quantal size below baseline. Interestingly, the second mutant contained an asparagine in place of the aspartate in TMD10, the residue implicated in proton translocation by the VMATs (see Section 4.6). This mutant may therefore act as a dominant negative by allowing the efflux of stored ACh down its concentration gradient.

Vesicular transport may influence quantal size through several distinct mechanisms. First, the transporters function at a relatively slow rate and synaptic vesicles have been reported to cycle within 20–90 seconds (367), raising the possibility that rapidly cycling vesicles may not have time to fill completely before they become available for a second round of exocytosis. To test this possibility, the dopamine content was compared in unstimulated midbrain cultures from VMAT2 +/– and wild-type mice. Under these conditions, the heterozygotes contained ~ 75% wild-type levels as compared to 50% with stimulation, supporting a role for the rate of recycling in the determination of quantal size but implicating additional factors as well since the heterozygotes still differed from wild-type (364). Secondly, increased transporter expression may recruit fusion-competent vesicles that otherwise contained no transporter. Under these circumstances, increased expression should increase frequency as well as quantal size and this was indeed observed in the *Xenopus* motor neuron–muscle co-culture system (366). Thirdly, transport activity may simply operate against a large, non-specific efflux of transmitter (368). Although a wasteful phenomenon, experiments indicate a substantial leak of transmitter from synaptic vesicles (369) and different levels of transporter expression would in this case also confer differences in quantal size. The relative importance of these three phenomena may well vary at different synapses. Conversely, it will also be very important to determine whether the state of filling influences the likelihood of exocytosis at certain synapses — in other words, is there a checkpoint that monitors vesicle filling?

4.7.3 Mechanisms of regulation

Classical studies have suggested that vesicular transport may indeed undergo physiological regulation. In sympathetic neurons, K^+-evoked depolarization increases VMAT expression at the transcriptional level (370, 371). At the post-translational level, cAMP appears to reduce VMAT activity (372) and phorbol esters protect against the inhibition of VAChT by vesamicol (373), but the molecular mechanisms responsible for these observations remain unknown. In addition, the heterotrimeric G protein G_{o2} inhibits vesicular monoamine transport in PC12 cells (374). Purified G protein exerts the inhibitory effect in semi-permeabilized cells and this does not appear to involve changes in subcellular location. The inhibition affects principally K_m and does not involve a change in ΔpH, but the nature of the interaction with the transporter remains unknown. Interestingly, a proteolytic fragment of fodrin inhibits both vesicular GABA and glutamate transport (375), raising questions about the relationship between vesicle filling, docking to the cytoskeleton, and availability for release. These observations clearly deserve further study, but the greatest potential for vesicular transport regulation appears to occur at the level of membrane trafficking.

Although closely related in sequence, the VMATs and VAChT show important differences in membrane trafficking. In PC12 cells, both VMATs sort preferentially to large dense core vesicles (LDCVs) (376) whereas VAChT localizes to synaptic-like microvesicles (301, 377). Since LDCVs occur in the cell body and dendrites as well as at nerve terminals, and release more slowly and in response to different stimuli from synaptic vesicles (SVs) (378), these differences in subcellular location have important implications for transmitter release. In addition, these transporters provide some of the first integral membrane protein markers unique to these specialized neurosecretory vesicles. Most SV proteins occur on LDCVs as well, and previous studies of LDCVs has focused on soluble, lumenal proteins that appear to sort by selective aggregation (379, 380). Integral membrane proteins may also sort through lumenal domains but have the potential to interact with sorting machinery in the cytoplasm.

Recent studies have begun to identify the sequences important for sorting VMATs and VAChT in PC12 cells. The analysis of chimeras has suggested that the cytoplasmic C-terminus of VAChT suffices for targeting VMAT2 to SVs (381). From this study, the VMAT2 signals for targeting to LDCVs appear less clear. Within the C-termini of both VMATs and VAChT, a dileucine-like motif similar to those found in many other proteins is both required for and suffices to confer internalization in PC12 cells and CHO fibroblasts (382). Interestingly, acidic residues at positions –4 and –5 relative to the dileucine often form an important part of the motif and the VMATs contain glutamates at these positions, but they are not required for endocytosis. However, dileucine motifs act at multiple sites in the secretory pathway, not simply endocytosis, suggesting that the acidic residues may influence other trafficking events such as sorting to LDCVs, an event generally considered to occur in the *trans*-Golgi network. Further, VAChT contains a serine at –5 and a glutamate at –4 relative to the dileucine, phosphorylation at the –5 position has been shown to control the internalization of other proteins (383, 384) and VAChT undergoes strongly regulated phosphorylation at this site (D. E. K., R. H. E., in preparation). Mutants mimicking or blocking phosphorylation at this site in VAChT also influence the steady state distribution of the transport protein. Thus, dileucine motifs appear very important for the trafficking of VMATs and VAChT, implicating an interaction with clathrin adaptor proteins AP-1, -2, or -3 (385, 386). A casein kinase 2 phosphorylation site at the C-terminus of VMAT2 (387) also appears to influence the trafficking of this protein, presumably through a distinct mechanism.

Sorting of the transporters in neurons differs slightly from that observed in PC12 cells. VAChT localizes preferentially to SVs in the brain (388) but VMAT2 appears in multiple vesicle populations, including somatodendritic tubulovesicular structures and SVs as well as LDCVs (389, 390). The tubulovesicular structures may indeed mediate the somatodendritic release of dopamine observed *in vivo* (391). Thus, VMAT2 differs in localization from VAChT in neurons as well as PC12 cells.

Recent work also suggests that neurons release multiple classical transmitters in addition to neural peptides. Dopamine neurons in primary midbrain culture form glutamatergic synapses (392) and cholinergic neurons in the basal forebrain also express the GABAergic markers glutamic acid decarboxylase and VGAT (393). It is particularly interesting that this co-expression involves one transporter from the ΔpH-coupled VMAT/VAChT family and the other from the more $\Delta\Psi$-coupled amino acid transport family. Although this storage within the same cell might be predicted to require distinct vesicular compartments expressing either the chemical or electrical components of $\Delta\tilde{\mu}_{H+}$, it is also possible that co-localization of the transport proteins in the same vesicle would actually prove synergistic. In particular, monoamine transport into vesicles would dissipate ΔpH, increasing $\Delta\Psi$, and promoting amino acid storage, whereas amino acid transport would dissipate $\Delta\Psi$, increase ΔpH, and promoting the storage of monoamines or ACh. Synergistic interactions of this sort have been observed in chromaffin ghosts between ATP and monoamines (318). Co-storage and release of monoamines with glutamate has also been demonstrated to occur transiently during development.

4.7.4 Development

Although the regulated exocytotic release of monoamines does not appear to be required for the normal development of monoamine cell populations and the projections to their targets (364), a series of interesting observations implicate monoamines in the segregation of cortical afferents during the critical period in development. Mice deficient in monoamine oxidase A (MAO-A) fail to metabolize serotonin and hence accumulate extremely high levels of this monoamine (394). In these mice, thalamic afferents representing the whiskers grow profusely within the somatosensory cortex and fail to segregate into the usual barrel fields visualized by staining for cytochrome oxidase (395). Although this could simply result from the excessive, essentially pharmacological stimulation of serotonin receptors, inhibition of serotonin biosynthesis reduces the size of barrels in both MAO-deficient and normal mice, suggesting that serotonin normally modulates the formation of barrel fields. In addition, the thalamocortical afferents accumulate large amounts of serotonin. Since they do not express the biosynthetic enzyme tryptophan hydroxylase, these cells must acquire from serotonin from more traditional serotonergic projections originating in brainstem raphe nuclei (396). The thalamocortical projections indeed express both the plasma membrane serotonin transporter and VMAT2 transiently during the critical period in development when barrel field formation depends on somatosensory input from the periphery. These cells also express the serotonin $5HT_{1b}$ autoreceptor, which can serve to inhibit transmitter release. Although the evidence remains largely circumstantial, these observations suggest several possibilities. First, serotonin storage in thalamocortical neurons may, like MAO, simply serve to reduce extracellular serotonin. Alternatively, the storage and release of serotonin may modulate the development of glutamatergic synapses in the cortex and the segregation of afferents. Future studies analysing the localization of VMAT2 and the vesicular glutamate transporter, together with physiological preparations and transporter knock-out mice, should help to resolve these questions.

5. Summary

Molecular analysis has begun to identify the proteins responsible for the transport of neurotransmitters at both the plasma membrane and secretory vesicles. However, we still understand very little about the structural mechanisms that couple transmitter accumulation to ionic gradients. In addition, we can only surmise the biological role for many of the proteins already identified. Perhaps most strikingly, the potential for these proteins to regulate synaptic transmission suggested by the pharmacology remains almost entirely unexplored.

Acknowledgements

We thank the NINDS, NIMH, NIDA, and Valley Foundation for their support and S. Amara for her thoughtful suggestions on the manuscript.

References

1. Axelrod, J., Weil-Malherbe, H., and Tomchick, R. (1959) The physiological disposition of ^{3}H-epinephrine and its metabolite metanephrine. *J. Pharm. Exp. Ther.* **127**, 251.
2. Axelrod, J., Whitby, L., and Hertting, G. (1961) Effect of psychotropic drugs on the uptake of ^{3}H-norepinephrine by tissues. *Science* **133**, 383.
3. Iversen, L. L. (1976) The uptake of biogenic amines. In *Handbook of psychopharmacology*, Vol. 3 (ed. S. D. Iversen and S. H. Snyder), pp. 381–442. Plenum Press, New York.
4. Frize, E. D. (1954) Mental depression in hypertensive patients treated for long periods with high doses of reserpine. *N. Engl. J. Med.* **251**, 1006.
5. Iversen, L. L. and Neal, M. J. (1968) The uptake of ^{3}H-GABA by slices of rat cerebral cortex. *J. Neurochem.* **15**, 1141.
6. Kanner, B. I. and Schuldiner, S. (1987) Mechanisms of storage and transport of neurotransmitters. *CRC Crit. Rev. Biochem.* **22**, 1.
7. Johnson, R. G. (1988) Accumulation of biological amines into chromaffin granules: a model for hormone and neurotransmitter transport. *Physiol. Rev.* **68**, 232.
8. Radian, R. and Kanner, B. I. (1985) Reconstitution and purification of the sodium- and chloride-coupled gamma-aminobutyric acid transporter from rat brain. *J. Biol. Chem.* **260**, 11859.
9. Guastella, J., Nelson, N., Nelson, H., Czyzyk, L., Kenyan, S., Miedel, M. C., *et al.* (1990) Cloning and expression of a rat brain GABA transporter. *Science* **249**, 1303.
10. Pacholczyk, T., Blakeley, R. D., and Amara, S. G. (1991) Expression cloning of a cocaine- and antidepressant-sensitive human noradrenaline transporter. *Nature* **350**, 350.
11. Pines, G., Dnabolt, N. C., Bjoras, M., Zhgang, Y., Bendahan, A., Eide, L., *et al.* (1992) Cloning and expression of a rat brain L-glutamate transporter. *Nature* **360**, 464.
12. Storck, T., Schulte, S., Hofmann, K., and Stoffel, W. (1992) Structure, expression and functional analysis of a Na^{+}-dependent glutamate/aspartate transporter from rat brain. *Proc. Natl. Acad. Sci. USA* **89**, 10955.
13. Kanai, Y. and Hediger, M. A. (1992) Primary structure and functional characterization of a high-affinity glutamate transporter. *Nature* **360**, 467.

14. Liu, Y., Roghani, A., and Edwards, R. H. (1992) Gene transfer of a reserpine-sensitive mechanism of resistance to MPP^+. *Proc. Natl. Acad. Sci. USA* **89**, 9074.
15. Liu, Y., Peter, D., Roghani, A., Schuldiner, S., Prive, G. G., Eisenberg, D., *et al.* (1992) A cDNA that supresses MPP^+ toxicity encodes a vesicular amine transporter. *Cell* **70**, 539.
16. Erickson, J. D., Eiden, L. E., and Hoffman, B. J. (1992) Expression cloning of a reserpine-sensitive vesicular monoamine transporter. *Proc. Natl. Acad. Sci. USA* **89**, 10993.
17. Alfonso, A., Grundahl, K., Duerr, J. S., Han, H.-P., and Rand, J. B. (1993) The *Caenorhabditis elegans unc-17* gene: a putative vesicular acetylcholine transporter. *Science* **261**, 617.
18. McIntire, S. L., Reimer, R. J., Schuske, K., Edwards, R. H., and Jorgensen, E. M. (1997) Identification and characterization of the vesicular GABA transporter. *Nature* **389**, 870.
19. Rudnick, G. and Clark, J. (1993) From synapse to vesicle: the reuptake and storage of biogenic amine neurotransmitters. *Bioc. Biop. Acta* **1144**, 249.
20. Amara, S. G. and Arriza, J. L. (1993) Neurotransmitter transporters: three distinct gene families. *Curr. Biol.* **3**, 337.
21. Palacin, M., Estevez, R., Bertran, J., and Zorzano, A. (1998) Molecular biology of the mammalian plasma membrane amino acid transporters. *Physiol. Rev.* **78**, 969.
22. Minelli, A., Brecha, N. C., Karschin, C., DeBiasi, S., and Conti, F. (1995) GAT-1, a high-affinity GABA plasma membrane transporter, is localized to neurons and astroglia in the cerebral cortex. *J. Neurosci.* **15**, 7734.
23. Conti, F., Melone, M., De Biasi, S., Minelli, A., Brecha, N. C., and Ducati, A. (1998) Neuronal and glial localization of GAT-1, a high-affinity gamma-aminobutyric acid plasma membrane transporter, in human cerebral cortex: with a note on its distribution in monkey cortex. *J. Comp. Neurol.* **396**, 51.
24. Borden, L. A., Murali-Dhar, T. G., Smith, K. E., Weinshank, R. L., Branchek, T. A., and Gluchowski, C. (1994) Tiagabine, SKandF 89976-A, CI-966, and NNC-711 are selective for the cloned GABA transporter GAT-1. *Eur. J. Pharmacol.* **269**, 219.
25. Lopez-Corcuera, B., Liu, Q. R., Mandiyan, S., Nelson, H., and Nelson, N. (1992) Expression of a mouse brain cDNA encoding novel gamma-aminobutyric acid transporter. *J. Biol. Chem.* **267**, 17491.
26. Borden, L. A., Smith, K. E., Hartig, P. R., Branchek, T. A., and Weinshank, R. L. (1992) Molecular heterogeneity of the gamma-aminobutyric acid (GABA) transport system. Cloning of two novel high affinity GABA transporters from rat brain. *J. Biol. Chem.* **267**, 21098.
27. Ikegaki, N., Saito, N., Hashima, M., and Tanaka, C. (1994) Production of specific antibodies against GABA transporter subtypes (GAT1, GAT2, GAT3) and their application to immunocytochemistry. *Brain Res.* **26**, 47.
28. Johnson, J., Chen, T. K., Rickman, D. W., Evans, C., and Brecha, N. C. (1996) Multiple gamma-aminobutyric acid plasma membrane transporters (GAT-1, GAT-2, GAT-3) in the rat retina. *J. Comp. Neurol.* **375**, 212.
29. Clark, J. A., Deutch, A. Y., Gallipoli, P. Z., and Amara, S. G. (1992) Functional expression and CNS distribution of a beta-alanine-sensitive neuronal GABA transporter. *Neuron* **9**, 337.
30. Clark, J. A. and Amara, S. G. (1994) Stable expression of a neuronal gamma-aminobutyric acid transporter, GAT-3, in mammalian cells demonstrates unique pharmacological properties and ion dependence. *Mol. Pharm.* **46**, 550.
31. Liu, Q. R., Lopez-Corcuera, B., Mandiyan, S., Nelson, H., and Nelson, N. (1993) Molecular characterization of four pharmacologically distinct gamma-aminobutyric acid transporters in mouse brain. *J. Biol. Chem.* **268**, 2106.

32. Minelli, A., Bebiasi, S., Brecha, N. C., Zuccarello, L. V., and Conti, F. (1996) GAT-3, a high-affinity GABA plasma membrane transporter is localized to astrocytic processes, and it is not confined to the vicinity of GABAergic synapses in the cerebral cortex. *J. Neurosci.* **16**, 6255.
33. Ribak, C. E., Tong, W. M., and Brecha, N. C. (1996) GABA plasma membrane transporters, GAT-1 and GAT-3, display different distributions in the rat hippocampus. *J. Comp. Neurol.* **367**, 595.
34. Liu, Q. R., Lopez-Corcuera, B., Nelson, H., Mandiyan, S., and Nelson, N. (1992) Cloning and expression of a cDNA encoding the transporter of taurine and beta-alanine in mouse brain. *Proc. Natl. Acad. Sci. USA* **89**, 12145.
35. Evans, J. E., Frostholm, A., and Rotter, A. (1996) Embryonic and postnatal expression of four gamma-aminobutyric acid transporter mRNAs in the mouse brain and leptomeninges. *J. Comp. Neurol.* **376**, 431.
36. Smith, K. E., Borden, L. A., Wang, C. H., Hartig, P. R., Branchek, T. A., and Weinshank, R. L. (1992) Cloning and expression of a high affinity taurine transporter from rat brain. *Mol. Pharm.* **42**, 563.
37. Uchida, S., Kwon, H. M., Yamauchi, A., Preston, A. S., Marumo, F., and Handler, J. S. (1992) Molecular cloning of the cDNA for an MDCK cell $Na(^{+})$- and $Cl(^{-})$-dependent taurine transporter that is regulated by hypertonicity. *Proc. Natl. Acad. Sci. USA* **89**, 8230.
38. Mayser, W., Schloss, P., and Betz, H. (1992) Primary structure and functional expression of a choline transporter expressed in the rat nervous system. *FEBS Lett.* **305**, 31.
39. Guimbal, C. and Kilimann, M. W. (1993) A $Na(^{+})$-dependent creatine transporter in rabbit brain, muscle, heart, and kidney. cDNA cloning and functional expression. *J. Biol. Chem.* **268**, 8418.
40. Adam-Vizi, V. (1992) External Ca^{++}-independent release of neurotransmitters. *J. Neurochem.* **58**, 395.
41. Attwell, D., Barbour, B., and Szatkowski, M. (1993) Nonvesicular release of neurotransmitter. *Neuron* **11**, 401.
42. Schwartz, E. A. (1987) Depolarization without calcium can release gamma-aminobutyric acid from a retinal neuron. *Science* **238**, 350.
43. Kilty, J. E., Lorang, D., and Amara, S. G. (1991) Cloning and expression of a cocaine-sensitive rat dopamine transporter. *Science* **254**, 578.
44. Shimada, S., Kitayama, S., Lin, C. L., Patel, A., Nanthakumar, E., Gregor, P., *et al.* (1991) Cloning and expression of a cocaine-sensitive dopamine transporter complimentary DNA. *Science* **254**, 576.
45. Giros, B., Mestikawy, S., Bertrand, L., and Caron, M. G. (1991) Cloning and functional characterization of a cocaine-sensitive dopamine transporter. *FEBS Lett.* **295**, 149.
46. Blakely, R. D., Berson, H. E., Fremeau, R. T., Caron, M. G., Peek, M. M., Prince, H. K., *et al.* (1991) Cloning and expression of a functional serotonin transporter from rat brain. *Nature* **354**, 66.
47. Hoffman, B. J., Mezey, E., and Brownstein, M. J. (1991) Cloning of a serotonin transporter affected by antidepressants. *Science* **254**, 579.
48. Gu, H., Wall, S. C., and Rudnick, G. (1994) Stable expression of the biogenic amine transporters reveals differences in inhibitor sensitivity, kinetics and ion dependence. *J. Biol. Chem.* **269**, 7124.
49. Apparsundaram, S., Moore, K. R., Malone, M. D., Hartzell, H. C., and Blakely, R. D. (1997) Molecular cloning and characterization of an L-epinephrine transporter from the sympathetic ganglia of the bullfrog Rana catesbiana. *J. Neurosci.* **17**, 2691.

50. Giros, B., Jaber, M., Jones, S. R., Wightman, R. M., and Caron, M. G. (1996) Hyper-locomotion and indifference to cocaine and amphetamine in mice lacking the dopamine transporter. *Nature* **379**, 606.
51. Kitayama, S., Shimada, S., and Uhl, G. R. (1992) Parkinsonism-inducing neurotoxin MPP^+: uptake and toxicity in nonneuronal COS cells expressing dopamine transporter cDNA. *Ann. Neurol.* **32**, 109.
52. Pifl, C., Giros, B., and Caron, M. G. (1993) Dopamine transporter expression confers cytotoxicity to low doses of the parkinsonism-inducing neurotoxin 1-methyl-4-phenylpyridinium. *J. Neurosci.* **13**, 4246.
53. Snyder, S., D'Amato, R., Nye, J., and Javitch, J. (1986) Selective uptake of MPP^+ by dopamine neurons is required for MPTP toxicity: studies in brain synaptosomes and PC12 cells. In *MPTP: a neurotoxin producing a Parkinsonian syndrome* (ed. S. Markey, N. Castagnoli, Jr., A. Trevor, and I. Kopin), pp. 191–201. Academic Press, New York.
54. Edwards, R. H. (1993) Neural degeneration and the transport of neurotransmitters. *Ann. Neurol.* **34**, 638.
55. Uhl, G. R. (1998) Hypothesis: the role of dopaminergic transporters in selective vulnerability of cells in Parkinson's disease. *Ann. Neurol.* **43**, 555.
56. Langston, J. W., Ballard, P., Tetrud, J. W., and Irwin, I. (1983) Chronic parkinsonism in humans due to a product of meperidine analog synthesis. *Science* **219**, 979.
57. Cerruti, C., Walther, D. M., Kuhar, M. J., and Uhl, G. R. (1993) Dopamine transporter mRNA expression is intense in rat midbrain neurons and modest outside midbrain. *Brain Res. Mol. Brain Res.* **18**, 181.
58. Fisher, J. F. and Cho, A. K. (1979) Chemical release of DA from striatal homogenates: evidence for an exchange diffusion model. *J. Pharm. Exp. Ther.* **208**, 203.
59. Rudnick, G. and Wall, S. C. (1992) The molecular mechanism of 'ecstasy' [3,4-methylenedioxymethamphetamin (MDMA)]: serotonin transporters are targets for MDMA-induced serotonin release. *Proc. Natl. Acad. Sci. USA* **89**, 1817.
60. Sulzer, D., Chen, T.-K., Lau, Y. Y., Kristensen, H., Rayport, S., and Ewing, A. (1995) Amphetamine redistributes dopamine from synaptic vesicles to the cytosol and promotes reverse transport. *J. Neurosci.* **15**, 4102.
61. Seiden, L. S., Sabol, K. E., and Ricaurte, G. A. (1993) Amphetamine: effects on catecholamine systems and behavior. *Annu. Rev. Pharmacol. Toxicol.* **32**, 639.
62. Jones, S. R., Gainetdinov, R. R., Wightman, R. M., and Caron, M. G. (1998) Mechanisms of amphetamine action revealed in mice lacking the dopamine transporter. *J. Neurosci.* **18**, 1979.
63. Fumagalli, F., Gainetdinov, R. R., Valenzano, K. J., and Caron, M. G. (1998) Role of dopamine transporter in methamphetamine-induced neurotoxicity: evidence from mice lacking the transporter. *J. Neurosci.* **18**, 4861.
64. Rocha, B. A., Fumagalli, F., Gainetdinov, R. R., Jones, S. R., Ator, R., Giros, B., *et al.* (1998) Cocaine self-administration in dop[amine-transporter knockout mice. *Nature Neurosci.* **1**, 132.
65. Bengel, D., Murphy, D. L., Andrews, A. M., Wichems, C. H., Feltner, D., Heils, A., *et al.* (1998) Altered brain serotonin homeostasis and locomotor insensitivity to 3,4-methylene-dioxymethamphetamine ('ecstasy') in serotonin transporter-deficient mice. *Mol. Pharm.* **53**, 649.
66. Lesch, K. P., Bengel, D., Heils, A., Sabol, S. Z., Greenberg, B. D., Petri, S., *et al.* (1996) Association of anxiety-related traits with a polymorphism in the serotonin transporter gene regulatory region. *Science* **274**, 1527.

67. Smith, K. E., Borden, L. A., Hartig, P. R., Branchek, T., and Weinshank, R. L. (1992) Cloning and expression of a glycine transporter reveal colocalization with NMDA receptors. *Neuron* **8**, 927.
68. Guastella, J., Brecha, N., Weigmann, C., Lester, H. A., and Davidson, N. (1992) Cloning, expression and localization of a rat brain high-affinity glycine transporter. *Proc. Natl. Acad. Sci. USA* **89**, 7189.
69. Liu, Q. R., Lopez-Corcuera, B., Mandiyan, S., Nelson, H., and Nelson, N. (1993) Cloning and expression of a spinal cord- and brain-specific glycine transporter with novel structural features. *J. Biol. Chem.* **268**, 22802.
70. Borowsky, B., Mezey, E., and Hoffman, B. J. (1993) Two glycine transporter variants with distinct localization in the CNS and peripheral tissues are encoded by a common gene. *Neuron* **10**, 851.
71. Zafra, F., Aragon, C., Olivares, L., Danbolt, N. C., Gimenez, C., and Storm-Mathisen, J. (1995) Glycine transporters are differentially expressed among CNS cells. *J. Neurosci.* **15**, 3952.
72. Fremeau, R. T. J., Caron, M. G., and Blakely, R. D. (1992) Molecular cloning and expression of a high affinity L-proline transporter expressed in putative glutamatergic pathways of rat brain. *Neuron* **8**, 915.
73. Velaz-Faircloth, M., Guadano-Ferraz, A., Henzi, V. A., and Fremeau, R. J. (1995) Mammalian brain-specific L-proline transporter. Neuronal localization of mRNA and enrichment of transporter protein in synaptic plasma membranes. *J. Biol. Chem.* **270**, 15755.
74. Liu, Q. R., Mandiyan, S., Lopez-Corcuera, B., Nelson, H., and Nelson, N. (1993) A rat brain cDNA encoding the neurotransmitter transporter with an unusual structure. *FEBS Lett.* **315**, 114.
75. Luque, J. M., Jursky, F., Nelson, N., and Richards, J. G. (1996) Distribution and sites of synthesis of NTT4, an orphan member of the $Na^+/Cl^{(-)}$-dependent neurotransmitter transporter family, in the rat CNS. *Eur. J. Neurosci.* **8**, 127.
76. El Mestikawy, S., Giros, B., Pohl, M., Hamon, M., Kingsmore, S. F., Seldin, M. F., *et al.* (1994) Characterization of an atypical member of the $Na^+/Cl^{(-)}$-dependent transporter family: chromosomal localization and distribution in GABAergic and glutamatergic neurons in the rat brain. *J. Neurochem.* **62**, 445.
77. El Mestikawy, S., Wehrle, R., Masson, J., Lombard, M. C., Hamon, M., and Sotelo, C. (1997) Distribution pattern and ultrastructural localization of Rxt1, an orphan $Na^+/Cl^{(-)}$-dependent transporter, in the central nervous system of rats and mice. *Neuroscience* **77**, 319.
78. Uhl, G. R., Kitayama, S., Gregor, P., Nanthakumar, E., Persico, A., and Shimada, S. (1992) Neurotransmitter transporter family cDNAs in a rat midbrain library: 'orphan transporters' suggest sizable structural variations. *Mol. Brain Res.* **16**, 353.
79. Inoue, K., Sato, K., Tohyama, M., Shimada, S., and Uhl, G. R. (1996) Widespread distribution of mRNA encoding the orphan neurotransmitter v7–3. *Brain Res. Mol. Brain Res.* **37**, 217.
80. Smith, K. E., Fried, S. G., Durkin, M. M., Gustafson, E. L., Borden, L. A., Branchek, T. A., *et al.* (1995) Molecular cloning of an orphan transporter. A new member of the neurotransmitter transporter family. *FEBS Lett.* **357**, 86.
81. Radian, R. and Kanner, B. I. (1983) Stoichiometry of sodium- and chloride-coupled gamma-aminobutyric acid transport by synaptic plasma membranes isolated from rat brain. *Biochemistry* **22**, 1236.
82. Gu, H., Wall, S., and Rudnick, G. (1996) Ion coupling stoichiometry for the norepinephrine transporter in membrane vesicles from stably transfected cells. *J. Biol. Chem.* **271**, 6911.

83. Rudnick, G. and Nelson, P. J. (1978) Platelet 5-hydroxytryptamine transport: an electroneutral mechanism coupled to potassium. *Biochemistry* **17**, 4739.
84. Nelson, P. J. and Rudnick, G. (1979) Coupling between platelet 5-hydroxytryptamine and potassium transport. *J. Biol. Chem.* **254**, 10084.
85. Keyes, S. R. and Rudnick, G. (1982) Coupling of transmembrane proton gradients to platelet serotonin transport. *J. Biol. Chem.* **257**, 1172.
86. Kavanaugh, M. P., Arriza, J. L., North, R. A., and Amara, S. G. (1992) Electrogenic uptake of γ-aminobutyric acid by a cloned transporter expressed in *Xenopus* oocytes. *J. Biol. Chem.* **267**, 22007.
87. Mager, S., Naeve, J., Quick, M., Labarca, C., Davidson, N., and Lester, H. A. (1993) Steady states, charge movements and rates for cloned GABA transporter expressed in *Xenopus* oocytes. *Neuron* **10**, 177.
88. Mager, S., Kleinberger-Doron, N., Keshet, G. I., Davidson, N., Kanner, B. I., and Lester, H. A. (1996) Ion binding and permeation at the GABA transporter GAT1. *J. Neurosci.* **16**, 5405.
89. Risso, S., DeFelice, L. J., and Blakely, R. D. (1996) Sodium-dependent GABA-induced currents in GAT1-transfected HeLa cells. *J. Physiol.* **490**, 691.
90. Sonders, S. and Amara, S. G. (1996) Channels in transporters. *Curr. Opin. Neurobiol.* **6**, 294.
91. Mager, S., Min, C., Henry, D. J., Chavkin, C., Hoffman, B. J., Davidson, N., *et al.* (1994) Conducting states of a mammalian serotonin transporter. *Neuron* **12**, 845.
92. Cammack, J. N., Rakhilin, S. V., and Schwartz, E. A. (1994) A GABA transporter operates asymmetrically and with variable stoichiometry. *Neuron* **13**, 949.
93. Sonders, M. S., Zhu, S.-J., Zahniser, N. R., Kavanaugh, M. P., and Amara, S. G. (1997) Multiple ionic conductances of the human dopamine transporter: the actions of dopamine and psychostimulants. *J. Neurosci.* **17**, 960.
94. Galli, A., deFelice, L. J., Duke, B.-J., Moore, K. R., and Blakely, R. D. (1995) Sodium-dependent norepinephrine-induced currents in norepinephrine-transporter-transfected HEK-293 cells blocked by cocaine and antidepressants. *J. Exp. Biol.* **198**, 2197.
95. Galli, A., Blakely, R. D., and deFelice, L. J. (1996) Norepinephrine transporters have channel modes of conduction. *Proc. Natl. Acad. Sci. USA* **93**, 8671.
96. Galli, A., Petersen, C. I., deBlaquiere, M., Blakely, R. D., and deFelice, L. J. (1997) *Drosophila* serotonin transporters have voltage-dependent uptake coupled to a serotonin-gated ion channel. *J. Neurosci.* **17**, 3401.
97. Lauger, P. (1991) *Electrogenic ion pumps.* Sinauer Assoc., Sunderland, Mass.
98. Su, A., Mager, S., Mayo, S. L., and Lester, H. A. (1996) A multi-substrate single-file model for ion-coupled transporters. *Biophys. J.* **70**, 762.
99. Cammack, J. N. and Schwartz, E. A. (1996) Channel behavior in a gamma-aminobutyric acid transporter. *Proc. Natl. Acad. Sci. USA* **93**, 723.
100. Lin, F., Lester, H. A., and Mager, S. (1996) Single-channel currents produced by the serotonin transporter and analysis of a mutation affecting ion permeation. *Biophys. J.* **71**, 3126.
101. Lester, H. A., Cao, Y., and Mager, S. (1996) Listening to neurotransmitter transporters. *Neuron* **17**, 807.
102. Haugh-Scheidt, L., Malchow, R. P., and Ripps, H. (1995) GABA transport and calcium dynamics in horizontal cells from the skate retina. *J. Physiol.* **488**, 565.
103. Bruss, M. R., Hammermann, R., Brimijoin, S., and Bonisch, H. (1995) Antipeptide antibodies confirm the topology of the human norepinephrine transporter. *J. Biol. Chem.* **270**, 9197.

104. Olivares, L., Aragon, C., Gimenez, C., and Zafra, F. (1995) The role of N-glycosylation in the targeting and activity of the GLYT1 glycine transporter. *J. Biol. Chem.* **270**, 9437.
105. Melikian, H. E., Ramamoorthy, S., Tate, C. G., and Blakely, R. D. (1996) Inability to N-glycosylate the human norepinephrine transporter reduces protein stability, surface trafficking and transport activity but not ligand recognition. *Mol. Pharm.* **50**, 266.
106. Bennet, E. R. and Kanner, B. I. (1997) The membrane topology of GAT-1, a (Na^+ and Cl^-)-coupled γ-aminobutyric acid transporter from rat brain. *J. Biol. Chem.* **272**, 1203.
107. Olivares, L., Aragon, C., Gimenez, C., and Zafra, F. (1997) Analysis of the transmembrane topology of the glycine transporter GLYT1. *J. Biol. Chem.* **270**, 9437.
108. Mabjeesh, N. J. and Kanner, B. I. (1993) Neither amino nor carboxyl termini are required for function of the sodium- and chloride-coupled g-aminobutyric acid transporter from rat brain. *J. Biol. Chem.* **267**, 2563.
109. Olivares, L., Aragon, C., Gimenez, C., and Zafra, F. (1994) Carboxyl terminus of the glycine transporter GLYT1 is necessary for correct processing of the protein. *J. Biol. Chem.* **269**, 28400.
110. Kanner, B. I., Bendahan, A., Pantanowitz, S., and Su, H. (1994) The number of amino acid residues in hydrophilic loops connecting transmembrane domains of the GABA transporter GAT-1 is critical for its function. *FEBS Lett.* **356**, 191.
111. Tamura, S., Nelson, H., Tamura, A., and Nelson, N. (1995) Short external loops as potential substrate binding site of γ-aminobutyric acid transporters. *J. Biol. Chem.* **270**, 28712.
112. Pantanowitz, S., Bendahan, A., and Kanner, B. I. (1993) Only one of the charged amino acids located in the transmembrane α-helices of the γ-aminobutyric acid transporter (subtype A) is essential for its activity. *J. Biol. Chem.* **268**, 3222.
113. Kleinberger-Doron, N. and Kanner, B. I. (1994) Identification of tryptophan residues critical for the function and targeting of the γ-aminobutyric acid transporter (subtype A). *J. Biol. Chem.* **269**, 3063.
114. Mager, S., Kleinberger-Doron, N., Keshet, G. I., Davidson, N., Kanner, B. I., and Lester, H. A. (1996) Ion binding and permeation at the GABA transporter GAT-1. *J. Neurosci.* **16**, 5405.
115. Keshet, G. I., Bendahan, A., Su, H., Mager, S., Lester, H. A., and Kanner, B. I. (1995) Glutamate-101 is critical for the function of the sodium- and chloride-coupled GABA transporter GAT-1. *FEBS Lett.* **371**, 39.
116. Sur, C., Betz, H., and Schloss, P. (1996) A single serine residue controls the cation dependence of substrate transport by the rat serotonin transporter. *Proc. Natl. Acad. Sci. USA* **94**, 7639.
117. Bismuth, Y., Kavanaugh, M. P., and Kanner, B. I. (1997) Tyrosine 140 of the γ-aminobutyric acid transporter GAT-1 plays a critical role in the neurotransmitter recognition. *J. Biol. Chem.* **272**, 16096.
118. Kitayama, S., Shimada, S., Xu, S., Markham, L., Donovan, D. M., and Uhl, G. R. (1992) Dopamine transporter site-directed mutations differentially alter substrate transport and cocaine binding. *Proc. Natl. Acad. Sci. USA* **89**, 7782.
119. Giros, B., Wang, Y. M., Suter, S., McLeskey, S. B., Pifl, C., and Caron, M. G. (1994) Delineation of discrete domains for substrate, cocaine, and tricyclic antidepressant interactions using chimeric dopamine-norepinephrine transporters. *J. Biol. Chem.* **269**, 15985.
120. Buck, K. J. and Amara, S. G. (1994) Chimeric dopamine-norepinephrine transporters delineate structural domains influencing selectivity for catecholamines and 1-methyl-4-phenylpyridinium. *Proc. Natl. Acad. Sci. USA* **91**, 12584.

121. Ferrer, J. V. and Javitch, J. A. (1998) Cocaine alters the accessibility of endogenous cysteines in putative extracellular and intracellular loops of the human dopamine transporter. *Proc. Natl. Acad. Sci. USA* **95**, 9238.
122. Chen, J. G., Liu-Chen, S., and Rudnick, G. (1998) Determination of external loop topology in the serotonin transporter by site-directed chemical labeling. *J. Biol. Chem.* **273**, 12675.
123. Chen, J. G., Sachpatzidis, A., and Rudnick, G. (1997) The third transmembrane domain of the serotonin transporter contains residues associated with substrate and cocaine binding. *J. Biol. Chem.* **272**, 28321.
124. Cao, Y., Li, M., Mager, S., and Lester, H. A. (1998) Amino acid residues that control pH modulation of transport-associated current in mammalian serotonin transporters. *J. Neurosci.* **18**, 7739.
125. Miller, K. J. and Hoffman, B. J. (1994) Adenosine A3 receptors regulate serotonin transport via nitric oxide and cGMP. *J. Biol. Chem.* **269**, 27351.
126. Corey, J. L., Davidson, N., Lester, H. A., Brecha, N., and Quick, M. W. (1994) Protein kinase C modulates the activity of a cloned gamma-aminobutyric acid transporter expressed in *Xenopus* oocytes via regulated subcellular redistribution of the transporter. *J. Biol. Chem.* **269**, 14759.
127. Quick, M. W., Corey, J. L., Davidson, N., and Lester, H. A. (1997) Second messengers, trafficking-related proteins and amino acid residues that contribute to the functional regulation of the rat brain GABA transporter GAT-1. *J. Neurosci.* **17**, 2967.
128. Kitayama, S., Dohi, T., and Uhl, G. R. (1994) Phorbol esters alter functions of the expressed dopamine transporter. *Eur. J. Pharmacol.* **268**, 115.
129. Qian, Y., Galli, A., Ramamoorthy, S., Risso, S., DeFelice, L. J., and Blakely, R. D. (1997) Protein kinase C activation regulates human serotonin transporters in HEK-293 cells via altered cell surface expression. *J. Neurosci.* **17**, 45.
130. Zhu, S. J., Kavanaugh, M. P., Sonders, M. S., Amara, S. G., and Zahniser, N. R. (1997) Activation of protein kinase C inhibits uptake, currents and binding associated with the human dopamine transproter expressed in *Xenopus* oocytes. *J. Pharm. Exp. Ther.* **282**, 1358.
131. Vaughan, R. A., Huff, R. A., Uhl, G. R., and Kuhar, M. J. (1997) Protein kinase C-mediated phosphorylation and functional regulation of dopamine transporters in striatal synaptosomes. *J. Biol. Chem.* **272**, 15541.
132. Ramamoorthy, S., Giovanetti, E., Qian, Y., and Blakely, R. D. (1998) Phosphorylation and regulation of antidepressant-sensitive serotonin transporters. *J. Biol. Chem.* **273**, 2458.
133. Zhu, M. Y., Blakely, R. D., Apparsundaram, S., and Ordway, G. A. (1998) Down-regulation of the human norepinephrine transporter in intact 293-hNET cells exposed to desipramine. *J. Neurochem.* **70**, 1547.
134. Pietrini, G., Suh, Y. J., Edelmann, L., Rudnick, G., and Caplan, M. J. (1994) The axonal gamma-aminobutyric acid transporter GAT-1 is sorted to the apical membranes of polarized epithelialc cells. *J. Biol. Chem.* **269**, 4668.
135. Ahn, J., Mundigl, O., Muth, T. R., Rudnick, G., and Caplan, M. J. (1996) Polarized expression of GABA transporters in Madin-Darby canine kidney cells and cultured hippocampal neurons. *J. Biol. Chem.* **271**, 6917.
136. Gu, H., Ahn, J., Caplan, M. J., Blakely, R. D., Levey, A. I., and Rudnick, G. (1996) Cell-specific sorting of biogenic amine transporters expressed in epithelial cells. *J. Biol. Chem.* **271**, 18100.
137. Schroeter, S., Levey, A. I., and Blakely, R. D. (1997) Polarized expression of the anti-

depressant-sensitive serotonin transporter in epinephrine-synthesizing chromaffin cells of the rat adrenal gland. *Mol. Cell. Neurosci.* **9**, 170.

138. Nirenberg, M. J., Vaughan, R. A., Uhl, G. R., Kuhar, M. J., and Pickel, V. M. (1996) The dopamine transporter is localized to dendritic and axonal plasma membranes of nigrostriatal dopaminergic neurons. *J. Neurosci.* **16**, 436.
139. Hersch, S. M., Yi, H., Heilman, C. J., Edwards, R. H., and Levey, A. I. (1997) Subcellular localization and molecular topology of the dopamine transporter in the striatum and substantia nigra. *J. Comp. Neurol.* **388**, 211.
140. Nirenberg, M. J., Chan, J., Vaughan, R. A., Uhl, G. R., Kuhar, M. J., and Pickel, V. M. (1997) Immunogold localization of the dopamine transporter: an ultrastructual study of the rat tegmental area. *J. Neurosci.* **17**, 4037.
141. Attwell, D. and Mobbs, P. (1994) Neurotransmitter transporters. *Curr. Opin. Neurobiol.* **4**, 353.
142. Kanner, B. I. and Sharon, I. (1978) Active transport of L-glutamate by membrane vesicles isolated from rat brain. *Biochemistry* **17**, 3949.
143. Barbour, B., Brew, H., and Attwell, D. (1988) Electrogenic glutamate uptake in glial cells is activated by intracellular potassium. *Nature* **335**, 433.
144. Kanner, B. I. and Bendahan, A. (1982) Binding order of substrates to the sodium and potassium ion-coupled L-glutamic acid transport from rat brain. *Biochemistry* **21**, 6327.
145. Bouvier, M., Szatkowski, M., Amato, A., and Attwell, D. (1992) The glial cell glutamate uptake carrier countertransports pH-changing anions. *Nature* **360**, 471.
146. Danbolt, N. C., Pines, G., and Kanner, B. I. (1990) Purification and reconstitution of the sodium- and potassium-coupled glutamate transport glycoprotein from rat brain. *Biochemistry* **29**, 6734.
147. Rothstein, J. D., Martin, L., Levey, A. I., Dykes-Hoberg, M., Jin, L., Wu, D., *et al.* (1994) Localization of neuronal and glial glutamate transporters. *Neuron* **13**, 713.
148. Chaudhry, F. A., Lehre, K. P., vanLookeren Campagne, M., Otterson, O. P., Danbolt, N. C., and Storm-Mathisen, J. (1995) Glutamate transporters in glial plasma membranes: highly differentiated localizations revealed by quantitative ultrastructural immunocytochemistry. *Neuron* **15**, 711.
149. Lehre, K. P., Levy, L. M., Ottersen, O. P., Storm-Mathisen, J., and Danbolt, N. C. (1995) Differential expression of two glial glutamate transporters in the rat brain: quantitative and immunocytochemical observations. *J. Neurosci.* **15**, 1835.
150. Mennerick, S., Dhond, R. P., Benz, A., Xu, W., Rothstein, J. D., Danbolt, N. C., *et al.* (1998) Neuronal expression of the glutamate transporter GLT-1 in hippocampal microcultures. *J. Neurosci.* **18**, 4490.
151. Robinson, M. B., Sinor, J. D., Dowd, L. A., and Kerwin, J. F. J. (1993) Subtypes of sodium-dependent high-affinity L-[^{3}H]glutamate transport activity: pharmacological specificity and regulation by sodium and potassium. *J. Neurochem.* **60**, 167.
152. Fairman, W. A., Vandenberg, R. J., Arriza, J. L., Kavanaugh, M. P., and Amara, S. G. (1995) An excitatory amino-acid transporter with properties of a ligand-gated chloride channel. *Nature* **375**, 599.
153. Nagao, S., Kwak, S., and Kanazawa, I. (1997) EAAT4, a glutamate transporter with properties of a chloride channel, is predominantly localized in Purkinje cell dendrites, and forms parasagittal compartments in rat cerebellum. *Neuroscience* **78**, 929.
154. Furuta, A., Martin, L. J., Lin, C. L., Dykes-Hoberg, M., and Rothstein, J. D. (1997) Cellular and synaptic localization of the neuronal glutamate transporters excitatory amino acid transporter 3 and 4. *Neuroscience* **81**, 1031.

155. Dehnes, Y., Chaudhry, F. A., Ullensvang, K., Lehre, K. P., Storm-Mathisen, J., and Danbolt, N. C. (1998) The glutamate transporter EAAT4 in rat cerebellar Purkinje cells: a glutamate-gated chloride channel concentrated near the synapse in parts of the dendritic membrane facing astroglia. *J. Neurosci.* **18**, 3606.
156. Arriza, J. L., Eliasof, S., Kavanaugh, M. P., and Amara, S. G. (1997) Excitatory amino acid transporter 5, a retinal glutamate transporter coupled to a chloride conductance. *Proc. Natl. Acad. Sci. USA* **94**, 4155.
157. Eliasof, S., Arriza, J. L., Leighton, B. H., Kavanaugh, M. P., and Amara, S. G. (1998) Excitatory amino acid transporters of the salamander retina: identification, localization and function. *J. Neurosci.* **18**, 698.
158. Sheng, M. (1996) PDZs and receptor/channel clustering: rounding up the latest suspects. *Neuron* **17**, 575.
159. Arriza, J. L., Kavanaugh, M. P., Fairman, W. A., Wu, Y.-N., Murdoch, G. H., North, R. A., *et al.* (1993) Cloning and expression of a human neutral amino acid transporter with structural similarity to the glutamate transporter gene family. *J. Biol. Chem.* **268**, 15329.
160. Utsunomiya-Tate, N., Endou, H., and Kanai, Y. (1996) Cloning and functional characterization of a system ASC-like Na^+-dependent neutral amino acid transporter. *J. Biol. Chem.* **271**, 14883.
161. Kekuda, R., Prasad, P. D., Fei, Y.-J., Torres-Zamorano, V., Sinha, S., Yang-Feng, T. L., *et al.* (1996) Molecular and functional characterization of intestinal Na^+-dependent, broad-scope, neutral amino acid transporter Bo from a human placental choriocarcinoma cell line. *J. Biol. Chem.* **271**, 18657.
162. Zerangue, N. and Kavanaugh, M. P. (1996) ASCT-1 is a neutral amino acid exchanger with chloride channel activity. *J. Biol. Chem.* **271**, 27991.
163. Arriza, J. L., Fairman, W. A., Wadiche, J. I., Murdoch, G. H., Kavanaugh, M. P., and Amara, S. G. (1994) Functional comparisons of three glutamate transporters cloned from human motor cortex. *J. Neurosci.* **14**, 5559.
164. Lebrun, B., Sakaitani, M., Shimamoto, K., Yasuda-Kamatani, Y., and Nakajima, T. (1997) New β-hydroxyaspartate derivatives are competitive blockers for the bovine glutamate/aspartate transporter. *J. Biol. Chem.* **272**, 20336.
165. Vandenberg, R. J., Mitrovic, A. D. C. M., Balcar, V. J., and Johnston, G. A. R. (1997) Contrasting modes of action of methylglutamate derivatives on the excitatory amino acid transporters, EAAT1 and EAAT2. *Mol. Pharmacol.* **51**, 809.
166. Kanner, B. I. (1993) Glutamate transporters from brain. A novel neurotransmitter transporter family. *FEBS Lett.* **325**, 95.
167. Kanai, Y., Nussberger, S., Romero, M. F., Boron, W. F., Herbert, S. C., and Hediger, M. A. (1995) Electrogenic properties of the epithelial and neuronal high affinity glutamate transporter. *J. Biol. Chem.* **270**, 16561.
168. Klockner, U., Storck, T., Conradt, M., and Stoffel, W. (1993) Electrogenic L-glutamate uptake in *Xenopus laevis* oocytes expressing a cloned rat brain L-glutamate/L-aspartate transporter (GLAST-1). *J. Biol. Chem.* **268**, 14594.
169. Zerangue, N. and Kavanaugh, M. P. (1995) Flux coupling in a neuronal glutamate transporter. *Nature* **383**, 634.
170. Levy, L. M., Warr, O., and Attwell, D. (1998) Stoichiometry of the glial glutamate transporter GLT-1 expressed inducibly in a chinese hamster ovary cell line selected for low endogenous Na+-dependent glutamate uptake. *J. Neurosci.* **18**, 9620.
171. Wadiche, J. I., Arriza, J. L., Amara, S. G., and Kavanaugh, M. P. (1995) Kinetics of a human glutamate transporter. *Neuron* **14**, 1019.

172. Zerangue, N. and Kavanaugh, M. P. (1996) Interaction of L-cysteine with a human excitatory amino acid transporter. *J. Physiol.* **493**, 419.
173. Tachibana, M. and Kaneko, A. (1988) L-glutamate-induced depolarization in solitary photoreceptors: a process that may contribute to the interaction between photoreceptors in situ. *Proc. Natl. Acad. Sci. USA* **85**, 5315.
174. Sarantis, M., Everett, K., and Attwell, D. (1988) A presynaptic action of glutamate at the cone output synapse. *Nature* **332**, 451.
175. Eliasof, S. and Werblin, F. (1993) Characterization of the glutamate transporter in retinal cones of the tiger salamander. *J. Neurosci.* **13**, 402.
176. Wadiche, J. I., Amara, S. G., and Kavanaugh, M. P. (1995) Ion fluxes associated with excitatory amino acid transport. *Neuron* **15**, 721.
177. Wadiche, J. I. and Kavanaugh, M. P. (1998) Macroscopic and microscopic properties of a cloned glutamate transporter/chloride channel. *J. Neurosci.* **18**, 7650.
178. Wahle, S. and Stoffel, W. (1996) Membrane topology of the high-affinity L-glutamate transporter (GLAST-1) of the central nervous system. *J. Cell Biol.* **135**, 1867.
179. Slotboom, D. J., Lolkema, J. S., and Konings, W. N. (1996) Membrane topology of the C-terminal half of the neuronal, glial and bacterial glutamate transporter family. *J. Biol. Chem.* **271**, 31327.
180. Grunewald, M., Bendahan, A., and Kanner, B. I. (1998) Biotinylation of single cysteine mutants of the glutamate transporter GLT-1 from rat brain reveals its topology. *Neuron* **21**, 623.
181. Vandenberg, R. J., Arriza, J. L., Amara, S. G., and Kavanaugh, M. P. (1995) Constitutive ion fluxes and substrate binding domains of human glutamate transporters. *J. Biol. Chem.* **270**, 17668.
182. Mitrovic, A. D., Amara, S. G., Johnston, G. A., and Vandenberg, R. J. (1998) Identification of functional domains of the human glutamate transporters EAAT1 and EAAT2. *J. Biol. Chem.* **273**, 14698.
183. Conradt, M. and Stoffel, W. (1995) Functional analysis of the high affinity, Na^+-dependent glutamate transporter GLAST-1 by site-directed mutagenesis. *J. Biol. Chem.* **270**, 25207.
184. Zhang, Y., Pines, G., and Kanner, B. I. (1994) Histidine 326 is critical for the function of GLT-1, a (Na^+ and K^+)-coupled glutamate transporter from rat brain. *J. Biol. Chem.* **269**, 19573.
185. Pines, G., Zhang, Y., and Kanner, B. I. (1995) Glutamate 404 is involved in the substrate discrimination of GLT-1, a Na^+ and K^+-coupled glutamate transporter from rat brain. *J. Biol. Chem.* **270**, 17093.
186. Kavanaugh, M. P., Bendahan, A., Zerangue, N., Zhang, Y., and Kanner, B. I. (1997) Mutation of an amino acid residue influencing potassium coupling in the glutamate transporter GLT-1 induces obligate exchange. *J. Biol. Chem.* **272**, 1703.
187. Zarbiv, R., Grunewald, M., Kavanaugh, M. P., and Kanner, B. I. (1998) Cysteine scanning of the surroundings of an alkali-ion binding site of the glutamate transporter GLT-1 reveals a conformationally sensitive residue. *J. Biol. Chem.* **273**, 14231.
188. Zhang, Y., Bendahan, A., Zarbiv, R., Kavanaugh, M. P., and Kanner, B. I. (1998) Molecular determinant of ion selectivity of a Na^+ and K^+-coupled rat brain glutamate transporter. *Proc. Natl. Acad. Sci. USA* **95**, 751.
189. Sahin-Toth, M. and Kaback, H. R. (1993) Properties of interacting aspartic and lysine residues in the lactose permease of *Escherichia coli*. *Biochemistry* **32**, 10027.
190. Haugeto, O., Ullensvang, K., Levy, L. M., Chaudhry, F. A., Honore, T., Nielsen, M., *et al.*

(1996) Brain glutamate transporter proteins from homomultimers. *J. Biol. Chem.* **271**, 27715.
191. Hestrin, S., Sah, P., and Nicoll, R. A. (1990) Mechanisms generating the time course of dual component excitatory synaptic currents recorded in hippocampal slices. *Neuron* **5**, 247.
192. Isaacson, J. S. and Nicoll, R. A. (1993) The uptake inhibitor L-trans-PDC enhances responses to glutamate but fails to alter the kinetics of excitatory synaptic currents in the hippocampus. *J. Neurophysiol.* **70**, 2187.
193. Sarantis, M., Ballerini, L., Miller, B., Silver, R. A., Edwards, M., and Attwell, D. (1993) Glutamate uptake from the synaptic cleft does not shape the decay of the non-NMDA component of the synaptic current. *Neuron* **11**, 541.
194. Takahashi, M., Kovalchuk, Y., and Attwell, D. (1995) Pre- and postsynaptic determinants of EPSC waveform at cerebellar climbing fiber and parallel fiber to Purkinje cell synapses. *J. Neurosci.* **15**, 5693.
195. Takahashi, M., Sarantis, M., and Attwell, D. (1996) Postsynaptic glutamate uptake in rat cerebellar Purkinje cells. *J. Physiol.* **497**, 523.
196. Mennerick, S. and Zorumski, C. F. (1994) Glial contributions to excitatory neurotransmission in cultured hippocampal cells. *Nature* **368**, 59.
197. Tong, G. and Jahr, C. E. (1994) Block of glutamate transporters potentiates postsynaptic excitation. *Neuron* **13**, 1195.
198. Diamond, J. S. and Jahr, C. E. (1997) Transporters buffer synaptically released glutamate on a submillisecond time scale. *J. Neurosci.* **17**, 4672.
199. Otis, T. S. and Jahr, C. E. (1998) Anion currents and predicted glutamate flux through a neuronal glutamate transporter. *J. Neurosci.* **18**, 7099.
200. Otis, T. S., Kavanaugh, M. P., and Jahr, C. E. (1997) Postsynaptic glutamate transport at the climbing fiber-Purkinje cell synapse. *Science* **277**, 1515.
201. Bergles, D. E. and Jahr, C. E. (1998) Glial contribution to glutamate uptake at Schaffer collateral-commissural synapses in the hippocampus. *J. Neurosci.* **18**, 7709.
202. Mennerick, S., Benz, A., and Zorumski, C. F. (1996) Components of glial responses to exogenous and synaptic glutamate in rat hippocampal cultures. *J. Neurosci.* **16**, 55.
203. Bergles, D. E., Dzubay, J. A., and Jahr, C. E. (1997) Glutamate transporter currents in bergmann glial cells follow the time course of extrasynaptic glutamate. *Proc. Natl. Acad. Sci. USA* **94**, 14821.
204. Diamond, J. S., Bergles, D. E., and Jahr, C. E. (1998) Glutamate release monitored with astrocyte transporter currents during LTP. *Neuron* **21**, 425.
205. Bergles, D. E., Dzubay, J. A., and Jahr, C. E. (1997) Glutamate transporter currents in bergmann glial cells follow the time course of extrasynaptic glutamate. *Proc. Natl. Acad. Sci. USA* **94**, 14821.
206. Luscher, C., Malenka, R. C., and Nicoll, R. A. (1998) Monitoring glutamate release during LTP with glial transporter currents. *Neuron* **21**, 335.
207. Picaud, S. A., Larsson, H. P., Grant, G. B., Lecar, H., and Werblin, F. S. (1995) Glutamate-gated chloride channel with glutamate-transporter-like properties in cone photoreceptors of the tiger salamander. *J. Neurophysiol.* **74**, 1760.
208. Larsson, H. P., Picaud, S. A., Werblin, F. S., and Lecar, H. (1996) Noise analysis of the glutamate-activated current in photoreceptors. *Biophys. J.* **70**, 733.
209. Picaud, S., Larsson, H. P., Wellis, D. P., Lecar, H., and Werblin, F. (1995) Cone photoreceptors respond to their own glutamate release in the tiger salamander. *Proc. Natl. Acad. Sci. USA* **92**, 9417.

210. Grant, G. B. and Dowling, J. E. (1995) A glutamate-activated chloride current in cone-driven ON bipolar cells of the white perch retina. *J. Neurosci.* **15**, 3852.
211. Hamberger, A. C., Chiang, G. H., Nylen, E. S., Scheff, S. W., and Cotman, C. W. (1979) Glutamate as a CNS transmitter. I. Evaluation of glucose and glutamine as precursors for the synthesis of preferentially released glutamate. *Brain Res.* **168**, 513.
212. Hamberger, A., Chiang, G. H., Sandoval, E., and Cotman, C. W. (1979) Glutamate as a CNS transmitter. II. Regulation of synthesis in the releasable pool. *Brain Res.* **168**, 531.
213. Ward, H. K., Thanki, C. M., and Bradford, H. F. (1983) Glutamine and glucose as precursors of transmitter amino acids: *ex vivo* studies. *J. Neurochem.* **40**, 855.
214. Raizen, D. M. and Avery, L. (1994) Electrical activity and behavior in the pharynx of *Caenorhabditis elegans*. *Neuron* **12**, 483.
215. Ni, B., Rosteck, P. R., Nadi, N. S., and Paul, S. M. (1994) Cloning and expression of a cDNA encoding a brain-specific Na^+-dependent inorganic phosphate cotransporter. *Proc. Natl. Acad. Sci. USA* **91**, 5607.
216. Ni, B., Wu, X., Yan, G.-M., Wang, J., and Paul, S. M. (1995) Regional expression and cellular localization of the Na^+-dependent inorganic phosphate cotransporter of rat brain. *J. Neurosci.* **15**, 5789.
217. Berger, A. J., Hart, A. C., and Kaplan, J. M. (1998) G alphas-induced neurodegeneration in *Caenorhabditis elegans*. *J. Neurosci.* **18**, 2871.
218. Bellocchio, E. E., Hu, H., Pohorille, A., Chan, J., Pickel, V. M., and Edwards, R. H. (1998) The localization of the brain-specific inorganic phosphate transporter suggests a specific presynaptic role in glutamatergic transmission. *J. Neurosci.* **18**, 8648.
219. Rothstein, J. D., Tsai, G., Kuncl, R. W., Clawson, L., Cornblath, D. R., Drachman, D. B., *et al.* (1990) Abnormal excitatory amino acid metabolism in amyotrophic lateral sclerosis. *Ann. Neurol.* **28**, 18.
220. Shaw, P. J., Forrest, V., Ince, P. G., Richardson, J. P., and Wastell, H. J. (1995) CSF and plasma amino acid levels in motor neuron disease: elevation of CSF glutamate in a subset of patients. *Neurodegeneration* **4**, 209.
221. Rothstein, J. D., Martin, L. J., and Kuncl, R. W. (1992) Decreased glutamate transport by the brain and spinal cord in amyotrophic lateral sclerosis. *N. Engl. J. Med.* **326**, 1464.
222. Rothstein, J. D., Jin, L., Dykes-Hoberg, M., and Kuncl, R. W. (1993) Chronic inhibition of glutamate uptake produces a model of slow neurotoxicity. *Proc. Natl. Acad. Sci. USA* **90**, 6591.
223. Rothstein, J. D., Van Kammen, M., Levey, A. I., Martin, L. J., and Kuncl, R. W. (1995) Selective loss of glial glutamate transporter GLT-1 in amyotrophic lateral sclerosis. *Ann. Neurol.* **38**, 73.
224. Rothstein, J. D., Dykes-Hoberg, M., Pardo, C. A., Bristol, L. A., Jin, L. A., Jin, L., *et al.* (1996) Knockout of glutamate transporters reveals a major role for astroglial transport in excitotoxicity and clearance of glutamate. *Neuron* **16**, 675.
225. Aoki, M., Lin, C. L. G., Rothstein, J. D., Geller, B. A., Hosler, B. A., Munsat, T. L., *et al.* (1998) Mutations in the glutamate transporter EAAT2 gene do not cause abnormal transcripts in ALS. *Ann. Neurol.* **43**, 645.
226. Bristol, L. A. and Rothstein, J. D. (1996) Glutamate transporter gene expression in amyotrophic lateral sclerosis motor cortex. *Ann. Neurol.* **39**, 676.
227. Lin, C. L., Bristol, L. A., Jin, L., Dykes-Hoberg, M., Crawford, T., Clawson, L., *et al.* (1998) Aberrant RNA processing in a neurodegenerative disease: the cause for absent EAAT2, a glutamate transporter, in amyotrophic lateral sclerosis. *Neuron* **20**, 589.
228. Tanaka, K., Watase, K., Manabe, T., Yamada, K., Watanabe, M., Takahashi, K., *et al.*

(1997) Epilepsy and exacerbation of brain injury in mice lacking the glutamate transporter GLT-1. *Science* **276**, 1699.

229. Takahashi, M., Billups, B., Rossi, D., Sarantis, M., Hamann, M., and Attwell, D. (1997) The role of glutamate transporters in glutamate homeostasis in the brain. *J. Exp. Biol.* **200**, 401.
230. Szatkowski, M., Barbour, B., and Attwell, D. (1990) Non-vesicular release of glutamate from glial cells by reversed electrogenic glutamate uptake. *Nature* **348**, 443.
231. Billups, B. and Attwell, D. (1996) Modulation of non-vesicular glutamate release by pH. *Nature* **379**, 171.
232. Peghini, P., Janzen, J., and Stoffel, W. (1997) Glutamate transporter EAAC-1-deficient mice develop dicarboxylic aminoaciduria and behavioral abnormalities but no neuro-degeneration. *EMBO J.* **16**, 3822.
233. Swanson, R. A., Liu, J., Miller, J. W., Rothstein, J. D., Farrell, K., Stein, B. A., *et al.* (1997) Neuronal regulation of glutamate transporter subtype expression in astrocytes. *J. Neurosci.* **17**, 932.
234. Schlag, B. D., Vondrasek, J. R., Munir, M., Kalandadze, A., Zelenaia, O. A., Rothstein, J. D., *et al.* (1998) Regulation of the glial Na^+-dependent glutamate transporters by cyclic AMP analogs and neurons. *Mol. Pharm.* **53**, 355.
235. Furuta, A., Rothstein, J. D., and Martin, L. J. (1997) Glutamate transporter protein subtypes are expressed differentially during rat CNS development. *J. Neurosci.* **17**, 8363.
236. Casado, M., Bendahan, A., Zafra, F., Danbolt, N. C., Aragon, C., Gimenez, C., *et al.* (1993) Phosphorylation and modulation of brain glutamate transporters by protein kinase C. *J. Biol. Chem.* **268**, 27313.
237. Davis, K. E., Straff, D. J., Weinstein, E. A., Bannerman, P. G., Correale, D. M., Rothstein, J. D., *et al.* (1998) Multiple signaling pathways regulate cell surface expression and activity of the excitatory amino acid carrier 1 subtype of Glu transporter in C6 glioma. *J. Neurosci.* **18**, 2475.
238. Barbour, B., Szatkowski, M., Ingledew, N., and Attwell, D. (1989) Arachidonic acid induces a prolonged inhibition of glutamate uptake into glial cells. *Nature* **342**, 918.
239. Chan, P. H., Kerlan, R., and Fishman, R. A. (1993) Reductions of γ-aminobutyric acid and glutamate uptake and (Na^+ and K^+)ATPase activity in brain slices and synaptosomes by arachidonic acid. *J. Neurochem.* **40**, 309.
240. Volterra, A., Trotti, D., and Racagni, G. (1994) Glutamate uptake is inhibited by arachidonic acid and oxygen radicals via two distinct and additive mechanisms. *Mol. Pharm.* **46**, 986.
241. Trotti, D., Volterra, A., Lehre, K. P., Rossi, D., Gjesdal, O., Racagni, G., *et al.* (1995) Arachidonic acid inhibits a purified and reconstituted glutamate transporter directly from the water phase and not via the phospholipid membrane. *J. Biol. Chem.* **270**, 9890.
242. Kataoka, Y., Morii, H., Watanabe, Y., and Ohmori, H. (1997) A postsynaptic excitatory amino acid transporter with chloride conductance functionally regulated by neuronal activity in cerebellar Purkinje cells. *J. Neurosci.* **17**, 7017.
243. Zerangue, N., Arriza, J. L., Amara, S. G., and Kavanaugh, M. P. (1995) Differential modulation of human glutamate transporter subtypes by arachidonic acid. *J. Biol. Chem.* **270**, 6433.
244. Fairman, W. A., Sonders, M. S., Murdoch, G. H., and Amara, S. G. (1998) Arachidonic acid elicits a substrate-gated proton current associated with the glutamate transporter EAAT4. *Nature Neurosci.* **1**, 105.
245. Tzingounis, A. V., Lin, C. L., Rothstein, J. D., and Kavanaugh, M. P. (1998) Arachidonic

acid activates a proton current in the rat glutamate transporter EAAT4. *J. Biol. Chem.* **273**, 17315.

246. Spiridon, M., Kamm, D., Billups, B., Mobbs, P., and Attwell, D. (1998) Modulation by zinc of the glutamate transporters in glial cells and cones isolated from the tiger salamander retina. *J. Physiol.* **506**, 363.
247. Volterra, A., Trotti, D., Tromba, C., Floridi, S., and Racagni, G. (1994) Glutamate uptake inhibition by oxygen free radicals in rat cortical astrocytes. *J. Neurosci.* **14**, 2924.
248. Trotti, D., Danbolt, N. C., and Volterra, A. (1998) Glutamate transporters are oxidant-vulnerable: a molecular link between oxidative and excitotoxic neurodegeneration? *Trends Pharm. Sci.* **19**, 328.
249. Trotti, D., Rossi, D., Gjesdal, O., Levy, L. M., Racagni, G., Danbolt, N. C., *et al.* (1996) Peroxynitrite inhibits glutamate transporter subtypes. *J. Biol. Chem.* **271**, 5976.
250. Stevens, T. H. and Forgac, M. (1997) Structure, function and regulation of the vacuolar H+-ATPase. *Annu. Rev. Cell Dev. Biol.* **13**, 779.
251. Knoth, J., Zallakian, M., and Njus, D. (1981) Stoichiometry of H^+-linked dopamine transport in chromaffin granule ghosts. *Biochemistry* **20**, 6625.
252. Daniels, A. J. and Reinhard, J. F. (1988) Energy-driven uptake of the neurotoxin 1-methyl-4-phenylpyridinium into chromaffine granules via the catcholamine transporter. *J. Biol. Chem.* **263**, 5034.
253. Scherman, D., Darchen, F., Desnos, C., and Henry, J. P. (1988) 1-Methyl-4-phenylpyridinium is a substrate of the vesicular monoamine transport system of chromaffin granules. *Eur. J. Pharmacol.* **146**, 359.
254. Liu, Y. and Edwards, R. H. (1997) The role of vesicular transport proteins in synaptic transmission and neural degeneration. *Annu. Rev. Neurosci.* **20**, 125.
255. Anderson, D. C., King, S. C., and Parsons, S. M. (1982) Proton gradient linkage to active uptake of ^{3}H-acetylcholine by Torpedo electric organ synaptic vesicles. *Biochemistry* **21**, 3037.
256. Nguyen, M. L., Cox, G. D., and Parsons, S. M. (1998) Kinetic parameters for the vesicular acetylcholine transporter: two protons are exchanged for one acetylcholine. *Biochemistry* **37**, 13400.
257. Parsons, S. M., Prior, C., and Marshall, I. G. (1993) Acetylcholine transport, storage and release. *Int. Rev. Neurobiol.* **35**, 279.
258. Carlson, M. D., Kish, P. E., and Ueda, T. (1989) Characterization of the solubilized and reconstituted ATP-dependent vesicular glutamate uptake system. *J. Biol. Chem.* **264**, 7369.
259. Maycox, P. R., Deckwerth, T., Hell, J. W., and Jahn, R. (1988) Glutamate uptake by brain synaptic vesicles. Energy dependence of transport and functional reconstitution in proteoliposomes. *J. Biol. Chem.* **263**, 15423.
260. Naito, S. and Ueda, T. (1983) Adenosine triphosphate-dependent uptake of glutamate into protein I-associated synaptic vesicles. *J. Biol. Chem.* **258**, 696.
261. Hartinger, J. and Jahn, R. (1993) An anion binding site that regulates the glutamate transporter of synaptic vesicles. *J. Biol. Chem.* **268**, 23122.
262. Tabb, J. S., Kish, P. E., Van Dyke, R., and Ueda, T. (1992) Glutamate transport into synaptic vesicles. *J. Biol. Chem.* **267**, 15412.
263. Hell, J. W., Maycox, P. R., and Jahn, R. (1990) Energy dependence and functional reconstitution of the gamma-aminobutyric acid carrier from synaptic vesicles. *J. Biol. Chem.* **265**, 2111.
264. Merickel, A. and Edwards, R. H. (1995) Transport of histamine by vesicular monoamine transporter-2. *Neuropharmacology* **34**, 1543.

265. Scherman, D. and Henry, J. P. (1984) Reserpine binding to bovine chromaffin granule membranes. Characterization and comparison with dihydrotetrabenazine binding. *Mol. Pharmacol.* **25**, 113.
266. Weaver, J. H. and Deupree, J. D. (1982) Conditions required for reserpine binding to the catecholamine transporter on chromaffin granule ghosts. *Eur. J. Pharmacol.* **80**, 437.
267. Rudnick, G., Steiner-Mordoch, S. S., Fishkes, H., Stern-Bach, Y., and Schuldiner, S. (1990) Energetics of reserpine binding and occlusion by the chromaffin granule biogenic amine transporter. *Biochemistry* **29**, 603.
268. Schuldiner, S., Liu, Y., and Edwards, R. H. (1993) Reserpine binding to a vesicular amine transporter expressed in chinese hamster ovary fibroblasts. *J. Biol. Chem.* **268**, 29.
269. Darchen, F., Scherman, E., and Henry, J. P. (1989) Reserpine binding to chromaffin granules suggests the existence of two conformations of the monoamine transporter. *Biochemistry* **28**, 1692.
270. Stern-Bach, Y., Greenberg-Ofrath, N., Flechner, I., and Schuldiner, S. (1990) Identification and purification of a functional amine transporter from bovine chromaffin granules. *J. Biol. Chem.* **265**, 3961.
271. Vincent, M. S. and Near, J. A. (1991) Identification of a [^{3}H]dihydrotetrabenazine-binding protein from bovine adrenal medulla. *Mol. Pharmacol.* **40**, 889.
272. Isambert, M. F., Gasnier, B., Botton, D., and Henry, J. P. (1992) Characterization and purification of the monoamine transporter of bovine chromaffin granules. *Biochemistry* **31**, 1980.
273. Langston, J. W., Irwin, I., Langston, E. G., and Forno, L. S. (1984) Pargyline prevents MPTP-induced parkinsonism in primates. *Science* **225**, 1480.
274. Heikkila, R. E., Manzino, L., Cabbat, F. S., and Duvoisin, R. C. (1984) Protection against the dopaminergic neurotoxicity of 1-methyl-4-phenyl-1,2,5,6-tetrahydropyridine by monoamine oxidase inhibitors. *Nature* **311**, 467.
275. Javitch, J., D'Amato, R., Nye, J., and Javitch, J. (1985) Parkinsonism-inducing neurotoxin, N-methyl-4-phenyl-1,2,3,6-tetrahydropyridine: uptake of the metabolite N-methyl-4-phenylpyridine by dopamine neurons explains selective toxicity. *Proc. Natl. Acad. Sci. USA* **82**, 2173.
276. Nicklas, W. J., Vyas, I., and Heikkila, R. E. (1985) Inhibition of NADH-linked oxidation in brain mitochondria by 1-methyl-4-phenylpyridine, a metabolite of the neurotoxin 1-methyl-4-phenyl-1,2,3,6-tetrahydropyridine. *Life Sci.* **36**, 2503.
277. Ramsay, R. R., Krueger, M. J., Youngster, S. K., Gluck, M. R., Casida, J. E., and Singer, T. P. (1991) Interaction of 1-methyl-4-phenylpyridinium (MPP^+) and its analogs with the rotenone/piericidin binding site of NADH dehydrogenase. *J. Neurochem.* **56**, 1184.
278. Nguyen, T. T., Postle, K., and Bertrand, K. P. (1983) Sequence homology between the tetracycline-resistance determinants of Tn10 and pBR322. *Gene* **25**, 83.
279. Neal, R. J. and Chater, K. F. (1987) Nucleotide sequence analysis reveals similarities between proteins determining methylenomycin A resistance in streptomyces and tetracycline resistance in eubacteria. *Gene* **58**, 229.
280. Neyfakh, A. A., Bidnenko, V. E., and Chen, L. B. (1991) Efflux-mediated multidrug resistance in Bacillus subtilis: similarities and dissimilarities with the mammalian system. *Proc. Natl. Acad. Sci. USA* **88**, 4781.
281. Kaneko, M., Yamaguchi, A., and Sawai, T. (1985) Energetics of tetracycline efflux system encoded by Tn10 in *Escherichia coli*. *FEBS Lett.* **193**, 194.
282. Sulzer, D. and Rayport, S. (1990) Amphetamine and other psychostimulants reduce pH

gradients in midbrain dopaminergic neurons and chromaffin granules: a mechanism of action. *Neuron* **5**, 797.

283. Sulzer, D., Maidment, N. T., and Rayport, S. (1993) Amphetamine and other weak bases act to promote reverse transport of dopamine in ventral midbrain neurons. *J. Neurochem.* **60**, 527.
284. Schuldiner, S., Steiner-Mordoch, S., Yelin, R., Wall, S. C., and Rudnick, G. (1993) Amphetamine derivatives interact with both plasma membrane and secretory vesicle biogenic amine transporters. *Mol. Pharm.* **44**, 1227.
285. Wall, S. C., Gu, H., and Rudnick, G. (1995) Biogenic amine flux mediated by cloned transporters stably expressed in cultured cell lines: amphetamine specificity for inhibition and efflux. *Mol. Pharm.* **47**, 544.
286. Maron, R., Stern, Y., Kanner, B. I., and Schuldiner, S. (1983) Functional asymmetry of the amine transporter from chromaffin granules. *J. Biol. Chem.* **258**, 11476.
287. Steele, T. D., McCann, U. D., and Ricaurte, G. A. (1994) 3,4-Methylenedioxymethamphetamine (MDMA, ecstasy): pharmacology and toxicology in animals and humans. *Addiction* **89**, 539.
288. Fischer, C., Hatzidimitriou, G., Wios, J., Katz, J., and Ricaurte, G. (1995) Reorganization of ascending 5-HT axon projections in animals previously exposed to the recreational drug 3,4-methylenedioxymethamphetamine (MDMA, 'ecstasy'). *J. Neurosci.* **15**, 5476.
289. Cubells, J. F., Rayport, S., Rajendran, G., and Sulzer, D. (1994) Methamphetamine neurotoxicity involves vacuolation of endocytic organells and dopamine-dependent intracellular oxidative stress. *J. Neurosci.* **14**, 2260.
290. Weihe, E., Schafer, M. K., Erickson, J. D., and Eiden, L. E. (1994) Localization of vesicular monoamine transporter isoforms (VMAT1 and VMAT2) to endocrine cells and neurons in rat. *J. Mol. Neurosci.* **5**, 149.
291. Peter, D., Liu, Y., Sternini, C., de Giorgio, R., Brecha, N., and Edwards, R. H. (1995) Differential expression of two vesicular monoamine transporters. *J. Neurosci.* **15**, 6179.
292. De Giorgio, R., Su, D., Peter, D., Edwards, R. H., Brecha, N. C., and Sternini, C. (1996) Vesicular monoamine transporter 2 expression in enteric neurons and enterochromaffin-like cells of the rat. *Neurosci. Lett.* **217**, 77.
293. Peter, D., Jimenez, J., Liu, Y., Kim, J., and Edwards, R. H. (1994) The chromaffin granule and synaptic vesicle amine transporters differ in substrate recognition and sensitivity to inhibitors. *J. Biol. Chem.* **269**, 7231.
294. Erickson, J. D., Schafer, M. K.-H., Bonner, T. I., Eiden, L. E., and Weihe, E. (1996) Distinct pharmacological properties and distribution in neurons and endocrine cells of two isoforms of the human vesicular monoamine transporter. *Proc. Natl. Acad. Sci. USA* **93**, 5166.
295. Finn, J. P. and Edwards, R. H. (1997) Individual residues contribute to multiple differences in ligand recognition between vesicular monoamine transporters 1 and 2. *J. Biol. Chem.* **272**, 16301.
296. Dimaline, R. and Struthers, J. (1996) Expression and regulation of a vesicular monoamine transporter in rat stomach: a putative histamine transporter. *J. Physiol.* **490**, 249.
297. Varoqui, H., Diebler, M.-F., Meunier, F.-M., Rand, J. B., Usdin, T. B., Bonner, T. I., *et al.* (1994) Cloning and expression of the vesamicol binding protein from the marine ray *Torpedo*. Homology with the putative vesicular acetylcholine transporter UNC-17 from *Caenorhabditis elegans*. *FEBS Lett.* **342**, 97.
298. Bejanin, S., Cervini, R., Mallet, J., and Berrard, S. (1994) A unique gene organization for

two cholinergic markers, choline acetyltransferase and a putative vesicular transporter of acetylcholine. *J. Biol. Chem.* **269**, 21944.

299. Roghani, A., Feldman, J., Kohan, S. A., Shirzadi, A., Gundersen, C. B., Brecha, N., *et al.* (1994) Molecular cloning of a putative vesicular transporter for acetylcholine. *Proc. Natl. Acad. Sci. USA* **91**, 10620.
300. Varoqui, H. and Erickson, J. D. (1996) Active transport of acetylcholine by the human vesicular acetylcholine transporter. *J. Biol. Chem.* **271**, 27229.
301. Liu, Y. and Edwards, R. H. (1997) Differential localization of vesicular acetylcholine and monoamine transporters in PC12 cells but not CHO cells. *J. Cell Biol.* **139**, 907.
302. Bahr, B. A. and Parsons, S. M. (1992) Purification of the vesamicol receptor. *Biochemistry* **31**, 5763.
303. Erickson, J. D., Varoqui, H., Schafer, M. D., Modi, W., Diebler, M. F., Weihe, E., *et al.* (1994) Functional identification of a vesicular acetylcholine transporter and its expression from a 'cholinergic' gene locus. *J. Biol. Chem.* **269**, 21929.
304. Alfonso, A., Grundahl, K., McManus, J. R., Asbury, J. M., and Rand, J. B. (1994) Alternative splicing leads to two cholinergic proteins in *Caenorhabditis elegans*. *J. Mol. Biol.* **241**, 627.
305. Kitamoto, T., Wang, W., and Salvaterra, P. M. (1998) Structure and organization of the *Drosophila* cholinergic locus. *J. Biol. Chem.* **273**, 2706.
306. Shimojo, M., Wu, D., and Hersh, L. B. (1998) The cholinergic gene locus is coordinately regulated by protein kinase A. *J. Neurochem.* **71**, 1118.
307. McIntire, S., Jorgensen, E., Kaplan, J., and Horvitz, H. R. (1993) The GABAergic nervous system of *Caenorhabditis elegans*. *Nature* **364**, 334.
308. McIntire, S., Jorgensen, E., and Horvitz, H. R. (1993) Genes required for GABA function in *Caenorhabditis elegans*. *Nature* **364**, 337.
309. Chaudhry, F. A., Reimer, R. J., Bellocchio, E. E., Danbolt, N. C., Osen, K. K., Edwards, R. H., *et al.* (1998) The vesicular GABA transporter VGAT localizes to synaptic vesicles in sets of glycinergic as well as GABAergic neurons. *J. Neurosci.* **18**, 9733.
310. Sagne, C., El Mestikawy, S., Isambert, M.-F., Hamon, M., Henry, J.-P., Giros, B., *et al.* (1997) Cloning of a functional vesicular GABA and glycine transporter by screening of genome databases. *FEBS Lett.* **417**, 177.
311. Hell, J. W., Maycox, P. R., Stadler, H., and Jahn, R. (1988) Uptake of GABA by rat brain synaptic vesicles isolated by a new procedure. *EMBO J.* **7**, 3023.
312. Jonas, P., Bischofberger, J., and Sandkuhler, J. (1998) Corelease of two fast neurotransmitters at a central synapse. *Science* **281**, 419.
313. Nicoll, R. A. and Malenka, R. C. (1998) A tale of two transmitters. *Science* **281**, 360.
314. Fischer, W. N., Kwart, M., Hummel, S., and Frommer, W. B. (1995) Substrate specificity and expression profile of amino acid transporters (AAPs) in *Arabidopsis*. *J. Biol. Chem.* **270**, 16315.
315. Brake, A. J. and Julius, D. (1996) Signaling by extracellular nucleotides. *Annu. Rev. Cell Dev. Biol.* **12**, 519.
316. Aberer, W., Kostron, H., Huber, E., and Winkler, H. (1978) A characterization of the nucleotide uptake by chromaffin granules of bovine adrenal medulla. *Biochem. J.* **172**, 353.
317. Luqmani, Y. A. and Giompres, P. (1981) On the specificity of uptake by isolated Torpedo synaptic vesicles. *Neurosci. Lett.* **23**, 81.
318. Bankston, L. A. and Guidotti, G. (1996) Characterization of ATP transport into chromaffin granule ghosts. Synergy of ATP and serotonin accumulation in chromaffin granule ghosts. *J. Biol. Chem.* **271**, 17132.

319. Stadler, H. and Fenwick, E. M. (1983) Cholinergic synaptic vesicles from Torpedo marmorata contain an atractyloside-binding protein related to the mitochondrial ADP/ATP carrier. *Eur. J. Biochem.* **136**, 377.
320. Gruninger, H. A., Apps, D. K., and Pillips, J. H. (1983) Adenine nucleotide and phosphoenolpyruvate transport by bovine chromaffin granule 'ghosts'. *Neuroscience* **9**, 917.
321. Gualix, J., Abal, M., Pintor, J., Garcia-Carmona, F., and Miras-Portugal, M. T. (1996) Nucleotide vesicular transporter of bovine chromaffin granules. Evidence for mnemonic regulation. *J. Biol. Chem.* **271**, 1957.
322. Perin, M. S., Fried, V. A., Slaughter, C. A., and Sudhof, T. C. (1988) The structure of cytochrome b561, a secretory vesicle-specific electron transport protein. *EMBO J.* **7**, 2697.
323. Fleming, P. J. and Kent, U. M. (1991) Cytochrome b561, ascorbic acid and transmembrane electron transfer. *Am. J. Clin. Nutr.* **54**, 1173S.
324. Tirrell, J. G. and Westhead, E. W. (1979) The uptake of ascorbic acid and dehydroascorbic acid by chromaffin granules of the adrenal medulla. *Neuroscience* **4**, 181.
325. Daniels, A. J., Dean, G., Viveros, O. H., and Diliberto, E. J. J. (1983) Secretion of newly taken up ascorbic acid by adrenomedullary chromaffin cells originates from a compartment different from the catecholamine storage vesicle. *Mol. Pharm.* **23**, 437.
326. Diliberto, E. J. J., Heckman, G. D., and Daniels, A. J. (1983) Characterization of ascorbic acid transport by adrenomedullary chromaffin cells. Evidence for Na^+-dependent cotransport. *J. Biol. Chem.* **258**, 12886.
327. Buckley, K. and Kelly, R. B. (1985) Identification of a transmembrane glycoprotein specific for secretory vesicles of neural and endocrine cells. *J. Cell Biol.* **100**, 1284.
328. Feany, M. B., Lee, S., Edwards, R. H., and Buckley, K. M. (1992) The synaptic vesicle protein SV2 is a novel type of transmembrane transporter. *Cell* **70**, 861.
329. Bajjalieh, S. M., Peterson, K., Shinghal, R., and Scheller, R. H. (1992) SV2, a brain synaptic vesicle protein homologous to bacterial transporters. *Science* **257**, 1271.
330. Gingrich, J. A., Andersen, P. H., Tiberi, M., El Mestikawy, S., Jorgensen, P. N., Fremeau, R. T. J., *et al.* (1992) Identification, characterization and molecular cloning of a novel transporter-like protein localized to the central nervous system. *FEBS Lett.* **312**, 115.
331. Nanavati, C. and Fernandez, J. M. (1993) The secretory granule matrix: a fast-acting smart polymer. *Science* **259**, 963.
332. Bajjalieh, S. M., Peterson, K., Linial, M., and Scheller, R. H. (1993) Brain contains two forms of synaptic vesicle protein 2. *Proc. Natl. Acad. Sci. USA* **90**, 2150.
333. Bajjalieh, S. M., Frantz, G. D., Weimann, J. M., McConnell, S. K., and Scheller, R. H. (1994) Differential expression of synaptic vesicle protein 2 (SV2) isoforms. *J. Neurosci.* **14**, 5223.
334. Schivell, A. E., Batchelor, R. H., and Bajjalieh, S. M. (1996) Isoform-specific, calcium-regulated interaction of the synaptic vesicle proteins SV2 and synaptotagmin. *J. Biol. Chem.* **271**, 27770.
335. Merickel, A., Rosandich, P., Peter, D., and Edwards, R. H. (1995) Identification of residues involved in substrate recognition by a vesicular monoamine transporter. *J. Biol. Chem.* **270**, 25798.
336. Strader, C. D., Sigal, I. S., Register, R. B., Candelore, M. R., Rands, E., and Dixon, R. A. F. (1987) Identification of residues required for ligand binding to the beta-adrenergic receptor. *Proc. Natl. Acad. Sci. USA* **84**, 4384.
337. Strader, C. D., Candelore, M. R., Hill, W. S., Sigal, I. S., and Dixon, R. A. F. (1989) Identification of two serine residues involved in agonist activation of the beta-adrenergic receptor. *J. Biol. Chem.* **264**, 13572.

338. Peter, D., Vu, T., and Edwards, R. H. (1996) Chimeric vesicular monoamine transporters identify structural domains that influence substrate affinity and sensitivity to tetrabenazine. *J. Biol. Chem.* **271**, 2979.
339. Finn, J. P. and Edwards, R. H. (1998) Multiple residues contribute independently to differences in ligand recognition between vesicular monoamine transporters 1 and 2. *J. Biol. Chem.* **273**, 3943.
340. Sievert, M. K., Thiriot, D. S., Edwards, R. H., and Ruoho, A. E. (1998) High-efficiency expression and characterization of the synaptic vesicle monoamine transporter from baculovirus-infected insect cells. *Biochem. J.* **330**, 959.
341. Sievert, M. K. and Ruoho, A. E. (1997) Peptide mapping of the ^{125}I-iodoazidoketanserin and ^{125}I-2-N-[(3′-iodo-4′-azidophenyl)propionyl]tetrabenazine binding sites for the synaptic vesicle monoamine transporter. *J. Biol. Chem.* **272**, 26049.
342. Merickel, A., Kaback, H. R., and Edwards, R. H. (1997) Charged residues in transmembrane domains II and XI of a vesicular monoamine transporter form a charge pair that promotes high affinity substrate recognition. *J. Biol. Chem.* **272**, 5403.
343. Steiner-Mordoch, S., Shirvan, A., and Schuldiner, S. (1996) Modification of the pH profile and tetrabenazine sensitivity of rat VMAT1 by replacement of aspartate 404 with glutamate. *J. Biol. Chem.* **271**, 13048.
344. Van der Kloot, W. (1990) The regulation of quantal size. *Prog. Neurobiol.* **36**, 93.
345. Van der Kloot, W. and Van der Kloot, T. E. (1985) Catecholamines, insulin and ACTH increase quantal size at the frog neuromuscular junction. *Brain Res.* **376**, 378.
346. Van der Kloot, W. (1993) Nicotinic agonists antagonize quantal size increases and evoked release at frog neuromuscular junction. *J. Physiol.* **468**, 567.
347. Van der Kloot, W. (1991) Down-regulation of quantal size at frog neuromuscular junctions: possible roles for elevated intracellular calcium and for protein kinase C. *J. Neurobiol.* **22**, 204.
348. Van der Kloot, W. (1987) Pretreatment with hypertonic solutions increases quantal size at the neuromuscular junction. *J. Neurophysiol.* **57**, 1536.
349. Van der Kloot, W. (1987) Ammonium inhibits the packing of acetylcholine into quanta. *FASEB J.* **1**, 298.
350. Colquhoun, D., Large, W. A., and Rang, H. P. (1977) An analysis of the action of a false transmitter at the neuromuscular junction. *J. Physiol.* **266**, 361.
351. Large, W. A. and Rang, H. P. (1978) Factors affecting the rate of incorporation of a false transmitter into mammalian motor nerve terminals. *J. Physiol.* **285**, 1.
352. Large, W. A. and Rang, H. P. (1978) Variability of transmitter quanta released during incorporation of a false transmitter into motor nerve terminals. *J. Physiol.* **285**, 25.
353. Bekkers, J. M., Richerson, G. B., and Stevens, C. F. (1990) Origin of variability in quantal size in cultured hippocampal neurons and hippocampal slices. *Proc. Natl. Acad. Sci. USA* **87**, 5359.
354. Liu, G. and Tsien, R. W. (1995) Properties of synaptic transmission at single hippocampal synaptic boutons. *Nature* **375**, 404.
355. Frerking, M., Borges, S., and Wilson, M. (1995) Variation in GABA mini amplitude is the consequence of variation in transmitter concentration. *Neuron* **15**, 885.
356. Frerking, M. and Wilson, M. (1996) Saturation of postsynaptic receptors at central synapses? *Curr. Opin. Neurobiol.* **6**, 395.
357. Pothos, E., Desmond, M., and Sulzer, D. (1996) L-3,4-dihydroxyphenylalanine increases the quantal size of exocytotic dopamine release *in vitro*. *J. Neurochem.* **66**, 629.
358. Pothos, E. N., Przedborski, S., Davila, V., Schmitz, Y., and Sulzer, D. (1998) D2-like

dopamine autoreceptor activation reduces quantal size in PC12 cells. *J. Neurosci.* **18**, 5575.
359. Zhang, B., Koh, Y. H., Beckstead, R. B., Budnik, V., Ganetzky, B., and Bellen, H. J. (1998) Synaptic vesicle size and number are regulated by a clathrin adaptor protein required for endocytosis. *Neuron* **21**, 1465.
360. Tamir, H., Piscopo, I., Liu, K. P., Hsiung, S. C., Adlersberg, M., Nicolaides, M., *et al.* (1994) Secretagogue-induced gating of chloride channels in the secretory vesicles of parafollicular cells. *Endocrinology* **135**, 2045.
361. Tamir, H., Liu, K. P., Adlersberg, M., Hsiung, S. C., and Gershon, M. D. (1996) Acidification of serotonin-containing secretory vesicles induced by a plasma membrane calcium receptor. *J. Biol. Chem.* **271**, 6441.
362. Reimer, R. J., Fon, E. A., and Edwards, R. H. (1998) Vesicular neurotransmitter transport and the presynaptic regulation of quantal size. *Curr. Opin. Neurobiol.* **8**, 405.
363. Takahashi, N., Miner, L. L., Sora, I., Ujike, H., Revay, R. S., Kostic, V., *et al.* (1997) VMAT2 knockout mice: heterozygous display reduced amphetamine-conditioned reward, enhanced amphetamine locomotion and enhanced MPTP toxicity. *Proc. Natl. Acad. Sci. USA* **94**, 9938.
364. Fon, E. A., Pothos, E. N., Sun, B.-C., Killeen, N., Sulzer, D., and Edwards, R. H. (1997) Vesicular transport regulates monoamine storage and release but is not essential for amphetamine action. *Neuron* **19**, 1271.
365. Wang, Y.-M., Gainetdinov, R. R., Fumagalli, F., Xu, F., Jones, S. R., Bock, C. B., *et al.* (1997) Knockout of the vesicular monoamine transporter 2 gene results in neonatal death and supersensitivity to cocaine and amphetamine. *Neuron* **19**, 1285.
366. Song, H.-j., Ming, G.-l., Fon, E., Bellocchio, E., Edwards, R. H., and Poo, M.-m. (1997) Expression of a putative vesicular acetylcholine transporter facilitates quantal transmitter packaging. *Neuron* **18**, 815.
367. Ryan, T. A., Reuter, H., Wendland, B., Schweizer, F. E., Tsien, R. W., and Smith, S. J. (1993) The kinetics of synaptic vesicle recycling measured at single presynaptic boutons. *Neuron* **11**, 713.
368. Williams, J. (1997) How does a vesicle know it is full? *Neuron* **18**, 683.
369. Floor, E., Leventhal, P. S., Wang, Y., Meng, L., and Chen, W. (1995) Dynamic storage of dopamine in rat brain synaptic vesicles *in vitro*. *J. Neurochem.* **64**, 689.
370. Desnos, C., Raynaud, B., Vidal, S., Weber, M. J., and Scherman, D. (1990) Induction of the vesicular monoamine transporter by elevated potassium concentration in cultures of rat sympathetic neurons. *Dev. Brain Res.* **52**, 161.
371. Krejci, E., Gasnier, B., Botton, D., Isambert, M. F., Sagne, C., Gagnon, J., *et al.* (1993) Expression and regulation of the bovine vesicular monoamine transporter gene. *FEBS Lett.* **335**, 27.
372. Nakanishi, N., Onozawa, S., Matsumoto, R., Hasegawa, H., and Yamada, S. (1995) Cyclic AMP-dependent modulation of vesicular monoamine transport in pheochromocytoma cells. *J. Neurochem.* **64**, 600.
373. Barbosa, J. J., Clarizia, A. D., Gomez, M. V., Romano-Silva, M. A., Prado, V. F., and Prado, M. A. (1997) Effect of protein kinase C activation on the release of [^{3}H]-acetylcholine in the presence of vesamicol. *J. Neurochem.* **69**, 2608.
374. Ahnert-Hilger, G., Nurnberg, B., Exner, T., Schafer, T., and Jahn, R. (1998) The heterotrimeric G protein G02 regulates catecholamine uptake by secretory vesicles. *EMBO J.* **17**, 406.
375. Ozkan, E. D., Lee, F. S., and Ueda, T. (1997) A protein factor that inhibits ATP-dependent

glutamate and γ-aminobutyric acid accumulation into synaptic vesicles: purification and initial characterization. *Proc. Natl. Acad. Sci. USA* **94**, 4137.

376. Liu, Y., Schweitzer, E. S., Nirenberg, M. J., Pickel, V. M., Evans, C. J., and Edwards, R. H. (1994) Preferential localization of a vesicular monoamine transporter to dense core vesicles in PC12 cells. *J. Cell Biol.* **127**, 1419.
377. Weihe, E., Tao-Cheng, J.-H., Schafer, M. K.-H., Erickson, J. D., and Eiden, L. E. (1996) Visualization of the vesicular acetylcholine transporter in cholinergic nerve terminals and its targeting to a specific population of small synaptic vesicles. *Proc. Natl. Acad. Sci. USA* **93**, 3547.
378. Martin, T. F. J. (1994) The molecular machinery for fast and slow neurosecretion. *Curr. Opin. Neurobiol.* **4**, 626.
379. Kelly, R. B. (1991) Secretory granule and synaptic vesicle formation. *Curr. Opin. Cell Biol.* **3**, 654.
380. Chanat, E. and Huttner, W. B. (1991) Milieu-induced, selective aggregation of regulated secretory proteins in the *trans*-Golgi network. *J. Cell Biol.* **115**, 1505.
381. Varoqui, H. and Erickson, J. D. (1998) The cytoplasmic tail of the vesicular acetylcholine transporter contains a synaptic vesicle targeting signal. *J. Biol. Chem.* **273**, 9094.
382. Tan, P. K., Waites, C., Liu, Y., Krantz, D. E., and Edwards, R. H. (1998) A leucine-based motif mediates the endocytosis of vesicular monoamine and acetylcholine transporters. *J. Biol. Chem.* **273**, 17351.
383. Shin, J., Dunbrack, R. L. J., Lee, S., and Strominger, J. L. (1991) Phosphorylation-dependent down-modulation of CD4 requires a specific structure within the cytoplasmic domain of CD4. *J. Biol. Chem.* **266**, 10658.
384. Dietrich, J., Hou, X., Wegener, A. M., and Geisler, C. (1994) CD3 gamma contains a phosphoserine-dependent di-leucine motif involved in down-regulation of the T cell receptor. *EMBO J.* **13**, 2156.
385. Kirchhausen, T., Bonifacino, J. S., and Riezman, H. (1997) Linking cargo to vesicle formation: receptor tail interactions with coat proteins. *Curr. Opin. Cell Biol.* **9**, 488.
386. Rapoport, T., Chen, Y. C., Cupers, P., Shoelson, S. E., and Kirchhausen, T. (1998) Dileucine-based sorting signals bind to the beta chain of AP-1 at a site distinct and regulated differently from the tyrosine-based motif-binding site. *EMBO J.* **17**, 2148.
387. Krantz, D. E., Peter, D., Liu, Y., and Edwards, R. H. (1997) Phosphorylation of a vesicular monoamine transporter by casein kinase II. *J. Biol. Chem.* **272**, 6752.
388. Gilmor, M. L., Nash, N. R., Roghani, A., Edwards, R. H., Yi, H., Hersch, S. M., *et al.* (1996) Expression of the putative vesicular acetylcholine transporter in rat brain and localization in cholinergic synaptic vesicles. *J. Neurosci.* **16**, 2179.
389. Nirenberg, M. J., Liu, Y., Peter, D., Edwards, R. H., and Pickel, V. M. (1995) The vesicular monoamine transporter-2 is present in small synaptic vesicles and preferentially localizes to large dense core vesicles in rat solitary tract nuclei. *Proc. Natl. Acad. Sci. USA* **92**, 8773.
390. Nirenberg, M. J., Chan, J., Liu, Y., Edwards, R. H., and Pickel, V. M. (1996) Ultrastructural localization of the vesicular monoamine transporter-2 in midbrain dopaminergic neurons: potential sites for somatodendritic storage and release of dopamine. *J. Neurosci.* **16**, 4135.
391. Kalivas, P. W. and Duffy, P. (1993) Time course of extracellular dopamine and behavioral sensitization to cocaine. II. Dopamine perikarya. *J. Neurosci.* **13**, 276.
392. Sulzer, D., Joyce, M. P., Lin, L., Geldwert, D., Haber, S. N., Hattori, T., *et al.* (1998) Dopamine neurons make glutamatergic synapses *in vitro*. *J. Neurosci.* **18**, 4588.

393. Tkatch, T., Baranauskas, G., and Surmeier, D. J. (1998) Basal forebrain neurons adjacent to the globus pallidus co-express GABAergic and cholinergic marker mRNAs. *Neuroreport* **9**, 1935.
394. Cases, O., Seif, I., Grimsby, J., Gaspar, P., Chen, K., Pournin, S., *et al.* (1995) Aggressive behavior and altered amounts of brain serotonin and norepinephrine in mice lacking MAO-A. *Science* **268**, 1763.
395. Cases, O., Vitalis, T., Seif, I., De Maeyer, E., Sotelo, C., and Gaspar, P. (1996) Lack of barrels in the somatosensory cortex of monoamine oxidase A-deficient mice: role of a serotonin excess during the critical period. *Neuron* **16**, 297.
396. Lebrand, C., Cases, O., Adelbrecht, C., Doye, A., Alvarez, C., El Mestikawy, S., *et al.* (1996) Transient uptake and storage of serotonin in developing thalamic neurons. *Neuron* **17**, 823.

6 Toxins that affect neurotransmitter release

CAHIR J. O'KANE, GIAMPIETRO SCHIAVO, and SEAN T. SWEENEY

1. Introduction

As a perceptive observer once remarked, given enough genetic variation, evolution can select for machine-gun resistance in dogs. Less surreally, one outcome of evolutionary arms races between predators and prey has been the evolution of toxins that can immobilize their victims effectively and in small amounts, by interfering with synaptic transmission. A happier outcome is that some of these toxins have greatly aided our understanding of the roles of several key synaptic proteins that function in neurotransmitter release.

(a) The most far-reaching insights to date have probably come from the clostridial neurotoxins, a family of proteases which cleave one or more of the three synaptic SNARE proteins, synaptobrevin, syntaxin, and SNAP-25.

(b) Excitatory synaptic toxins cause neurotransmitter release in the absence of an action potential. Their mechanism is less well understood, although one of them, α-latrotoxin, is starting to reveal new insights into the control of the neurotransmitter release machinery.

(c) Calcium channel blockers have been useful in dissecting the roles of various calcium channels.

(d) Presynaptic phospholipases are neurotoxins isolated from arthropod and snake venoms which cause a persistent blockade of neurotransmitter release in a variety of synapses of the CNS. Their inhibitory mechanism of action is still not completely understood and possibly not related to their well known enzymatic activity on phospholipids.

2. Clostridial neurotoxins

Clostridial bacteria are obligate anaerobes that include pathogens which can colonize either vertebrate (e.g. 1, 2) or invertebrate (3, 4) hosts. The species *Clostridium botulinum* and *Clostridium tetani* can produce neurotoxins which lead to paralysis of their host, in severe cases death, and a field day for the pathogen, which is effectively a predator that has created an anaerobic fermenter for itself. Bacterial survival is then ensured by formation of spores that are hardy and relatively resistant to abuse. For a concise but wide-ranging review of earlier work on *Clostridia* and their neurotoxins, see ref. 2.

2.1 The phenomenology of toxicity

C. botulinum infection of humans, or exposure to botulinum neurotoxins (BoNTs), normally occurs by ingestion of contaminated food. At least seven different serotypes are known, A to G, each of which produces a different neurotoxin. The toxin is taken up across the gut epithelium by transcytosis (5–7) enters the circulatory system, and is taken up by motor neurons at the neuromuscular junction (NMJ). This leads to failure of synaptic transmission at the NMJ, and flaccid paralysis.

C. tetani penetration into the organism normally occurs through wounds. In contrast with the presence of distinct botulinum neurotoxin serotypes, only one type of tetanus neurotoxin (TeNT) is known. It is taken up peripherally by motor neurons, and transported back along axons into the cell body (8). Within the spinal cord, TeNT migrates trans-synaptically into coupled inhibitory interneurons across the synaptic cleft (9, 10) and it blocks the release of inhibitory neurotransmitters (11, 12), thus causing spastic paralysis. Although this pathway accounts for the main clinical symptoms of tetanus, TeNT ascends to the CNS not only through motor neurons, but also through sensory and adrenergic fibres (13).

Poisoning of individual synapses by botulinum neurotoxins lasts for several weeks, and recovery is accompanied by *de novo* sprouting of the motorneuron terminal at the NMJ (14–16). These NMJ sprouts are then re-absorbed when the functionality of the old NMJ is recovered (16).

2.2 Heavy and light chains

Each clostridial neurotoxin is encoded by a single gene, and post-translational processing in the bacteria or in the host gives rise to an active di-chain toxin composed by a heavy chain (HC) and a light chain (LC), which are joined by a disulfide bridge (17–19) and non-covalent forces (see Fig. 1; reviewed in ref. 20). The roles of the heavy and light chains have been investigated in at least two experimental paradigms.

First, the effects of toxin LCs have been studied in defined neurons of the mollusc *Aplysia*, where it is possible to stimulate presynaptically and record postsynaptically. Intracellular injection of the LCs of TeNT, BoNT/A, or BoNT/B leads to loss of transmission. BoNT serotypes A and B also require either extracellular or intracellular application of HC for toxicity (21–23). The basis for this observation is not clear, but subsequent work has shown that native or recombinant LCs of the

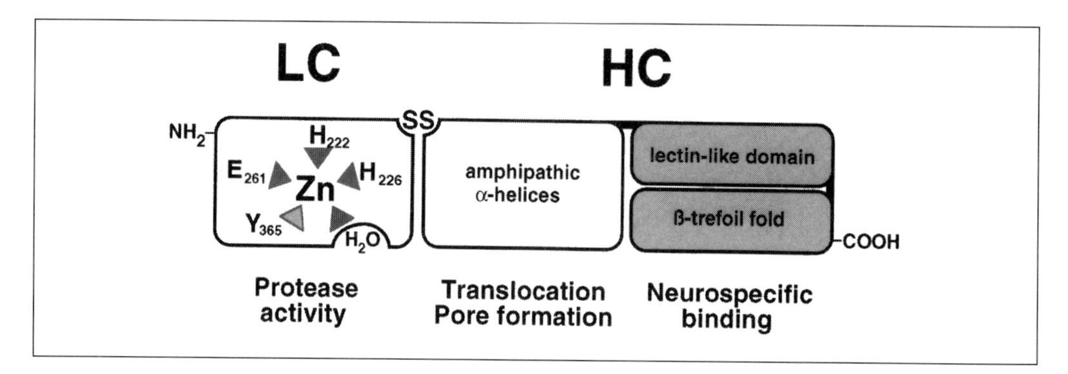

Fig. 1 Structure of clostridial neurotoxins. The crystal structure of the di-chain BoNT/A (28) has confirmed early biochemical observations that the neurotoxin is composed of three independently folded domains. The carboxy-terminal portion of the heavy chain (HC) is responsible for the neuro-specific binding and is subdivided in two parts, both having structural homology with sugar-interacting proteins. The amino-terminus of the HC contains a pair of very long amphipathic α-helices, likely to play a key role in the translocation of the neurotoxin through the endosomal membrane and having a pore-forming activity. The light chain (LC) is a metallo-endopeptidase and contains an atom of zinc essential for the enzymatic activity of the neurotoxin. This zinc atom is co-ordinated by two histidines (H222 and H226 in BoNT/A) and one glutamate residue (E261). A second glutamate residue (E223) co-ordinates a water molecule responsible for the hydrolysis. A tyrosine (Y365), likely to be involved in the interaction with the substrate and secondary bonding networks, completes the active site.

botulinum toxins are sufficient for degradation of their proteolytic targets in mammalian cells (see Section 2.3). Secondly, the LCs of TeNT, and BoNT serotypes A, B, and E, are sufficient to prevent regulated release from mammalian neuro-secretory cells (24).

When applied extracellularly to *Aplysia*, the toxins show cell type specificity of action not unlike that observed in vertebrates: BoNTs inhibit cholinergic synapses preferentially, whereas TeNT is more toxic to non-cholinergic ones. Reconstitution experiments that involve swapping of TeNT and BoNT heavy and light chains demonstrate that the cell type specificity of toxin action is determined by the HC (25). Hence, the LCs are required for intracellular toxicity, while the HCs are responsible for the neuro-specific binding and internalization.

A large number of studies highlight the importance of two classes of molecules in the binding of clostridial neurotoxins: a class of glycolipids containing multiple sialic acid residues, termed polysialogangliosides, and surface proteins (reviewed in ref. 26). The results of studies favouring a lipid receptor for clostridial neurotoxins are briefly summarized below:

(a) BoNTs and TeNT interact with polysialogangliosides, particularly members of the G1b series.

(b) Pre-incubation of toxins with polysialogangliosides partially prevents the BoNT poisoning of the NMJ and the retroaxonal transport of TeNT.

(c) Incubation of cultured chromaffin cells with polysialogangliosides increases their sensitivity to TeNT and BoNT/A.

(d) Treatment of membranes with neuraminidase, which removes sialic acid residues, decreases toxin binding.

However, experiments carried out with cells in culture or purified synaptosomes from brain have indicated that cell surface proteins may be involved in toxin binding (26). Protection experiments (27) and the presence of both lectin-like and protein binding domains in the carboxy-terminal portion of TeNT and BoNT/A (28, 29) suggest that clostridial neurotoxins may bind strongly and specifically to the presynaptic membrane because they display multiple interactions with sugar and protein ligands (30). Recent experiments provided strong evidence in favour of such a model by showing that BoNTs binds to the intralumenal domain of the synaptic vesicle protein synaptotagmin in the presence of polysialogangliosides (31, 32).

After binding to the neuronal cell surface, clostridial neurotoxins enter the lumen of vesicular structures in a temperature- and energy-dependent process (33, 34) and, in hippocampal neurons, TeNT was found to enter synaptic terminals inside the lumen of a compartment indistinguishable from synaptic vesicles (35). Therefore, clostridial neurotoxins may use recycling synaptic vesicles as Trojan horses to enter CNS neurons. In order to elicit their inhibitory activity, LCs must cross the hydrophobic barrier of the vesicle membrane to reach the cytosol. The different trafficking of TeNT and BoNTs at the NMJ clearly indicates that internalization is not directly followed by membrane translocation. There is compelling evidence that TeNT and BoNTs have to be exposed to a low pH step for nerve intoxication to occur (35–37). Acidic pH does not induce a direct activation of the toxin via a structural change, since the introduction of a non-acid treated LC in the cytosol is sufficient to block exocytosis (22, 24); however it is instrumental in the process of membrane translocation of the LC from the vesicle lumen into the cytosol, and possibly for the dissociation of LC from HC. In this respect, TeNT and BoNTs appear to behave similarly to the other bacterial protein toxins characterized by a structure composed of three distinct portions (38). Membrane interaction studies have shown that low pH induces TeNT and BoNTs to undergo a conformational change from a water soluble neutral form to an acid form with surface exposed hydrophobic segments. This hydrophobicity enables the penetration of both the H and L chains in the hydrocarbon core of the lipid bilayer (reviewed in ref. 38). The crystallographic structure of BoNT/A reveals that the amino-terminal portion of HC consists of a bundle of α-helices buried in the domain core (28), and resembles the structural organization of another bacterial toxin, colicin Ia (39). There is a general consensus that this hydrophobic core is related to the process of translocation of the L domain across the vesicle membrane into the nerve cytosol. However, the molecular details of this mechanism are still unknown.

The isolated carboxy-terminal portion of the HC of TeNT is responsible for the uptake, retrograde axonal transport, and transcytosis of the entire molecule and could be exploited for the targeting of exogenous substances to the CNS (40, 41). A similar process of transcytosis may mediate uptake of BoNTs across gut epithelium (6).

2.3 Mechanism of light chain action

Clues to the mechanism of clostridial neurotoxin action came from the identification of the intracellular active regions (23, 24), followed by the observation that one of the conserved regions of the LC contains the putative HExxH metalloprotease zinc binding motif, and the demonstration of a zinc requirement for TeNT activity (42). Identification of proteolytic substrates of the toxins followed rapidly, by using conventional biochemical techniques, such as Western blotting with specific antibodies against synaptic proteins, or treating *in vitro* translated synaptic proteins with the toxins. Proteolytic activity was, as predicted, confined to the LC of each toxin. Cleavage sites were identified by amino-terminal sequencing of the cleavage products. The results of these studies are summarized in Table 1 and Fig. 2.

Strikingly, clostridial neurotoxins cleave distinct peptide bonds on three synaptic membrane-bound proteins, VAMP/synaptobrevin, syntaxin, and SNAP-25. These proteins have been termed SNAREs on the basis of their ability to act as receptors for two soluble proteins (NSF and αSNAP) shown to be required for vesicle fusion in an *in vitro* intra-Golgi transport assay (43, 44). These three proteins can form an SDS-resistant 7S complex, and a larger 20S complex with NSF and αSNAP (44, 45). In the synapse, synaptobrevin is found on synaptic vesicles, whereas syntaxin and SNAP-25 are found predominantly in the plasma membrane (46–48). Some of the main features of the synaptic SNARE proteins are summarized in Fig. 2.

Hence, the interaction of the SNAREs with NSF and αSNAP, together with the

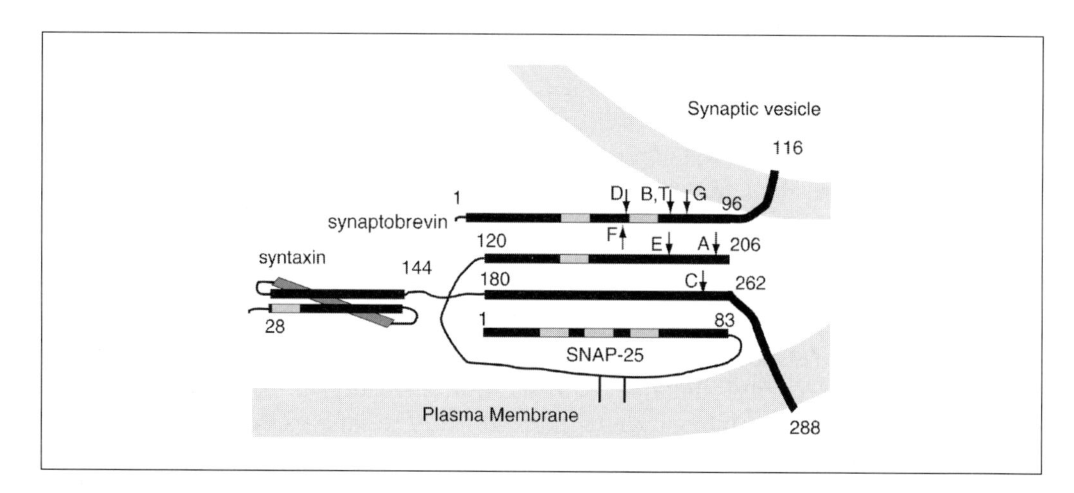

Fig. 2 Main features of SNARE proteins (not to scale). Known α-helical regions are shown as thick lines, as are the possibly α-helical transmembrane regions of synaptobrevin and syntaxin; unstructured regions are shown as thin lines (64, 65, 101). The four α-helical regions that form the SNARE complex are shown in parallel to each other; they are actually wound around each other to form a left-handed quadruple helix. The N-terminal region of syntaxin comprises mainly three α-helical regions of about 40 amino acid residues each (second helix in dark grey, for clarity). Numbering indicates amino acid residues. A, B, C, D, E, F, G, T, next to arrows, indicate the cleavage sites of BoNT types A–G and of TeNT respectively; precise bonds cleaved are listed in Table 1. Light grey boxes indicate proposed SNARE motifs (Section 2.4); only those within known α-helical regions are shown. Several palmitoyl side groups of SNAP-25 anchor this protein in the plasma membrane.

Table 1 Botulinum toxins, their targets, and cleavage sites

Toxin	Target	Cleaves at	Minimal substrate[a]		References
TeNT	Synaptobrevin	Q76–F77	47–116	–	55, 81, 152, 153
			39–88	+	
BoNT/A	SNAP-25	Q197–R198	187–202	+	154–159
BoNT/B	Synaptobrevin	Q76–F77	60–94	+	58, 81
BoNT/C	Syntaxin	K253–A254			160, 161
	SNAP-25	R198–A199	SNAP-25 in intact cells		52, 162, 163, 164
BoNT/D	Synaptobrevin	K59–L60	47–116	–	55, 77, 155
			27–116	+	
BoNT/E	SNAP-25	R180–I181			154, 155, 157
BoNT/F	Synaptobrevin	Q58–K59	47–116	–	55, 77, 165
			27–116	+	
BoNT/G	Synaptobrevin	A81–A82	54–COOH	+	56, 166
			60–COOH	±	

[a] –, not cleaved; +, cleaved.

blockage of synaptic transmission by the clostridial toxins, suggests that the toxins interfere with proteins that play a key role in calcium-dependent vesicle fusion at the synaptic plasma membrane. The nature of this role, and the way in which the toxins have helped shed light on it, are discussed in more depth in Section 2.6.

2.4 Factors affecting toxin–target recognition

One striking feature of clostridial neurotoxins is their specificity of action. For example, TeNT action in *Drosophila* does not disrupt any functions required for cell viability, muscle function, neuronal development, or synapse formation (49, 50). TeNT and most BoNTs can cause long-lasting inhibition of mammalian synaptic transmission, without obviously affecting neuronal viability; in contrast, BoNT/C causes neuronal cell death in culture (51, 52). This correlates with an apparent requirement of one of its targets, syntaxin, for cell viability (53, 54); however, other targets of BoNT/C cannot presently be excluded.

While the sequences of the cleavage sites have some influence on their toxin sensitivity (55–57), they alone cannot easily account for this high degree of specificity. Furthermore, at least BoNT/B has some flexibility in the specificity of its peptide bond cleavage — it normally cleaves at Q76–F77 of synaptobrevin, but can also cleave (Q/N/A)–F or Q–(F/Y) peptide bonds at the same position (58). In addition, efficient toxin cleavage requires quite long target peptides in most cases (Table 1). Together, these data support the notion that toxin–substrate interaction requires some other feature of SNARE sequence or structure.

Such a structural feature has been proposed by Montecucco and colleagues, and designated the 'SNARE motif' (59–63). This motif was suggested to have the sequence xh- -xh-xhp (x, any amino acid; h, hydrophobic residue; -, acidic residue; p, polar residue) and form an α-helix which has a hydrophobic face and a negatively charged face. Multiple copies of this motif are present in syntaxin, SNAP-25, and VAMP/synaptobrevin. As presented in Fig. 2, the majority of them are included in the four-helix bundle structure of the synaptic SNARE complex with their hydrophobic faces to the interior and the acidic faces to the exterior, or (in one case) in the three-helix bundle at the N-terminus of syntaxin (64, 65). While direct structural evidence, for example co-crystallization of a LC with its substrate, is so far lacking, antibodies raised to a SNARE motif of one target protein can partially inhibit the toxin-mediated cleavage of other SNARE proteins, and some combinations of neurotoxins show cross-inhibition of cleavage (59, 61). Toxin resistance is sometimes associated with mutations in SNARE motifs, or with deviations from the consensus — *Drosophila* ubiquitously expressed synaptobrevin lacks one of the three acidic residues in each SNARE motif, and is not cleaved by TeNT (49). At least some of these mutations do not simply disrupt overall secondary or tertiary structure — some amino acid substitutions in the SNARE motifs of rat synaptobrevin-2 that confer toxin resistance have no effect on the ability of the protein to restore Ca^{2+}-regulated insulin exocytosis in TeNT-treated pancreatic cells (57). More direct evidence of the importance of this motif might come from, say, introducing amino acid substitutions into the resistant *Drosophila* variant and testing whether it now becomes susceptible to TeNT cleavage — to our knowledge this experiment has not yet been done.

2.5 Do the toxins have other targets or other activities?

While the above model is now supported by a large body of evidence and is now accepted wisdom, there are some suggestions in the literature that the toxins may have other targets or activities. The existence of such alternative targets could well affect the interpretation of many experiments using the toxins. It is therefore worth reviewing the evidence on this question.

2.5.1 Arguments for SNARE specificity

1. Very many studies, in a wide range of organisms and cell types, show that the clostridial neurotoxins inhibit calcium-dependent transmitter release. Several studies show no effects of them on plasma membrane calcium currents (66, 67) or presynaptic calcium influx (68) in *Aplysia*, or in mammalian sympathetic neurons, or motor endplates (69, 70). Also, treatments which would bypass a block prior to calcium influx (calcium ionophores, permeabilization, or calcium mimics) do not relieve the block of transmitter release caused by several of the toxins in contexts as diverse as the mammalian hippocampus (71, 72), chromaffin cells (73, 74), or *Torpedo* cholinergic nerve endings (75). However, inhibition of synaptic transmission by BoNT/A and BoNT/C in hippocampal cultures can be partially rescued by large calcium influxes; in contrast, TeNT inhibition cannot be rescued (76). This suggests that one effect of these two botulinal toxins is on the response of the SNARE complex to the calcium sensing machinery in the synapse.

2. Does the blockage of calcium-dependent vesicle release depend on protease activity, or do the toxins have other enzymatic activities? This question can be addressed by asking whether mutations affecting the protease active site of TeNT or BoNT retain toxicity (and not merely the ability to cleave their respective substrates). The results of such experiments are largely, but not exclusively, in favour of the protease activity being a requirement for blockage of neurotransmitter release. Two main kinds of mutant change which abolish protease activity have been made in the toxin active sites. First, mutation of the histidine residues in the active site (H232 and H237 in TeNT) abolishes zinc binding (77). A caveat with this class of mutations is that there is no direct evidence for maintenance of the overall tertiary structure of the toxin. Secondly, mutation of the glutamate residue thought to co-ordinate a water molecule important in catalysis (E234 in TeNT), leaves both zinc binding and substrate binding of TeNT intact (77–79), and hence presumably has no drastic effects on overall tertiary structure of the toxin. An E234Q variant of reconstituted TeNT recombinant LC with TeNT HC has no detectable toxicity when assayed using muscle tension after nerve stimulation in mouse (78); TeNT-E234Q LC shows no inhibition of catecholamine secretion in permeabilized PC12 cells (80); and TeNT-E234Q LC has no obvious effect on viability when expressed throughout the *Drosophila* embryonic nervous system (49). In contrast, an E234A mutant of reconstituted TeNT showed some inhibition of release after injection into *Aplysia* neurons (79). The different toxin and assay preparations used in these experiments, and the difficulty of comparing local intracellular concentrations of toxin, must add caveats to any comparisons. Nevertheless, in the *Aplysia* preparation one must invoke mechanisms such as a secondary activity of the protease-inactive LC (perhaps stimulation of an endogenous transglutaminase; see Section 2.5.2), an effect of the inactive LC in the presence of HC, or competitive inhibition of interactions between synaptobrevin and other proteins by binding of high concentrations of the inactive toxin. In conclusion, while some preparations suggest no protease-independent activity of TeNT involved in the inhibition of transmission, investigators would be wise to include protease-negative mutant controls in their experiments before drawing any conclusions about the mechanism by which any clostridial toxin brings about any effect observed.

3. The inhibition by TeNT of transmission in *Aplysia* neurons is delayed by injection of peptides that contain the cleavage site of TeNT (81). The intracellular concentration of toxin used here (8 nM versus 10 nM), the nature of toxin (holotoxin, albeit not reconstituted), and the class of synapse (cholinergic) were similar to those used by Ashton *et al.* (79), but the conclusions are at variance with those of the latter experiment, which suggested a protease-independent activity of the toxin (79).

4. Antibodies raised against rat synaptobrevin effectively block the inhibition of transmission by extracellularly applied TeNT and BoNT/B in *Aplysia*. In contrast, synaptobrevin antibodies do not block the effect of BoNT/A, which cleaves SNAP-25, and unrelated control antibody preparations showed no blockage of BoNT/B (82).

5. The effects of neuronal expression of TeNT LC (49) and of mutating neuronal synaptobrevin (83) on the *Drosophila* NMJ are virtually identical. There is no sign of any developmental defect, evoked synaptic transmission is abolished, and there is a reduction in the frequency of spontaneous vesicle fusion.

6. Ca^{2+}-dependent insulin secretion from pancreatic cells be inhibited by TeNT. This inhibition can be rescued by synaptobrevin, or its ubiquitously expressed homologue cellubrevin, which have been made resistant to TeNT proteolysis (57). Hence, synaptobrevin (or a closely related homologue) is the only target of TeNT which is relevant for blockage of exocytosis in this assay system, and TeNT toxicity can be ascribed to proteolysis of synaptobrevin.

7. Leech synaptobrevin is cleaved by TeNT, whereas leech SNAP-25 is resistant to BoNT/A (84). This correlates with the inhibition of synaptic transmission by injected TeNT LC, and the lack of effect of injected BoNT/A LC.

8. Non-neuronal toxicity can often be accounted for by susceptibility of homologues of the synaptic SNAREs to the toxins. For example, the vertebrate synaptobrevin homologue, cellubrevin, is ubiquitously expressed, and is sensitive to TeNT (85); the ubiquitously expressed *Drosophila* synaptobrevin isoform is not susceptible to TeNT (49). This correlates with the toxic effects of TeNT in mammalian non-neuronal cells (86), and the apparent lack of non-neuronal defects in *Drosophila* (49).

2.5.2 Arguments against SNARE specificity

1. An E234A protease-deficient mutant of reconstituted TeNT showed some inhibition of neurotransmission after injection into *Aplysia* neurons (79). See Section 2.5.1 for a discussion of these results.

2. Another reported activity of TeNT is to stimulate cellular transglutaminase, which can crosslink glutamine residues with primary amino groups which may be on different protein molecules (87). One substrate of transglutaminase is synapsin, at least *in vitro* (88). However, in contrast to the large body of work that correlates proteolytic cleavage of SNARE proteins *in vivo* with the loss of neurotransmission, the *in vivo* evidence for a role of transglutaminase is conflicting. While one group has found some effect of transglutaminase inhibitors on TeNT action in *Aplysia* synapses (79) and synaptosomes (89), two other studies in mammalian neurons failed to find any effect on toxicity of either transglutaminase inhibitors or other conditions that reduced transglutaminase activity (90, 91). In summary, while there is evidence of an effect of tetanus toxin on transglutaminase activity, the contribution of any such effect to blockage of neurotransmitter release is controversial and remains to be clarified.

3. Disruption of actin or microtubule cytoskeleton reduces TeNT toxicity in mammalian synaptosomes (89). One interpretation of these results is an effect of the toxin or its substrate on movement of vesicles from the cytoskeleton-linked reserve pool into the readily releasable pool at the plasma membrane. However, a requirement of

the cytoskeleton for some components of toxin action (e.g. for transport of toxin molecules or toxin-containing vesicles from their uptake sites) is also a possibility.

4. Several of the neurotoxins are substrates for src protein kinase *in vitro*, and at least BoNT/A LC is tyrosine-phosphorylated in PC12 cells *in vivo* (92). This modification of serotypes A and E dramatically increases both their catalytic activity and thermal stability, while dephosphorylation reverses the effect, suggesting that a biologically significant form of the neurotoxins inside neurons is phosphorylated. This finding also raises the possibility that phosphorylated toxins could interfere with intracellular signal transduction pathways by binding endogenous SH2 domain proteins.

5. Even inactive clostridial neurotoxins may still bind their substrates (78). Hence, the toxins also have potential to act as competitive inhibitors of the normal interactions between their substrates and other proteins — especially when applied at concentrations that may be well above the minimal levels required for toxicity by the normal route.

6. While toxin-resistant SNAREs can rescue toxin-induced blockage of calcium-dependent insulin secretion (57), no similar rescue of the toxin-induced blockage of synaptic transmission has yet been reported.

2.5.3 Assessment of toxin specificity

In conclusion, there is overwhelming evidence that the clostridial neurotoxins are highly specific metalloproteases which cleave synaptic SNARE proteins or their homologues, and that this is the primary mechanism of toxicity. Non-synaptic effects of the toxins can often be accounted for by either by roles of the synaptic SNAREs outside the synapse, or by cleavage of closely related homologues of the synaptic SNAREs by the toxins. However, it is impossible to rule out either the possibility of other substrates, or other activities of the toxins. Such alternative targets or activities are unlikely to make a major contribution to the blockage of synaptic transmission, but under certain circumstances might conceivably produce measurable effects on cell physiology.

2.6 Using clostridial neurotoxins to study the function of their targets

The identification of the sole or major target molecules of the clostridial neurotoxins as SNAREs has established the central role of SNAREs in synaptic vesicle exocytosis. However, the precise nature of this role is still a matter for debate. Much of the debate has centred around the question of whether synaptobrevin, syntaxin, and SNAP-25 are involved in determining the specificity of the vesicle–plasma membrane interaction (often paraphrased as the specificity of 'docking'), or whether they are more intimately involved in the actual fusion event. While these roles are in principle distinct, they are by no means mutually exclusive.

2.6.1 SNAREs and vesicle–plasma membrane docking

Eukaryotic cells have many classes of membrane-bound compartments, and traffic of vesicles between these compartments is highly directional, highly regulated, and occurs only between specific combinations of vesicles and target membranes. The requirement of NSF/SNAP for vesicle–target membrane fusion, the identification of molecules such as synaptobrevin, syntaxin, and SNAP-25 as membrane receptors for NSF/SNAP (SNAREs) (43, 44), and the roles of different, but structurally homologous classes of SNAREs on different membrane compartments in yeast (93–96) have suggested SNAREs as the major determinants of specificity in the interaction between vesicles and their target membranes. However, treatment of squid synapses with TeNT (68) or BoNT/C (97), or treatment of *Drosophila* synapses with TeNT or removal of syntaxin by genetic means (50), does not prevent 'docking' of vesicles to plasma membrane active zones, as assayed by electron microscopy.

However, the synapse active zone is a relatively complex macromolecular structure, in which many components of the vesicle fusion machinery are assembled (reviewed in ref. 98). While SNARE-like molecules may well be major determinants of the specificity of membrane interactions in some circumstances, it appears that vesicle localization at the presynaptic membrane is a consequence of the series of interactions that lead to an assembled active zone. This process can give rise to vesicles in close apposition with the presynaptic membrane that appear 'docked' by electron microscopy, but is not sufficient to produce fusion-competent vesicles. Hence, the complete absence of vesicle fusion when SNAREs are removed, despite the close proximity of the synaptic vesicles with plasma membrane, points to a key role of the SNAREs in the fusion process itself. Very recently, evidence supporting the hypothesis of SNAREs being the minimal machinery for membrane fusion has been provided by a series of experiments showing the clostridial neurotoxin-sensitive heterotypic fusion of artificial liposomes containing synaptic SNAREs (99).

2.6.2 Clostridial neurotoxins and the SNARE complex

How might the SNARE proteins function during the membrane fusion process? The three SNAREs that are targets of the clostridial neurotoxins can form a complex that is extremely stable, being resistant to dissociation by SDS (45). This complex, known as the SNARE complex, can exist in a 7S form, and as a larger 20S complex with NSF and αSNAP, the structures of which have been solved at a low resolution (100). Recent work (65, 101) (summarized in Fig. 2) has shown that the SNARE complex consists of four α-helical regions, which are wound in parallel around each other to form a quadruple left-handed helix, and which are derived from:

- most of the cytoplasmic domain of synaptobrevin (residues 1–96)
- the C-terminal portion of the cytoplasmic domain of syntaxin (residues 180–262)
- an N-terminal portion of SNAP-25 (residues 1–83)
- a C-terminal portion of SNAP-25 (residues 120–206).

The N-terminus of synaptobrevin and the longer N-terminus of syntaxin are both cytoplasmic, with their C-terminal regions anchored in the vesicle membrane and plasma membrane respectively. The N-terminal portion of syntaxin does not take part in the SNARE complex, but largely comprises a bundle of three long α-helices, which is likely to be involved in interactions with other proteins (64). The N- and C-terminal portions of SNAP-25 are separated by a relatively unstructured linker of about 37 amino acids, which includes the Cys linked palmitoyl groups that anchor this protein to the plasma membrane.

The conformation and stability of the SNARE complex and its components suggest a model in which formation of the complex, and the large free energy thus made available, would bring the membranes in which the proteins are anchored in to close proximity. This would then drive, or at least facilitate, membrane bilayer fusion (102). Such a model is strongly supported by the finding that synthetic lipid vesicles with reconstituted synaptobrevin can interact and subsequently fuse *in vitro* with vesicles into which syntaxin and SNAP-25 have been inserted (99). This model would also require dissociation of the stable complex at some subsequent stage during the vesicle recycling process; indeed NSF and αSNAP, in the presence of ATP, can make the stable complex susceptible to dissociation (103). While there are some lines of evidence that support this model, we will focus here on the information that the clostridial neurotoxins have yielded on the mechanism of vesicle fusion.

1. The effects of the toxins on assembly and disassembly of the stable SNARE complex support the idea that a cycle of assembly and disassembly is a key process in the process of exocytosis. For example, synaptobrevin cleavage by BoNT D or F, or SNAP-25 cleavage by BoNT A or E, or syntaxin cleavage by BoNT/C (SNAP-25 is not cleaved *in vitro*), do not prevent assembly of a synaptobrevin–syntaxin–SNAP-25 complex; however, the complex is no longer SDS-resistant, suggesting that formation of the most stable form of the complex is an important step which is blocked by toxin action (45, 104). Further, SNARE complexes (both SDS-resistant and SDS-sensitive conformations) can be dissociated by NSF and αSNAP in the presence of ATP; the N-terminal fragments of SNAP-25 produced by BoNT/A and E allow more efficient dissociation, whereas the N-terminal fragment of synaptobrevin produced by TeNT cleavage did not permit dissociation of the ternary complex (103). Taken as a whole, these results also suggest that the blockade of vesicle fusion caused by the clostridial neurotoxins does not impair the recognition between synaptic SNAREs, but affects a later event in the neuroexocytosis process which involves formation of the stable form of the SNARE complex.

2. SNARE proteins are resistant to cleavage by clostridial toxins when assembled in the SNARE complex *in vitro* (45, 105). At least for TeNT, and BoNT serotypes B, C, D, E, and G, this is consistent with the target bonds being close to the central axis of the four-helix bundle of the SNARE complex, and it is also likely that some or all of the extended toxin recognition sites are also inaccessible (65). However, the completeness of the inhibition of vesicle fusion by most of the neurotoxins suggests that their target proteins pass through a long phase in which they are not in the stable complex.

This sensitive period extends even past the step of ATP-dependent 'priming' (which includes dissociation of stable complexes by NSF and αSNAP) in the case of SNAP-25 cleavage by BoNT/E, also supporting the model that the SNAREs do not form the stable complex until a late stage in the fusion process (106, 107).

3. The toxin resistance of the components of the stable SNARE complex can be used as an assay for the effects of manipulating other steps of the vesicle release pathway. For example, the small GTPase Rab3A negatively regulates the probability of vesicle release (108). A clue to the possible mechanism of this effect comes from the observation that a constitutively activated mutant of Rab3 delays the onset of toxin sensitivity of neurotransmitter release in *Aplysia* neurons (109), suggesting that Rab3 activation favours formation of a stable complex (or inhibits its dissociation by NSF and αSNAP), and may inhibit progression to a fusion-competent state.

4. Cleavage of SNAREs by the toxins *in vivo* can be used as a very convenient proxy for deletion mutants of the SNAREs. For example, cleavage of SNAP-25 with BoNT/A, which removes only the 9 C-terminal amino acids, does not inhibit exocytosis entirely, but BoNT/E, which removes the 26 C-terminal amino acids, does do so (107, 110). In contrast to BoNT/A, BoNT/E proteolysis induces fast replacement of cleaved SNAP-25 at the NMJ (110a). These results suggest a distinctive role for the C-terminal portion of SNAP-25, possibly in coupling the fusion machinery to the calcium sensing machinery and in SNAP-25 turnover.

5. A large body of evidence suggests a highly structured arrangement of the vesicle fusion machinery in close proximity to plasma membrane calcium channels in the presynaptic active zone (see also Section 3.2). One interaction that contributes to this arrangement is that between calcium channels and syntaxin (111–115). An obvious implication of this proximity is that the vesicle fusion machinery is rapidly exposed to elevated calcium levels after depolarization — but it also provides a framework for more complex interactions. For example, calcium channels in the large calyx-type nerve terminals of the chick ciliary ganglion are regulated by G proteins; this regulation is abolished by BoNT/C, probably because of its cleavage of syntaxin.

3. Latrotoxins and excitatory toxins

A further class of toxins that affect neurotransmitter release are excitatory toxins — these cause a drastic increase in neurotransmitter release. Some such toxins do not act directly on synaptic vesicle release — the effects of many can be blocked by tetrodotoxin, which blocks sodium channels and hence action potentials. However, some appear to act on the vesicle release machinery, causing massive release by overriding the controls that normally prevent this in the absence of an action potential. The best understood excitatory toxin, α-latrotoxin (αLTX), has started to be useful in identifying the ways in which the vesicle fusion process can be controlled.

3.1 α-Latrotoxin

The venom of the black widow spider *Latrodectus mactans* contains components that cause uncontrolled vesicle release in synapses over a broad phylogenetic range. Cloning of the genes that encode three purified components of the venom has shown that it contains structurally homologous proteins that are active in different phyla, and probably have similar mechanisms of action (116–118). Shared sequence features include two predicted transmembrane domains and a series of ankyrin repeats (Fig. 3).

Of the cloned black widow toxins, αLTX and its mechanism of action has been best characterized. The toxin can lead to formation of non-specific cation channels in artificial lipid bilayers and in cell membranes of synapses and neurosecretory cells; the subsequent entry of calcium can plausibly contribute to trigger vesicle release (119, 120). However, there are both calcium-dependent and calcium-independent components to toxicity (discussed in more detail below).

Biochemical approaches have led to identification of two receptors for αLTX on synaptic membranes: neurexin Iα (121), which binds the toxin in a calcium-dependent manner (122), and a calcium-independent receptor, latrophilin (or CIRL) which is a member of the G protein-coupled receptor superfamily (123, 124). However, the simplistic model of neurexin Iα binding mediating the calcium-dependent actions, and latrophilin binding mediating the calcium-independent actions, does not stand up well to further scrutiny, and the effects of αLTX turn out to be quite complex.

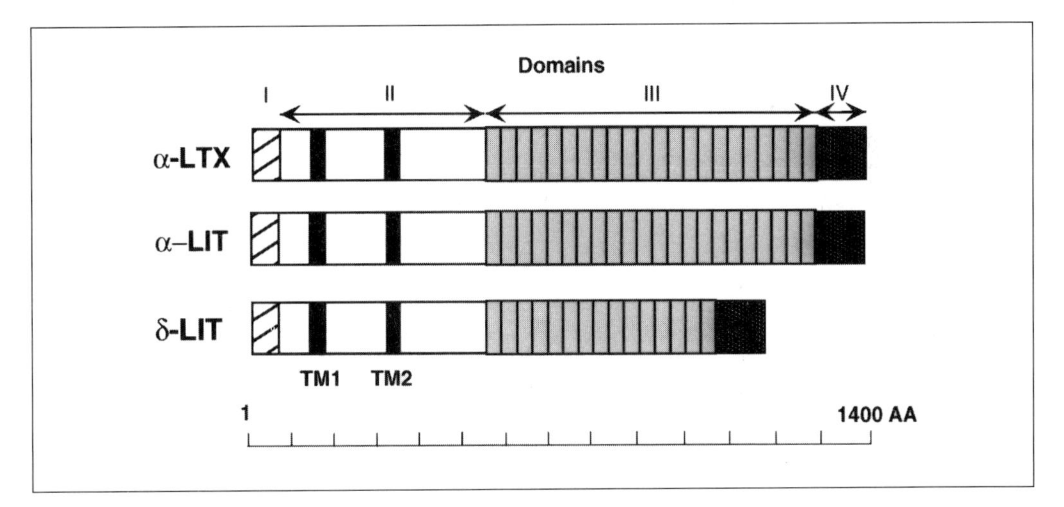

Fig. 3 Structure of latrotoxins. Four domains can be highlighted in the structure of the three cloned members of the latrotoxin family, α-latrotoxin (α-LTX), α-latroinsectotoxin (α-LIT) and δ-latroinsectotoxin (δ-LIT) (116–118). The signal sequence (domain I, hatched area) is followed by a conserved amino-terminal domain (II), containing two putative transmembrane domains (TM1 and 2) and by a portion of variable length, rich in ankyrin repeats (domain III, grey). The carboxy-terminal part (domain IV) is not present in the mature toxin and is not essential for its activity.

1. The full latrophilin-dependent effects of the toxin do actually require calcium. For example, transfection of bovine chromaffin cells with a latrophilin gene increases their sensitivity to the effects of αLTX. As predicted from the calcium-independence of αLTX–latrophilin binding, this effect can occur when toxin binding is allowed to take place in the absence of extracellular calcium — but calcium must subsequently be added in order to see a rise in exocytosis (125).

2. The increased sensitivity of latrophilin-transfected cells to αLTX (125) clearly implicates latrophilin in toxin action. Surprisingly then, insertion of four extra amino acid residues between domains II and III (Fig. 3) eliminates the ability of a recombinant αLTX to evoke vesicle release from synaptosomes, whereas it still binds normally to latrophilin (126). Also, a truncated latrophilin protein consisting of only the N-terminal extracellular domain and the first transmembrane domain, which lacks the regions that are likely to interact with G proteins, increased the sensitivity of PC12 cells to αLTX just as efficiently as did a full-length latrophilin (127). These paradoxical results suggest that stimulation of signalling through latrophilin is not relevant to the effect of αLTX, and that the role of latrophilin may simply be to recruit the toxin to synapses, where it evokes vesicle release by interacting with a different protein.

3. The calcium-dependence of the αLTX effect is not due primarily to the ability of the toxin to form cation channels. Blockage of αLTX-induced calcium influx (by removal of magnesium, which is required for this calcium influx), does not prevent the neurotransmitter release elicited from synaptosomes by the combination of αLTX and calcium. Intracellular chelation of calcium, or pharmacological depletion of intracellular calcium stores, does however inhibit this release (128). Hence, it appears that αLTX acts to stimulate release of calcium from intracellular stores, and that this release accounts for the calcium-dependent component of αLTX-elicited vesicle release. Consistent with this, an inhibitor of phospholipase C (the enzyme that makes inositol-(1,4,5)-trisphosphate, which can cause release of calcium from one class of intracellular calcium stores) also inhibits the effect of 1 nM αLTX, without affecting αLTX-induced calcium influx. While the phenomena in these experiments have not shown to be latrophilin-dependent, they can be elicited under conditions in which neurexin Iα does not bind αLTX.

4. At least in synaptosomes exposed to higher concentrations (> 1 nM) of αLTX, about 40% of the αLTX-elicited transmitter release does not require extracellular calcium (128). This appears to be due to formation of pores in the synapse membrane which allow passage of low molecular weight compounds, including fluorescein, aspartate, and calcium. Consistent with this model, reserpine (which blocks uptake of norepinephrine from cytoplasm into vesicles) causes an increase in the calcium-independent αLTX-elicited release, but decreases the calcium-dependent release which is due to synaptic vesicle exocytosis (see next point). Further, the calcium-independent release is not blocked by BoNT/C or BoNT/D, suggesting that it is independent of vesicle fusion. While αLTX can form pores in artificial lipid bilayers

(119), this may not be the mechanism of formation of these pores; binding of αLTX to synaptosomes and to latrophilin-expressing non-neuronal cells allowed pore formation (measured by calcium influx) only in the synaptosomes, suggesting that pore formation is due to an endogenous cell type-specific component that is activated by αLTX. Again, the calcium-independent effects in synaptosomes have not been shown directly to be mediated by latrophilin binding of αLTX, but they occur under conditions when there is no detectable binding to neurexin Iα.

5. The calcium-dependent vesicle release elicited by αLTX in synaptosomes is eliminated by treatment with BoNT/C or BoNT/D (128). Hence, αLTX must activate a step in exocytosis upstream of the formation of the stable SNARE complex. αLTX also causes an increase in the ratio of the GDP/GTP-bound forms of Rab3A (129). Since Rab3A (probably the GTP-bound form) inhibits SNARE-mediated membrane fusion (108, 109), the effect of αLTX is also upstream of Rab3A. Strikingly, syntaxin co-purifies with latrophilin on an α-latrotoxin affinity column, suggesting a close association of latrophilin with the vesicle fusion machinery (124). However, it is unclear whether this association reflects a control of syntaxin function by latrophilin, or is merely a mechanism for anchoring latrophilin in the vicinity of other proteins in the active zone with which it must interact.

6. While neurexin Iα is not absolutely required for the calcium-dependent effects of αLTX, it does make some contribution. αLTX elicits a reduced amount of neurotransmitter release in synaptosomes from neurexin Iα-deficient mice, compared to wild-type synaptosomes (130).

In conclusion, there are still some unsolved puzzles about the mechanism of αLTX action. These include:

(a) If latrophilin is not the cellular partner of αLTX that mediates its effect on vesicle release, then what is?
(b) How does neurexin binding of αLTX potentiate its toxicity, and what endogenous cellular mechanisms are affected?
(c) Why does αLTX have ankyrin repeats, whose role is presumably to bind the toxin intracellularly to the cytoskeleton?

4. Calcium channel toxins

Other predators have evolved toxins that disable their prey by blocking calcium channels. These include the ω-agatoxins, a diverse class of peptide toxins from the funnel web spider *Agelenopsis aperta*, and the ω-conotoxins, another diverse class of peptide toxins from the marine snail genus *Conus*, which both inhibit specific classes of voltage gated calcium channels irreversibly. *Conus* venoms are a complex potion of toxins, of which the ω-conotoxins are only one class. A high rate of toxin sequence divergence outside a few highly conserved residues is suggestive of selection for diverse target specificity within a species. It is also conceivable that it reflects an evolutionary 'arms race' between predator and prey, with continual selection in the prey for toxin-resistant targets, and selection in the predator for peptides that are active against novel target variants (131). The resulting toxin diversity means that conotoxins will offer a rich source of useful ligands for some time to come.

Voltage gated calcium channels form a large family, whose members differ in their gating, kinetic, conductivity, and pharmacological properties (reviewed in ref. 132). The properties of each receptor are an adaptation to its particular cellular role, and can be used to classify the receptors into several classes; the clearest distinctions can be made pharmacologically (e.g. see Section 4.1). The two classes that are most important in mediating presynaptic calcium influx are the N-type and P-type channels. A class of Q-type channels has also been proposed, but it is unclear how different these are from the P-type channels.

4.1 Distinguishing the roles of different calcium channels

The simplest use of the calcium channel toxins has been as pharmacological tools, to distinguish between the functions of the various calcium channels that are found in neurons (reviewed in ref. 132). For example, ω-agatoxin IV-A is an inhibitor of P-type calcium channels (133, 134), and various ω-conotoxins inhibit N-type calcium channels (135).

The use of such specific inhibitors has demonstrated the requirement for these calcium channels to mediate the calcium influx that subsequently triggers exocytosis, and has shown that the pattern of N-type and P-type calcium channel usage at synapses is complex. Either single or multiple channel types can mediate release at a single class of synapse, and different synapses can differ in the predominant class of calcium channel used (136–139).

The reasons for the complexity are less clear. Presumably it contributes to the flexibility with which both calcium influx, and the effect of depolarization on vesicle fusion, can be regulated.

4.2 Using calcium channel toxins to probe synaptic structure

An elegant study has used gold-labelled ω-conotoxin-GVIA to examine the arrangement of calcium channels in presynaptic membranes of calyx-type terminals in the chick ciliary ganglion (140). By using atomic force microscopy to localize the gold particles, they could visualize regions with a high density of channels, presumably corresponding to the presynaptic active zones. Within these clusters their measurements also provided evidence for a preferred channel separation of 40 nm, and frequent linear arrays of 3–10 channels. These observations reinforce the idea that synapses are not just bags of vesicles which respond to a uniform calcium influx, but that the regions of calcium influx are localized, and the vesicle release machinery is arranged so that it will be exposed rapidly to a localized calcium influx after neuronal stimulation (reviewed in ref. 114).

5. Phospholipase toxins

Phospholipase A_2 (PLA_2) toxins are major toxic components of many arthropod and snake venoms (141). They are disulfide-rich small toxins which are characterized by a phospholipase activity specific for the *sn*-2 fatty acid-acyl ester bond of 1,2-phosphoglycerides (141). They constitute a very large protein family in excess of 700 members, showing similarities with secreted mammalian phospholipases, enzymes important in inflammatory and digestive processes (142). Some of these toxins (about 50 distinct proteins) are endowed with a well characterized presynaptic toxicity which causes a persistent blockade of neurotransmitter release in a variety of synapses of the CNS. Despite their low sequence conservation, all low molecular weight phospholipases present a very similar three-dimensional structure characterized by six or seven disulfide bridges, which are crucial for the stability of the protein in the extracellular space. Their toxicity is modest when compared with that of clostridial neurotoxins (in mice, 5–500 μg/kg versus 0.1 ng/kg of BoNT/A) and, despite the massive amount of data available, a strict correlation between PLA_2 activity and toxicity is not apparent (141). It is clear that also other factors play a major role in determining the toxicity of these proteins.

5.1 Structure–function relationships

Four classes can be distinguished on the basis of their structure (143).

(a) Class I: single chain toxins of low molecular weight (13–15 kDa) containing seven disulfide bridges. Representative examples of this class are notexin (144) and ammodytoxin (145).

(b) Class II: dimeric toxins composed of non-covalently linked subunits that show significant homology to each other (146).

(c) Class III: dimeric toxins composed of non-covalently linked non-homologous subunits. One of the members of this class, β-bungarotoxin, is by far the most studied PLA_2 neurotoxin (147, 148).

(d) Class IV: non-covalently linked multimers of homologous subunits. In taipoxin, a member of this class, the active trimer is composed of a PLA_2 toxic subunit, a non-toxic component, and another protein with PLA_2 activity but completely atoxic (147).

The active site of PLA_2 neurotoxins contains a calcium ion essential for hydrolysis of the ester bond linking the fatty acid residue to the phosphoglycerol moiety (143). This calcium ion plays a double role in the catalysis: it interacts with the substrate and it polarizes the target bond, promoting the entry of water. All PLA_2s show a higher activity on phospholipids inserted in biological membranes, compared with monomeric lipids. This is due to the higher efficiency of interfacial catalysis, which depends on adsorption of the enzyme onto the lipid–water interface.

5.2 Presynaptic activity

PLA_2 neurotoxins have been studied both *in vivo* and *in vitro* on isolated nerve–muscle preparations (149). This system allows convenient study of the presynaptic inhibition, allowing the direct and homogeneous comparison of different toxins. PLA_2 neurotoxins decrease the size of evoked endplate potentials (EPPs), decreasing at the same time the frequency of the miniature endplate potentials (MEPPs) without affecting their size (149). The inhibitory activity is preceded by a lag phase, independent of the dose, which is reduced by nerve stimulation. At the beginning, the inhibition of ACh release is very mild, can be blocked by antitoxin antibody, and is reversible, whilst in a later phase, the inhibition is complete and irreversible. This phase is characterized by the complete depletion of synaptic vesicles from the nerve terminal with the appearance of swollen mitochondria and large endosomes, together with clathrin-coated endocytic vesicles incompletely pinched-off from the synaptic membrane. Although no conclusive experiments are available at the moment, the kinetic and the morphological changes induced by the PLA_2 neurotoxins at the neuromuscular junction suggest that they may inhibit retrieval of the synaptic vesicles after fusion. Endocytosis appears to be blocked at a stage following formation of the clathrin scaffold, but before the closure of the vesicle, which requires dynamin, amphiphysin, and adaptor proteins (150). To date, attempts to correlate this inhibition with PLA_2 enzymatic activity have been unsuccessful. In fact, the level of enzymatic activity of these neurotoxins and the amounts of hydrolysed phospholipids are not sufficient to account for their toxicity and intensity of their effects at synaptic terminals. These neurotoxins may possess yet unknown pharmacological sites and/or activity(ies), which partially overlap with the PLA_2 active site. An interesting hypothesis is that presynaptic PLA_2 neurotoxins might be not only hydrolases, but could also transfer fatty acids to proteins, in analogy to diphtheria toxin which is both a NAD^+ glycohydrolase and an ADP-ribosyltransferase (151). Acylation of selected residues of the PLA_2 neurotoxin target could lead to an impairment or modification of its function, thus causing a blockade of neurotransmitter release. Alternatively, the binding properties of the toxins might direct their PLA_2 activity toward subsites of the presynapse of critical importance for neurotransmitter release, where a minimal alteration of the phospholipid composition could account for the final irreversible inhibitory effect. In this regard, the characterization of the high affinity surface receptors of these toxins will be instrumental to clarify their mechanism of action.

6. Conclusions

Toxins affecting synaptic vesicle release have been important tools in identifying and studying the protein components of this process. Current techniques of *in vitro* mutagenesis and generation of mutant organisms may have superseded the uses of toxins for some purposes, when it is necessary (or possible) to study the role of a defined protein, or a defined region of a protein, in an organism in which genetic manipulation is possible. However, the mechanisms of some classes of toxins are still not well understood, and when we are trying to identify important components of the synaptic vesicle cycle, it would be philistine in the extreme to ignore the long experience of nature at doing the same thing.

References

1. Hatheway, C. L. (1995) Botulism: the present status of the disease. *Curr. Top. Microbiol. Immunol.*, **195**, 55.
2. Montecucco, C. and Schiavo, G. (1995) Structure and function of tetanus and botulinum neurotoxins. *Q. Rev. Biophys.*, **28**, 423.
3. Edwards, D. D., McFeters, G. A., and Venkatesan, M. I. (1998) Distribution of *Clostridium perfringens* and fecal sterols in a benthic coastal marine environment influenced by the sewage outfall from McMurdo Station, Antarctica. *Appl. Environ. Microbiol.*, **64**, 2596.
4. Bauer, J. C. and Agerter, C. J. (1994) Isolation of potentially pathogenic bacterial flora from tropical sea urchins in selected West Atlantic and East Pacific sites. *Bull. Marine Sci.*, **55**, 142.
5. Wolochow, H., Hildebrand, G. J., and Lamanna, C. (1966) Translocation of microorganisms across the intestinal wall of the rat: effect of microbial size and concentration. *J. Infect. Dis.*, **116**, 523.
6. Maksymowych, A. B. and Simpson, L. L. (1998) Binding and transcytosis of botulinum neurotoxin by polarized human colon-carcinoma cells. *J. Biol. Chem.*, **273**, 21950.
7. Lamanna, C., Hillowalla, R. A., and Alling, C. C. (1967) Buccal exposure to botulinal toxin. *J. Infect. Dis.*, **117**, 327.
8. Schwab, M. E. and Thoenen, H. (1978) Selective binding, uptake, and retrograde transport of tetanus toxin by nerve terminals in the rat iris: An electron microscope study using colloidal gold as a tracer. *J. Cell Biol.*, **77**, 1.
9. Schwab, M. E. and Thoenen, H. (1976) Electron microscopic evidence for a transsynaptic migration of tetanus toxin in spinal cord motoneurons: an autoradiographic and morphometric study. *Brain Res.*, **105**, 213.
10. Schwab, M. E., Suda, K., and Thoenen, H. (1979) Selective retrograde transsynaptic transfer of a protein, tetanus toxin, subsequent to its retrograde axonal transport. *J. Cell Biol.*, **82**, 798.
11. Curtis, D. R., Game, C. J., Lodge, D., and McCulloch, R. M. (1976) A pharmacological study of Renshaw cell inhibition. *J. Physiol. (London)*, **258**, 227.
12. Brooks, V. B., Curtis, D. R., and Eccles, J. C. (1955) Mode of action of tetanus toxin. *Nature*, **175**, 120.
13. Stoeckel, K., Schwab, M., and Thoenen, H. (1975) Comparison between the retrograde axonal transport of nerve growth factor and tetanus toxin in motor, sensory and adrenergic neurons. *Brain Res.*, **99**, 1.
14. Juzans, P., Comella, J. X., Molgo, J., Faille, L., and Angaut-Petit, D. (1996) Nerve terminal sprouting in botulinum type-A treated mouse levator auris longus muscle. *Neuromuscular Disorders*, **6**, 177.
15. Bonner, P. H., Friedli, P. F., and Baker, R. S. (1994) Botulinum-A toxin stimulates neurite branching in nerve-muscle cultures. *Dev. Brain Res.*, **79**, 39.
16. dePaiva, A., Meunier, F. A., Molgo, J., Aoki, K. R., Dolly, J. O. (1999) Functional repair of motor endplates after botulinum neurotoxin type A poisoning: biphasic switch of synaptic activity between nerve sprouts and their parent terminals. *Proc. Natl. Acad. Sci. USA*, **96**, 3200.

17. Antharavally, B., Tepp, W., and DasGupta, B. R. (1998) Status of Cys residues in the covalent structure of botulinum neurotoxin types A, B, and E. *J. Protein Chem.*, **17**, 187.
18. Krieglstein, K., Henschen, A., Weller, U., and Habermann, E. (1990) Arrangement of disulphide bridges and positions of sulfhydryl groups in tetanus toxin. *Eur. J. Biochem.*, **188**, 39.
19. Krieglstein, K. G., DasGupta, B. R., and Henschen, A. H. (1994) Covalent structure of botulinum neurotoxin type A: location of sulfhydryl groups, and disulphide bridges and identification of C-termini of light and heavy chains. *J. Protein Chem.*, **13**, 49.
20. DasGupta, B. R. (1990) Structure and biological activity of botulinum neurotoxin. *J. Physiol. (Paris)*, **84**, 220.
21. Mochida, S., Poulain, B., Weller, U., Habermann, E., and Tauc, L. (1989) Light chain of tetanus toxin intracellularly inhibits acetylcholine release at neuro-neuronal synapses, and its internalisation is mediated by heavy chain. *FEBS Lett.*, **253**, 47.
22. Poulain, B., Tauc, L., Maisey, E. A., Wadsworth, J. D. F., Mohan, P. M., and Dolly, J. O. (1988) Neurotransmitter release is blocked intracellularly by botulinum neurotoxin, and this requires uptake of both toxin polypeptides by a process mediated by the larger chain. *Proc. Natl. Acad. Sci. USA*, **85**, 4090.
23. Poulain, B., Wadsworth, J. D., Maisey, E. A., Shone, C. C., Melling, J., Tauc, L., *et al.* (1989) Inhibition of transmitter release by botulinum neurotoxin A. Contribution of various fragments to the intoxication process. *Eur. J. Biochem.*, **185**, 197.
24. Bittner, M. A., DasGupta, B. R., and Holz, R. W. (1989) Isolated light chains of botulinum neurotoxins inhibit exocytosis: Studies in digitonin-permeabilized chromaffin cells. *J. Biol. Chem.*, **264**, 10354.
25. Poulain, B., Mochida, S., Weller, U., Hogy, B., Habermann, E., Wadsworth, J. D. F., *et al.* (1991) Heterologous combinations of heavy and light chains from botulinum toxin A and tetanus toxin inhibit neurotransmitter release in Aplysia. *J. Biol. Chem.*, **266**, 9580.
26. Halpern, J. L. and Neale, E. A. (1995) Neurospecific binding, internalisation, and retrograde axonal transport. *Curr. Topics Microbiol. Immunol.*, **195**, 221.
27. Lalli, G., Herreros, J., Osborne, S. L., Montecucco, C., Rossetto, R. and Schiavo, G. (1999) Functional characterisation of tetanus and botulinum neurotoxins binding domains. *J. Cell Sci.*, (in press).
28. Lacy, D. B., Tepp, W., DasGupta, B. R., and Stevens, R. C. (1998) Crystal structure of Botulinum Neurotoxin Type A and implication for toxicity. *Nature Struct. Biol.*, **5**, 898.
29. Umland, T. C., Wingert, L. M., Swaminathan, S., Furey, W. F., Schmidt, J. J., and Sax, M. (1997) Structure of the receptor binding fragment Hc of tetanus toxin. *Nature Struct. Biol.*, **4**, 788.
30. Montecucco, C. (1986) How do tetanus and botulinum toxins bind to neuronal membranes? *Trends Biochem. Sci.*, **11**, 315.
31. Nishiki, T., Tokuyama, Y., Kamata, Y., Nemoto, Y., Yoshida, A., Sato, K., *et al.* (1996) The high-affinity binding of clostridium botulinum type-B neurotoxin to synaptotagmin-II associated with gangliosides G(T1B)/G(D1A). *FEBS Lett.*, **378**, 253.
32. Li, L. and Singh, B. R. (1998) Isolation of synaptotagmin as a receptor for type A and type E botulinum neurotoxin and analysis of their comparative binding using a new microtiter plate assay. *J. Nat. Toxins*, **7**, 215.
33. Black, J. D. and Dolly, J. O. (1986) Interaction of 125I-labeled botulinum neurotoxins with nerve terminals. II. Autoradiographic evidence for its uptake into motor nerves by acceptor-mediated endocytosis. *J. Cell Biol.*, **103**, 535.

34. Black, J. D. and Dolly, J. O. (1986) Interaction of 125I-labeled botulinum neurotoxins with nerve terminals. I. Ultrastructural autoradiographic localization and quantitation of distinct membrane acceptors for types A and B on motor nerves. *J. Cell Biol.*, **103**, 521.
35. Matteoli, M., Verderio, C., Rossetto, O., Iezzi, N., Coco, S., Schiavo, G., *et al.* (1996) Synaptic vesicle endocytosis mediates the entry of tetanus neurotoxin into hippocampal neurons. *Proc. Natl. Acad. Sci. USA*, **93**, 13310.
36. Simpson, L. L. (1983) Ammonium chloride and methylamine hydrochloride antagonize clostridial neurotoxins. *J. Pharmacol. Exp. Therapeut.*, **225**, 546.
37. Simpson, L. L., Coffield, J. L., and Bakry, N. (1994) Inhibition of vacuolar adenosine triphosphatase antagonizes the effects of clostridial neurotoxins but not phospholipase A2 neurotoxins. *J. Pharmacol. Exp. Therapeut.*, **269**, 256.
38. Montecucco, C., Papini, E., and Schiavo, G. (1994) Bacterial protein toxins penetrate cells via a four-step mechanism. *FEBS Lett.*, **346**, 92.
39. Wiener, M., Freymann, D., Ghosh, P., and Stroud, R. M. (1997) Crystal structure of colicin Ia. *Nature*, **385**, 461.
40. Coen, L., Osta, R., Maury, M., and Brulet, P. (1997) Construction of hybrid proteins that migrate retrogradely and transynaptically into the central nervous system. *Proc. Natl. Acad. Sci. USA*, **94**, 9400.
41. Cabot, J. B., Mennone, A., Bogan, N., Carroll, J., Evinger, C., and Erichsen, J. T. (1991) Retrograde, transsynaptic and transneuronal transport of fragment-C of tetanus toxin by sympathetic preganglionic neurons. *Neuroscience*, **40**, 805.
42. Schiavo, G., Poulain, B., Rossetto, O., Benfenati, F., Tauc, L., and Montecucco, C. (1992) Tetanus toxin is a zinc protein and its inhibition of neurotransmitter release and protease activity depend on zinc. *EMBO J.*, **11**, 3577.
43. Söllner, T., Whiteheart, S. W., Brunner, M., Erdjument-Bromage, H., Geromanos, S., Tempst, P., *et al.* (1993) SNAP receptors implicated in vesicle targeting and fusion. *Nature*, **362**, 318.
44. Söllner, T., Bennett, M. K., Whiteheart, S. W., Scheller, R. H., and Rothman, J. E. (1993) A protein assembly-disassembly pathway in vitro that may correspond to sequential steps of vesicle docking, activation and fusion. *Cell*, **75**, 409.
45. Hayashi, T., McMahon, H., Yamasaki, S., Binz, T., Hata, Y., and Südhof, T. C. (1994) Synaptic vesicle membrane fusion complex: action of clostridial neurotoxins on assembly. *EMBO J.*, **13**, 5051.
46. Trimble, W. S., Cowan, D. M., and Scheller, R. H. (1988) VAMP-1: a synaptic vesicle-associated integral membrane protein. *Proc. Natl. Acad. Sci. USA*, **85**, 4538.
47. Walch-Solimena, C., Blasi, J., Edelmann, L., Chapman, E. R., von Mollard, G. F., and Jahn, R. (1995) The t-SNAREs syntaxin 1 and SNAP-25 are present on organelles that participate in synaptic vesicle recycling. *J. Cell Biol.*, **128**, 637.
48. Baumert, M., Maycox, P. R., Navone, F., De Camilli, P., and Jahn, R. (1989) Synaptobrevin: an integral membrane protein of 18,000 daltons present in small synaptic vesicles of rat brain. *EMBO J.*, **8**, 379.
49. Sweeney, S. T., Broadie, K., Keane, J., Niemann, H., and O'Kane, C. J. (1995) Targeted expression of tetanus toxin light chain in *Drosophila* specifically eliminates synaptic transmission and causes behavioural defects. *Neuron*, **14**, 341.
50. Broadie, K., Prokop, A., Bellen, H. J., O'Kane, C. J., Schulze, K. L., and Sweeney, S. T. (1995) Syntaxin and synaptobrevin function downstream of vesicle docking in *Drosophila*. *Neuron*, **15**, 663.
51. Williamson, L. C. and Neale, E. A. (1998) Syntaxin and 25 kDa synaptosomal-associated

protein: differential effects of botulinum neurotoxins C1 and A on neuronal survival. *J. Neurosci. Res.*, **52**, 569.

52. Osen-Sand, A., Staple, J. K., Naldi, E., Schiavo, G., Rossetto, O., Petitpierre, S., *et al.* (1996) Common and distinct fusion proteins in axonal growth and neurotransmitter release. *J. Comp. Neurol.*, **367**, 222.
53. Burgess, R. W., Deitcher, D. L., and Schwarz, T. L. (1997) The synaptic protein syntaxin 1 is required for cellularization of *Drosophila* embryos. *J. Cell Biol.*, **138**, 861.
54. Schulze, K. L. and Bellen, H. J. (1996) *Drosophila* syntaxin is required for cell viability and may function in membrane formation and stabilization. *Genetics*, **144**, 1713.
55. Yamasaki, S., Baumeister, A., Binz, T., Blasi, J., Link, E., Cornille, F., *et al.* (1994) Cleavage of members of the synaptobrevin/VAMP family by types D and F botulinum neurotoxins and tetanus toxin. *J. Biol. Chem.*, **269**, 12764.
56. Yamasaki, S., Binz, T., Hayashi, T., Szabo, E., Yamasaki, N., Eklund, M., *et al.* (1994) Botulinum neurotoxin type G proteolyses the Ala81-Ala82 bond of rat synaptobrevin 2. *Biochem. Biophys. Res. Commun.*, **200**, 829.
57. Regazzi, R., Sadoul, K., Meda, P., Kelly, R. B., Halban, P. A., and Wollheim, C. B. (1996) Mutational analysis of VAMP domains implicated in Ca^{2+}-induced insulin exocytosis. *EMBO J.*, **15**, 6951.
58. Shone, C. C. and Roberts, K. A. (1994) Peptide-substrate specificity and properties of the zinc-endopeptidase activity of botulinum type-B neurotoxin. *Eur. J. Biochem.*, **225**, 263.
59. Rossetto, O., Schiavo, G., Montecucco, C., Poulain, B., Deloye, F., Lozzi, L., *et al.* (1994) SNARE motif and neurotoxins. *Nature*, **372**, 415.
60. Rossetto, O., Deloye, F., Poulain, B., Pellizari, R., Schiavo, G., and Montecucco, C. (1995) The metalloproteinase activity of tetanus and botulism neurotoxins. *J. Physiol. (Paris)*, **89**, 43.
61. Pellizari, R., Rossetto, O., Lozzi, L., Giovedi, S., Johnson, E., Shone, C. C., *et al.* (1996) Structural determinants of the specificity for synaptic vesicle-associated membrane protein/synaptobrevin of tetanus and botulinum type-B and type-G neurotoxins. *J. Biol. Chem.*, **271**, 20353.
62. Pellizari, R., Mason, S., Shone, C. C., and Montecucco, C. (1997) The interaction of synaptic vesicle-associated membrane protein synaptobrevin with botulinum neurotoxins D and F. *FEBS Lett.*, **409**, 339.
63. Washbourne, P., Pellizari, R., Baldini, G., Wilson, M. C., and Montecucco, C. (1997) Botulinum neurotoxin types A and E require the SNARE motif in SNAP-25 for proteolysis. *FEBS Lett.*, **418**, 1.
64. Fernandez, I., Ubach, J., Dulubova, I., Zhang, X., Südhof, T. C., and Rizo, J. (1998) Three-dimensional structure of an evolutionarily conserved N-terminal domain of syntaxin 1A. *Cell*, **94**, 841.
65. Sutton, R. B., Fasshauer, D., Jahn, R., and Brunger, A. T. (1998) Crystal structure of a SNARE complex involved in synaptic exocytosis at 2.4 Å resolution. *Nature*, **395**, 347.
66. Llinás, R., Sugimori, M., Chu, D., Morita, M., Blasi, J., Herreros, J., *et al.* (1994) Transmission at the squid giant synapse was blocked by tetanus toxin by affecting synaptobrevin, a vesicle-bound protein. *J. Physiol. (London)*, **477**, 129.
67. Marsal, J., Ruiz-Montasell, B., Blasi, J., Moreira, J. E., Contreras, D., Sugimori, M., *et al.* (1997) Block of transmitter release by botulinum C1 action on syntaxin at the squid giant synapse. *Proc. Natl. Acad. Sci. USA*, **94**, 14871.
68. Hunt, J. M., Bommert, K., Charlton, M. P., Kistner, A., Habermann, E., Augustine, G. J., *et*

al. (1994) A postdocking role for synaptobrevin in synaptic vesicle fusion. *Neuron*, **12**, 1269.

69. Mochida, S., Saisu, H., Kobayashi, H., and Abe, T. (1995) Impairment of syntaxin by botulinum neurotoxin C-1 or antibodies inhibits acetylcholine release but not Ca^{2+} channel activity. *Neuroscience*, **65**, 905.
70. Dreyer, F., Mallart, A., and Brigant, J. L. (1983) Botulinum-A toxin and tetanus toxin do not affect presynaptic membrane currents in mammalian motor-nerve endings. *Brain Res.*, **270**, 373.
71. Capogna, M., Gahwiler, B. H., and Thompson, S. M. (1996) Presynaptic inhibition of calcium-dependent and calcium-independent release elicited with ionomycin, gadolinium and alpha-latrotoxin in the hippocampus. *J. Neurophysiol.*, **75**, 2017.
72. Trudeau, L.-E., Emery, D. G., and Haydon, P. G. (1996) Direct modulation of the secretory machinery underlies PKA-dependent synaptic facilitation in hippocampal neurons. *Neuron*, **17**, 789.
73. Marxen, P., Bartels, F., Ahnert-Hilger, G., and Bigalke, H. (1991) Distinct targets for tetanus and botulinum-A neurotoxins within the signal transducing pathway in chromaffin cells. *Naunyn-Schmiedebergs Arch. Pharmakol.*, **344**, 387.
74. Glenn, D. E. and Burgoyne, R. D. (1996) Botulinum neurotoxin light-chains inhibit both Ca^{2+}-induced and CTP analog-induced catecholamine release from permeabilized adrenal chromaffin cells. *FEBS Lett.*, **386**, 137.
75. Lopez-Alonso, E., Canaves, J., Arribas, M., Casanova, A., Marsal, J., Gonzales-Ros, J. M., *et al.* (1995) Botulinum toxin type-A inhibits Ca^{2+}-dependent transport of acetylcholine in reconstituted giant liposomes made from presynaptic membranes from cholinergic nerve terminals. *Neurosci. Lett.*, **196**, 37.
76. Capogna, M., McKinney, R. A., O'Connor, V., Gahwiler, B. H., and Thompson, S. M. (1997) Ca^{2+} or Sr^{2+} partially rescues synaptic transmission in hippocampal cultures treated with botulinum toxin-A and toxin-C, but not tetanus toxin. *J. Neurosci.*, **17**, 7190.
77. Yamasaki, S., Hu, Y., Binz, T., Kalkuhl, A., Kurazono, H., Tamura, T., *et al.* (1994) Synaptobrevin/vesicle associated membrane protein (VAMP) of *Aplysia californica*: Structure and proteolysis by tetanus toxin, and botulinal neurotoxins type D and F. *Proc. Natl. Acad. Sci. USA*, **91**, 4688.
78. Li, Y., Foran, P., Fairweather, N. F., dePaiva, A., Weller, U., Dougan, G., *et al.* (1994) A single mutation in the recombinant light chain of tetanus toxin abolishes its proteolytic activity and removes the toxicity seen after reconstitution with native heavy chain. *Biochemistry*, **33**, 7014.
79. Ashton, A. C., Li, Y., Doussau, F., Weller, U., Dougan, G., Poulain, B., *et al.* (1995) Tetanus toxin inhibits neuroexocytosis even when its Zn^{2+}-dependent protease activity is removed. *J. Biol. Chem.*, **270**, 31386.
80. Höhne-Zell, B., Stecher, B., and Gratzl, M. (1993) Functional characterization of the catalytic site of the tetanus toxin light chain using permeabilized adrenal chromaffin cells. *FEBS Lett.*, **336**, 175.
81. Schiavo, G., Benfenati, F., Poulain, B., Rossetto, O., Polverino De Laureto, P., DasGupta, B. R., *et al.* (1992) Tetanus and botulinum-B neurotoxins block neurotransmitter release by proteolytic cleavage of synaptobrevin. *Nature*, **359**, 832.
82. Poulain, B., Rossetto, O., Deloye, F., Schiavo, G., Tauc, L., and Montecucco, C. (1993) Antibodies against rat brain vesicle-associated membrane protein (synaptobrevin) prevent inhibition of acetylcholine release by tetanus toxin or botulinum neurotoxin type-B. *J. Neurochem.*, **61**, 1175.

83. Deitcher, D. L., Ueda, A., Stewart, B. A., Burgess, R. W., Kidokoro, Y., and Schwarz, T. L. (1998) Distinct requirements for evoked and spontaneous release of neurotransmitter are revealed by mutations in the *Drosophila* gene neuronal-synaptobrevin. *J. Neurosci.*, **18**, 2028.
84. Bruns, D., Engers, S., Yang, C., Ossig, R., Jeromin, A., and Jahn, R. (1997) Inhibition of neurotransmitter release correlates with the proteolytic activity of tetanus toxin and botulinus toxin A in individual cultured synapses of Hirudo medicinalis. *J. Neurosci.*, **17**, 1898.
85. McMahon, H. T., Ushkaryov, Y. A., Edelmann, L., Link, E., Binz, T., Niemann, H., *et al.* (1993) Cellubrevin is a ubiquitous tetanus-toxin substrate homologous to a putative synaptic vesicle fusion protein. *Nature*, **364**, 346.
86. Eisel, U., Reynolds, K., Riddick, M., Zimmer, A., Niemann, H., and Zimmer, A. (1993) Tetanus toxin light chain expression in Sertoli cells of transgenic mice causes alterations of the actin cytoskeleton and disrupts spermatogenesis. *EMBO J.*, **12**, 3365.
87. Facchiano, F. and Luini, A. (1992) Tetanus toxin potently stimulates tissue transglutaminase. *J. Biol. Chem.*, **267**, 13267.
88. Facchiano, F., Benfenati, F., Valtorta, F., and Luini, A. (1993) Covalent modification of synapsin I by a tetanus toxin-activated transglutaminase. *J. Biol. Chem.*, **268**, 4588.
89. Ashton, A. C. and Dolly, J. O. (1997) Microtubules and microfilaments participate in the inhibition of synaptosomal noradrenaline release by tetanus toxin. *J. Neurochem.*, **68**, 649.
90. Coffield, J. A., Considine, R. V., Jeyapaul, J., Maksymowych, A. B., Zhang, R., and Simpson, L. L. (1994) The role of transglutaminase in the mechanism of action of tetanus toxin. *J. Biol. Chem.*, **269**, 24454.
91. Gobbi, M., Frittoli, E., and Mennini, T. (1996) Role of transglutaminase in [H-3]5-HT release from synaptosomes and in the inhibitory effect of tetanus toxin. *Neurochem. Int.*, **29**, 129.
92. Ferrer-Montiel, A. V., Canaves, J. M., DasGupta, B. R., Wilson, M. C., and Montal, M. (1996) Tyrosine phosphorylation modulates the activity of clostridial neurotoxins. *J. Biol. Chem.*, **271**, 18322.
93. Nichols, B. J. and Pelham, H. R. B. (1998) SNAREs and membrane-fusion in the Golgi-apparatus. *Biochim. Biophys. Acta [Mol. Cell Res.]*, **1404**, 9.
94. Lewis, M. J., Rayner, J. C., and Pelham, H. R. B. (1997) A novel SNARE complex implicated in vesicle fusion with the endoplasmic reticulum. *EMBO J.*, **16**, 3017.
95. Edwardson, J. M. (1998) Membrane-fusion—all done with SNAREpins. *Curr Biol.*, **8**, R 390.
96. Götte, M. and von Mollard, G. F. (1998) A new beat for the SNARE drum. *Trends Cell Biol.*, **8**, 215.
97. O'Connor, V., Heuss, C., De Bello, W. M., Dresbach, T., Charlton, M. P., Hunt, J. H., *et al.* (1997) Disruption of syntaxin-mediated protein interactions blocks neurotransmitter secretion. *Proc. Natl. Acad. Sci. USA*, **94**, 12186.
98. Matthews, G. (1996) Neurotransmitter release. *Annu. Rev. Neurosci.*, **19**, 219.
99. Weber, T., Zemelman, B. V., McNew, J. A., Westermann, B., Gmachl, M., Parlati, F., *et al.* (1998) SNAREpins: minimal machinery for membrane fusion. *Cell*, **92**, 759.
100. Hanson, P. I., Roth, R., Morisaki, H., Jahn, R., and Heuser, J. E. (1997) Structure and conformational changes in NSF and its membrane receptor complexes visualized by quick-freeze/deep-etch electron microscopy. *Cell*, **90**, 523.
101. Poirier, M. A., Xiao, W. Z., MacOsko, J. C., Chan, C., Shin, Y. K., and Bennett, M. K. (1998) The synaptic SNARE complex is a parallel 4-stranded helical bundle. *Nature Struct. Biol.*, **5**, 765.

102. Rizo, J. and Südhof, T. C. (1998) Mechanics of membrane fusion. *Nature Struct. Biol.*, **5**, 839.
103. Hayashi, T., Yamasaki, S., Nauenberg, S., Binz, T., and Niemann, H. (1995) Disassembly of the reconstituted synaptic vesicle membrane fusion complex in vitro. *EMBO J.*, **14**, 2317.
104. Pellegrini, L. L., O'Connor, V., Lottspeich, F., and Betz, H. (1995) Clostridial neurotoxins compromise the stability of a low energy SNARE complex mediating NSF activation of synaptic vesicle fusion. *EMBO J.*, **14**, 4705.
105. Pellegrini, L. L., O'Connor, V., and Betz, H. (1994) Fusion complex formation protects synaptobrevin against proteolysis by tetanus toxin light chain. *FEBS Lett.*, **353**, 319.
106. Banerjee, A., Kowalchyk, J. A., DasGupta, B. R., and Martin, T. F. J. (1996) SNAP-25 is required for a late postdocking step in Ca^{2+}-dependent exocytosis. *J. Biol. Chem.*, **271**, 20227.
107. Xu, T., Binz, T., Niemann, H., and Neher, E. (1998) Multiple kinetic components of exocytosis distinguished by neurotoxin sensitivity. *Nature Neurosci.*, **1**, 192.
108. Geppert, M., Goda, Y., Stevens, C. F., and Südhof, T. C. (1997) The small GTP-binding protein Rab3A regulates a late step in synaptic vesicle fusion. *Nature*, **387**, 810.
109. Johannes, L., Doussau, F., Clabecq, A., Henry, J. P., Darchen, F., and Poulain, B. (1996) Evidence for a functional link between Rab3 and the SNARE complex. *J. Cell Sci.*, **109**, 2875.
110. Lawrence, G., Foran, P., and Dolly, J. (1996) Distinct exocytotic responses of intact and permeabilized chromaffin cells after cleavage of the 25-kDa synaptosomal-associated protein (SNAP-25) or synaptobrevin by botulinum toxin-A or toxin-B. *Eur. J. Biochem.*, **236**, 877.
110a. Eleopra, R., Tugnoli, V., Rossetto, O., De Grandis, D. and Montecucco, C. (1998) Botulinum neurotoxin serotype A and E in human: evidence of a different temporal profile in the neuromuscular block induced. *Neurosci. Lett.*, **224**, 91.
111. Bezprozvanny, I., Scheller, R. H., and Tsien, R. W. (1995) Functional impact of syntaxin on gating of N-type and Q-type calcium channels. *Nature*, **378**, 623.
112. Martin-Moutot, N., Charvin, N., Leveque, C., Sato, K., Nishiki, T., Kozaki, S., *et al.* (1996) Interaction of SNARE complexes with P/Q-type calcium channels in rat cerebellar synaptosomes. *J. Biol. Chem.*, **271**, 6567.
113. Stanley, E. F. and Mirotznik, R. R. (1997) Cleavage of syntaxin prevents G-protein regulation of presynaptic calcium channels. *Nature*, **385**, 340.
114. Stanley, E. F. (1997) The calcium channel and the organisation of the presynaptic release face. *Trends Neurosci.*, **20**, 404.
115. Wiser, O., Bennett, M. K., and Atlas, D. (1996) Functional interaction of syntaxin and SNAP-25 with voltage-sensitive L- and N-type Ca^{2+} channels. *EMBO J.*, **15**, 4100.
116. Kiyatkin, N., Dulubova, I., and Grishin, E. (1993) Cloning and structural analysis of alpha-latroinsectotoxin cDNA—abundance of ankyrin-like repeats. *Eur. J. Biochem.*, **213**, 121.
117. Dulubova, I. E., Krasnoperov, V. G., Khvotchev, M. V., Pluzhnikov, K. A., Volkova, T. M., Grishin, E. V., *et al.* (1996) Cloning and structure of δ-latroinsectotoxin, a novel insect-specific member of the latrotoxin family. *J. Biol. Chem.*, **271**, 7535.
118. Kiyatkin, N. I., Dulubova, I. E., Chekhovskaya, I. A., and Grishin, E. V. (1990) Cloning and structure of cDNA encoding alpha-latrotoxin from black widow spider venom. *FEBS Lett.*, **270**, 127.
119. Finkelstein, A., Rubin, L. L., and Tzeng, M.-C. (1976) Black widow spider venom: effect of purified toxin on lipid bilayer membranes. *Science*, **193**, 1009.

120. Grasso, A., Alemà, S., Rufini, S., and Senni, M. I. (1980) Black widow spider toxin-induced calcium fluxes and transmitter release in a neurosecretory cell line. *Nature*, **283**, 774.
121. Ushkaryov, Y. A., Petrenko, A. G., Geppert, M., and Südhof, T. C. (1992) Neurexins: synaptic cell surface proteins related to the α-latrotoxin receptor and laminin. *Science*, **257**, 50.
122. Davletov, B. A., Krasnoperov, V., Hata, Y., Petrenko, A. G., and Südhof, T. C. (1995) High-affinity binding of α-latrotoxin to recombinant neurexin Iα. *J. Biol. Chem.*, **270**, 23903.
123. Lelianova, V. G., Davletov, B. A., Sterling, A., Rahman, M. A., Grishin, E. V., Totty, N. F., *et al.* (1997) α-Latrotoxin receptor, latrophilin, is a novel member of the secretin family of G protein-coupled receptors. *J. Biol. Chem.*, **272**, 21504.
124. Krasnoperov, V. G., Bittner, M. A., Beavis, R., Kuang, Y., Salnikow, K. V., Chepurny, O. G., *et al.* (1997) α-Latrotoxin stimulates exocytosis by the interaction with a neuronal G-protein-coupled receptor. *Neuron*, **18**, 925.
125. Bittner, M. A., Krasnoperov, V. G., Stuenkel, E. L., Petrenko, A. G., and Holz, R. W. (1998) A Ca^{2+}-independent receptor for α-latrotoxin, CIRL, mediates effects on secretion via multiple mechanisms. *J. Neurosci.*, **18**, 2914.
126. Ichtchenko, K., Khvotchev, M., Kiyatkin, N., Simpson, L., Sugita, S., and Südhof, T. C. (1998) α-Latrotoxin action probed with recombinant toxin: receptors recruit α-latrotoxin but do not transduce an exocytotic signal. *EMBO J.*, **17**, 6188.
127. Sugita, S., Ichtchenko, K., Khvotchev, M., and Südhof, T. C. (1998) α-Latrotoxin receptor CIRL/Latrophilin 1 (CL1) defines an unusual family of ubiquitous G-protein-linked receptors. *J. Biol. Chem.*, **273**, 32715.
128. Davletov, B. A., Meunier, F. A., Ashton, A. C., Matsushita, H., Hirst, W. D., Lelianova, V. G., *et al.* (1998) Vesicle exocytosis stimulated by α-latrotoxin is mediated by latrophilin and requires both external and stored Ca^{2+}. *EMBO J.*, **17**, 3209.
129. Stahl, B., von Mollard, G. F., Walch-Solimena, C., and Jahn, R. (1994) GTP cleavage by the small GTP-binding protein Rab3A is associated with exocytosis of synaptic vesicles induced by alpha-latrotoxin. *J. Biol. Chem.*, **269**, 24770.
130. Geppert, M., Khvotchev, M., Krasnoperov, V., Goda, Y., Missler, M., Hammer, R. E., *et al.* (1998) Neurexin Iα is a major α-latrotoxin receptor that cooperates in α-latrotoxin action. *J. Biol. Chem.*, **273**, 1705.
131. Olivera, B. M., Rivier, J., Clark, C., Ramilo, C. A., Corpuz, G. P., Abogadie, F. C., *et al.* (1990) Diversity of Conus neuropeptides. *Science*, **249**, 257.
132. Dunlap, K., Luebke, J. I., and Turner, T. J. (1995) Exocytotic Ca^{2+} channels in mammalian central neurons. *Trends Neurosci.*, **18**, 89.
133. Turner, T. J., Adams, M. E., and Dunlap, K. (1992) Calcium channels coupled to glutamate release identified by omega-Aga-IVA. *Science*, **258**, 310.
134. Mintz, I. M., Venema, V. J., Swiderek, K. M., Lee, T. D., Bean, B. P., and Adams, M. E. (1992) P-type calcium channels blocked by the spider toxin omega-Aga-IVA. *Nature*, **355**, 827.
135. Olivera, B. M., Gray, W. M., Zeikus, R., McIntosh, J. M., Varga, J., Rivier, J., *et al.* (1985) Peptide neurotoxins from fish-hunting cone snails. *Science*, **230**, 1338.
136. Reuter, H. (1995) Measurements of exocytosis from single presynaptic nerve terminals reveal heterogeneous inhibition by Ca^{2+}-channel blockers. *Neuron*, **14**, 773.
137. Turner, T. J., Adams, M. E., and Dunlap, K. (1993) Multiple Ca^{2+} channel types coexist to regulate neurotransmitter release. *Proc. Natl. Acad. Sci. USA*, **90**, 9518.
138. Luebke, J. I., Dunlap, K., and Turner, T. J. (1993) Multiple calcium-channel types control glutamatergic synaptic transmission in the hippocampus. *Neuron*, **11**, 895.

139. Mintz, I. M., Sabatini, B. L., and Regehr, W. G. (1995) Calcium control of transmitter release at a cerebellar synapse. *Neuron*, **15**, 675.
140. Haydon, P. G., Henderson, E., and Stanley, E. F. (1994) Localization of individual calcium channels at the release face of a presynaptic nerve terminal. *Neuron*, **13**, 1275.
141. Rappuoli, R. and Montecucco, C. (ed.) (1997) *Protein toxins and their use in cell biology*. Oxford University Press, Oxford.
142. Tischfield, J. A. (1997) A reassessment of the low molecular weight phospholipase A2 gene family in mammals. *J. Biol. Chem.*, **272**, 17247.
143. Arni, R. K. and Ward, R. J. (1996) Phospholipase A2–a structural review. *Toxicon*, **34**, 827.
144. Westerlund, B., Nordlund, P., Uhlin, U., Eaker, D., and Eklund, H. (1992) The three-dimensional structure of notexin, a presynaptic neurotoxic phospholipase A2 at 2.0 Å resolution. *FEBS Lett.*, **301**, 159.
145. Pungercar, J., Kordis, D., Strukelj, B., Liang, N. S., and Gubensek, F. (1991) Cloning and nucleotide sequence of a cDNA encoding ammodytoxin A, the most toxic phospholipase A2 from the venom of long-nosed viper (*Vipera ammodytes*). *Toxicon*, **29**, 269.
146. Bibier, A. L., Bon, C., and Faure, G. (1997) Rattlesnake venom neurotoxins: crotoxin-related proteins. In *Protein toxins and their use in cell biology* (ed. R. Rappuoli and C. Montecucco), p. 221. Oxford University Press, Oxford.
147. Dolly, J. O. (1992) Peptide toxins that alter neurotransmitter release. In *Handbook of experimental pharmacology. Selective neurotoxicity* (ed. H. Herken and F. Hucho), Vol. 102, p. 683. Springer–Verlag, Berlin.
148. Mebs, D. (1997) β-bungarotoxin. In *Protein toxins and their use in cell biology* (ed. R. Rappuoli and C. Montecucco), p. 219. Oxford University Press, Oxford.
149. Harris, J. B. (1991) Phospholipases in snake venoms and their effects on nerve and muscle. In *Snake toxins* (ed. A. L. Harvey), p. 91. Pergamon Press, Oxford.
150. Cremona, O. and De Camilli, P. (1997) Synaptic vesicle endocytosis. *Curr. Opin. Neurobiol.*, **7**, 323.
151. Choe, S., Bennett, M. J., Fujii, G., Curmi, P. M., Kantardjieff, K. A., Collier, R. J., *et al.* (1992) The crystal structure of diphtheria toxin. *Nature*, **357**, 216.
152. Soleil-Hac, J. M., Cornille, F., Martin, L., Lenoir, C., Fournie-Zaluski, M. C., and Roques, B. P. (1996) A sensitive and rapid fluorescence-based assay for determination of tetanus toxin peptidase activity. *Anal. Biochem.*, **241**, 120.
153. Link, E., Edelman, L., Chou, J. H., Binz, T., Yamasaki, S., Eisel, U., *et al.* (1992) Tetanus toxin action: inhibition of neurotransmitter release linked to synaptobrevin proteolysis. *Biochem. Biophys. Res. Commun.*, **189**, 1017.
154. Schiavo, G., Santucci, A., DasGupta, B. R., Metha, P. P., Jontes, J., Benfenati, F., *et al.* (1993) Botulinum neurotoxins serotypes A and E cleave SNAP-25 at distinct COOH-terminal peptide bonds. *FEBS Lett.*, **335**, 99.
155. Schiavo, G., Rossetto, O., Catsicas, S., DeLaureto, P. P., DasGupta, B. R., Benfenati, F., *et al.* (1993) Identification of the nerve terminal targets of botulinum neurotoxin serotypes A, D, and E. *J. Biol. Chem.*, **268**, 23784.
156. Blasi, J., Chapman, E. R., Link, E., Binz, T., Yamasaki, S., DeCamilli, P., *et al.* (1993) Botulinum neurotoxin A selectively cleaves the synaptic protein SNAP-25. *Nature*, **365**, 160.
157. Binz, T., Blasi, J., Yamasaki, S., Baumeister, A., Link, E., Südhof, T., *et al.* (1994) Proteolysis of SNAP-25 by types E and A botulinal neurotoxins. *J. Biol. Chem.*, **269**, 1617.
158. Hallis, B., James, B. A. F., and Shone, C. C. (1996) Development of novel assays for

botulinum type-A and type-B neurotoxins based on their endopeptidase activities. *J. Clin. Microbiol.*, **34**, 1934.

159. Schmidt, J. J. and Bostian, K. A. (1995) Proteolysis of synthetic peptides by type-A botulinum neurotoxin. *J. Prot. Chem.*, **14**, 703.
160. Blasi, J., Chapman, E. R., Yamasaki, S., Binz, T., Niemann, H., and Jahn, R. (1993) Botulinum neurotoxin C1 blocks neurotransmitter release by means of cleaving HPC-1/syntaxin. *EMBO J.*, **12**, 4821.
161. Schiavo, G., Shone, C. C., Bennett, M. K., Scheller, R. H., and Montecucco, C. (1995) Botulinum neurotoxin type-C cleaves a single lys-ala bond within the carboxyl-terminal region of syntaxins. *J. Biol. Chem.*, **270**, 10566.
162. Foran, P., Lawrence, G. W., Shone, C. C., Foster, K. A., and Dolly, J. O. (1996) Botulinum neurotoxin C1 cleaves both syntaxin and SNAP-25 in intact and permeabilized chromaffin cells—correlation with its blockade of catecholamine release. *Biochemistry*, **35**, 2630.
163. Williamson, L. C., Halpern, J. L., Montecucco, C., Brown, J. E., and Neale, E. A. (1996) Clostridial neurotoxins and substrate proteolysis in intact neurons—botulinum neurotoxin acts on synaptosomal-associated protein of 25 kDa. *J. Biol. Chem.*, **271**, 7694.
164. Vaidyanathan, V. V., Yoshino, K., Jahnz, M., Dorries, C., Bade, S., Nauenburg, S., *et al.* (1999) Proteolysis of SNAP-25 isoforms by botulinum neurotoxin types A, C, and E: Domains and amino acid residues controlling the formation of enzyme-substrate complexes and cleavage. *J. Neurochem.*, **72**, 327.
165. Schiavo, G., Shone, C., Rossetto, O., Alexander, F. C. G., and Montecucco, C. (1993) Botulinum neurotoxin type F is a zinc endopeptidase specific for VAMP/synaptobrevin. *J. Biol. Chem.*, **268**, 11516.
166. Schiavo, G., Malizio, C., Trimble, W. S., DeLaureto, P. P., Milan, G., Sugiyama, H., *et al.* (1994) Botulimum-G neurotoxin cleaves VAMP/synaptobrevin at a single ala-ala peptide bond. *J. Biol. Chem.*, **269**, 20213.

7 Functional studies of presynaptic proteins at the squid giant synapse

MARIE E. BURNS and GEORGE J. AUGUSTINE

1. Introduction

The work of Katz (1) and colleagues established that synaptic transmission arises from the secretion of quantal packets of neurotransmitter from the presynaptic terminals of neurons. More recent work has revealed that release of neurotransmitters is due to the influx of calcium ions into presynaptic terminals and the subsequent localized elevation of calcium concentration within these terminals (2). This rise in presynaptic calcium concentration triggers the fusion of synaptic vesicles with the presynaptic plasma membrane and results in exocytotic discharge of the neurotransmitters stored within the vesicles (3). Exocytosis is only one reaction in a series of synaptic vesicle trafficking events that leads to the local recycling of vesicle components within the presynaptic terminal (see Chapter 1).

While this view of synaptic transmitter release as a form of calcium-regulated secretion is well established, many important questions remain about the molecular basis of this process. In particular, while many presynaptic proteins have been identified (4), the functions of these proteins are largely unclear. In order to identify the proteins and protein–protein interactions responsible for neurotransmitter release and the other steps of the vesicle trafficking cycle, it is necessary to combine analyses of the physiology, biochemistry, and microscopy of the presynaptic terminal. Here we describe the application of such approaches at the squid giant synapse. This synapse has the largest known presynaptic terminal (5), an advantage that makes it possible to microinject into presynaptic terminals molecular probes that perturb the function of individual proteins and then evaluate the consequences of these molecular manipulations upon neurotransmitter release. Further, these terminals possess many thousands of active zones, the sites of neurotransmitter release, making it practical to use electron microscopy to evaluate the structural consequences of perturbing the function of presynaptic proteins. The goal of this chapter is to provide some of the theoretical underpinnings of this approach and some practical details that will be useful for those considering experimentation upon the squid giant synapse.

2. Overview of the squid giant synapse

The large dimensions of the squid giant synapse have long made this synapse a classical preparation for the study of presynaptic function (1, 6). We therefore begin with a brief overview of the extraordinary anatomy of this synapse and a historical description of how useful this synapse has been in elucidating presynaptic mechanisms. The presynaptic neuron of the squid giant synapse is the second of three elements that constitute a motor circuit involved in producing the rapid escape movements of this animal (5). The cell body of this second-order neuron lies within the palliovisceral ganglion of the squid and its axon travels within the pallial nerve to innervate the stellate ganglion. Once entering the stellate ganglion, this second-order axon forms 10–15 finger-like branches; each of these branches is a giant presynaptic terminal. In a typical specimen of *Loligo pealei* (perhaps 15 cm in length), the most distal presynaptic terminal — hereafter referred to as the giant presynaptic terminal — is on the order of 1 mm in length and approximately 50 μm in diameter (Fig. 1). These dimensions make the terminal approximately one million times larger in volume than a typical, 1 μm diameter presynaptic terminal of the mammalian nervous system.

Each of the presynaptic terminals of the second-order neuron innervates even larger third-order motor neurons. The most distal of these third-order neurons has received considerable attention as the 'giant' axon preparation so useful in elucidating the ionic basis of action potential generation (7). Synaptic contacts are formed between the giant presynaptic terminal and fine dendritic processes emerging from the postsynaptic third-order neuron. Approximately 10000–40000 of these contacts are found in a typical giant synapse (8, 9), yielding (by far) the largest number of active zones known to be present in any presynaptic terminal. Despite their large number, each individual active zone is not remarkably different from their counterparts in mammalian neurons, making it possible to use this synapse to make general conclusions about the relationship between presynaptic structure and function.

The large size of the squid giant synapse has proven to be of decisive advantage for studies of presynaptic mechanisms (1, 6). For example, the large size of the presynaptic terminal permitted the first direct measurements of presynaptic action potentials (10), a feat that has been replicated in only a handful of preparations in the intervening 40 years. Current-clamp studies at the squid presynaptic terminal demonstrated the distinctive bell-shaped relationship between presynaptic membrane potential and neurotransmitter release, which served as the key observation behind the 'calcium hypothesis' of neurotransmitter release (11). Microinjection of calcium ions into this terminal was the initial test of the hypothesis that intracellular calcium concentration determines the rate of transmitter release (12), an observation that has been extended at this synapse through some of the earliest applications of 'caged' calcium compounds (13, 14) and by microinjection of calcium chelators (15).

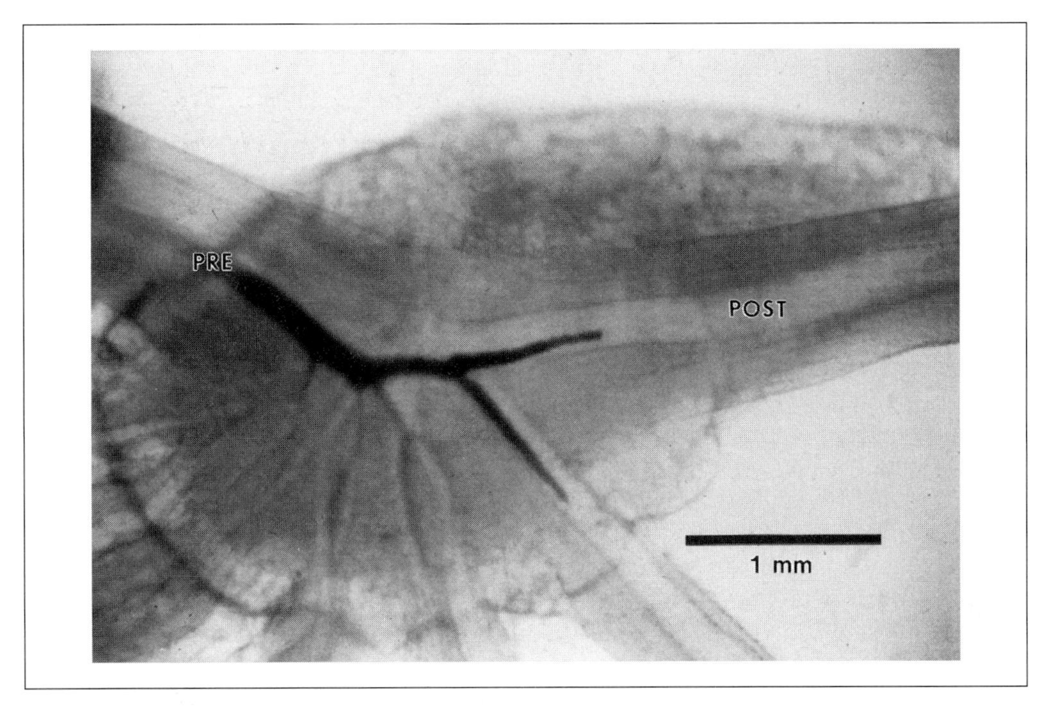

Fig. 1 Bright-field micrograph of a squid stellate ganglion. The giant second-order neuron (PRE) has been microinjected with a red dye. Upon entering the ganglion, the axon of this neuron forms 11 finger-like presynaptic terminals that each innervate a single third-order motor neuron. The largest of these third-order axons is labelled POST and is visible as a translucent structure lying under the most distal presynaptic terminal. The region of overlap between the two neurons is the giant synapse whose large dimensions can be appreciated by comparison to the calibration bar, which is approximately 1 mm in length. From an unpublished experiment by Dr Stephen J. Smith (Stanford University Medical Center).

Further, the first optical measurements of presynaptic calcium levels were performed at this synapse (16–18) and documented the accumulation of calcium during transmitter release. Subsequent imaging studies with fluorescent calcium indicators such as fura-2 (19) demonstrated that calcium influx is restricted to the terminal region of the presynaptic neuron, suggesting that calcium channels are specifically targeted to presynaptic terminals (20, 21). The first voltage-clamp analysis of presynaptic calcium currents (22–24) established the gating properties of the calcium channels of this presynaptic terminal and served as the predecessor for subsequent experiments demonstrating that the 'co-operative' triggering of neurotransmitter release is due to a non-linear relationship between calcium entry and transmitter release (25, 26). Measurements of presynaptic membrane capacitance performed at this synapse documented that neurotransmitter release is associated with a transient increase in the surface area of the presynaptic plasma membrane, as expected from an exocytotic mechanism for neurotransmitter release (27). Last, injection of synapsin into the squid giant presynaptic terminal provided the initial demonstration of the physiological importance of synapsin (28) and was the precursor to the many contemporary studies of the function of presynaptic proteins that serve as the focus of this chapter.

3. Functional studies of presynaptic proteins

To determine the physiological roles of presynaptic proteins in neurotransmitter release, one must be able to monitor the amount of neurotransmitter released while perturbing the protein's function. Microinjection acutely introduces reagents that perturb the function of a protein directly into the squid giant nerve terminal while neurotransmitter release is simultaneously monitored by measuring the electrical responses evoked in the large postsynaptic cell. Additional sophisticated physiological analyses of presynaptic function, such as measurements of calcium influx and presynaptic membrane capacitance, can be valuable in elucidating the mechanism of action of an injected reagent. Correlation of molecular function with presynaptic structure is also valuable and requires the ability to examine the ultrastructure of microinjected terminals. Once a reagent has inhibited neurotransmitter release, the giant synapse can be rapidly fixed for electron microscopy (EM). This can provide information about the stage of the vesicle trafficking cycle that is disrupted by the injected reagent and, thereby, help to define the reaction in which the protein of interest is involved.

In this section, we will discuss several important aspects of these microinjection experiments. We begin with a brief description of some of the practical details important in such experiments, then discuss the various reagents used to perturb presynaptic proteins, the rationale behind their design and interpretation of their actions, and conclude with an overview of our current understanding of the molecular basis of neurotransmitter release based on applications of the microinjection approach.

3.1 Using the squid giant synapse for microinjection studies

A comprehensive description of the simple dissection of the squid giant synapse can be found in the summary by Stanley (29). While squid are available at many locales around the world, the most reliable supply is available at the Marine Biological Laboratory in Woods Hole, Massachusetts. This laboratory provides a continuous supply of *Loligo pealei* from May until November each year and has sufficient infrastructural support to make microinjection experiments feasible.

Microinjection of a reagent into the presynaptic terminal is best achieved by pressure injection, which permits fine control of the rate and extent of injection. To prevent osmotic trauma to the terminal, the reagent to be microinjected is dissolved in a solution that mimics the composition of squid axoplasm (200 mM KCl, 100 mM taurine, 200 mM K isethionate, 50 mM K Hepes pH 7.4). Because the reagent must be soluble in this medium, microinjection is restricted to reagents that are soluble in this solution, such as cytoplasmic proteins or the cytoplasmic domains of integral proteins. The reagent solution is placed within an electrode (5–20 MΩ resistance) that is inserted directly in the presynaptic terminal. This electrode can both inject the reagent and record the presynaptic action potential evoked by an electrical stimulus. It is very important to monitor the rate of injection throughout the course of an experiment, because overly rapid injections will damage the terminal and result in a decrease in neurotransmitter release that is usually irreversible. To estimate the amount of reagent injected, the reagent is either fluorescently tagged or is mixed with an inert fluorescent tracer, such as dextran conjugated to fluorescein or some other dye. The presynaptic terminal is then imaged with a fluorescence microscope and the fluorescence of the terminal is quantified by comparison to a calibration standard. Because of variations in terminal geometry, these estimates of reagent concentration are accurate to within a factor of two to three.

When a reagent is microinjected, it will diffuse throughout the presynaptic terminal and, eventually, into the rest of the giant presynaptic neuron. This diffusion causes the reagent to be diluted by a factor of 10–100, meaning that the concentration of reagent within the injection pipette must be 10–100 times more concentrated than is necessary to affect transmitter release. This can be a problem for reagents that are not highly soluble in the axoplasm-like injection buffer. However, one advantage of the large presynaptic axon acting as a diffusion sink is that following injection, the concentration of reagents injected into the presynaptic terminal will decrease and result in reversal of the effects of the reagent. Reversibility allows each injected terminal to serve as its own control and can yield some insight into the mechanism of action of the reagent, such as the specificity and affinity of the reagent for its target.

To evoke neurotransmitter release, action potentials are generated in the presynaptic axon by using an intracellular stimulating electrode (~ 5 MΩ resistance and filled with 3 M KCl) to pass current into the presynaptic axon. Typically release is measured by placing a third microelectrode (~ 5 MΩ; filled with 3 M KCl) in the postsynaptic cell within the region that is in contact with the presynaptic terminal. This electrode can record the postsynaptic potential (PSP) resulting from presynaptic neurotransmitter release or the postsynaptic cell can be voltage-clamped to measure the postsynaptic current (25). PSPs are typically measured because voltage-clamping requires a fourth microelectrode. These PSPs are strong enough to elicit a postsynaptic action potential, so transmitter release is usually quantified by measuring the initial slope of the PSP at a time before the postsynaptic action potential occurs.

The only serious limitation of this synapse for electrophysiological studies is that it is quite difficult to record spontaneous 'miniature' postsynaptic potentials. Measurement of these responses is necessary to determine the size of the quanta that serve as the unitary secretion event and to determine the number of these quanta that are released by a presynaptic stimulus (30). However, the large size of the postsynaptic neuron causes its electrical input resistance to be very low, on the order of 25–50 KΩ. This low resistance causes spontaneous miniature potentials to be very small in amplitude, typically on the order of 1–10 μV, and challenges the sensitivity of intracellular recording methods. None the less, it has been possible to detect such events by using synapses from small squid, which have relatively smaller postsynaptic axons (31, 32), or by using a low-noise axial wire electrode (33). Another approach to quantifying these spontaneous events has been to measure the macroscopic postsynaptic depolarizations that arise from large increases in the frequency of miniature responses (12, 18, 34, 35) and/or to use fluctuation analysis to estimate the amplitude and time course of miniature potentials (12, 33, 36). Although these methods can yield useful information about spontaneous miniature potentials, it has not been practical to measure spontaneous transmitter release characteristics routinely during microinjection experiments.

One of the great advantages of the squid giant synapse is that individual terminals microinjected with a molecular reagent can be examined with electron microscopy (EM) to examine the ultrastructure of active zones in these terminals. Such an analysis of the structural consequences of molecular perturbation can help assign a defined function for the protein within the synaptic vesicle trafficking cycle. To sample the active zones from various distances from the injection site, cross-sections should be cut at regular intervals along the entire length of the terminal. For example, sectioning the terminal every 50 μm along its length typically yields ten or more sections per terminal and, with 10–20 active zones in each cross-section, 150–200 active zones for analysis. This yields a large and statistically meaningful sample of the structure of the injected terminal.

3.2 Molecular probes of presynaptic proteins

Perhaps the most important aspect of the microinjection approach to protein perturbation is the design of the reagents that are microinjected into the presynaptic terminal. These reagents need to fulfil three important criteria:

(a) They must be specific for the protein or protein–protein interaction of interest.

(b) They must be soluble.

(c) Because of dilution upon microinjection, they must be effective at relatively low concentrations relative to their solubility limit.

Several different types of reagents, each with distinct advantages and caveats, have been used in microinjection experiments. We next discuss these different types of reagents and give an example of how each has contributed to our current understanding of neurotransmitter release mechanisms.

3.2.1 Proteins

One type of reagent that can be used for microinjection experiments is proteins. These proteins can be native proteins, purified from sources such as squid optic lobe or mammalian brain, or recombinant proteins produced in bacterial expression systems. Squid and mammalian proteins usually are very similar in their primary structure and, in our experience, usually have very similar effects when microinjected into the squid giant presynaptic terminal. Protein microinjection can be thought of as an acute form of a genetic overexpression experiment, with the expectation being that the injected protein will mimic the action of the endogenous protein in the terminal. If the endogenous protein is present in rate-limiting quantities, the injected protein will then exaggerate the normal function of the protein (Fig. 2A, B). Suitable controls for such injections can include inactive mutated forms of recombinant proteins or heat denatured native proteins, though the latter are often difficult to microinject because they tend to form precipitates that clog injection pipettes.

While interpretation of protein injection experiments is conceptually straightforward, two types of errors can arise. First, if the endogenous protein already is present in excess, then microinjection of additional protein should have no effect and yield a negative result (Fig. 2C). However, such an outcome can be informative if one is confident that the injected protein is biologically active and that the quantities injected are comparable to those already present within the presynaptic terminal. A more insidious error in interpretation can occur if an excess of injected protein disrupts the usual order of protein–protein interactions involved in neurotransmitter release (Fig. 2D). Such an outcome would not reflect the normal function of the protein but instead yield a physiological phenotype that is opposite to that expected from the native protein, analogous to a 'dominant-negative' mutation. Therefore, it is important to establish whether the action of the exogenous protein is via exaggeration of its normal function, or due to a loss-of-function phenotype.

While there is no evidence to suggest that such false positive results account for any of the published actions of microinjected proteins, two tests can be performed to assess the likelihood of such actions. First, because such effects might be expected to require relatively high concentrations of protein, the concentration of injected protein can be varied systematically during an experiment to determine whether low and high concentrations produce qualitatively similar effects on transmitter release. Secondly, the actions of injected proteins can be compared to those of other injected reagents, such as binding site peptides. If perturbing a protein–protein interaction with these reagents interferes with transmitter release, then microinjecting the full-length protein might be expected to have a stimulatory effect upon release. For example, while injection of a number of peptide fragments of the soluble NSF-attachment protein (SNAP) inhibits transmitter release, injection of full-length SNAP stimulates release (37). This logical correspondence between the actions of the two different classes of reagents indicates that the action of endogenous SNAP is to promote neurotransmitter release and provides a general paradigm for sorting out the physiological functions of other presynaptic proteins.

A number of different proteins have been injected into the squid giant presynaptic terminal (Table 1). The classic example of such experiments is the microinjection of synapsin by Llinas, Greengard, and colleagues (28, 38). In these experiments, synapsin I was purified from mammalian brain and injected into the squid giant synapse. It was found that dephosphorylated synapsin inhibited transmitter release evoked by presynaptic depolarizations, while phosphorylated synapsin had little or no effect. A fluorescent tag (Texas red) was attached to the synapsin to observe diffusion of injected protein throughout the presynaptic terminal. The degree of inhibition of transmitter release was proportional to the spread of the injected dephosphorylated

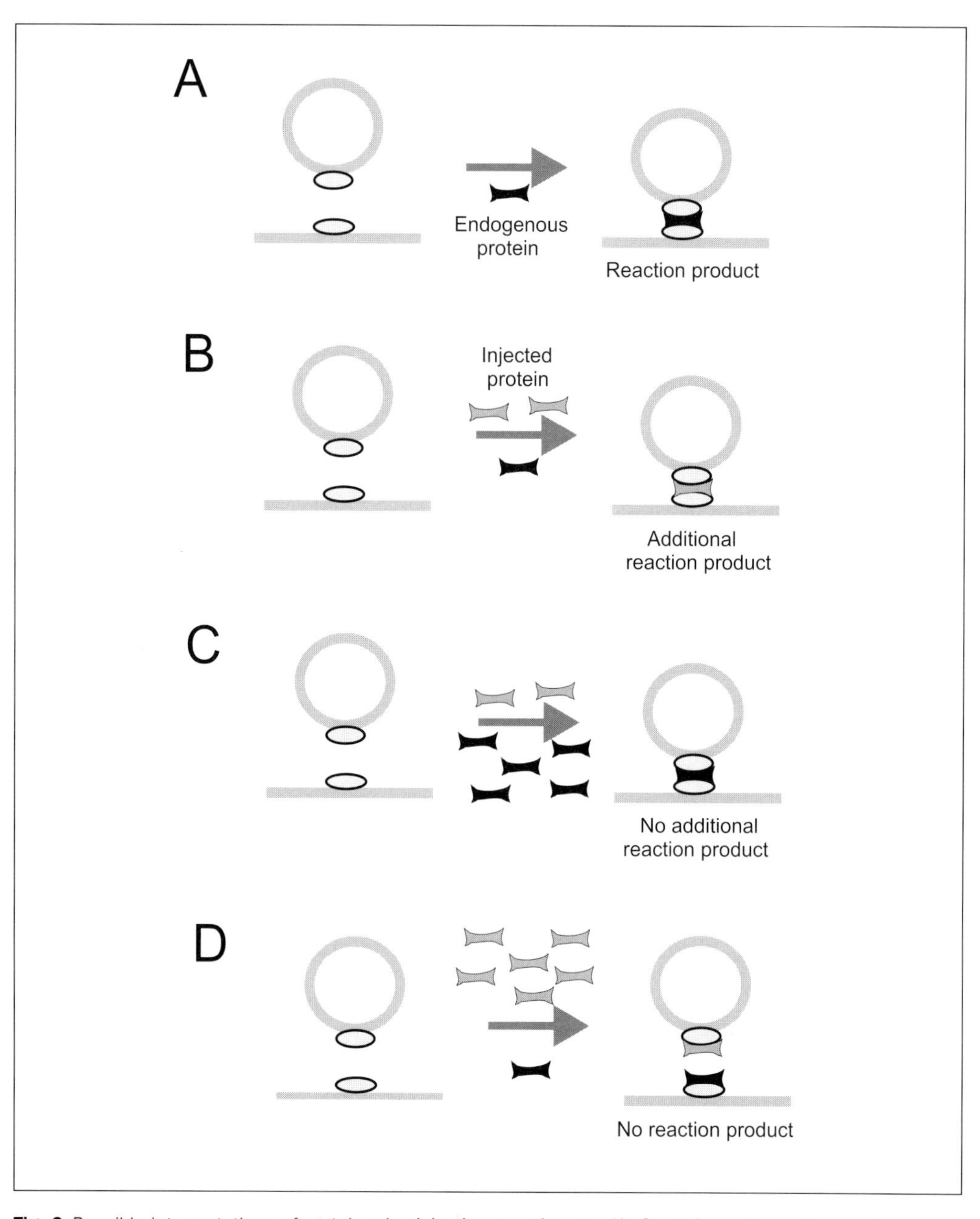

Fig. 2 Possible interpretations of protein microinjection experiments. (A) Control condition. Binding of an endogenous protein to its partners, in this hypothetical case proteins on both synaptic vesicle and the plasma membranes, results in a reaction that is important for neurotransmitter release. (B) If the endogenous protein is present in rate-limiting quantities, injection of additional protein will increase the rate of the reaction and increase the reaction product. (C) If the endogenous protein is present in excess within the terminal, injection of additional protein has no effect on the reaction nor its product. (D) If a large excess of protein is injected, it may bind inappropriately and act as a competitive inhibitor of the normal reaction and prevent generation of the normal reaction product.

Table 1 Effects of microinjected proteins on neurotransmitter release and presynaptic structure

Protein	Effects on release	Structural effects	Reference
Synapsin	Decrease		38
Calmodulin kinase II	Increase		28, 38
Mss4	Increase		44
αSNAP	Increase		37
Squid SNAP	Increase		37
γSNAP	Increase		37
Yeast SNAP (Sec17)	None		37
NSF	None		W. DeBello, F. Schweizer, and G. Augustine, unpublished
Rab3A	None		35
SNAP-25	None		45
Tyrosine kinase	Increase[a]		46
Squid-sec1	Decrease	None	47
Rabphilin-3A	Decrease	None	35
Complexin-1	Decrease		48
Complexin-2	Decrease		48
Squid complexin	Decrease		48

[a] Also increased presynaptic calcium current.

synapsin within the terminal, suggesting that the amount of inhibition reflects the number of active zones exposed to the injected synapsin. Simultaneous voltage-clamping of the presynaptic terminal, under conditions where presynaptic sodium channels and potassium channels were blocked, revealed that synapsin injection had no effect on presynaptic calcium currents but instead reduced the amount of transmitter released by a given influx of calcium through these channels. Although the fine-structure of synapsin-injected terminals was not examined, the results are consistent with the proposal that synapsin regulates the number of vesicles available for release by acting as a phosphorylation-sensitive link between synaptic vesicles and the presynaptic cytoskeleton. This hypothesis has been largely borne out by more recent studies of the actions of injected synapsin fragments into the squid giant terminal (39), injection of antibodies into lamprey giant axons (40), and by studies of synaptic function in synapsin knock-out mice (41–43).

3.3.2 Binding site peptides

A second type of microinjection reagent is based upon peptides that mimic the sites of protein–protein interactions. These peptides can be recombinant fragments of presynaptic proteins or smaller synthetic peptides. Because they can bind to a target protein but should possess none of the other biological activity of a full-length protein, such peptides should act as competitive inhibitors of the interaction between native proteins within the terminal (Fig. 3A, B). Suitable controls include 'scrambled' peptides that contain the same amino acids in an order that no longer resembles the

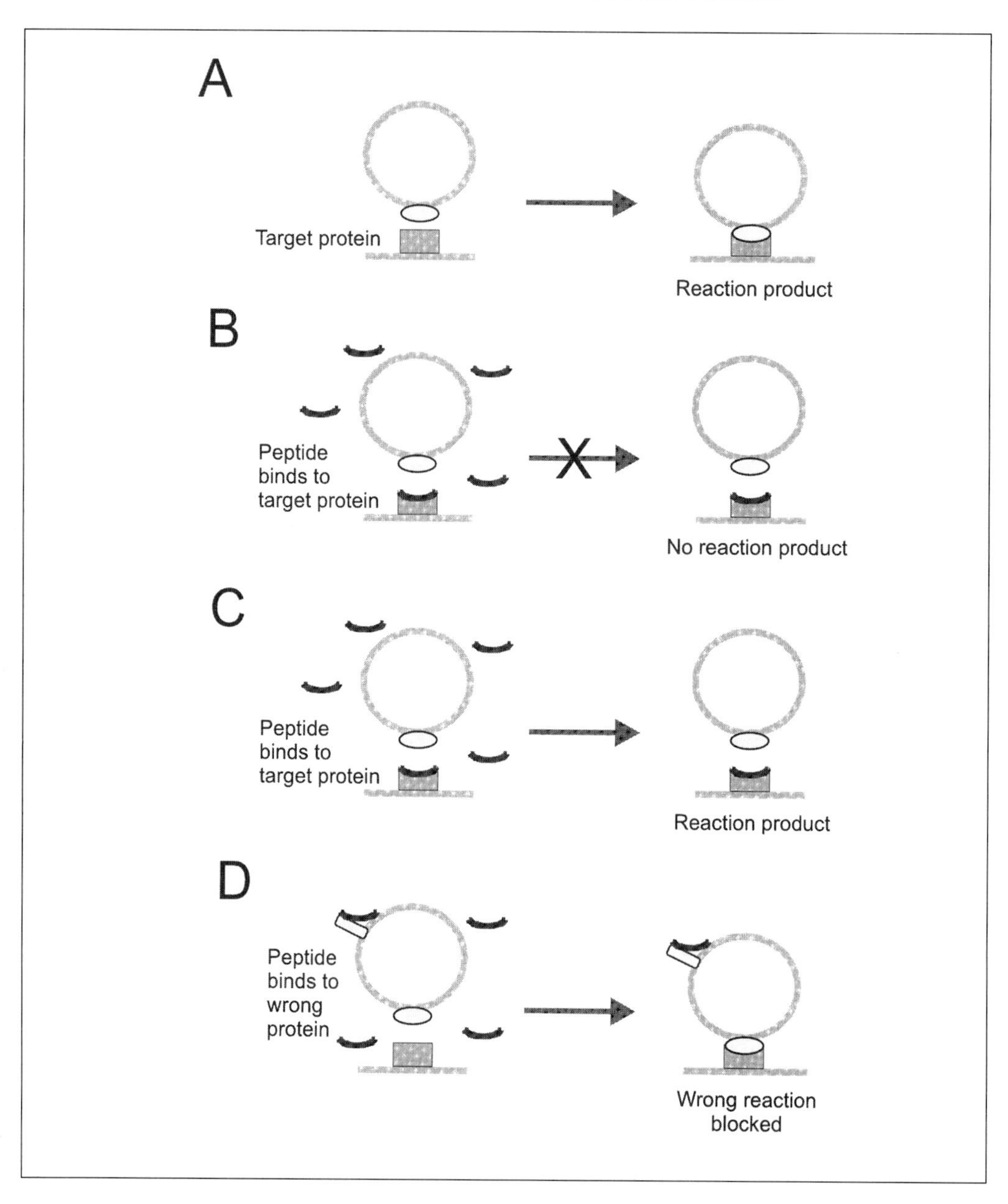

Fig. 3 Possible interpretations of peptide microinjection experiments. (A) Control condition. Binding of an endogenous protein, in this hypothetical case a vesicular protein, to its partner on the plasma membrane results in a reaction that is important for neurotransmitter release. (B) A peptide that mimics the binding domain of the endogenous (vesicular) protein binds to the usual target protein, preventing the endogenous protein from binding. The antagonistic action of the peptide thereby prevents generation of the normal reaction product. (C) If binding of the injected peptide to partner protein is sufficient to mediate the reaction that normally requires the endogenous, full-length protein, then the product of the reaction will be generated. (D) If the injected peptide is not specific for the protein–protein interaction of interest, it may bind to other presynaptic proteins and inhibit their interactions and resultant products. This will cause the function of the protein of interest to be misassigned.

binding site or point mutant peptides having single amino acid substitutions which, ideally, will abolish binding activity.

In principle, these binding site peptides can serve as a means of probing the function of defined protein–protein interactions in neurotransmitter release and have proven very useful for such applications (Table 2). However, there are several limitations that must be kept in mind when evaluating the results of such experiments. One limitation is that binding sites have been identified for only a fraction of the known protein–protein interactions. To circumvent this problem, the earliest binding site peptides were based on computer analysis of the amino acid sequences of presynaptic proteins (e.g. ref. 49). These analyses identified regions of the protein that were predicted to be surface exposed and were conserved across species and isoforms, common features of regions involved in binding to other proteins. Approximately one-third of the peptides developed with this strategy had physiological effects when injected into the giant presynaptic terminal, suggesting that this approach is capable of identifying binding sites. For example, one peptide produced via this approach (Pep15) (49) has subsequently been shown by crystallography to

Table 2 Effects of active microinjected peptides on neurotransmitter release and presynaptic structure

Peptide	Peptide source	Effects on release	Structural effects	Reference
Pep20	Synaptotagmin C2B	Decrease	Increased docked vesicles	49
Pep15	Synaptotagmin C2A	Decrease		49
Synaptobrevin cytoplasmic domain	Synaptobrevin	Decrease		60
SS19	Squid SNAP	Decrease	Increased docked vesicles and decreased total vesicles	37
SS24	Squid SNAP	Decrease		37
MS20	Bovine αSNAP	Decrease		37
NSF-1	Squid NSF	Decrease		59
NSF-2	Squid NSF	Decrease[a]	Decrease total vesicles	59
NSF-3	Squid NSF	Decrease[a]		59
TAX-86	Mammalian syntaxin H3 domain	Decrease	None	53
TAX-74	Squid syntaxin H3 domain	Decrease		53
Secpep3	Squid Sec-1	Decrease	Increased docked vesicles	47
Rabphilin N-terminus	Bovine rabphilin	Decrease	Decrease total vesicles	35
Rabphilin C-terminus	Bovine rabphilin	Decrease	None	35
Domain E peptide	Squid synapsin	Decrease[a]	Decrease total vesicles	39
Domain E peptide	Mammalian synapsin	Decrease[a]		39
Domain E peptide	*Drospohila* synapsin	Decrease[a]		39
sSBD-2	Squid complexin	Decrease	None	48
rSBD-2	Rat complexin	Decrease		48
AP180 peptide	Squid, mouse AP180	Decrease	Decrease total vesicles	64
AP2 peptide	Mouse AP2	Decrease		64

[a] Also slows down the kinetics of neurotransmitter release.

correspond to an apparent binding pocket in the C2A domain of synaptotagmin (50, 51).

More recent studies have used *in vitro* assays of protein binding to directly assess the ability of binding site peptides to prevent specific protein–protein interactions and to otherwise improve the design of binding site peptides. For example, mutational analyses of recombinant proteins have been very useful in identifying the sites of protein–protein interactions of syntaxin (52), which led to the development of syntaxin binding site peptides for microinjection experiments (53). The recent use of X-ray crystallography (54–56) or solution NMR methods (57, 58) to describe the structure of the binding sites or complexes of presynaptic proteins should further improve our understanding of the interactions among presynaptic proteins and will be extremely helpful in designing the next generation of binding site peptides for microinjection studies.

The main limitation in interpreting the action of these binding site peptides is that they could act as either an antagonist (Fig. 3B) or as an agonist (Fig. 3C) of the interaction of interest. Biochemical assays of the ability of these peptides to disrupt a defined protein–protein interaction are therefore necessary to define the biological activity of these peptides and are becoming a standard part of this approach (47, 53, 59). Thus far, all of the peptides analysed in this way by our laboratory appear to act as antagonists of protein–protein interactions but it is likely that agonist peptides eventually will be identified. It is worth noting that either type of peptide can be useful in microinjection experiments, as long as its action is known with certainty.

Another general limitation of the binding site peptide approach is that these reagents tend to have a fairly low affinity for binding to their partners. For example, while the n-sec1 protein binds to syntaxin with nanomolar affinity, a peptide derived from n-sec1 half-maximally binds to syntaxin at approximately 50 μM (47). The reduced affinity of peptides, relative to their protein of origin, probably arises because the structures of the peptides are generally much less constrained than the same region in the intact protein. The lack of structural constraint therefore allows peptides to assume multiple conformations and spend only a fraction of their time in the conformation needed to disrupt the protein–protein interaction of interest. Consistent with this explanation, there is some correlation between peptide size and the concentration required to inhibit neurotransmitter release. For example, 50–100 amino acid long fragments of SNARE proteins inhibit neurotransmitter release at low micromolar concentrations (45, 53, 60) while smaller peptides (15–25 amino acids in length) typically act at concentrations on the order of 100 μM (37, 47, 59).

The relatively low affinity of peptides for their targets may cause two problems. One practical problem is that the peptides must be highly soluble, at concentrations on the order of 10 mM in the injection buffer, to allow for the dilution that occurs following injection into the giant presynaptic terminal. In our experience, approximately 80% of the synthetic peptides that we have produced meet this solubility criterion. Secondly, the need for relatively high peptide concentrations increases the possibility that a peptide produces physiological effects that are not specific to the protein–protein interaction of interest (Fig. 3D). Although a low affinity does not mandate low specificity, it is always important to consider the possibility of such non-specific physiological effects. The starting point for such an analysis is to consider the biological effects of control peptides. Point mutants are generally considered an optimal control for peptide specificity and should be used whenever practical. The main limitation is that detailed mutagenesis has been performed for very few binding sites for presynaptic proteins. One noteworthy exception is a peptide from NSF, termed NSF-3, which inhibits transmitter release and slows release kinetics (59). A point mutation in comatose, a *Drosophila* NSF gene, similarly inhibits synaptic function (61, 62). The amino acid residue affected by this mutation is found within the NSF-3 peptide, allowing the corresponding residue change to produce an inactive peptide that served as a control for specificity. When point mutations that abolish function are not known, scrambled peptides can evaluate the importance of physical factors, such as charge and hydrophobicity, in determining the activity of these peptides. Any biological actions shared by scrambled peptides are unlikely to result from disrupting the specific protein–protein interaction of interest.

Recent studies of a peptide from the syntaxin binding domain of complexin, termed SBD-2 (48), illustrate two means of assessing the specificity of a binding site peptide *in vitro*. First, immobilized SBD-2 peptide was found to bind to syntaxin but to none of the many other proteins present in squid optic lobe homogenates. Secondly, concentrations of SBD-2 that completely blocked the interaction of these two proteins had no effect upon the interaction of syntaxin with other proteins, such as synaptotagmin, SNAP-25, or synaptobrevin. This indicates that SBD-2 is highly specific for the interaction of complexin with syntaxin and that even when SBD-2 is bound to syntaxin, syntaxin still can interact with its numerous other binding partners. While this level of analysis has not been achieved for other binding site peptides, it should be possible to do so in future studies to establish the specificity of peptide action.

The binding site peptide approach has been used to examine the function of many protein–protein interactions at the squid synapse (Table 2) and one of the best examples of the utility of this approach is the study of SNAP function by DeBello *et al.* (37). A number of peptides derived from both squid and mammalian SNAPs inhibited neurotransmitter release reversibly when injected into the squid giant terminal, whereas other peptides from other regions of SNAP had no effect on transmitter release. In contrast, injection of full-length mammalian αSNAP or squid SNAP proteins enhanced neurotransmitter release, suggesting that the active SNAP peptides were disrupting a SNAP-dependent protein–protein interaction required for neurotransmitter release.

At present, the binding target of SS-19 and the other SNAP peptides is unknown because biochemical analysis of the interaction of these peptides with the known binding partners is incomplete. SS-19, one of the most potent of these SNAP peptides, does not prevent the interaction of αSNAP or squid SNAP with NSF, one of the binding partners of these SNAPs (J. Morgan and G. J. Augustine, unpublished). Thus, SS-19 presumably is disrupting the interaction of SNAPs with other binding partners, such as the membrane SNAP receptor (SNARE) proteins. Although the binding partner for SS-19 is not clear, *in vitro* experiments suggest that the action of the peptide is very specific. Measurements of the fusion of Golgi membranes, a process known to require SNAP (63), demonstrated that SNAP peptides which inhibited neurotransmitter release were also capable of inhibiting Golgi membrane fusion. Conversely, those SNAP peptides that had no effects on release also had no effect on Golgi membrane fusion. This strong similarity between the biological actions of the SNAP peptides in the two systems strongly suggests that the peptides act specifically to prevent the function of SNAP.

Injection of the SS-19 peptide was used to answer three questions about the presynaptic function of SNAP. First, Ca imaging revealed that SS-19 had no effect on Ca influx during presynaptic action potentials, suggesting that SNAP does not regulate presynaptic Ca channels. Secondly, blockade of neurotransmitter release occurs within four seconds of injecting SS-19, showing that SNAP is involved in a reaction that occurs no more than four seconds away from the membrane fusion reaction. This rules out the possibility that this peptide inhibits transmitter release by blocking the slower processes of endocytosis and vesicle recycling. In fact, because of delays associated with peptide diffusion within the giant terminal, it is possible that SNAP could work even closer to the time of vesicle fusion. Finally, EM analysis of terminals injected with SS-19 revealed that inhibition of neurotransmitter release is correlated with an accumulation of synaptic vesicles that were attached to (within 50 nm of) the presynaptic plasma membrane. This indicated that blockade of vesicle fusion by SS-19 is likely to be due to inhibition of a SNAP-dependent reaction that occurs after synaptic vesicles dock at the plasma membrane. The peptide also reduced the number of synaptic vesicle by approximately half, suggesting an additional role for SNAP in the endocytotic pathway. These conclusions of SNAP function are in general agreement with subsequent work done in permeabilized chromaffin cells (65) and in temperature-sensitive mutations in the *Drosophila* gene for NSF, one of the binding partners of SNAP (66).

3.2.3 Antibodies

Antibodies are a third class of reagent employed in microinjection studies at the squid giant synapse (Table 3). The attraction of antibodies is that they can, in favourable cases, bind to their target proteins with very high affinity and specificity. However, antibodies are substantially larger than most of the presynaptic proteins they are intended to study. As a result, the binding of antibodies to their target proteins may create steric hindrance that non-selectively inhibits presynaptic function. Thus, careful control experiments must be designed to evaluate such concerns. Ideal controls are antibodies that bind to a second epitope that is present, at comparable density, at locations near the epitope of interest. Such control antibodies should have no effect when microinjected. A more pragmatic concern is that antibodies tend to recognize their presynaptic antigens in a species-specific manner; for example, most of the antibodies generated against mammalian SNAP-25 do not recognize the squid homologue of this protein. Thus, it is often necessary to generate antibodies that specifically recognize squid presynaptic proteins.

Table 3 Effect of microinjected toxins and antibodies on neurotransmitter release and presynaptic structure

Reagent	Target	Effects on release	Structural effects	Reference
Toxins				
Botulinum B	Synaptobrevin	Decrease		60
Tetanus	Synaptobrevin	Decrease	Increased docked vesicles	60, 74
Botulinum C1	Syntaxin	Decrease	None	53
		Decrease	Increased vesicle number	75
Botulinum A	SNAP-25	None		45
Botulinum E	SNAP-25	None		45
α-Latrotoxin	Neurexin	None		H. Betz, K. Bommert, M. P. Charlton, and G. J. Augustine, unpublished
Antibodies				
Anti-synaptotagmin	Squid C2A domain	Decrease	Increased vesicle number	67
Anti-synaptotagmin	Squid C2B domain	Decrease during repetitive activity	Reduced vesicle number	68
Anti-SNAP-25	Mammalian SNAP-25	None		45
Anti-syntaxin	Mammalian syntaxin	Inhibits	Increased vesicle number	76

The most compelling example of the microinjection of antibodies into the squid giant presynaptic terminal is the study of Mikoshiba *et al.* (67) and Fukuda *et al.* (68), who injected antibodies directed against the two C2 domains of squid synaptotagmin. They found that presynaptic injection of antibodies generated against these two domains had different effects; while an anti-C2A domain antibody blocked transmitter release evoked by single presynaptic action potentials, an anti-C2B antibody did not have this effect. In response to trains of action potentials, the anti-C2B antibody produced a time-dependent depression of synaptic transmission while the anti-C2A antibody did not. Further, EM analysis suggested that the two antibodies had different effects on the number of synaptic vesicles: the anti-C2A antibody caused an accumulation of synaptic vesicles while the anti-C2B antibody depleted synaptic vesicles. From a technical standpoint, the differential actions of these two antibodies provides a good argument that the actions of each antibody are not due to steric effects. From the standpoint of vesicle trafficking, the results suggest that synaptotagmin is a multifunctional protein, with the C2A domain involved in vesicle fusion and the C2B domain involved in endocytosis. These conclusions are consistent with peptide injection experiments (49, 51) and results obtained from genetic mutation of synaptotagmin in *Drosophila*, *C. elegans*, and mice (69–73).

3.2.4 Toxins

A number of neurotoxins have defined presynaptic targets and have therefore proven valuable as molecular probes of neurotransmitter release (Table 3). Among these, clostridial toxins that act as metalloproteases have attracted special attention because these proteases cleave SNARE proteins important for neurotransmitter release (77, 78). The simplest scenario is that injection of such toxins will selectively destroy only the target of interest (Fig. 4B), revealing the normal function of this protein (Fig. 4A). However, some toxins may have multiple substrates (Fig. 4C); for example, Botulinum toxin, serotype C1, is known to cleave both syntaxin and SNAP-25 (79). As is the case for all other microinjection reagents, any loss of selectivity limits the conclusions that can be drawn from the response to the reagent. Controls can include heat treated toxin samples or inactive mutant recombinant toxins.

Microinjection of clostridial toxins into the squid giant presynaptic terminal has

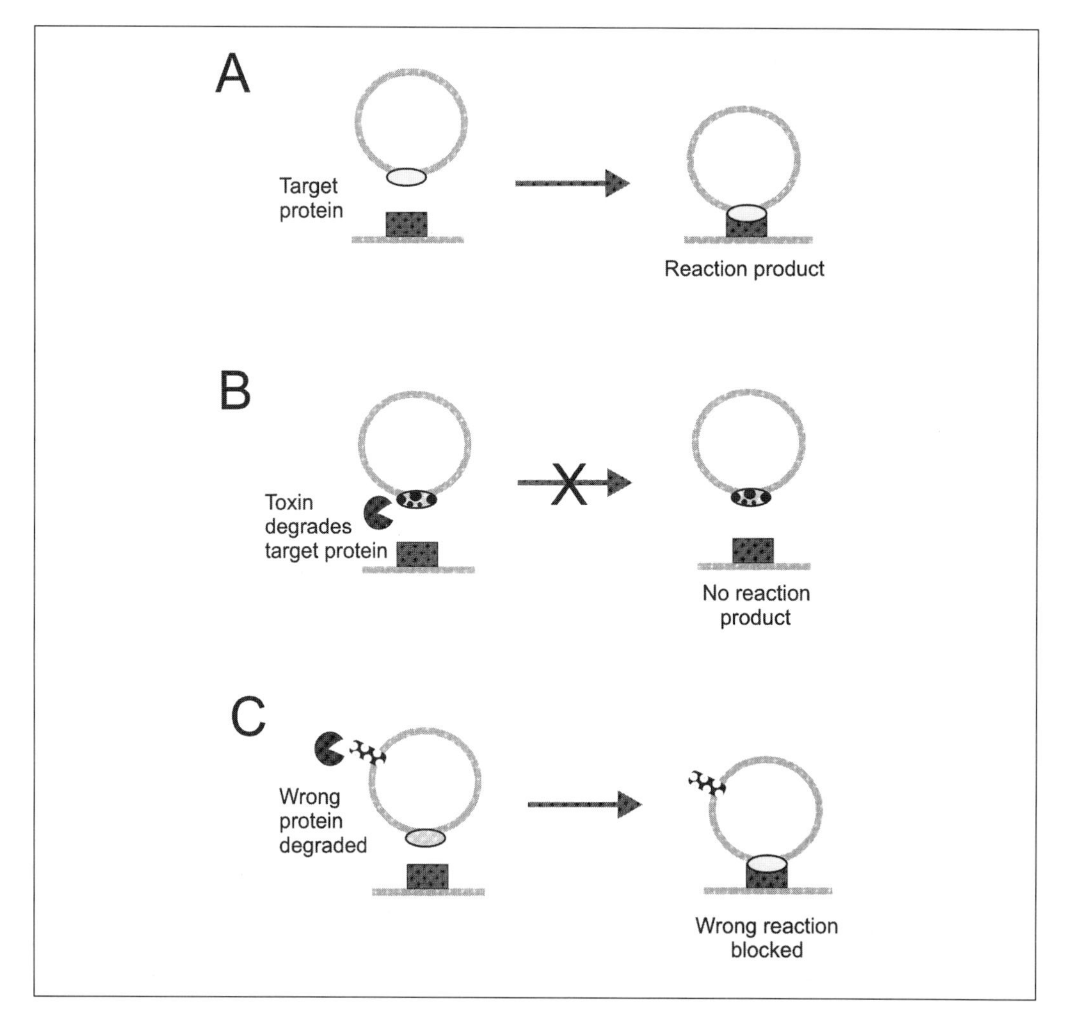

Fig. 4 Possible interpretations of neurotoxin microinjection experiments. (A) Control condition. An endogenous protein, in this hypothetical case a vesicular protein, mediates a reaction that is important for neurotransmitter release. (B) Injection of neurotoxin can proteolyse the vesicle protein, preventing it from participating in its usual reaction and producing its normal product. (C) Injected neurotoxin could proteolyse a different presynaptic protein and impair some other reaction, which will be wrongly attributed to the theoretical target of the neurotoxin.

proven effective in some cases. The light chains of both Botulinum toxin, serotype B, and tetanus toxin inhibit transmitter release when injected into the squid giant presynaptic terminal (60, 74). These toxins cleave squid synaptobrevin and cleavage of this protein is presumably their mode of action. One line of support for this conclusion is that the inhibitory effects of tetanus toxin are prevented by co-injection of a peptide that mimics the toxin cleavage site of synaptobrevin (60). Active zones of terminals injected with tetanus toxin have an accumulation of synaptic vesicles, including docked vesicles, suggesting that cleavage of synaptobrevin blocks vesicle trafficking at a step that follows vesicle docking but precedes fusion. This conclusion is consistent with the outcome of other studies involving genetic targeting of tetanus toxin to *Drosophila* synapses (80) or studies of null mutations in the *Drosophila* synaptobrevin gene (81).

Likewise, presynaptic injection of Botulinum toxin, serotype C1, cleaves squid syntaxin and inhibits transmitter release (53, 75). Ultrastructural analyses suggest either a normal number of synaptic vesicles (53) or an increased number of vesicles (75). In both cases, there was no depletion of docked vesicles, again indicating that cleavage of syntaxin blocks vesicle trafficking at a step that follows vesicle docking but precedes fusion. This result is consistent with the results of null (80) or temperature-sensitive (66) mutations in a *Drosophila* syntaxin gene.

Finally, Botulinum toxins serotypes A and E, which cleave mammalian SNAP-25, are unable to cleave the squid homologue of this protein, presumably due to variations in the proteolytic cleavage site of squid SNAP-25. In addition, neither of these toxins inhibit transmitter release when injected into the squid giant presynaptic terminal (45). This is consistent with the conclusion that the inhibitory actions of these toxins at mammalian synapses is due to cleavage of SNAP-25. In summary, clostridial toxins have been useful as tools to study the function of SNARE proteins at the squid giant synapse but are limited by the fact that squid SNAREs are not substrates for some of these toxins.

3.2.5 Other reagents

In addition to the protein-selective reagents described above, a number of other reagents with a broader range of action have been microinjected into the squid giant presynaptic terminal (Table 4). Among these, the best example is the use of non-hydrolysable guanine nucleotides to study the function of presynaptic GTP binding proteins (82). It was found that both GTPγS and GDPβS inhibited evoked neurotransmitter release, while neither aluminium fluoride nor GDP had an effect. This pattern of inhibition is consistent with the notion that a small molecular weight GTP binding protein, rather than a heterotrimeric GTP binding protein, promotes transmitter release when bound to GTP. Terminals injected with GTPγS had a reduced number of synaptic vesicles except for those docked at the plasma membrane (82), while terminals injected with GDPβS had a selective reduction in docked vesicles but no change in the number of vesicles elsewhere (83). Thus at least one role of GTP binding proteins in neurotransmitter release is to promote the docking of synaptic vesicles.

Table 4 Effect of miscellaneous reagents on neurotransmitter release and presynaptic structure

Reagent	Effects on release	Structural effects	Reference
ATPγS	Decrease		W. DeBello, G. J. F. Schweizer, and G. Augustine, unpublished
GTPγS	Decrease	Increased docked vesicles	82
GDPβS	Decrease	Decreased docked vesicles	82, 83
NEM	Decrease		W. DeBello, F. Schweizer, and G. J. Augustine, unpublished
Phorbol esters	Increase		84
IP4, IP5, IP6	Decrease		85
Pervanadate	Decrease[a]		46

[a] Decrease in postsynaptic response was accompanied by an increase in presynaptic Ca currents.

3.3 Electron microscopy defines protein action at the level of synaptic vesicle trafficking

One of the main challenges of the perturbation approach is to interpret a physiological phenotype in terms of changes in synaptic vesicle trafficking. While measurements of presynaptic electrical and ionic events can be useful in this regard, the most valuable approach to date has been to analyse the effects of microinjected reagents on the ultrastructure of presynaptic terminals. However, in order to draw meaningful conclusions about the structural changes induced by microinjection, it is necessary to assess variability both within a given terminal and between experimental and control terminals. For example, in our study of the effects of microinjecting Rabphilin-3A reagents, we found that there was more structural variability between synapses in a given terminal that there was on average between terminals within a given treatment group (35). Thus, it is important to quantify structural properties and to compare these parameters both within and across treatment groups. Two different types of quantitative analysis have been used for this purpose.

(a) Determination of the spatial distribution of synaptic vesicles. This analysis defines both the number of vesicles and their static position at the presynaptic active zone by measuring the distance between the presynaptic plasma membrane and the centre of each synaptic vesicle. Such measurements are not too labour-intensive and have been employed in many studies (Tables 1–4).

(b) Measurements of the membrane surface areas of the organelles involved in vesicle trafficking (see Chapter 1), including synaptic vesicles, plasma membrane, coated vesicles, and endosomes, can provide a glimpse into the function of the entire vesicle trafficking cycle (86). This analysis is done by using image processing software to determine the linear amount of membrane present in a given tissue section and then converting these values to membrane areas based on the thickness of the section. These measurements are very time-consuming and, at present, have been combined with microinjection only in the study of Rabphilin-3A function by Burns *et al.* (35).

We will briefly summarize our current view of the role of various presynaptic proteins in synaptic vesicle trafficking to demonstrate how structural changes produced by microinjected reagents can be used to deduce the function of proteins in synaptic vesicle trafficking (Figure 5). When interpreting the structural consequences of perturbing neurotransmitter release at the squid giant presynaptic terminal, we use the model of vesicle trafficking proposed by Heuser and Reese (86). While there have been many challenges to this model in the intervening 25 years (see Chapter 1), in our opinion it remains the most likely explanation of synaptic vesicle trafficking.

3.3.1 Vesicle mobilization

The Heuser and Reese (86) model postulates that synaptic vesicles within the interior of the presynaptic terminal are not directly involved in neurotransmitter release; more recently these vesicles have been proposed to serve as a 'reserve pool' of vesicles that is mobilized to replenish the vesicles that undergo fusion (40). Perturbation of this reserve pool should cause a selective loss of synaptic vesicles in the interior of the terminal. Microinjection of peptides from the E domain of synapsin have such an effect (39), consistent with the hypothesis that synapsin is involved in the maintenance of this reserve pool (Fig. 5). Domain E peptides also prevent docked synaptic vesicles from fusing and, therefore, indicate an additional action of synapsin on docked vesicles (not shown in Fig. 5).

3.3.2 Vesicle docking

Mobilized vesicles attach to the presynaptic plasma membrane during the process of docking. Inhibition of proteins needed for vesicle docking should prevent vesicles from attaching to the plasma membrane. We have found that injection of GDPβS causes a selective loss of vesicles attached to the plasma membrane, suggesting that a GTP binding protein is important for vesicle docking (83). Consistent with this notion, a peptide from Rabphilin-3A, a protein that binds to the GTP binding protein Rab3A, causes a relative accumulation of vesicles that are near (50–100 nm away from) the plasma membrane but are not attached to it (35). Thus, Rab3A or some other GTP binding protein must be bound to GTP to permit vesicles to dock (Fig. 5).

3.3.3 Vesicle priming

It is thought that docked vesicles must undergo ATP-dependent priming reactions before they are competent to undergo fusion with the plasma membrane (87, 88). Perturbing vesicle priming should prevent docked vesicles from fusing with the plasma membrane and cause vesicles to accumulate at the plasma membrane. Two reagents that cause such changes when microinjected into the squid presynaptic terminal are peptides from SNAP (37) and NSF (59). This is consistent with indications that these proteins must form a 20S complex that dissociates during priming (88) and suggests that NSF and SNAP are involved in vesicle priming (Fig. 5). Presumably the function of these proteins is to dissociate the membrane SNARE proteins in preparation for fusion, though it remains possible that they are important after fusion (66, 89).

3.3.4 Vesicle fusion

The observation that membrane fusion occurs after formation of a complex of the membrane SNARE proteins (syntaxin, SNAP-25, and synaptobrevin) has led to the conclusion that these proteins are bound to each other at the time that synaptic vesicle fuse (90). Disrupting the formation of this SNARE complex would therefore be expected to prevent docked vesicles from fusing. Microinjection of clostridial toxins (53, 60) and several other reagents that prevent SNAREs from interacting (45, 47, 53) produce such a phenotype. Thus, interaction among the membrane SNAREs is not required for vesicle docking but is required for fusion (Fig. 5).

In vitro, the interactions among the SNARE proteins are not sensitive to calcium ions. Because neurotransmitter release requires calcium, one or more calcium binding proteins must confer calcium sensitivity to SNARE-dependent membrane fusion. We have found that injection of a synaptotagmin peptide causes an

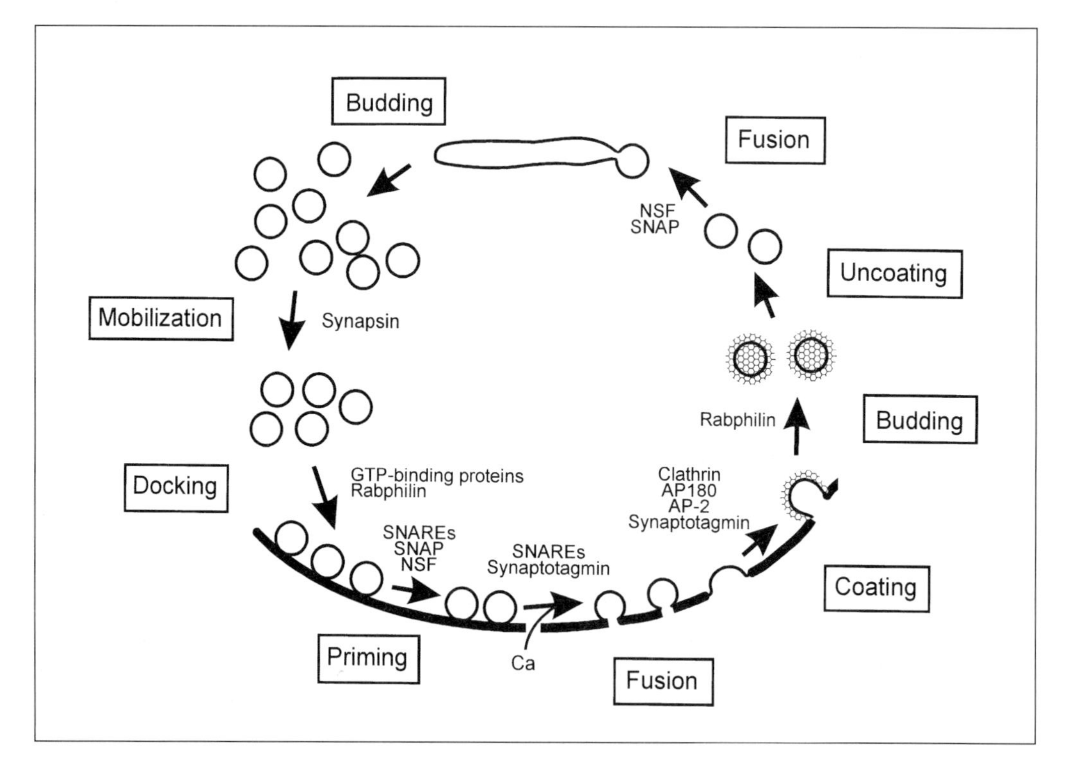

Fig. 5 Presynaptic vesicle trafficking reactions and their molecular mediators, as deduced from the results of squid microinjection experiments.

accumulation of docked vesicles that cannot fuse, suggesting that synaptotagmin may confer calcium sensitivity to neurotransmitter release (Fig. 5) (49). Injection of antibodies that bind to the C2A domain of synaptotagmin has a qualitatively similar effect (67). Though these results are consistent with evidence obtained via other approaches (69–72), questions still remain. For example, it is not yet clear how interaction of calcium-bound synaptotagmin with the SNARE proteins could trigger vesicle fusion (91).

3.3.5 Endocytosis and vesicle recycling

Heuser and Reese (86) proposed that vesicular membrane which has fused with the plasma membrane is retrieved by a budding process that requires the formation of clathrin-coated pits and coated vesicles. Inhibition of this endocytotic process should reduce the number of coated pits and vesicles and should trap vesicular membrane in the plasma membrane. Peptides from two clathrin assembly proteins, AP180 and AP2, prevent clathrin assembly *in vitro*. Further, when microinjected into the squid terminal these peptides severely reduce the number of synaptic vesicles, reduce the number of coated vesicles, and increase the surface area of the presynaptic terminal (64). These results are consistent with a null mutation of the *Drosophila* AP180 gene (92) and suggest that assembly of clathrin by these proteins is needed for the budding of vesicular membrane during endocytosis (Fig. 5). Microinjection of a Rabphilin-3A peptide into the presynaptic terminal produces very similar structural changes, while full-length protein has the opposite effect (35). This suggests that Rabphilin-3A may also regulate membrane budding during endocytosis (Fig. 5).

Heuser and Reese (86) also suggested that endocytosed vesicular membrane transits through other intermediates, such as endosomes, before being re-formed into synaptic vesicles. The temporary incorporation of this endocytosed membrane into endosomes requires fusion between the retrieved vesicles and the endosomes, so that blockade of this process should block endocytosis. Peptides from both SNAP (37) and NSF (59) substantially reduce the number of synaptic vesicles, suggesting that endocytosis requires a fusion process based on SNARE protein complexes (Fig. 5).

4. Summary and future prospects

In this chapter we have summarized the types of reagents that are used for microinjection experiments and some of the insights into the molecular basis of neurotransmitter release that have resulted from application of this approach at the squid giant synapse. Though we have restricted our discussion to examples from this synapse, similar experiments can be performed at other synapses with presynaptic terminals that permit microinjection, such as those of *Aplysia* central ganglia, crayfish neuromuscular synapses, cultured *Xenopus* neuromuscular synapses, cultured mammalian hippocampal neurons, and giant calyx synapses in the auditory system of vertebrates. Likewise the reagents we have described can be applied to other sorts of secretory cells that can be permeabilized or dialysed via whole-cell patch-clamp methods. Thus, this style of experimentation has fairly general applicability to studies of secretory systems.

For studies of secretory protein function, the main alternative to the approach we have described is to use genetic methods to knock-out or overexpress these proteins (see Chapters 9 and 10). The genetic approach currently is more fashionable, in part because it offers distinct advantages. The most obvious advantage is specificity; by manipulating only a single gene, there is a higher chance that the resulting change in secretion is due to a change in a single protein. However, because such genetic manipulations are often chronic they can also produce secondary effects — such as aberrant organismal development or compensatory changes in gene expression — that seriously confound interpretation of experimental results. Another advantage of genetic approaches is that they largely obviate the need to worry about reagent solubility, permitting study of integral membrane proteins and of insoluble domains of other proteins.

Microinjection experiments have their own advantages as well. Because microinjection at the squid synapse is acute, it neatly side-steps many of the possible complications associated with chronic genetic manipulations. Further, because many presynaptic proteins have multiple isoforms encoded by multiple genes, it may be difficult to knock-out all functional protein with genetic methods. However, careful design should permit production of microinjection reagents that specifically perturb all isoforms of a given protein and yield a clearer analysis of the function of the protein. For example, the presence of multiple synapsin genes may explain why genetic knock-out of one or two of these genes produces very mild synaptic phenotypes (41–43), while microinjection of synapsin reagents always has been found to produce dramatic changes in synaptic function (28, 38–40).

In summary, their differing strengths and weakness make microinjection and genetic approaches quite complementary. As we have tried to emphasize throughout this chapter, the results from microinjection experiments generally are in very good agreement with results obtained via genetic manipulations. This strong consistency suggests that both methods are reliable ways to study the function of presynaptic proteins and should motivate the continued application of both in the future.

Future microinjection experiments should take better advantage of the unique attributes of this acute perturbation strategy. For example, one underutilized benefit of microinjection is that it permits measurement of the rate at which the microinjected reagent affects transmitter release. Such kinetic studies may yield information about the time at which individual proteins are important relative to the time of neurotransmitter release. For example, the observation that the SS-19 peptide works within four seconds of the time that it is injected into the presynaptic terminal suggests that SNAP works at a step that is temporally proximal to synaptic vesicle fusion (37). The time resolution of the approach presently is on the order of seconds, which is several orders of magnitude more rapid than chronic genetic manipulations and compares favourably with the speed at which temperature-sensitive mutations can be effected. Photochemical approaches may permit even higher temporal resolution to be achieved (13, 14). Likewise, the ability to microinject fluorescently-tagged proteins or even organelles (93) can be further exploited to determine the location or timing of protein–protein interactions. Finally, the ability to microinject multiple reagents at defined times may be quite useful in sorting out the order in which proteins act within the vesicle trafficking pathway. In conclusion, although the squid giant synapse has been used as an experimental system for more than 50 years, its uniquely large presynaptic terminal continues to make it valuable for contemporary studies of the molecular biology of neurotransmitter release.

References

1. Katz, B. (1969) *The release of neural transmitter substances.* Liverpool University Press.
2. Augustine, G. J., Charlton, M. P., and Smith, S. J. (1987) Calcium action in synaptic transmitter release. *Annu. Rev. Neurosci.* **10**, 633.
3. Heuser, J. E., Reese, T. S., Dennis, M. J., Jan, Y., Jan, L., and Evans, L. (1979) Synaptic vesicle exocytosis captured by quick freezing and correlated with quantal transmitter release. *J. Cell Biol.* **81**, 275.
4. Südhof, T. C. (1995) The synaptic vesicle cycle: a cascade of protein-protein interactions. *Nature* **375**, 645.
5. Young, J. Z. (1939) Fused neurons and synaptic contacts in the giant nerve fibers of cephalopods. *Phil. Trans. R. Soc. Lond. B* **229**, 465.
6. Llinas, R. R. (1984) The squid giant synapse. *Curr. Top. Memb. Transport* **22**, 519.
7. Hodgkin, A. L., Huxley, A. F., and Katz, B. (1952) Measurement of the current-voltage relations in the membrane of the giant axon of Loligo. *J. Physiol.* **116**, 424.
8. Pumplin, D. W., Reese, T. S., and Llinás, R. (1981) Are the presynaptic membrane particles the calcium channels? *Proc. Natl. Acad. Sci. USA* **78**, 7210.
9. Martin, R. and Miledi, R. (1986) The form and dimensions of the giant synapse of squids. *Phil. Trans. R. Soc. Lond. B* **312**, 355.
10. Bullock, T. H. and Hagiwara, S. (1957) Intracellular recording from the giant synapse of the squid. *J. Gen. Physiol.* **40**, 565.
11. Katz, B. and Miledi, R. (1967) A study of synaptic transmission in the absence of nerve impulses. *J. Physiol.* **192**, 407.
12. Miledi, R. (1973) Transmitter release induced by injection of calcium ions into nerve terminals. *Proc. R. Soc. Lond. Ser. B* **183**, 421.
13. Delaney, K. R. and Zucker, R. S. (1990) Calcium released by photolysis of DM-nitrophen stimulates transmitter release at squid giant synapse. *J. Physiol.* **426**, 473.
14. Hsu, S. F., Augustine, G. J., and Jackson, M. B. (1996) Adaptation of Ca^{2+}-triggered exocytosis in presynaptic terminals. *Neuron* **17**, 501.
15. Adler, E. M., Augustine, G. J., Duffy, S. N., and Charlton, M. P. (1991) Alien intracellular calcium chelators attenuate neurotransmitter release at the squid giant synapse. *J. Neurosci.* **11**, 1496.
16. Llinás, R., Blinks, J. R., and Nicholson, C. (1972) Calcium transient in presynaptic terminal of squid giant synapse: detection with aequorin. *Science* **176**, 1127.
17. Miledi, R. and Parker, I. (1981) Calcium transients recorded with arsenazo III in the presynaptic terminal of the squid giant synapse. *Proc. R. Soc. Lond. Ser. B* **212**, 197.
18. Charlton, M. P., Smith, S. J., and Zucker, R. S. (1982) Role of presynaptic calcium ions and channels in synaptic facilitation and depression at the squid giant synapse. *J. Physiol.* **323**, 173.

19. Grynkiewicz, G., Poenie, M., and Tsien, R. Y. (1985) A new generation of Ca^{2+} indicators with greatly improved fluorescence properties. *J. Biol. Chem.* **260**, 3440.
20. Smith, S. J., Osses, L. R., and Augustine, G. J. (1988) Fura-2 imaging of localized calcium accumulation within squid 'giant' presynaptic terminal. In *Calcium and ion channel modulation* (ed. A. D. Grinnell, D. Armstrong, and M. B. Jackson), pp. 147–55. Plenum Press, New York.
21. Smith, S. J., Buchanan, J., Osses, L. R., Charlton, M. P., and Augustine, G. J. (1993) The spatial distribution of calcium signals in squid presynaptic terminals. *J. Physiol.* **472**, 573.
22. Llinás, R., Steinberg, I. Z., and Walton, K. (1976) Presynaptic calcium currents and their relation to synaptic transmission: voltage clamp study in squid giant synapse and theoretical model for the calcium gate. *Proc. Natl. Acad. Sci. USA* **73**, 2913.
23. Llinás, R., Steinberg, I. Z., and Walton, K. (1981) Presynaptic calcium currents in squid giant synapse. *Biophys. J.* **33**, 289.
24. Augustine, G. J., Charlton, M. P., and Smith, S. J. (1985) Calcium entry into voltage-clamped presynaptic terminals of squid. *J. Physiol.* **367**, 143.
25. Augustine, G. J., Charlton, M. P., and Smith, S. J. (1985) Calcium entry and transmitter release at voltage-clamped nerve terminals of squid. *J. Physiol.* **367**, 163.
26. Augustine, G. J. and Charlton, M. P. (1986) Calcium-dependence of presynaptic calcium current and post-synaptic response at the squid giant synapse. *J. Physiol.* **381**, 619.
27. Gillespie, J. I. (1979) The effect of repetitive stimulation on the passive electrical properties of the presynaptic terminal of the squid giant synapse. *Proc. R. Soc. Lond. Ser. B* **206**, 293.
28. Llinás, R., McGuinness, T. L., Leonard, C. S., Sugimori, M., and Greengard, P. (1985) Intraterminal injection of synapsin I or calcium/calmodulin-dependent protein kinase II alters neurotransmitter release at the squid giant synapse. *Proc. Natl. Acad. Sci. USA* **82**, 3035.
29. Stanley, E. F. (1990) The preparation of the squid giant synapse for electrophysiological investigation. In *Squid as experimental animals* (ed. D. L. Gilbert, W. J. Adelman, and J. M. Arnold), pp. 171–92. Plenum, New York.
30. del Castillo, J. and Katz, B. (1954) Quantal components of the end-plate potential. *J. Physiol.* **124**, 560.
31. Miledi, R. (1967) Spontaneous synaptic potentials and quantal release of transmitter in the stellate ganglion of the squid. *J. Physiol.* **192**, 379.
32. Mann, D. W. and Joyner, R. W. (1978) Miniature synaptic potentials at the squid giant synapse. *J. Neurobiol.* **9**, 329.
33. Augustine, G. J. and Eckert, R. (1984) Divalent cations differentially support transmitter release at the squid giant synapse. *J. Physiol.* **346**, 257.
34. Augustine, G. J., Charlton, M. P., and Horn, R. (1988) Role of calcium-activated potassium channels in transmitter release at the squid giant synapse. *J. Physiol.* **398**, 149.
35. Burns, M. E., Sasaki, T., Takai, Y., and Augustine, G. J. (1998) Rabphilin-3A: a multifunctional regulator of synaptic vesicle traffic. *J. Gen. Physiol.* **111**, 243.
36. Lin, J. W., Sugimori, M., Llinás, R. R., McGuinness, T. L., and Greengard, P. (1990) Effects of synapsin I and calcium/calmodulin-dependent protein kinase II on spontaneous neurotransmitter release in the squid giant synapse. *Proc. Natl. Acad. Sci. USA* **87**, 8257.
37. DeBello, W. M., O'Connor, V., Dresbach, T., Whiteheart, S. W., Wang, S. S., Schweizer, F. E., *et al.* (1995) SNAP-mediated protein-protein interactions essential for neurotransmitter release. *Nature* **373**, 626.

38. Llinás, R., Gruner, J. A., Sugimori, M., McGuinness, T. L., and Greengard, P. (1991) Regulation by synapsin I and Ca($^{2+}$)-calmodulin-dependent protein kinase II of the transmitter release in squid giant synapse. *J. Physiol.* **436**, 257.
39. Hilfiker, S., Schweizer, F. E., Kao, H.-T., Czernik, A. J., Greengard, P., and Augustine, G. J. (1998) Two sites of action for synapsin domain E in regulating neurotransmitter release. *Nature Neurosci.* **1**, 29.
40. Pieribone, V. A., Shupliakov, O., Brodin, L., Hilfiker-Rothenfluh, S., Czernik, A. J., and Greengard, P. (1995) Distinct pools of synaptic vesicles in neurotransmitter release. *Nature* **375**, 493.
41. Rosahl, T. W., Spillane, D., Missler, M., Herz, J., Selig, D. K., Wolff, J. R., *et al.* (1995) Essential functions of synapsins I and II in synaptic vesicle regulation. *Nature* **375**, 488.
42. Li, L., Chin, L. S., Shupliakov, O., Brodin, L., Sihra, T. S., Hvalby, O., *et al.* (1995) Impairment of synaptic vesicle clustering and of synaptic transmission, and increased seizure propensity, in synapsin I-deficient mice. *Proc. Natl. Acad. Sci. USA* **92**, 9235.
43. Takei, Y., Harada, A., Takeda, S., Kobayashi, K., Terada, S., Noda, T., *et al.* (1995) Synapsin I deficiency results in the structural change in the presynaptic terminals in the murine nervous system. *J. Cell Biol.* **131**, 1789.
44. Burton, J. L., Burns, M. E., Gatti, E., Augustine, G. J., and De Camilli, P. (1994) Specific interactions of Mss4 with members of the Rab GTPase subfamily. *EMBO J.* **13**, 5547.
45. Burns, M. E. (1996) Molecular mechanisms of neurotransmitter release: the functions of Rab3A, Rabphilin, and SNAP-25. Ph.D. Dissertation, Duke University, Durham, NC.
46. Llinás, R., Moreno, H., Sugimori, M., Mohammadi, M., and Schlessinger, J. (1997) Differential pre- and postsynaptic modulation of chemical transmission in the squid giant synapse by tyrosine phosphorylation. *Proc. Natl. Acad. Sci. USA* **94**, 1990.
47. Dresbach, T., Burns, M. E., O'Connor, V., DeBello, W. M., Betz, H., and Augustine, G. J. (1998) A neuronal Sec1 homolog regulates neurotransmitter release at the squid giant synapse. *J. Neurosci.* **18**, 2923.
48. Tokumaru, H., Pelligrini, L. L., Ishizuka, T., Umayahara, K., Saisu, H., Betz, H., *et al.* (1999) Interactions between synaphin/complexin and syntaxin are essential for neurotransmitter release. (submitted).
49. Bommert, K., Charlton, M. P., DeBello, W. M., Chin, G. J., Betz, H., and Augustine, G. J. (1993) Inhibition of neurotransmitter release by C2-domain peptides implicates synaptotagmin in exocytosis. *Nature* **363**, 163.
50. Sutton, R. B., Davletov, B. A., Berghuis, A. M., Sudhof, T. C., and Sprang, S. R. (1995) Structure of the first C2 domain of synaptotagmin I: a novel Ca^{2+}/phospholipid-binding fold. *Cell* **80**, 929.
51. Thomas, D. M. and Elferink, L. A. (1998) Functional analysis of the C2A domain of synaptotagmin 1: implications for calcium-regulated secretion. *J. Neurosci.* **18**, 3511.
52. Kee, Y., Lin, R. C., Hsu, S. C., and Scheller, R. H. (1995) Distinct domains of syntaxin are required for synaptic vesicle fusion complex formation and dissociation. *Neuron* **14**, 991.
53. O'Connor, V., Heuss, C., De Bello, W. M., Dresbach, T., Charlton, M. P., Hunt, J. H., *et al.* (1997) Disruption of syntaxin-mediated protein interactions blocks neurotransmitter secretion. *Proc. Natl. Acad. Sci. USA* **94**, 12186.
54. Esser, L., Wang, C. R., Hosaka, M., Smagula, C. S., Sudhof, T. C., and Deisenhofer, J. (1998) Synapsin I is structurally similar to ATP-utilizing enzymes. *EMBO J.* **17**, 977.
55. Lenzen, C. U., Steinmann, D., Whiteheart, S. W., and Weis, W. I. (1998) Crystal structure of the hexamerization domain of *N*-ethylmaleimide-sensitive fusion protein. *Cell* **94**, 525.

56. Sutton, R. B., Fasshauer, D., Jahn, R., and Brunger, A. T. (1998) Crystal structure of a SNARE complex involved in synaptic exocytosis at 2.4 Å resolution. *Nature* **395**, 347.
57. Shao, X., Li, C., Fernandez, I., Zhang, X., Sudhof, T. C., and Rizo J. (1997) Synaptotagmin-syntaxin interaction: the C2 domain as a Ca^{2+}-dependent electrostatic switch. *Neuron* **18**, 133.
58. Fernandez, I., Ubach, J., Dulubova, I., Zhang, X., Sudhof, T. C., and Rizo, J. (1998) Three dimensional structure of an evolutionarily conserved N-terminal domain of syntaxin 1A. *Cell* **94**, 841.
59. Schweizer, F. E., Dresbach, T., DeBello, W. M., O'Connor, V., Augustine, G. J., and Betz, H. (1998) Regulation of neurotransmitter release kinetics by NSF. *Science* **279**, 1203.
60. Hunt, J. M., Bommert, K., Charlton, M. P., Kistner, A., Habermann, E., Augustine, G. J., *et al.* (1994) A post-docking role for synaptobrevin in synaptic vesicle fusion. *Neuron* **12**, 1269.
61. Pallanck, L., Ordway, R. W., and Ganetzky, B. (1995) A *Drosophila* NSF mutant. *Nature* **376**, 25.
62. Kawasaki, F., Mattiuz, A. M., and Ordway, R. W. (1998) Synaptic physiology and ultrastructure in comatose mutants define an *in vivo* role for NSF in neurotransmitter release. *J. Neurosci.* **18**, 10241.
63. Rothman, J. E. (1994) Mechanisms of intracellular protein transport. *Nature* **372**, 55.
64. Morgan, J. R., Zhao, X., Womack, M., Prasad, K., Augustine, G. J., and Lafer, E. M. (1999) A role for the clathrin assembly domain of AP180 in synaptic vesicle endocytosis (submitted).
65. Burgoyne, R. D., Morgan, A., Barnard, R. J., Chamberlain, L. H., Glenn, D. E., and Kibble, A. V. (1996) SNAPs and SNAREs in exocytosis in chromaffin cells. *Biochem. Soc. Trans.* **24**, 653.
66. Littleton, J. T., Chapman, E. R., Kreber, R., Garment, M. B., Carlson, S. D., and Ganetzky, B. (1998) Temperature-sensitive paralytic mutations demonstrate that synaptic exocytosis requires SNARE complex assembly and disassembly. *Neuron* **21**, 401.
67. Mikoshiba, K., Fukuda, M., Moreira, J. E., Lewis, F. M. T., Sugimori, M., Niiobe, M., *et al.* (1995) Role of the C2A domain of synaptotagmin in transmitter release as determined by specific antibody injection into the squid giant synapse preterminal. *Proc. Natl. Acad. Sci. USA* **92**, 10703.
68. Fukuda, M., Moreira, J. E., Lewis, F. M., Sugimori, M., Niinobe, M., Mikoshiba, K., *et al.* (1995) Role of the C2B domain of synaptotagmin in vesicular release and recycling as determined by specific antibody injection into the squid giant synapse preterminal. *Proc. Natl. Acad. Sci. USA* **92**, 10708.
69. Littleton, J. T., Stern, M., Schulze, K., Perin, M., and Bellen, H. J. (1993) Mutational analysis of *Drosophila* synaptotagmin demonstrates its essential role in Ca^{2+}-activated neurotransmitter release. *Cell* **74**, 1125.
70. Littleton, J. T., Stern, M., Perin, M., and Bellen, H. J. (1994) Calcium dependence of neurotransmitter release and rate of spontaneous vesicle fusions are altered in *Drosophila* synaptotagmin mutants. *Proc. Natl. Acad. Sci. USA* **91**, 10888.
71. Broadie, K., Bellen, H. J., DiAntonio, A., Littleton, J. T., and Schwarz, T. L. (1994) Absence of synaptotagmin disrupts excitation-secretion coupling during synaptic transmission. *Proc. Natl. Acad. Sci. USA* **91**, 10727.
72. Geppert, M., Goda, Y., Hammer, R. E., Li, C., Rosahl, T. W., Stevens, C. F., *et al.* (1994) Synaptotagmin I: a major Ca^{2+} sensor for transmitter release at a central synapse. *Cell* **79**, 717.

73. Jorgensen, E. M., Hartwieg, E., Schuske, K., Nonet, M. L., Jin, Y., and Horvitz, H. R. (1995) Defective recycling of synaptic vesicles in synaptotagmin mutants of *Caenorhabditis elegans*. *Nature* **378**, 196.
74. Llinás, R., Sugimori, M., Chu, D., Morita, M., Blasi, J., Herreros, J., *et al.* (1994) Transmission at the squid giant synapse was blocked by tetanus toxin by affecting synaptobrevin, a vesicle-bound protein. *J. Physiol.* **477**, 129.
75. Marsal, J., Ruiz-Montasell, B., Blasi, J., Moreira, J. E., Contreras, D., Sugimori, M., *et al.* (1997) Block of transmitter release by botulinum C1 action on syntaxin at the squid giant synapse. *Proc. Natl. Acad. Sci. USA* **94**, 14871.
76. Sugimori, M., Tong, C. K., Fukuda, M., Moreira, J. E., Kojima, T., Mikoshiba, K., *et al.* (1998) Presynaptic injection of syntaxin-specific antibodies blocks transmission in the squid giant synapse. *Neuroscience* **86**, 39.
77. Montecucco, C. and Schiavo, G. (1993) Tetanus and botulism neurotoxins: a new group of zinc proteases. *Trends Biochem. Sci.* **18**, 324.
78. Niemann, H., Blasi, J., and Jahn, R. (1994) Clostridial neurotoxins: new tools for dissecting exocytosis. *Trends Cell Biol.* **4**, 179.
79. Williamson, L. C., Halpern, J., Montecucco, C., Brown, J. E., and Neale, E. A. (1996) Clostridial neurotoxins and substrate proteolysis in intact neurons: botulinum C acts on synaptosomal-associated protein of 25 kDa. *J. Biol. Chem.* **271**, 7694.
80. Broadie, K., Prokop, A., Bellen, H. J., O'Kane, C. J., Schulze, K. L., and Sweeney, S. T. (1995) Syntaxin and synaptobrevin function downstream of vesicle docking in *Drosophila*. *Neuron* **15**, 663.
81. Deitcher, D. L., Ueda, A., Stewart, B. A., Burgess, R. W., Kidokoro, Y., and Schwartz, T. L. (1998) Distinct requirements for evoked and spontaneous release of neurotransmitter are revealed by mutations in the *Drosophila* gene neuronal-synaptobrevin. *J. Neurosci.* **18**, 2028.
82. Hess, S. D., Doroshenko, P. A., and Augustine, G. J. (1993) A functional role for GTP-binding proteins in synaptic vesicle cycling. *Science* **259**, 1169.
83. Doroshenko, P. A., Burns, M. E., and Augustine, G. J. (1999) Tethering of synaptic vesicles by a GTP-binding protein (in preparation).
84. Osses, L. R., Barry, S. R., and Augustine, G. J. (1989) Protein kinase C activators enhance transmission at the squid giant synapse. *Biol. Bull.* **177**, 146.
85. Llinás, R., Sugimori, M., Lang, E. J., Morita, M., Fukuda, M., Niinobe, M., *et al.* (1994) The inositol high-polyphosphate series blocks synaptic transmission by preventing vesicular fusion: a squid giant synapse study. *Proc. Natl. Acad. Sci. USA* **91**, 12990.
86. Heuser, J. E. and Reese, T. S. (1973) Evidence for recycling of synaptic vesicle membrane during transmitter release at the frog neuromuscular junction. *J. Cell Biol.* **57**, 315.
87. Parsons, T. D., Coorssen, J. R., Horstmann, H., and Almers, W. (1995) Docked granules, the exocytic burst, and the need for ATP hydrolysis in endocrine cells. *Neuron.* **15**, 1085.
88. Banerjee, A., Barry, V. A., DasGupta, B. R., and Martin, T. F. J. (1996) N-Ethylmaleimide-sensitive factor acts at a prefusion ATP-dependent step in Ca^{2+}-activated exocytosis. *J. Biol. Chem.* **271**, 20223.
89. Tolar, L. A. and Pallanck, L. (1998) NSF function in neurotransmitter release involves rearrangement of the SNARE complex downstream of synaptic vesicle docking. *J. Neurosci.* **18**, 10250.
90. Weber, T., Zemelman, B. V., McNew, J. A., Westermann, B., Gmachl, M., Parlati, F., *et al.* (1998) SNAREpins: Minimal machinery for membrane fusion. *Cell* **92**, 759.
91. Hilfiker, S., Greengard, P., and Augustine, G. J. (1999) Coupling calcium to SNARE-mediated synaptic vesicle fusion. *Nature Neurosci.* **2**, 104.

92. Zhang, B., Koh, Y. H., Beckstead, R. B., Budnik, V., Ganetsky, B., and Bellen, H. J. (1998) Synaptic vesicle size and number are regulated by a clathrin adaptor protein required for endocytosis. *Neuron* **21**, 1465.
93. Llinás, R., Sugimori, M., Lin, J. W., Leopold, P. L., and Brady, S. T. (1989) ATP-dependent directional movement of rat synaptic vesicles injected into the presynaptic terminal of squid giant synapse. *Proc. Natl. Acad. Sci. USA* **86**, 5656.

8 Studying mutants that affect neurotransmitter release in *C. elegans*

MICHAEL L. NONET

1. Introduction

Sydney Brenner initiated the study of the nematode *Caenorhabditis elegans* in the early 1960s with the specific intent of developing the organism as a model to study the nervous system. Primarily through the isolation and characterization of mutants that disrupt the functioning of the nervous system at the behavioural level, studies of the nematode have contributed to our understanding of the molecular mechanisms underlying a wide variety of neuronal processes including axonal outgrowth (1), sensory reception (2, 3), and synaptic transmission (4). The strength of the nematode as a model organism is undoubtedly the ability to apply powerful genetics to a biological problem. However, in the long-term, the dissection of the role of molecules in a complex process such as synaptic transmission requires the synthesis of results obtained using a variety of experimental approaches. Here, I first review the approaches available in *C. elegans* which have been utilized to characterize the molecules underlying the regulated release of neurotransmitter. Subsequently, I examine the complexity of the synaptic transmission apparatus in *C. elegans* through a discussion of mutants in a variety of genes which perturb the process. Finally, I discuss the avenues of research where the study of *C. elegans* is most likely to provide future insight that is applicable to understanding the machinery operating at synapses in all organisms.

2. *C. elegans* as a model system

Caenorhabditis elegans is a small hermaphroditic nematode that feeds on bacteria and grows to about a millimetre in length. In the laboratory, it usually is propagated on small agar plates covered with a 'lawn' of bacteria (5, 6). Genetic manipulation of the organism is particularly simple and fast because the hermaphrodite is self-fertile with a generation time of only three days (7). In addition, nematode strains can be stored in a frozen state, greatly simplifying the maintenance of large numbers of mutant strains (6). Its small size and simple nervous system present many advantages in genetic manipulations of the organism, but also have limited the development of biochemical and physiological techniques.

2.1 Overview of the nervous system and characteristics of synapses

The nervous system of the *C. elegans* hermaphrodite consists of only 302 neurons (8). The anatomy and connectivity of these neurons is known in great detail from reconstruction of the nervous system from serial section electron micrographs (9). The neurons have a very simple morphology, many extending only a single process that serves a dual role as axon and dendrite. Cell bodies are very small, ranging around 3 μm in diameter, and processes are thin, averaging around 0.5 μm in the adult (10). Axons are not myelinated, although some sensory neurons are surrounded by glial-like support cells (11). In their analysis of the nervous system, White and colleagues (9) organized the neurons into 118 distinct classes based on morphology and connectivity. Although only a few individual cells are dedicated to distinct functions, the classes define mechano-, thermo-, and chemosensory neurons, interneurons, as well as motorneurons. Different classes of neurons express a wide variety of the classically defined neurotransmitters including glutamate, acetylcholine, GABA, serotonin, dopamine, octopamine, and neuropeptides (12–18). Thus, although the nervous system is small, it contains a wide repertoire of neuronal cell types, and these cells exhibit many fundamental properties associated with vertebrate neurons.

The nervous system is organized into three major regions of neuropil and several associated ganglia in which neuronal cell bodies reside (Fig. 1A). Since most synaptic connections in *C. elegans* are formed *en passant*, the vast majority of synapses are located in the three neuropils: the nerve ring, the ventral nerve cord, and the dorsal nerve cord. The nerve ring, a bundle of approximately 160 processes, is the major site of interneuronal synapses. Sensory neuron endings are concentrated in the head and virtually all neurons send processes into the nerve ring where they synapse onto interneurons. The cell bodies of most sensory and interneurons are positioned in two ganglia just anterior and posterior to the nerve ring (Fig. 1A). Motor output is regulated by a small set of command interneurons with processes running down the ventral nerve cord where they synapse onto motor neurons that innervate the body wall muscles. The cell bodies of virtually all motor neurons are situated in the ventral nerve cord (Fig. 1A). These consist of both excitatory cholinergic and inhibitory GABAergic neurons. *C. elegans* body wall muscle cells are located in each of the four dorso-lateral and ventro-lateral quadrants (Fig. 1B). Rather than the neurons extending processes to muscle, the muscle cells extend process-like 'muscle arms' to meet motor neurons (Fig. 1B). As a result, neuromuscular junctions are located at the edge of the ventral cord and dorsal cord process bundles. Head muscles also extend process-like arms to the periphery of the nerve ring bundle where they contact motor neurons and form neuromuscular junctions. Thus, virtually all the roughly 7000 synapses in *C. elegans* are concentrated in the ventral nerve cord, dorsal nerve cord, and nerve ring.

Interneuronal and neuromuscular synapses share many morphological features in *C. elegans*. Both types of synapses usually form *en passant*. The presynaptic axon swells to accommodate a cluster of up to several hundred vesicles in the largest synapses. Synaptic vesicles at most synapses are 30–35 nm in diameter and have clear cores. Dense core vesicles of 37–57 nm are also found in certain neurons (9). A

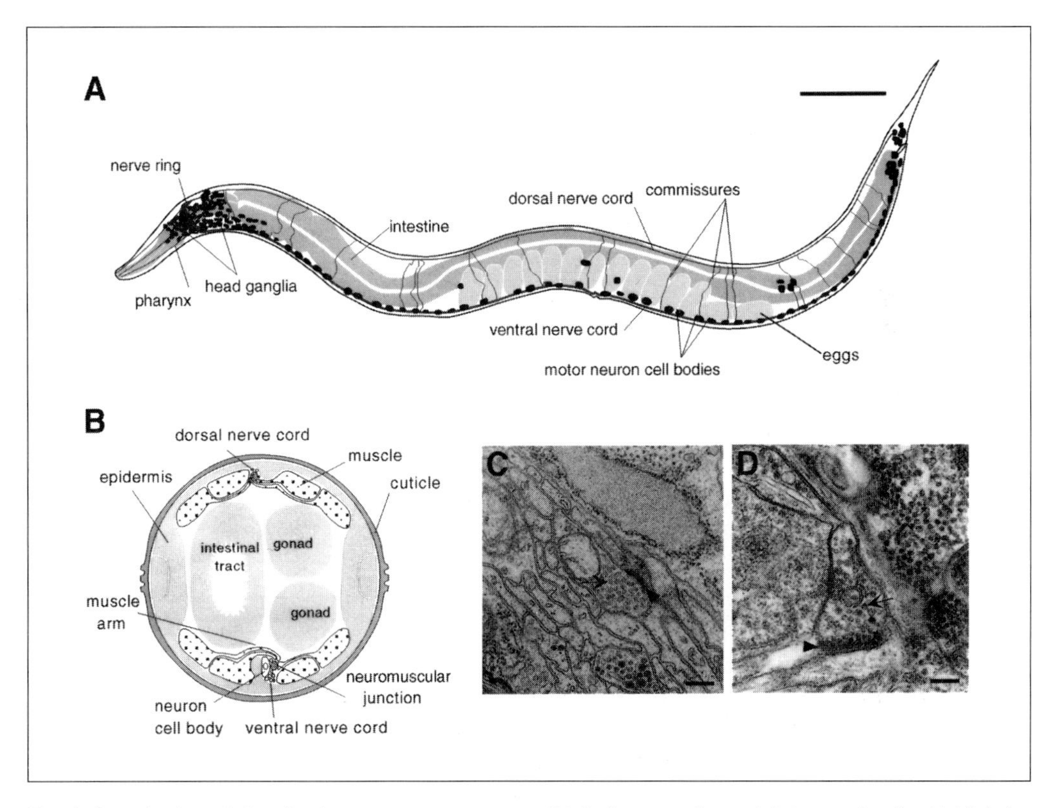

Fig. 1 Organization of the *C. elegans* nervous system. (A) A diagram of an adult hermaphrodite highlighting nervous system structures. Neuronal cell bodies are represented as black ovals. Major processes bundles, commissures, and representative sensory processes in the head are depicted as dark grey lines, but minor laterally-positioned process bundles are omitted for clarity. The pharynx, the intestine, and eggs are shown in light grey. Several other neuro-anatomical landmarks are also labelled. Adapted from ref. 179. (B) A diagram of a cross-section through the mid-body of an adult animal showing the position of muscles, the ventral nerve cord, and the dorsal nerve cord. Minor process bundles are omitted from the diagram for clarity. A basal lamina is present between the epidermis and the pseudocoelom and surrounding the muscle quadrants. Adapted from ref. 179. (C and D) Ultrastructure of synapses in *C. elegans*. (C) A transverse section through the mid-body of a wild-type adult hermaphrodite showing a typical neuromuscular junction in the ventral nerve cord. (D) A longitudinal section through the nerve ring showing an interneuronal synapse. An arrowhead in each micrograph identifies the electron dense presynaptic density. An arrow identifies a representative synaptic vesicle at each synapse. Clear and dense core vesicles from other synaptic specializations are visible in both micrographs. Micrographs courtesy of Dr David Hall. Scale bars: 200 nm.

large thick presynaptic density is present at most synapses (Fig. 1C and D). At neuromuscular junctions, the vesicle cluster is centred around, and extends about 250 nm from the density (19). Components of the density have not been identified, and no mutants lacking it have been described. By contrast with the extensive presynaptic density, the postsynaptic density is very subtle and often difficult to detect in micrographs. Other features seen at vertebrate neuromuscular junctions such as junctional folds and sub-synaptic nuclei are absent at *C. elegans* synapses. Indeed, the major identifiable difference between interneuronal and neuromuscular synapses in *C. elegans* is that a basal lamina is only present at the neuromuscular junction.

2.2 Identification of genes regulating synaptic transmission

2.2.1 Behavioural screens

The majority of genes known to regulate synaptic transmission in *C. elegans* were first identified in genetic screens for mutants. The earliest screens were simply designed to identify mutants with behavioural defects (5). Brenner's initial screen for mutants with uncoordinated locomotion identified a collection of 75 different *unc* genes. Although these screens did not specifically target synaptic components, genes such as *unc-13* and *unc-18* that encode synaptic regulators were isolated in this manner. More recent genetic screens for animals with more subtle behavioural defects have also identified mutants that disrupt synaptic components. For example, extensive genetic screens have been performed for mutants with egg laying defects (*egl* mutants) (20), feeding defects (*eat* mutants) (21), and for defecation defects (*aex* and *exp* mutants) (22). In fact, alleles of many synaptic genes were identified in multiple different screens for behavioural defects. The major limitation of these types of behavioural screens in specifically identifying mutants perturbing synaptic transmission is that such screens are not selective for lesions which disturb synaptic or even neuronal functions. For example, many genes acting in muscle are also represented among these mutants.

2.2.2 Selecting for mutants using pharmacological agents

The incorporation of pharmacological agents into genetic screens has provided further selectivity for mutants that disrupt the synaptic transmission process. The most successful application of this approach has been the identification of mutants resistant to acetylcholinesterase inhibitors such as aldicarb and trichlorfon (23–25). Much of motor function in *C. elegans* is regulated by cholinergic transmission. In the presence of millimolar concentrations of aldicarb, wild-type animals slowly become hypercontracted and paralysed over the course of several hours. The vast majority of these animals eventually die. Aldicarb resistance has been the basis of several genetic screens, generating a large collection of mutants that are capable of surviving and reproducing in the continuous presence of aldicarb (Table 1). These mutants include a subset of the *unc*, *egl*, *aex*, and *eat* mutants originally isolated in behavioural screens, but also mutants that were not isolated (24). As discussed in detail in later sections,

Table 1 Genes identified by selecting for aldicarb-resistant mutants

Gene	Molecular description of gene product	Alleles[a]	References
unc-13	Phorbol ester/DAG binding protein	29	72
unc-41	Similarity to *Drosophila stonedB*	22	b, 4
ric-1	Not cloned	19	4, 121
snt-1	Synaptotagmin	17	28
ric-3	Not cloned	15	4, 121
unc-18	Syntaxin binding protein	14	29
unc-11	AP180 adaptor protein	6	80
unc-17	Vesicular acetylcholine transporter	6	180
unc-75	Not cloned	6	4, 121
egl-10	RGS	5	136
cha-1	Choline acetyltransferase	4	16
ric-4	SNAP-25	4	c, 4, 121
unc-104	Kinesin-like molecule	4	181
unc-10	Rim	3	d, 4, 121
aex-3	Rab3 exchange factor	1	90
egl-30	$G\alpha_q$	1	135
unc-2	α subunit of Ca^{2+} channel	1	66
unc-64	Syntaxin	1	26
ric-8	Uncloned	1	24
unc-26	Synaptojanin	1	e, 4, 121
unc-31	CAPS	1	125–127
snb-1	Synaptobrevin	1	32

[a] Data from refs 24 and 32.
[b] J. B. Rand, unpublished data.
[c] J. Lee, Y. Lee, M. L. Nonet, J. B. Rand, and B. J. Meyer, in preparation.
[d] L. Wei, J. Staunton, G. Hadwiger, and M. L. Nonet, in preparation.
[e] E. M. Jorgensen, unpublished data.

all the genes lesioned in these mutants encode components that participate in regulating synaptic transmission, or are required indirectly in neurons for synaptic function. The technical advantages of these screens to isolate mutants disrupting synaptic components are twofold. First, the screens specifically identify mutants that disrupt cholinergic transmission (and in many cases all types of transmission). Secondly, aldicarb kills animals that do not contain lesions in these genes, and hence provides a selection, rather than a screen, for mutants of interest. The ability to select for a large number of distinct mutants using aldicarb also allows the isolation of vast numbers of mutations in a gene. Indeed, over 20 distinct alleles of *unc-13* were isolated in one such screen (24). When the appropriate mutagen (e.g. ethyl methanesulfonate) is used, most of the mutations are single base substitutions, and many are useful in structure–function analysis of gene products (26). A significant limitation of this drug selection approach is that only lesions reducing cholinergic transmission can be isolated. Hence, genes that function to negatively regulate synaptic transmission likely will be missed.

By definition, all genetic screens that depend on isolating a viable animal have the limitation that genes essential for survival of the organism will be overlooked. Two factors compensate for this problem in *C. elegans*. First, *C. elegans* is very tolerant to lesions of the nervous system, probably because the animal need not move either to reproduce or to feed (27). Thus, many mutations which result in lethal consequences in other organisms are viable in *C. elegans*. For example, both *C. elegans* synaptotagmin (28) and *unc-18* (29) null mutants are viable as homozygotes, while *Drosophila* mutants in the analogous genes have lethal phenotypes (30, 31). Secondly, the ability to select for and characterize a large number of mutants partially compensates for this problem because rare viable hypomorphic alleles occasionally are isolated. For example, hypomorphic lesions in both the synaptobrevin and syntaxin genes were isolated as aldicarb-resistant mutants (26, 32). However, complete loss-of-function mutations in both of these genes result in a lethal phenotype. An additional approach to identify mutations in essential genes that has been used more extensively in *Drosophila* is to identify conditional alleles such as temperature-sensitive lethal or paralysed mutants (33–37). Regardless of the means by which the initial alleles of essential genes are isolated, once a single allele is in hand, additional alleles that disrupt gene function can readily be acquired using classical genetic techniques such as non-complementation screens. Hence, the ability to screen through vast numbers of animals to isolate mutants is one of the most powerful aspects of *C. elegans* genetics.

2.2.3 Suppressor genetics and identifying interacting genes

Another strength of *C. elegans* genetics is the feasibility of screening for recessive enhancers or suppressors of a phenotype. This complex genetic approach has been extremely successful in identifying additional components acting in cellular pathways including the ras (38–41) and the dosage compensation pathways (42). Several laboratories recently have begun to use this approach to study the regulation of synaptic transmission. Miller and Rand have isolated a group of suppressors of *ric-8;* Saifee and Nonet have been characterizing suppressors of syntaxin hypomorphs (K. Miller, J. B. Rand, O. Saifee, and M. L. Nonet, unpublished data). Interestingly, many suppressors isolated in both of these screens exhibit a novel phenotype: hypersensitivity to aldicarb. Screens directed at isolating additional members of this growing class of genes, which include the Ca^{2+} calmodulin-dependent protein kinase II gene (O. Saifee and M. L. Nonet, unpublished data), offer a potentially very powerful approach to identify genes that behave as negative regulators of synaptic transmission.

In summary, classical genetic studies using a combination of behavioural and pharmacological assays have led to the identification of many *C. elegans* genes involved in synaptic transmission. The approaches have been highly effective at isolating mutants in both previously identified and novel synaptic components. Molecular characterization of these genes is incomplete. In fact, over a dozen different aldicarb-resistant mutants remain uncloned. Although none of the genetic screens have been performed to saturation and additional mutants clearly remain to be isolated, their identification will become more time-consuming as the odds of isolating novel genes continues to decrease. More sophisticated genetic approaches which have been used successfully to dissect other cellular processes remain largely untapped. In time, these genetic approaches will undoubtedly identify additional regulators of synaptic function in *C. elegans*.

2.3 Molecular manipulation of genes

Molecular tools to clone and manipulate genes are well developed and provide a powerful complementary tool to the genetics available in *C. elegans*. *C. elegans* is the first and only multicellular organism whose genomic constitution has been determined virtually in its entirety (43). Preliminary analysis of the sequence suggests that the 97 megabase *C. elegans* genome encodes approximately 19100 genes. Gene density on the chromosomes is very high and most genes are smaller than 10 kb in size, hence small enough to be easily manipulated in plasmids (44). Furthermore, comprehensive collections of cosmid and yeast artificial chromosome clones covering the entire genome are available for molecular studies (45, 46). As a result, experiments that incorporate transgenic technology are relatively easy to perform in *C. elegans*.

2.3.1 Analysis of genomic sequence

The genomic sequence data and the molecular reagents generated during both the genomic and expressed sequence tag (EST) sequencing projects provide a unique set of tools for the *C. elegans* researcher. The most useful information the genomic sequence provides is a precise description of the genes present in the organism. Determining if *C. elegans* contains a gene with similarity to any vertebrate gene is now simple and straightforward. In fact, it often possible to precisely define the gene structure of such homologues when EST data is also available. This type of analysis has revealed the extent of conservation of synaptic components between *C. elegans* and vertebrates (Table 2). In short, the vast majority of identified vertebrate synaptic proteins are present in *C. elegans* and homologues of all components implicated in synaptic function in *C. elegans* are found in vertebrates. In most cases, only a single homologue of a vertebrate protein family is present in *C. elegans*. For example, while four rab3 isoforms (rab3A, B, C, and D) are found in vertebrates, only a single *rab-3* gene is present in *C. elegans* (19). This suggests that, in most cases, gene redundancy is unlikely to complicate the analysis of synaptic mutants in *C. elegans*. In cases where an extended family of related proteins has been identified in vertebrates, the number of genes encoding recognizable homologues are much fewer in *C. elegans* and the products are usually much more divergent. For example, in the case of the synaptotagmin family, the *C. elegans* gene *snt-1* is most closely related to synaptotagmin I in vertebrates (74% identity in the C2A and C2B domains) (28). In the nematode, the two closest *snt-1* homologues are C08G5.4 and F42G9.2 which exhibit 41% and 52% identity, respectively, in the conserved C2A and C2B domains. By comparison, nine synaptotagmin family members identified to date in the rat share 43% to 85% identity in these domains with synaptotagmin I. Although the genomic sequence of *C. elegans* is already a very valuable resource for all biologists, the age of genomics is

Table 2 Components of the transmission apparatus in *C. elegans*

Protein[a]	Identity to rat protein	Gene	Expression[b]	Synaptic defects[c]	Comments and references
CSP	40%	K02G10.8			d
rab-3	76%	*rab-3*	N	Very mild	19
SCAMP	36%	M01D7.2			d, h
SVOP	48%	ZK637.1			162
Synapsin	56%	Y38C1_7.c			h, i
Synaptobrevin	68%	*snb-1*	N, S, G	Essential	32
Synaptogyrin	30%	*sng-1*	N		79
Synaptophysin	23%	F42G8.11			f, g
Synaptotagmin	54%	*snt-1*	N, S	Moderate	28
Syntaxin	63%	*unc-64*	N, I, S	Essential	26
SNAP-25	51%	*ric-4*	N	Essential	m
CAPS	54%	*unc-31*	N	Moderate	125, 126
HRS-2	32%	C07G1.5			d
MINT	37%	*lin-10*	N, E	Mild	71
NSF	55%	ZK1014.1			e
Tomosyn	32%	M01A10.2A			d
SNAP	50%	D1014.3			d
UNC-13	50%	*unc-13*	N	Essential	72
UNC-18	57%	*unc-18*	N	Severe	29
RAB-3 GAP p130	24%	F20D1.6			d
rab GDI	66%	*gdi-1*			e
rab-3 NEF	35%	*aex-3*	N, I	Mild	90
Rabphilin	41%	*rbf-1*	N	None	j, l
rim	27%	*unc-10*	N	Moderate	j, n
Amphiphysin	33%	F58G6.1			d
AP180	45%	*unc-11*	N, cc, I, S	Strong	80
Dynamin	62%	*dyn-1*	N	Essential	37
Synaptojanin	37%	*unc-26*		Strong	k

[a] Genes that have not been molecularly characterized were identified by analysis of *C. elegans* genomic sequence and the wormpep predicted peptide database using the programs *BLAST* and *clustalW*. Most predicted protein sequences available at `http://www.sanger.ac.uk/Projects/C_elegans/`.

[b] Expression in neurons (N); the intestine (I); coelomocytes (cc); non-neuronal secretory cells (S); spermatheca (G).

[c] Mutants divided into classes based on behavioural and pharmacological defects.

[d] EST data supports predicted sequence.

[e] EST data confirms entire predicted sequence.

[f] No EST data available.

[g] Hydrophobicity plots also very similar.

[h] Only a partial open reading frame predicted from databases. Similarity based on that interval only.

[i] From preliminary *C. elegans* genome sequence DNA.

[j] Homology restricted to domains with higher similarity than overall score.

[k] T. Harris and E. Jorgensen, in preparation.

[l] J. Staunton, B. Ganetzky, and M. L. Nonet, in preparation.

[m] J. Lee, Y. Lee, M. L. Nonet, J. B. Rand, and B. J. Meyer, in preparation.

[n] L. Wei, J. Staunton, G. Hadwiger, and M. L. Nonet, in preparation.

just beginning and its potential to transform the approaches we use to study neurons is still unclear.

2.3.2 Transgenic animals

Transformation of *C. elegans* is relatively simple and efficient; when utilized in conjunction with mutants, transformation serves well for the analysis of gene function at the molecular level. Transformation is achieved by direct injection into the germline of a mixture of a plasmid DNA of interest and a marker plasmid such as *rol-6(su1006)*, a collagen mutation which leads to a dominant rolling phenotype (47–49). In the germline, the injected plasmids concatenate via recombination into megabase-sized mini-chromosomes called extrachromosomal arrays. These mini-chromosomes are semi-stable: they are effectively transmitted from generation to generation, but not all germline or somatic cells receive the arrays because they segregate inefficiently during mitosis and meiosis. Transformants are easily identified as mutant animals expressing the injected marker and can be isolated with a few hours of labour spread over the course of a week. For many applications, these transformants are very useful. However, mosaicism in expression levels and spatial expression pattern make some types of analyses difficult. Truly stable transgenic lines that behave more reproducibly are derived from the semi-stable strains by integration of the extrachromosomal element into the genome using gamma rays (48). This process is more time-consuming, requiring a small genetic screen to identify the integration events. A drawback of transformation in *C. elegans* that is shared with most other transgenic systems is that it is difficult to control copy number of the injected gene precisely, and thus expression levels can vary dramatically among lines and are often higher than the native gene. Despite these limitations, transformation procedures in *C. elegans* are robust and expedient, and enable one to test large numbers of constructs that might be required in experiments such as structure–function studies.

2.3.3 Reverse genetics

In the last few years, several efficient techniques have been developed to isolate lesions in specific genes in *C. elegans*. Although homologous integration has only rarely been observed in *C. elegans*, the ability to grow large numbers of worms has permitted two distinct approaches for isolating small (mostly under 3 kb) randomly positioned deletions in a gene of interest. Briefly, both approaches are based on growing large numbers of mutagenized nematodes and using PCR to identify an individual population containing some animals carrying a lesion of interest (50–53). From this complex pool of animals, a single animal carrying the lesion of interest is isolated through repetitive rounds of subdivision and PCR analysis of pools of lower and lower complexity. Because *C. elegans* is a self-fertile hermaphrodite, even mutations that result in lethality when homozygous can be efficiently propagated and isolated as heterozygous animals using these techniques. Over 200 genes now have been knocked-out using these approaches (R. Barstead and R. Plasterk, personal communication) and several groups have initiated projects aimed at the large scale isolation of mutants in thousands of genes identified by the genome project. Still missing is the ability to lesion a specific base pair in the genome. However, until methodologies for homologous recombination are developed, a suitable alternative is the introduction of a specifically lesioned gene (using transgenic methods) into a strain containing a deletion of the chromosomal copies of the gene of interest.

2.3.4 RNA-mediated interference

Another unique method for disrupting gene function has recently been developed in *C. elegans*. RNA-mediated interference is based on the injection of double-stranded RNA into an adult animal (54). For many genes tested, progeny of the injected animal exhibit a phenotype very similar to the loss-of-function phenotype of the gene. Although the exact mechanism of this gene silencing is unknown, it provides an extremely quick method for determining the likely null mutant phenotype of a gene. One current hypothesis is that the injected double-stranded RNA acts catalytically to stimulate the degradation of native mRNA, resulting in an epigenetic phenocopy of the loss-of-function lesion in the gene. Unfortunately, late expressed neuronal genes seem to be some of the most resistant to this interference method (J. Fleenor, H. Zhao, L. Wei, M. Nonet, A. Fire, unpublished data). Hence, in practical terms, this technique provides only limited usefulness for the study of synaptic transmission until a molecular understanding of the mechanism of action permits refinement of the technology.

2.4 Behavioural and pharmacological analysis

Many of the arguments that certain *C. elegans* genetic lesions specifically disrupt synaptic transmission are based upon indirect behavioural and pharmacological assays. Although these correlations between behavioural deficits and synaptic abnormalities may not seem compelling, many homologues of these *C. elegans* genes have been identified in vertebrates, and such homologues are implicated in synaptic transmission. Several distinct assays have been used to assess the nature of functional defects in mutants including examination of behaviours regulated by specific neurotransmitters and the analysis of responses to cholinergic and GABAergic pharmacological agents.

2.4.1 Behavioural analysis

A variety of simple behaviours have been studied in *C. elegans*. The commonly assayed behaviours include locomotion, feeding and defecation (21, 22, 55), foraging (56), egg laying (20), responses to both soft and harsh touch (2, 57), chemotaxis and thermotaxis (3), and male mating (58–60). These behaviours provide quantitative assays to assess the relative strength of defects in different mutants, as well as different alleles of a gene. Furthermore, the circuits which regulate many of these behaviours are at least partially defined and the neurotransmitter phenotype of many of the neurons regulating specific behaviours is known. As a result, it has been possible to correlate certain behavioural deficits with loss-of-function of a specific transmitter. For example, mutants in the *unc-25* glutamic acid decarboxylase gene (61) have specific behavioural abnormalities including defecation defects and altered foraging behaviour (14, 56). Mutants with defects in glutamatergic transmission have distinct behavioural defects in response to harsh stimuli (62, 63), whereas GABA-deficient *unc-25* mutants have normal responses to such stimuli. Hence, analysis of the behavioural deficits of a mutant also provides a method for determining if neuronal deficits perturb the function of neurons using distinct transmitters.

2.4.2 Pharmacological assays

Pharmacological assays complement the behavioural assays by providing a simple assay for the likely site(s) of action of a gene product. Specifically, cholinergic agents have been used to assess whether genes disrupt pre- or postsynaptic functions at the neuromuscular junction (24, 28). In mutants that are resistant to the acetyl-cholinesterase inhibitor aldicarb, the deficit is likely to be presynaptic if the animals respond normally to cholinergic receptor agonists like levamisole (64, 65). Similar arguments can be made using muscimol for GABAergic transmission (14, 56), and imipramine for serotonergic transmission (20). Pharmacological agents have also been used to examine long-term modulation of synaptic function. Schafer and Kenyon screened for mutants that failed to adapt to chronic serotonin and dopamine exposure. They identified the α-subunit of the N-type calcium channel as a mediator of this adaptation process (66). An unusual aspect of pharmacological experiments in *C. elegans* is that, in many cases, 100- to 1000-fold higher concentrations of drug are required to produce an effect on living animals than in mammalian cells (67). However, in most cases this simply reflects limited permeability of the drugs across the *C. elegans* cuticle.

2.5 Biochemical and cell biological analysis

2.5.1 Biochemistry

By contrast with the rich history of genetic analysis in *C. elegans*, only a very limited number of biochemical studies have been undertaken in *C. elegans*. Under most circumstances, biochemical analysis of nervous system components is feasible, although impractical. The greatest limitation to biochemistry in *C. elegans* is the ability to obtain suitable starting materials for fractionation. Although *C. elegans* can be grown in multi-kilogram quantities in large fermentors, in a standard laboratory situation the isolation of nematodes in 10–100 gram quantities is more realistic (68). From these cultures of worms it is not possible to fractionate or dissect the nervous system from the rest of the organism. In an adult, the nervous system represents less than 2% of the mass of the animal (less than 1% of the nervous system is synaptic vesicles). To further complicate matters, the cuticle of *C. elegans* is also very durable requiring either a French press or the crushing of frozen tissue with a mortar and pestle to effectively lyse worms for biochemical fractionation (69). Thirdly, the nematode culture is contaminated with *E. coli*, the food source used to grow the worms.

Thus, at best it is difficult to isolate more than 10 grams of nervous system tissue and that representation of tissue will be diluted in 500 grams of gut, germline, muscle, cuticle, and *E. coli*. By comparison, standard synaptosomal preparations are typically started with approximately 200 grams of rat brain. Harada *et al.* (1994) have described a simple fractionation protocol starting with whole worms that may enrich for synaptic vesicles (70). Although the preparation contains some vesicles in the 30 nm range, the purity of this prep was not assessed using cellular markers. Cell culture does not represent a viable option for biochemical analysis as no immortalized *C. elegans* tissue culture cell line has been described. Perhaps the most practical alternative to examine certain interactions is expressing *C. elegans* components in a heterologous systems as recently described for the LIN-2, LIN-7, LIN-10 PDZ protein complexes (71).

In vitro analysis of biochemical interactions among vertebrate synaptic proteins using bacterially-expressed sources of protein has contributed greatly to our current understanding of the nerve terminal (see Chapter 3). Although few studies have been performed with *C. elegans* proteins, the organismal source of the proteins has very little influence on the experimental protocol in these *in vitro* assays. Indeed, early *in vitro* studies of the UNC-13 C1 domain documented that the protein bound phorbol esters (72, 73) suggesting the protein was acting as a signal transduction molecule at the synapse before UNC-13 homologues were identified in vertebrates. In those instances where biochemical interactions have been tested in *C. elegans*, they have largely agreed with vertebrate studies. Syntaxin binds avidly to UNC-18 (74) and stable SNARE complexes even can be formed with mixtures of *C. elegans* and vertebrate SNAREs (G. Hadwiger, L. Wei and M. Nonet, in preparation). While replicating interactions identified among vertebrate proteins using the *C. elegans* homologues will be less enlightening, analysis of the effects of missense lesions identified in *C. elegans* on *in vitro* interactions may be quite fruitful, especially in studying proteins like syntaxin that form complex sets of interactions.

2.5.2 Cell biology

C. elegans is translucent and thus very amenable to analysis using optical techniques that are easy to integrate with genetic approaches. It has become relatively straightforward to tag proteins *in vivo* with green fluorescent protein (GFP) (75) and characterize the subcellular localization of these protein fusions (76–79). Transformation and phenotypic rescue of mutants provides a stringent test for the functionality of these tagged proteins. In cases where the localization of fusion proteins has been compared to the localization of native protein using immunohistochemical techniques, the functional GFP fusion relatively faithfully reproduces the native pattern of localization when expressed at appropriate levels (79). These GFP markers provide a powerful assay for dissecting the processes responsible for regulating expression, trafficking, and subcellular localization of synaptic components. Targeting domains can be identified by assessing the ability of transgenic deletion constructs to localize efficiently. Indeed, these assays can be combined with genetics to identify molecules responsible for these targeting functions in neurons. For example, the *C. elegans* AP180 homologue *unc-11* is required for the proper synaptic localization of synaptobrevin (80), the *C. elegans* Mint homologue *lin-10* is required for localization of the glutamate receptor GLR-1 (78), and ODR-4 is required for the localization of olfactory receptors to dendrites (76). A limitation to these types of studies is the fact that most synapses are located in bundles of processes, thus preventing the visualization of individual synaptic sites. However, limiting the expression of molecules to subsets of neurons provides an suitable alternative to permit visualization of individual varicosities (79, 81).

2.5.3 Ultrastructural analysis

Valuable insights to the function of a synaptic component can be attained by characterizing the changes in synaptic morphology that are associated with disruption of that component. In *C. elegans*, these morphological studies have usually been performed on neuromuscular junctions in the ventral nerve cord. Detailed analysis of junctions have identified alterations in synaptic connectivity (72, 82), in vesicle distributions at active sites (77, 83), of mean vesicle diameter (80), and of vesicle density (19) in mutants lacking specific synaptic components. Unfortunately, these studies have not been extended to immunolocalization of components at the ultrastructural level. The major stumbling block in this analysis is combining manipulation of small animals with fixation techniques that preserve immunoreactivity. A thoughtful discussion of approaches to both traditional and immunoelectron microscopy in *C. elegans* has recently been published (84).

2.6 Physiological tools

Despite the powerful genetics available in *C. elegans*, characterization of the role of molecules in regulating synaptic transmission has been hampered by the lack of sophisticated physiological recording preparations. The morphological properties of specific circuits such as those controlling body wall muscle are very similar in the much larger nematode *Ascaris suum* (85). Physiological analysis of *Ascaris* has been used to infer the likely properties of *C. elegans* neurons, but direct analysis of the circuits in *C. elegans* has not been performed. In *C. elegans*, two physiology techniques have been developed.

2.6.1 Electropharyngeograms (EPGs)

The pharynx is the muscular feeding organ that the nematode uses to feed on bacteria. The pharynx acts both as an efficient filter to trap bacteria and as a grinder to rupture bacteria. The co-ordinated pumping action responsible for feeding behaviour is under neuronal control. Raizen and Avery (86) have carefully characterized the nature of capacitive currents that can be measured from extracellular recording of pharyngeal muscle. In this preparation, the head of a live animal, or a dissected head, is sucked into a recording pipette. Endogenous muscle currents and synaptic currents can be recorded in this configuration (Fig. 2). A variety of evidence including the analysis of genetic mutants (86–88), uncaging of glutamate (12), and laser killing of specific neurons (86, 87) provide compelling evidence that the activity of two distinct neuronal classes can be identified in these extracellular recordings. Briefly, activity of both the MC and M3 neurons can be detected in a typical recording (Fig. 2). The pair of cholinergic MC neurons regulates the initiation of pharyngeal pumps (although pharyngeal muscle has its own intrinsic rhythm and will pump even if uninnervated) (27). In many synaptic mutants, pharyngeal pumping rates are reduced and an MC transient is often absent at the initiation of pharyngeal pumps (Fig. 2). Furthermore, spontaneous bursts of transients that do not elicit a pharyngeal pump often are observed in interpump intervals in synaptic transmission mutants. Raizen, Lee and Avery (1995) have demonstrated that this subthreshold activity requires MC function in synaptotagmin mutants (87). M3s are a pair of inhibitory glutamatergic neurons which hasten the repolarization of muscle (12, 88, 89). The amplitude, frequency, and spacing of M3 IPSPs is relatively constant in wild-type animals (Fig. 2A). In severe synaptic transmission mutants, M3 IPSPs often are absent (26, 32), while in less severe mutants M3 IPSP frequency and amplitude frequently are reduced in magnitude and M3 spacing or timing often is irregular (Fig. 2B–D) (19, 26, 90). Analysis of the activities of the two neurons currently represents the most direct measure of synaptic activity in *C. elegans*. The assay provides a quantitative method of comparing the severity of defects in different mutants. However, the inability to stimulate MC or M3 limits its practical application to dissecting the mechanistic defects in release in synaptic mutants.

2.6.2 Whole-cell recordings from identified neurons

Goodman and colleagues have developed a preparation where they expose neurons of the nerve ring ganglia and record from these neurons *in situ* with a whole-cell patch-clamp (10). Recordings from a sensory neuron suggest that these cells have high input resistance, do not fire sodium-based action potentials, and have a wide

---

Fig. 2 Physiological analysis of *C. elegans* synaptic mutants. Representative electropharyngeograms from wild-type and mutant strains of *C. elegans*. (A) The electrical activity associated with two typical pharyngeal pumps of the wild-type. The large positive transient and negative transient, labelled E and R in (A) only, are associated with the contraction (depolarization) and relaxation (repolarization) of muscles of the metacorpus bulb of the pharynx. The activity of the MC neuron, which regulates the initiation of pharyngeal pumping, is seen as a small positive transient (arrows) that usually precedes an E transient. The activity of the inhibitory motor neuron M3 is seen as negative transients (marked by a dot) in the E to R interval when the pharynx is contracted. (B) A recording from the RAB-3 nucleotide exchange factor mutant *aex-3* illustrating a typical defect associated with weak synaptic function mutants. Both MC (arrows) and M3 transient amplitudes are reduced. Furthermore, the M3 transients are often broader and multiphasic suggesting M3 transmission is less synchronous. Subthreshold MC transients that fail to elicit a pharyngeal pump also often observed in the interpump intervals. (C) Recordings from syntaxin *unc-64(e246)* illustrating a mutant with altered kinetics of transmission. Note that the initial M3 inhibitory transient is delayed relative to the onset of the pharyngeal contraction (E transient) and that successive M3 transients increase in amplitude. Nevertheless, after the initial delay M3 transients are evenly spaced and well defined. (D) Recordings from the synaptobrevin mutant *snb-1(js17)*. In this mutant, the first few M3 transients after depolarization are well defined, but the both the amplitude and synchrony of subsequent transients are markedly reduced.

dynamic range in which they are sensitive to small current input. Although the authors did observe occasional spontaneous activity, they do not deduce whether the activity is synaptic in nature. In principle, recording from any identified neuron in the nerve ring ganglia should be feasible if the neurons can be identified with a selective promoter GFP fusion. A single recording electrode synaptic physiology preparation could in theory be developed by recording from an interneuron and stimulating sensory neurons with mechanical stimuli or with identified odorants.

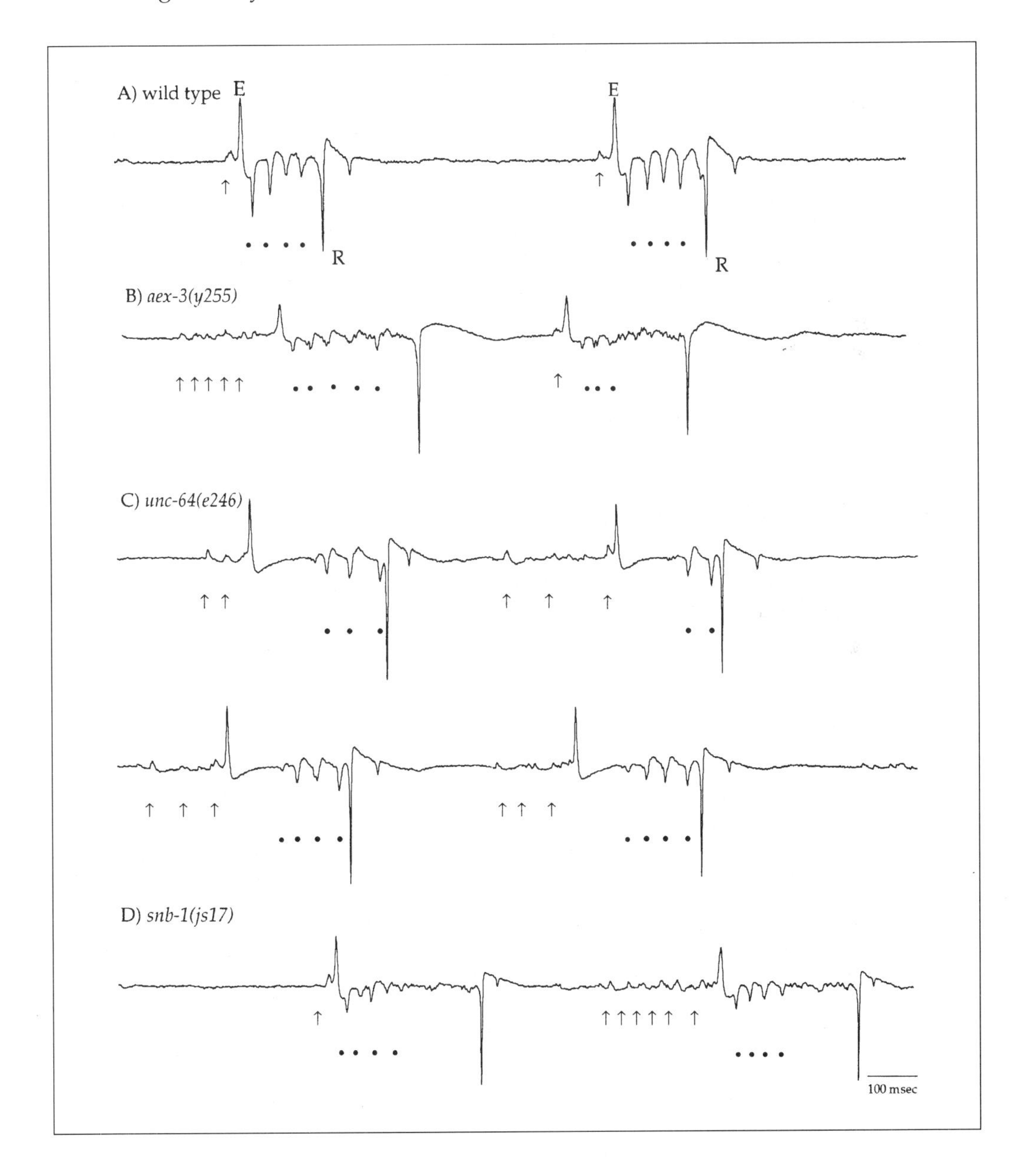

2.6.3 Prospects for a synaptic physiology preparation

Although the neuronal whole-cell patch preparation may be useful in developing a synaptic physiological preparation, body wall muscle is probably a more accessible target. Recently, Richmond and Jorgensen have reported preliminary results regarding the development of a neuromuscular preparation (Fig. 3A) (Richmond and Jorgensen, personal communication). In their preparation, application of either GABA or the cholinergic agonist levamisole leads to inward currents in voltage-clamped body muscle of filleted, immobilized, adult animals (Fig. 3A and B). Furthermore, spontaneous activity from body wall muscle is observed in their recordings (Fig. 3C). Although they have yet to report the ability to evoke release by

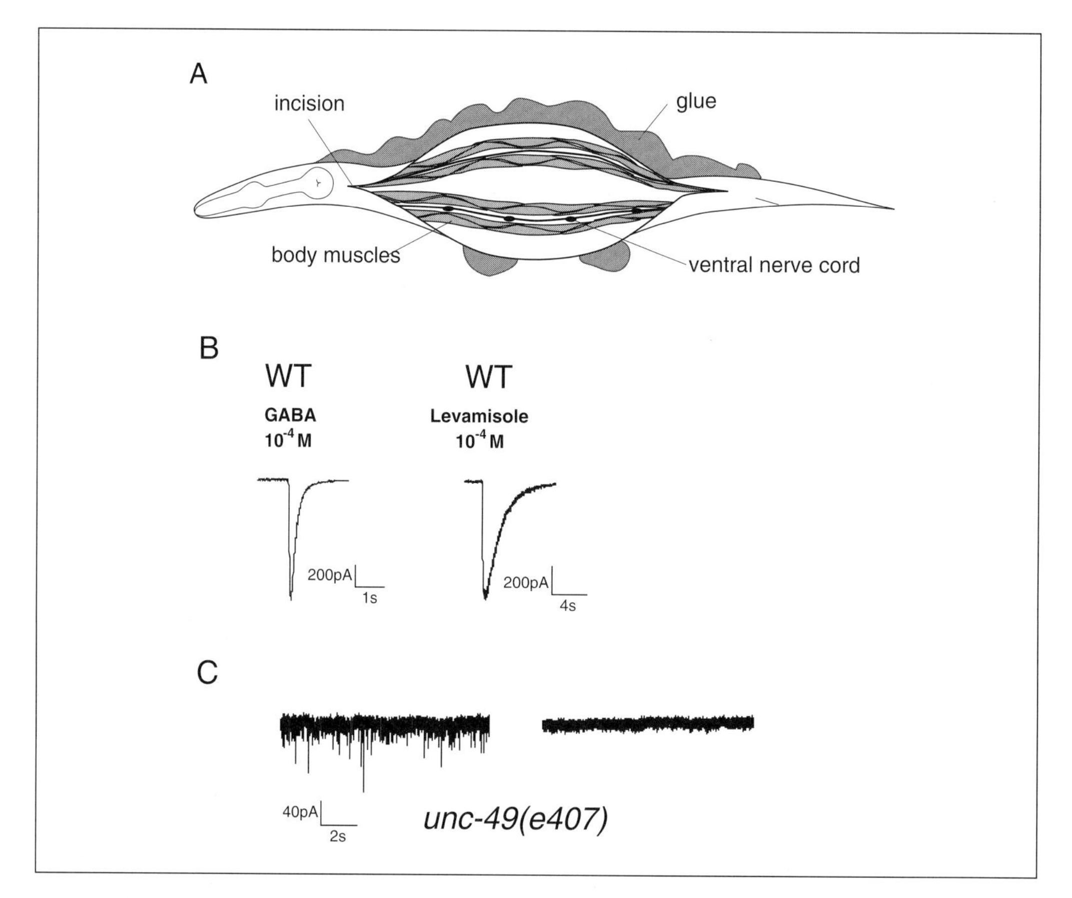

Fig. 3 Intracellular recording from *C. elegans* body wall muscle. (A) A diagram of the body wall muscle recording preparation. Adult *C. elegans* are glued down and an incision is made to expose body wall muscles that are innervated by both GABAergic and cholinergic motor neurons. (B) Application of GABA or the cholinergic agonist levamisole elicited inward currents in voltage-clamped body wall muscle cell. The holding potential used was more hyperpolarized than the reversal potential for chloride. (C) Spontaneous cholinergic activity recorded from an *unc-49* GABA receptor mutant (56) (B. Bamber and E. Jorgensen, in preparation). The spontaneous activity was completely blocked by the application of d-tubocurare (right recording). Recordings kindly provided by J. Richmond and E. Jorgensen.

stimulation of motor neurons, their preliminary findings show promise to a much needed physiological assay.

3. An overview of *C. elegans* synaptic mutants

Genetic and molecular analysis of a variety of behavioural and aldicarb-resistant mutants has identified over 30 genes that operate at the synapse. Below, I discuss the general properties of mutants in these genes, dividing them arbitrarily into classes. I emphasize the unique findings that these mutants have contributed to our understanding of the synaptic transmission apparatus.

3.1 The fusion machinery

3.1.1 v- and t-SNAREs

Vesicle and target soluble *N*-ethylmaleimide-sensitive factor attachment protein receptors (v- and t-SNAREs) constitute essential components of the fusion apparatus at the synapse. In *C. elegans*, mutants have been isolated in all three SNAREs: the v-SNARE synaptobrevin, and the t-SNAREs syntaxin and SNAP-25 (26, 32, 74) (J. Lee, Y. Lee, M. L. Nonet, J. B. Rand, and B. J. Meyer, in preparation). As predicted from genetic and biochemical studies on SNAREs in various systems, complete loss-of-function lesions in any of the three synaptic SNARE genes result in a lethal phenotype. Although detailed physiology is not available, the behavioural defects of the mutants are consistent with severe transmission abnormalities. The most severe phenotype is that of syntaxin loss-of-function mutations; the animals arrest and die as paralysed larvae just after completing embryogenesis (26). Mutants in SNAP-25 and synaptobrevin retain some limited capabilities for movement, but arrest shortly after hatching (32) (J. Lee, Y. Lee, M. L. Nonet, J. B. Rand, and B. J. Meyer, in preparation).

Hypomorphic mutations also have been isolated in all three genes and these have been much more informative about the role of the SNAREs in regulating release. Most of the hypomorphic mutations are missense lesions on the hydrophobic face of amphipathic helical domains involved in forming the four-stranded coiled-coil-like SNARE complex (91–93). The recent crystal structure of the SNARE complex revealed a set of stacked planar contacts between the amino acids on the hydrophobic face of the amphipathic helices of each protein in the complex (Fig. 4A) (92). Analysis of double mutant combinations of missense lesions in synaptobrevin and syntaxin supports interactions predicted by the crystal structure: when lesions disrupting distinct planes of these stacked interactions are combined, the double mutants exhibit a strong synergistic increase in the severity of the phenotype (Fig. 4B). In contrast, double mutant combinations that disrupt the same hydrophobic plane show little phenotypic enhancement (26) (Fig. 4C). These observations illustrate the use of genetics to provide *in vivo* evidence to corroborate *in vitro* biochemical and structural observations.

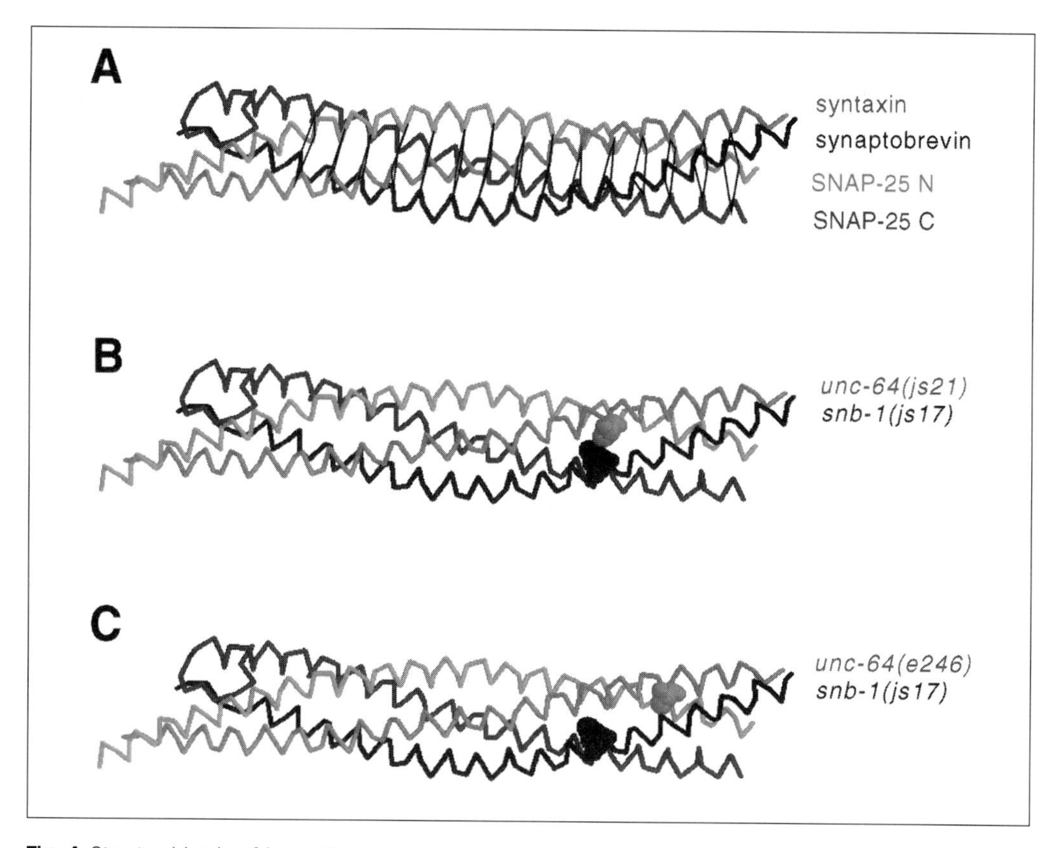

Fig. 4 Structural basis of interactions between lesions in the *C. elegans* SNAREs. Backbone drawings of the amphipathic four-strand helical synaptic core complex. Hydrophobic contacts between the chains are organized into planar layers (92) that are marked in (A) as black virtual bonds between the α carbons of the amino acids of each chain contributing to the interaction. The position of lesions in the *C. elegans* synaptobrevin and syntaxin genes (26, 32) were mapped onto the vertebrate SNARE complex structure and highlighted by representing the affected amino acid using a space-filling representation. Lesions in synaptobrevin and syntaxin which disrupt the same planar interaction, such as those illustrated in (B), showed minimal genetic interactions. By contrast, lesions that disrupted distinct planar interactions, such as those illustrated in (C), showed strong synergistic genetic interactions resulting in much more severe behavioural and physiological defects.

Particular lesions have implicated the transmembrane domains of syntaxin and synaptobrevin as more complex than simple membrane-anchoring domains. First, *C. elegans* expresses three forms of syntaxin that differ only in their transmembrane domain. Although two of these forms are expressed in neurons, a termination codon late in the coding sequence for one of the transmembrane domains is sufficient to cause lethality (26). This observation suggests that syntaxin isoforms that differ only in transmembrane sequences are not functionally equivalent. Interestingly, two vertebrate syntaxin genes also express isoforms with distinct transmembrane domains (94, 95). One possible role of these different transmembrane domains is to regulate subcellular localization through specific association of these sequences with other components of the transmission apparatus. However, direct evidence supporting this idea is still lacking.

Biochemical studies and analysis of genetic lesions in the *C. elegans* synaptobrevin gene point to both anchoring and non-anchoring roles for the transmembrane domain of this SNARE. The *C. elegans* allele *md247* has a mutation that alters the last half of the synaptobrevin transmembrane domain without altering the synaptic localization of synaptobrevin. This mutant is viable, but shows moderate behavioural and synaptic abnormalities indicating that the anchoring function of the transmembrane domain is insufficient to provide complete synaptobrevin activity (32). *In vitro* biochemical data from Laage and Langosch (96) demonstrates that the transmembrane domain dimerizes and suggests that the synaptobrevin transmembrane domain plays roles in addition to acting as a membrane anchor. *C. elegans* provides an ideal system to assess the functional requirements for dimerization or subcellular localization *in vivo*, as transgenic animals with specific lesions in the transmembrane domain can be assayed for behavioural, cellular, and physiological defects.

Studies of *C. elegans* SNARE mutants have also revealed a potential role of the SNARE complex in mediating volatile anaesthetic actions. Many lesions in the *C. elegans* SNARE genes exhibit altered sensitivity to volatile anaesthetics including halothane and isoflurane (97). Alleles in each of the three SNAREs that cause mild behavioural defects result in hypersensitivity to anaesthetics. These observations are consistent with presynaptic actions of volatile anaesthetics which have been observed in vertebrates (98). Of particular interest is the *md130* allele of syntaxin which results in the expression of several syntaxin products that are truncated within the H3 helical domain that mediates interactions with other SNAREs (97). This mutant behaves as a recessive loss-of-function mutation that reduces synaptic transmission as assayed by electropharyngeograms, behavioural assays, and pharmacological assays (26). However, the same lesion behaves as a strong dominant lesion for resistance to halothane (97). In sum, the *md130* mutant implicates syntaxin, a syntaxin containing complex, or a syntaxin interacting protein as a target for volatile anaesthetics in *C. elegans* and perhaps other metazoans.

3.1.2 UNC-18 and Mint

unc-18 was initially identified by mutation in *C. elegans* (5, 99) and subsequently linked to SNAREs by the biochemical identification of strong syntaxin binding partners in vertebrate homologues (100, 101). In fact, both *C. elegans* and mammalian UNC-18s have a much higher affinity for syntaxin than the other individual SNAREs (74, 100, 102). Unlike the SNARE mutants, all *unc-18* mutants are viable, including mutants that express no protein or RNA (29, 99). However, *unc-18* mutants are virtually paralysed, indicating that the *unc-18* gene is critical for synaptic function. Although UNC-18s role remains uncertain, several groups have suggested that it acts to mediate docking of v-SNAREs to syntaxin. Since, UNC-18 does not associate directly with synaptic vesicles, biochemical binding partners of UNC-18, such as the *m*ammalian UNC-18 *int*eracting *p*roteins (Mint1 and Mint2) could act as intermediates in this docking function (103). Mints, in turn, could interact with vesicles via their phosphatidylinositol bisphosphate binding domain, forming a biochemical link from synaptic vesicles to syntaxin that could be regulated by UNC-18–syntaxin interactions. However, mutants in the *C. elegans* Mint gene *lin-10* exhibit no behavioural phenotypes indicative of general synaptic dysfunction (104). *lin-10* mutants do exhibit a variety of defects in non-neuronal cells and defects in localization of postsynaptic glutamate receptors, but show no deficits in many behaviours mediated by neurotransmitters other than glutamate (78, 104). The mild phenotype of *lin-10* mutants underscores the need for genetic analysis to complement biochemical studies. At least in *C. elegans*, genetic lesions suggest that Mint plays a minor role, if any, in regulating transmitter release.

3.1.3 SNAP and NSF

Analysis of genomic and EST sequences indicates that only a single SNAP gene and a single NSF gene are present in *C. elegans*. As expected, both genes are very similar to their vertebrate counterparts (Table 2). The presence of only a single SNAP gene in *C. elegans* is similar to yeast, but contrasts with vertebrates that express three isoforms of SNAP, one of which binds to synaptotagmin (105) and is specifically expressed in neurons (106). Mutants have not been isolated in either SNAP or NSF. Since these genes almost certainly function in all cells of the organism, it is unlikely that mutants would easily be isolated in genetic screens that have been performed for *C. elegans* synaptic mutants. Recent findings that NSF binds to glutamate receptors and is much more abundant in the postsynaptic dendrites than in the presynaptic terminal indicates that ascribing specific defects in NSF (or SNAP) mutants to a presynaptic function may be very complicated (107–109).

3.2 Calcium signalling at the nerve terminal

An essential feature of synaptic transmission is its regulation by calcium influx into the nerve terminal. Identification of synaptic calcium sensors is crucial to a comprehensive understanding of neurotransmission at the molecular level. In *C. elegans*, mutants have been identified in calcium channels and also in other potential calcium sensors. The diversity of synaptic molecules that bind calcium suggests the calcium sensing machinery at the synapse is very complex.

3.2.1 Calcium channels

The calcium channel subunit repertoire in *C. elegans* appears to consist of genes encoding five α-subunits, two $\alpha 2/\delta$-subunits, and two β-subunits. Of the *C. elegans* α-subunits, *unc-2* is the most similar to the vertebrate α_{1B}-subunit of N-type channels and the best candidate to be present at release sites (66). Both *unc-2* and the *unc-36* $\alpha 2/\delta$-subunits are expressed in the nervous system (110). Lesions in each of these genes cause severe locomotion defects, but do not completely disrupt synaptic transmission (66, 110). Which additional calcium channels function in neurons has not been determined, although some evidence suggests the L-type *egl-19* gene may also function in neurons (110). From electrophysiological studies, Goodman and colleagues (10) have argued that only 30 calcium channels are active in the identified sensory neuron (ASE) whose physiology they have best characterized. Since, the ASE neuron forms 25 synapses (9), this would argue that only a single calcium channel is present at a release site. Although washout in these whole-cell physiological recordings may result in an underestimate of calcium channel numbers, they suggest that calcium channel density may be very low in *C. elegans* neurons.

Recently, vertebrate N- and P/Q-type calcium channels have been shown to interact with syntaxin and SNAP-25 in a calcium-dependent manner through a region of the calcium channel known as the 'synprint' (111, 112). This small region is located between the second and third repeated domain of these calcium channels and is well conserved among vertebrate N- and P/Q-type channels, but not among vertebrate L-type channels. Sequences homologous to the synprint domain are not present in any of the *C. elegans* calcium channel α-subunits. In fact, the equivalent intracellular loop of the *unc-2* α-subunit is much smaller than that of the vertebrate α_{1B}-subunit. Hence, even if regulation of calcium channels by the release machinery (and vice versa) occurs in *C. elegans,* it likely occurs using a distinct mechanism.

The *egl-19* gene shows most similarity to L-type channel α-subunits and is likely to be the functional homologue of vertebrate L-type channels (113). This channel is expressed strongly in muscle and animals lacking the channel arrest when muscle contraction is required for elongation of the developing embryo. The L-type *egl-19* α-subunit is also expressed in neurons and may also play a role in regulating transmission (113) since *egl-19* mutants show genetic interactions with *unc-2* (110). The three other α-subunits identified in *C. elegans* have not been well characterized. One is most similar to the vertebrate T-type channel (114), while the other two are very similar to each other but are not easily classified into one of the vertebrate families. It is unclear whether these α-subunits are expressed in neurons in *C. elegans.*

3.2.2 C2 domain proteins

The best characterized protein that fits the role of a calcium sensor at the nerve terminal is synaptotagmin (115). The phospholipid-dependent calcium binding C2 domains found in synaptotagmin turn out to be present in a number of other proteins that reside at the nerve terminal. In fact, most nerve terminal-associated proteins which bind calcium under physiological conditions contain at least one C2 domain. The synaptic proteins synaptotagmin (115), DOC2 (116), UNC-13 (72, 117), Rim (118), and rabphilin (119) all contain at least one C2 domain, although not all of the proteins have been documented to bind calcium. In both vertebrates and in *C. elegans* multiple different members of some of these families have been identified (120). Genetic and biochemical analysis is just beginning to sort out the roles of these proteins. In *C. elegans,* mutants have been isolated in four of these genes: the synaptotagmin homologue *snt-1* (28), *unc-13* (72), the Rim homologue *unc-10* (121), and the rabphilin homologue *rbf-1* (J. Staunton, B. Ganetzky, and M. Nonet, in preparation). Rim and rabphilin interact with rab3 and are discussed in Section 3.3.

Synaptotagmins and DOCs (for double C2 domain) both contain a closely spaced pair of C2 domains (120). In the synaptotagmins, these domains are linked to the membrane by an N-terminal hydrophobic anchor, while DOCs lack this anchor. Analysis of genomic sequences of *C. elegans* indicate that at least six proteins belong the synaptotagmin/DOC family. The lack of amino acid conservation outside of the C2 domains make assigning the genes to either family difficult except in two cases where cDNA sequences indicate the presence of N-terminal hydrophobic domain. Among these, only one, *snt-1*, exhibits above 50% similarity to a vertebrate counterpart in the conserved C2 domains (28). *C. elegans snt-1* shares 72% identity to vertebrate synaptotagmin I in the C2 domains and is expressed throughout the *C. elegans* nervous system. Synaptotagmin (*snt-1*) deletion mutants in *C. elegans* survive to adulthood and have surprisingly mild phenotypes (28). They remain capable of substantial movement and respond to touch. Although this mutant provided an early indication that synaptotagmin was unlikely to be the sole calcium sensor at the nerve terminal in *C. elegans*, identifying other molecules that might be involved has not been simple. One confounding issue is the likely involvement of synaptotagmin in the endocytosis pathway as evidenced by vertebrate synaptotagmin's association with the adaptor complex AP2 (122) and the depletion of synaptic vesicles from neuromuscular junctions in *C. elegans* (77). A second complication is the presence of the other *C. elegans* two C2 domain genes. Although these other genes have not been characterized, it is likely that some are expressed in *C. elegans* neurons. Whether multiple C2 domain proteins share roles in regulating calcium sensitivity with synaptotagmin at the nerve terminal will require a comprehensive analysis of mutants. Given the complexity of the C2 domain protein family, interactions between genes at a physiological level will likely be first addressed in *C. elegans* where isolation of mutants in multiple family members is practical.

unc-13 encodes a large protein containing domains similar to both the C1 and C2 domains found in calcium-activated protein kinase C (72). Recombinant *C. elegans* UNC-13 protein binds to phorbol esters in a calcium- and phospholipid-dependent manner (72, 73). Similar observations have been made for the vertebrate homologues (123). The *C. elegans unc-13* gene is complex expressing several distinct isoforms in neurons (72) (J. B. Rand and I. Maruyama, unpublished data). Although many lesions that disrupt neuronal isoforms survive as homozygotes and exhibit phenotypes very similar to *unc-18* mutants, other lesions in the gene result in a lethal phenotype. Interestingly, the mammalian UNC-13s have been demonstrated to bind the same domain of syntaxin (124) that interacts with UNC-18, suggesting that *unc-13* may directly regulate some aspect of vesicle fusion.

3.2.3 Regulation of dense core vesicle fusion by *unc-31*

Secretion of dense core vesicles is thought to be regulated differently from that of small clear vesicles. *unc-31* encodes a gene homologous to the biochemically characterized CAPS (calcium-dependent activator protein for secretion) (125, 126). Mutants in the *unc-31* gene exhibit phenotypes that are distinct from most synaptic uncoordinated mutants (127). The animals exhibit food-dependent sluggish locomotion, altered regulation of pharyngeal pumping and egg laying, as well as other behavioural abnormalities. The defects have been interpreted as altering behaviours which are regulated by serotonin (127). The identification of vertebrates CAPS as a component of the dense core vesicle secretion apparatus fits well with the observed *unc-31* behavioural abnormalities (128). *unc-31* also exhibits genetic interactions with both *aex-3*, a rab3 nucleotide exchange factor (discussed in Section 3.3), and specific alleles of syntaxin (90). Genetic screens for other mutants that show similar interactions with *aex-3* may provide one method of identifying other genes that regulate secretion of dense core vesicles.

3.3 Components of the rab3 pathway

rab-3 encodes the *C. elegans* homologue of a highly conserved small GTPase that is associated with synaptic vesicles (19, 129). One of the most unexpected findings from the analysis of synaptic transmission mutants in *C. elegans* has been that mutants which disrupt genes encoding components of the rab3 pathway only cause mild perturbations of neuronal function. This contrasts markedly with the essential nature of rab proteins in the secretion pathway in yeast (130). Biochemical studies have identified a number of vertebrate proteins that interact with rab3 including Rim (118), rabphilin (119), the rab3 nucleotide exchange factor (90, 131), rab-GDI (132), and two subunits of a rab3-specific GTPase (133, 134). A clear homologue of each of these genes is present in *C. elegans* and mutants have been isolated in four of the genes: Rim (*unc-10*), rabphilin (*rbf-1*), rab3 EF (*aex-3*), and *rab-3* itself. Although *rab-3* (19), *aex-3* (90), and *unc-10* (121) (L. Wei, J. Staunton, G. Hadwiger, and M. L. Nonet, in preparation), mutants all exhibit some behavioural, pharmacological, and physiological abnormalities, none of these genes are essential for synaptic transmission. In fact, rabphilin mutants do not even lead to a detectable decrease in synaptic function (J. Staunton, B. Ganetzky, and M. L. Nonet, in preparation). A second piece of evidence suggesting that the rab3 pathway is not essential for synaptic transmission is that none of the double and triple mutant combinations of *aex-3*, *rab-3*, and *unc-10* exhibit any synergistic phenotypic enhancement (90) (J. Staunton and M. L. Nonet, unpublished data). The mild phenotypes of these double and triple mutants reduce the possibility that other molecules (e.g. rab8 or rab11) compensate for absence of the individual components and mask the regular role of the proteins at the synapse.

3.4 G-protein signalling pathways

Many synaptic modulators like opiates alter synaptic transmission through G-protein coupled pathways. The nematode may be particularly useful in studying these genes because they differ from most of the other synaptic components in that the genes are neither essential for viability nor for synaptic function and thus are extremely amenable to genetic analysis. Mutants in several genes encoding different G-protein signalling molecules have been isolated that cause decreases in synaptic function suggesting that conserved mechanisms operate to modulate activity of the synaptic apparatus in *C. elegans*. *C. elegans egl-10* and *egl-30* mutants encode a Gα_q subunit and an RGS protein, respectively, that are both highly expressed in neurons (135, 136). Mutants in these genes are resistant to aldicarb and have a variety of behavioural deficits. Phospholipase Cβ, another component of the G-protein signalling pathway, is encoded by the *egl-8* gene, (K. Miller and J. B. Rand, in preparation). Additionally, several other ric genes including the *ric-8* gene have similar phenotypes to *egl-30* and are likely to encode components in this pathway which modulates synaptic function in *C. elegans*.

3.5 Vesicular transporters and ATPases

A critical function of the release apparatus is to efficiently load neurotransmitter into synaptic vesicles. This involves the creation of an electrochemical or pH gradient and the transport of transmitter into vesicles via specific transporters using the gradient as an energy source. Mutants in at least three transporters have been isolated in *C. elegans*. In two of the three cases, the initial isolation of the *C. elegans* gene preceded and led directly to the subsequent isolation of the vertebrate molecules which had long eluded molecular characterization using biochemical methods. The *unc-17* gene encodes the vesicular acetylcholine transporter. This essential gene is expressed co-ordinately with the choline acetyltransferase gene in a complex locus. Interestingly, the mechanism of co-ordinate regulation of these two genes has apparently been preserved in the mammalian lineage (137, 138). The *unc-47* gene encodes the GABA transporter (139), and *cat-1* encodes the vesicular monoamine transporter (140). Neither of these two genes are essential, but each mutant phenotype resembles the phenotype observed in animals missing the synthetic enzyme for the respective transmitter (18, 56). A few homologues of the *unc-47* transporter are present in *C. elegans* and these may represent other transmitter transporters (e.g. the glutamate transporter). A component of the multisubunit vacuolar pump required for loading of transmitter has also been isolated genetically. *unc-32* encodes a molecule similar to the 116 kDa subunit of the vacuolar ATPase proton pump (141) (D. Thierry-Meig, unpublished data). The strongest mutants in this gene arrest with phenotypes similar to syntaxin mutants; the phenotype predicted for a gene required for loading synaptic vesicles.

3.6 Endocytosis genes

The cyclical nature of synaptic vesicle biogenesis complicates the interpretation of many experiments which aim to localize the site of action of synaptic regulators. Since endocytosis is required for exocytosis and vice versa, mutants that block endocytosis at the synapse could exhibit very similar phenotypes as mutants that lesion the exocytic arm of the cycle. Indeed, in *C. elegans*, most mutants that have been isolated in components thought to participate in endocytosis exhibit defects in synaptic transmission and were isolated in mutants screens for aldicarb-resistant mutants. The main criteria that has been used to classify these mutants as defective in the endocytic arm of the synaptic vesicle cycle is an alteration in synaptic vesicle morphology or populations at the synapse.

3.6.1 Clathrin-associated proteins

A group of four aldicarb-resistant mutants which have similar behavioural phenotypes have all been implicated in functioning in endocytosis. *snt-1*, *unc-11*, *unc-26*, and *unc-41* are all small, slow growing, chronically starved and strongly aldicarb-resistant animals, but each retains a surprisingly robust capacity for jerky locomotion (5, 28, 121). *snt-1*, which encodes synaptotagmin, is implicated in endocytosis because of the depletion of synaptic vesicles observed in these mutants (77). Biochemical interactions between vertebrate synaptotagmin and AP2 support a role for the protein in endocytosis (122). *unc-11* encodes a homologue of AP180, a protein also implicated in endocytosis from its *in vitro* AP2 and clathrin binding properties (142–145). The mean size of synaptic vesicles and the synaptic localization of synaptobrevin are altered in *unc-11* mutants suggesting that, *in vivo*, the protein regulates both clathrin assembly and synaptobrevin sorting (80). *unc-26* encodes the *C. elegans* homologue of synaptojanin (146), an inositol polyphosphate 5-phosphatase (T. Harris and E. Jorgensen, in preparation). Alterations in phosphatidylinositol phospholipid composition that likely occur in *unc-26* mutants may disrupt the function of multiple components of the endocytosis apparatus that bind to phospholipid ligands including AP180 (147), synaptotagmin (115), and AP2 (148). Finally, *unc-41* encodes a molecule with some similarity to one of the products of the *Drosophila stoned* gene (149, 150). Both *stonedB* and the *unc-41* gene (J. B Rand, unpublished data) contain domains with similarity to the $\mu1$, $\mu2$, and $\mu3$ subunits of the AP complexes which have been implicated in recognition of cargo during clathrin coat formation (151). Thus, *unc-41* also may regulate aspects of synaptic vesicle membrane trafficking. More detailed cellular analysis of these genes should aid in the refining the emerging picture of the endocytosis pathway.

A single gene for each of the clathrin light and heavy chains is present in *C. elegans* as well as the α, $\beta1$, $\beta2$, $\beta3$, δ, and γ subunits of the three adaptor complexes (see Chapter 11). Most of these genes are known solely from genomic and EST sequences and have not been thoroughly characterized. In vertebrates, neuronal-specific isoforms of clathrin and adaptor subunits have been identified. However, only one isoform of the clathrin light chain is present among the light chain EST cDNAs, although an intron/exon boundary is present at the precise site of alternative splicing in vertebrates (M. L. Nonet, unpublished data). Additionally, only a single $\mu3$ gene has been identified in *C. elegans* suggesting that neuronal-specific isoforms of this gene are also absent. Two distinct genes which code for $\mu1$ subunits are found in *C. elegans* and *unc-101* mutants defines lesions in one of the genes (152). *unc-101* exhibit a variety of locomotory, egg laying, and defecation defects as well as non-neuronal defects. However, the cellular focus of the neuronal defect is unclear since *unc-101* mutants also exhibit altered neuronal morphology suggestive of axonal outgrowth defects.

3.6.2 Dynamin

The dynamin gene *shibire* in *Drosophila* was implicated in endocytosis of synaptic vesicles because temperature-sensitive alleles of *shibire* block endocytosis resulting in depletion of vesicles from synaptic terminals (153, 154). In *C. elegans*, one sub-viable temperature-sensitive mutant has been isolated in the *dyn-1* gene (37). Similar to the *Drosophila* mutant, the mutant results in a quickly reversible temperature-sensitive paralysis. Surprisingly, this dynamin mutant is hypersensitive to aldicarb suggesting that *dyn-1* mutant synapses may not be depleted of vesicles. *dyn-1* is very likely to function outside the nervous system since the only other dynamin family member found in *C. elegans* is T12E12.4, a gene that is closely related to Drp1 which has been implicated in regulating mitochondrial distribution (155). If more subtle defects in endocytosis remain at the permissive temperature, this could result in elevated levels of cholinergic receptors on the postsynaptic plasma membrane and thus account for the unexpected sensitivity to aldicarb. Additional analysis of *dyn-1* mutants, including ultrastructural analysis, will be required to resolved the molecular basis that underlies the unusual aldicarb hypersensitive phenotype of the *dyn-1* mutant.

3.7 Other components

Genes have also been identified in *C. elegans* for a variety of other components thought to localize to synaptic vesicles or function at the synapse. Mutants have not been identified in any of these genes, although some of them may be among the poorly characterized aldicarb-resistant mutants presently under analysis in several *C. elegans* laboratories. Genes similar to those encoding the vertebrate synaptic vesicle-associated proteins synaptogyrin (156), synaptophysin (157), synapsin (158), SCAMP (159, 160), and cysteine string protein (161) are all present in *C. elegans* (Table 2). The synaptogyrin homologue, *sng-1* is known to be expressed in neurons and appears to localize to synaptic sites (79), but the expression pattern of most of the other genes has not been well characterized. Although no SV2 homologue has been identified to date, Janz and colleagues cleverly used the *C. elegans* and *Drosophila* genome databases to identify SVOP (162), a protein with similarity to SV2 (163, 164). Among the synapse-associated proteins, genes encoding molecules with similarity to tomosyn (165), synucleins (166), VAP-33 (167), CIRL (168), and hrs-2 (169) have been identified in *C. elegans* (Table 2). In short, a homologue of virtually every molecule identified biochemically in vertebrates is present in *C. elegans*.

3.8 Are there additional synaptic mutants to be identified?

Additional *C. elegans* synaptic regulators remain unmined. Foremost, a number of well-characterized aldicarb-resistant mutants remain uncloned including *ric-1*, *ric-3*, *ric-6-10*, *ric-12*, *ric-14-16*, *sup-6*, *unc-65*, and *unc-75*. Many of these genes map to physical locations distinct from the positions of homologues of biochemically identified vertebrate synaptic proteins suggesting novel synaptic regulators remain to be identified in *C. elegans*. A second class of mutants that will be harder to isolate and characterize are lesions in components that are essential for either exocytosis or endocytosis of membranes in *C. elegans*. Mutants have only been isolated in a small fraction of the genes that we predict encode essential components. For example, mutants have not been identified in the genes encoding the clathrin light and heavy chains, subunits of the AP complexes, SNAP, or NSF. Mutants in these genes will likely only come from directed attempts to isolate null alleles using reverse genetics or by fortuitous isolation of rare hypomorphic or temperature-sensitive alleles. Since so few of the known 'essential' components have been defined by mutation, it is also very likely that novel genes that play essential roles in regulating synaptic function also remain unidentified. Lastly, components which provide a redundant function at the synapse will be extremely difficult to identify using classical genetic approaches. A combination of biochemical and reverse genetic methods will provide the most efficient way of characterizing the role of these proteins. However, genes encoding redundant functions also may be identified as suppressors or enhancers of synaptic mutants. Despite some limitations, the methodical characterization of mutants with behavioural or pharmacological phenotypes characteristic of synaptic deficits is certain to identify a variety of additional molecular components of the transmission apparatus.

4. A brief comparison of phenotypic consequences of mutations in different organisms

Drosophila, *C. elegans*, and mouse represent the three most widely studied organisms in which genetic approaches aimed at studying synaptic transmission are presently amenable. Although there exist only a few cases in which mutants in a synaptic component have been isolated and characterized in all three organisms, similarities have usually been apparent in the phenotypes of mutants disrupting the homologous gene product in at least two organisms. For example *rab-3* mutants in both *C. elegans* (19) and the mouse (170) lead to relatively minor perturbations of the synaptic apparatus and (m)UNC-18 mutants cause dramatic defects in all three systems (30, 99, 103). Probably more informative are the distinctions that can be drawn among mutants isolated in the different organisms. For, example synaptotagmin mutants have been isolated in mouse, *Drosophila*, and *C. elegans*. In each case, transmission is dramatically reduced but not eliminated. However, the morphology of synapses differs dramatically in the three mutants. Synaptic terminals of mouse synaptotagmin mutants are unchanged (171), while terminals of *Drosophila* synaptotagmin animals contain reduced numbers of docked vesicles and elevated numbers of larger vesicles (172), and those of *C. elegans* are greatly depleted of vesicles (77). Does this imply that synaptotagmin functions differently in each system? It more likely reflects the complex nature of the functions synaptotagmin regulates in interacting with calcium (115), phospholipids (115), syntaxin (173), βSNAP (105), the AP2 receptor (122), and probably other proteins as well. The net phenotypic outcome we observe depends on how processes synaptotagmin regulates are perturbed by removing synaptotagmin at the synapse that is examined, the modulatory effects of potential compensation mechanisms (e.g. other synaptotagmins, etc.), and the nature of the genetic lesion analysed. The phenotype we observe in each system probably reflects the fact that a different role of synaptotagmin is 'rate'-limiting in each system. Hence, each system will focus attention on different aspects of the roles of proteins in regulating the transmission machinery.

5. Future contributions from the analysis of *C. elegans*

The molecular identification of the genes encoding components of the synaptic machinery will not, by itself, provide an outline of the role these molecules play in regulating the release of transmitter. The roles of many long-identified synaptic components including SV2 and synaptophysin remain almost as elusive today as the moment they were first cloned. In the case of synaptophysin, even genetic lesions in the mouse have not provided much insight into the role of the protein in regulating transmitter release (174). At first glance, one must conclude that synaptophysin plays no substantial role in the transmission process. However, this does not provide a satisfactory answer to explain why the protein has been conserved throughout the evolutionary divergence of *C. elegans* and mouse. Indeed, the use of genetics is limited by the ability of the researchers to identify the defects associated with lesioning a gene. Characterization of mutants in such genes in multiple different organisms will probably play a crucial role in determining the role of these genes. Just as synaptotagmin mutants result in distinct abnormalities in different organisms, the requirement for synaptophysin may well differ in mouse, *Drosophila*, and *C. elegans*. Likewise, different assays to identify abnormalities have been developed in the three systems. *C. elegans* genetic approaches complement mouse approaches to the analysis of these genes because it affords the capability to isolate rare mutations with unusual properties. For example, recessive lesions in several ion channel genes in *C. elegans* have no apparent phenotype, but rare dominant lesions have severe abnormalities (175–177). Other lesions such as *sup-1* and *sup-8* specifically suppress a missense mutation in the vesicular acetylcholine transporter, but have no detectable phenotypes on their own (J. Hodgkin, unpublished data). These types of rare mutations can identify roles for genes that cannot be deduced from the characterization of simple knock-out mutants.

The synapse presents us with a wealth of intriguing biological questions that can be addressed using the unusual experimental capabilities provided by this small nematode. There exist many attributes in *C. elegans* that make them stand out as particularly amenable to study synaptic function. First, although the use of pharmacological agents to isolate genetic lesions which reduce synaptic transmission have been used extensively, acetylcholinesterase inhibitors have not been exploited to identify genes that negatively regulate the transmission process through the isolation of hypersensitive mutants. Secondly, the robust transformation techniques and compact genomic organization of genes in *C. elegans* make the dissection of structure–function relationships *in vivo* uniquely accessible in *C. elegans*. Thirdly, the translucent nature of *C. elegans* offers an unique advantage to study the genetic constituents which regulate the targeting of components to the subcellular domains of the neuron including the synapse. No other organism currently affords the capabilities to use classical genetics to study subcellular targeting in neurons. Finally, although perhaps outside the direct scope of synaptic transmission, *C. elegans* will offer a simple system to study the mechanisms that co-ordinate regulation of the expression of synaptic components.

Classical genetic analysis has provided the foundation for the use of *C. elegans* to study many aspects of the function and development the nervous system. Numerous genes present in vertebrates have been molecularly characterized and functionally implicated in transmission from the analysis of *C. elegans* mutants including the acetylcholine and GABA vesicular transporters, *aex-3*, *unc-13*, and *unc-18*. The uncloned aldicarb-resistant mutants and suppressor and enhancer mutants that are now being characterized will identify additional genes. However, within a few years, all genes present in metazoans will be identified (178) and neuroscientists studying synaptic transmission will concentrate on defining the roles these proteins play in the neuron. As physiology assays are improved in *C. elegans*, the rich collection of lesions in fundamental components of the nerve terminal will permit these mutants to be used to dissect subtle features of the roles of proteins in regulating quantal size, pool size, potentiation, probability of release, and other fundamental properties of the nerve terminal. The goal of the *C. elegans* community must be a balanced combination of further exploiting the genetic strengths of the system while simultaneously developing assays to compensate for the present weaknesses of the system. Each organism that neuroscientists use to study synaptic transmission offers unique advantages for physiological, genetic, biochemical, or ultrastructural analysis of the synapse, but also has distinct limitations. However, used in conjunction, these diversified approaches provide a much more powerful arsenal to dissect the intricate molecular details of the synaptic machinery.

Acknowledgements

I thank Janet Richmond and Erik Jorgensen for generously providing unpublished data. Furthermore, I thank Sandhya Koushika, Maya Kunkel, James Rand, and Anneliese Schaefer for critical comments on the manuscript. Work in my laboratory on synaptic transmission is funded by a grant from the United States Public Health Service.

References

1. Antebi, A., Norris, C. R., Hedgecock, E. M., and Garriga, G. (1997) Cell and growth cone migrations. In *C. elegans II* (ed. D. L. Riddle, T. Blumenthal, B. J. Meyer, and J. R. Priess), p. 583. Cold Spring Harbor Laboratory Press, Cold Spring Harbor.

2. Driscoll, M. and Kaplan, J. (1997) Mechanosensation. In *C. elegans II* (ed. D. L. Riddle, T. Blumenthal, B. J. Meyer, and J. R. Priess), p. 645. Cold Spring Harbor Laboratory Press, Cold Spring Harbor.
3. Bargmann, C. I. and Mori, I. (1997) Chemotaxis and thermotaxis. In *C. elegans II* (ed. D. L. Riddle, T. Blumenthal, B. J. Meyer, and J. R. Priess), p. 717. Cold Spring Harbor Laboratory Press, Cold Spring Harbor.
4. Rand, J. B. and Nonet, M. L. (1997) Synaptic transmission. In *C. elegans II* (ed. D. L. Riddle, T. Blumenthal, B. J. Meyer, and J. R. Priess), p. 611. Cold Spring Harbor Laboratory Press, Cold Spring Harbor.
5. Brenner, S. (1974) The genetics of *Caenorhabditis elegans*. *Genetics*, **77**, 71.
6. Sulston, J. and Hodgkin, J. (1988) Methods. In *The nematode Caenorhabditis elegans* (ed. W. B. Wood), p. 587. Cold Spring Harbor Laboratory, Cold Spring Harbor, New York.
7. Herman, R. K. and Horvitz, H. R. (1980) Genetic analysis of *Caenorhabditis elegans*. In *Nematodes as biological models, volume 1: Behavioural and developmental models* (ed. B. M. Zuckerman), p. 227. Academic Press, New York.
8. Sulston, J. E. and Horvitz, H. R. (1977) Post-embryonic cell lineages of the nematode *Caenorhabditis elegans*. *Dev. Biol.*, **56**, 110.
9. White, J. G., Southgate, E., Thomson, J. N., and Brenner, S. (1986) The structure of the nervous system of *Caenorhabditis elegans*. *Phil. Trans. R. Soc. Lond. [Biol]*, **314**, 1.
10. Goodman, M. B., Hall, D. H., Avery, L., and Lockery, S. R. (1998) Active currents regulate sensitivity and dynamic range in *C. elegans* neurons. *Neuron*, **20**, 763.
11. Ward, S., Thomson, N., White, J. G., and Brenner, S. (1975) Electron microscopical reconstruction of the anterior sensory anatomy of the nematode *Caenorhabditis elegans*. *J. Comp. Neurol.*, **160**, 313.
12. Li, H., Avery, L., Denk, W., and Hess, G. (1997) Identification of chemical synapses in the pharynx of *Caenorhabditis elegans*. *Proc. Natl. Acad. Sci. USA*, **94**, 5912.
13. Schinkman, K. and Li, C. (1992) Localization of FMRFamide-like peptides in *Caenorhabditis elegans*. *J. Comp. Neurol.*, **316**, 251.
14. McIntire, S. L., Jorgensen, E., Kaplan, J., and Horvitz, H. R. (1993) The GABAergic nervous system of *Caenorhabditis elegans*. *Nature*, **364**, 337.
15. Horvitz, H. R., Chalfie, M., Trent, C., Sulston, J. E., and Evans, P. D. (1982) Serotonin and octopamine in the nematode *Caenorhabditis elegans*. *Science*, **216**, 1012.
16. Alfonso, A., Grundahl, K., McManus, J. R., and Rand, J. B. (1994) Cloning and characterization of the choline acetyltransferase structural gene (*cha-1*) from *C. elegans*. *J. Neurosci.*, **14**, 2290.
17. Rand, J. B. and Nonet, M. L. (1997) Neurotransmitter assignments for specific neurons. In *C. elegans II* (ed. D. L. Riddle, T. Blumenthal, B. J. Meyer, and J. R. Priess), p. 1049. Cold Spring Harbor Laboratory Press, Cold Spring Harbor.
18. Sulston, J., Dew, M., and Brenner, S. (1975) Dopaminergic neurons in the nematode *Caenorhabditis elegans*. *J. Comp. Neurol.*, **163**, 215.
19. Nonet, M. L., Staunton, J., Kilgard, M. P., Fergestad, T., Hartweig, E., Horvitz, H. R., *et al.* (1997) *C. elegans rab-3* mutant synapses exhibit impaired function and are partially depleted of vesicles. *J. Neurosci.*, **17**, 8021.
20. Trent, C., Tsung, N., and Horvitz, H. R. (1983) Egg-laying defective mutants of the nematode *Caenorhabditis elegans*. *Genetics*, **104**, 619.
21. Avery, L. (1993) The genetics of feeding in *Caenorhabditis elegans*. *Genetics*, **133**, 897.
22. Thomas, J. H. (1990) Genetic analysis of defecation in *Caenorhabditis elegans*. *Genetics*, **124**, 855.

23. Rand, J. B. and Russell, R. L. (1985) Molecular basis of drug-resistance mutations in the nematode *Caenorhabditis elegans*. *Psychopharm. Bull.*, **21**, 623.
24. Miller, K. G., Alfonso, A., Nguyen, M., Crowell, J. A., Johnson, C. D., and Rand, J. B. (1996) A genetic selection for *Caenorhabditis elegans* synaptic transmission mutants. *Proc. Natl. Acad. Sci. USA*, **93**, 12593.
25. Hosono, R., Sassa, T., and Kuno, S. (1989) Spontaneous mutations of trichlorfon resistance in the nematode *Caenorhabditis elegans*. *Zool. Sci.*, **6**, 697.
26. Saifee, O., Wei, L. P., and Nonet, M. L. (1998) The *C. elegans unc-64* gene encodes a syntaxin which interacts with genetically with synaptobrevin. *Mol. Biol. Cell*, **9**, 1235.
27. Avery, L. and Horvitz, H. R. (1989) Pharyngeal pumping continues after laser killing of the pharyngeal nervous system of *C. elegans*. *Neuron*, **3**, 473.
28. Nonet, M. L., Grundahl, K., Meyer, B. J., and Rand, J. B. (1993) Synaptic function is impaired but not eliminated in *C. elegans* mutants lacking synaptotagmin. *Cell*, **73**, 1291.
29. Gengyo-Ando, K., Kamiya, Y., Yamakawa, A., Kodaira, K., Nishiwaki, K., Miwa, J., *et al.* (1993) The *C. elegans unc-18* gene encodes a protein expressed in motor neurons. *Neuron*, **11**, 703.
30. Schulze, K. L., Littleton, J. T., Salzberg, A., Halachmi, N., Stern, M., Lev, Z., *et al.* (1994) rop, a *Drosophila* homolog of yeast Sec1 and vertebrate n-Sec1/Munc-18 proteins, is a negative regulator of neurotransmitter release *in vivo*. *Neuron*, **13**, 1099.
31. DiAntonio, A., Burgess, R. W., Chin, A. C., Deitcher, D. L., Scheller, R. H., and Schwarz, T. L. (1993) Identification and characterization of *Drosophila* genes for synaptic vesicle proteins. *J. Neurosci.*, **13**, 4924.
32. Nonet, M. L., Saifee, O., Zhao, H., Rand, J. B., and Wei, L. (1998) Synaptic transmission deficits in *C. elegans* synaptobrevin mutants. *J. Neurosci.*, **18**, 70.
33. Siddiqi, O. and Benzer, S. (1976) Neurophysiological defects in temperature-sensitive paralytic mutants of *Drosophila melanogaster*. *Proc. Natl. Acad. Sci. USA*, **73**, 3253.
34. Suzuki, D. T. (1970) Temperature-sensitive mutations in *Drosophila melanogaster*. *Science*, **170**, 695.
35. Grigliatti, T. A., Hall, L., Rosenbluth, R., and Suzuki, D. T. (1973) Temperature-sensitive mutations in *Drosophila melanogaster*. XIV. A selection of immobile adults. *Mol. Gen. Genet.*, **120**, 107.
36. Pallanck, L., Ordway, R. W., and Ganetzky, B. (1995) A *Drosophila* NSF mutant. *Nature*, **376**, 25.
37. Clark, S. G., Shurland, D. L., Meyerowitz, E. M., Bargmann, C. I., and van der Bliek, A. M. (1997) A dynamin GTPase mutation causes a rapid and reversible temperature-inducible locomotion defect in *C. elegans*. *Proc. Natl. Acad. Sci. USA*, **94**, 10438.
38. Kornfeld, K., Hom, D. B., and Horvitz, H. R. (1995) The *ksr-1* gene encodes a novel protein kinase involved in Ras-mediated signaling in *C. elegans*. *Cell*, **83**, 903.
39. Sundaram, M. and Han, M. (1995) The *C. elegans ksr-1* gene encodes a novel Raf-related kinase involved in Ras-mediated signal transduction. *Cell*, **83**, 889.
40. Singh, N. and Han, M. (1995) *sur-2*, a novel gene, functions late in the *let-60* ras-mediated signaling pathway during *Caenorhabditis elegans* vulval induction. *Genes Dev.*, **9**, 2251.
41. Kornfeld, K., Guan, K. L., and Horvitz, H. R. (1995) The *Caenorhabditis elegans* gene *mek-2* is required for vulval induction and encodes a protein similar to the protein kinase MEK. *Genes Dev.*, **9**, 756.
42. Villeneuve, A. M. and Meyer, B. J. (1990) The regulatory hierarchy controlling sex determination and dosage compensation in *Caenorhabditis elegans*. *Adv. Genet.*, **27**, 117.

43. The *C. elegans* Sequencing Consortium (1998) Genome sequence of the nematode *C. elegans*: A platform for investigating Biology. *Science*, **282**, 2012.
44. Wilson, R., Ainscough, R., Anderson, K., Baynes, C., Berks, M., Bonfield, J., *et al.* (1994) 2.2 Mb of contiguous nucleotide sequence from chromosome III of *C. elegans*. *Nature*, **368**, 32.
45. Coulson, A., Sulson, J., Brenner, S., and Karn, J. (1986) Toward a physical map of the *Caenorhabditis elegans* genome. *Proc. Natl. Acad. Sci. USA*, **83**, 7821.
46. Coulson, A., Waterston, R., Kiff, J., Sulston, J., and Kohara, Y. (1988) Genome linking with yeast artificial chromosomes. *Nature*, **335**, 184.
47. Mello, C. C., Kramer, J. M., Stinchcomb, D., and Ambros, V. (1991) Efficient gene transfer in *C. elegans*: extrachromosomal maintenance and integration of transforming sequences. *EMBO J.*, **10**, 3959.
48. Mello, C. and Fire, A. (1995) DNA transformation. In *Caenorhabditis elegans: Modern biological analysis of an organism* (ed. H. F. Epstein and D. C. Shakes), Vol. 48, p. 451. Academic Press, San Diego.
49. Kramer, J. M., French, R. P., Park, E. C., and Johnson, J. J. (1990) The *Caenorhabditis elegans rol-6* gene, which interacts with the *sqt-1* collagen gene to determine organismal morphology, encodes a collagen. *Mol. Cell Biol.*, **10**, 2081.
50. Zwaal, R. R., Broeks, A., van Meurs, J., Groenen, J. T., and Plasterk, R. H. (1993) Target-selected gene inactivation in *Caenorhabditis elegans* by using a frozen transposon insertion mutant bank. *Proc. Natl. Acad. Sci. USA*, **90**, 7431.
51. Dernburg, A. F., McDonald, K., Moulder, G., Barstead, R., Dresser, M., and Villeneuve, A. M. (1998) Meiotic recombination in *C. elegans* initiates by a conserved mechanism and is dispensable for homologous chromosome synapsis. *Cell*, **94**, 387.
52. Plasterk, R. H. (1995) Reverse genetics: from gene sequence to mutant worm. In *Caenorhabditis elegans: Modern biological analysis of an organism* (ed. H. F. Epstein and D. C. Shakes), Vol. 48, p. 59. Academic Press, San Diego.
53. Jansen, G., Hazendonk, E., Thijssen, K., and Plasterk, R. (1997) Reverse genetics by chemical mutagenesis in *Caenorhabditis elegans*. *Nat. Genet.*, **17**, 119.
54. Fire, A., Xu, S., Montgomery, M. K., Kostas, S. A., Driver, S. E., and Mello, C. C. (1998) Potent and specific genetic interference by double-stranded RNA in *Caenorhabditis elegans*. *Nature*, **391**, 806.
55. Avery, L. and Thomas, J. (1997) Feeding and defecation. In *C. elegans II* (ed. D. L. Riddle, T. Blumenthal, B. J. Meyer, and J. R. Priess), p. 679. Cold Spring Harbor Laboratory Press, Cold Spring Harbor.
56. McIntire, S. L., Jorgensen, E., and Horvitz, H. R. (1993) Genes required for GABA function in *Caenorhabditis elegans*. *Nature*, **364**, 334.
57. Chalfie, M. and Au, M. (1989) Genetic control of differentiation of the *Caenorhabditis elegans* touch receptor neurons. *Science*, **243**, 1027.
58. Liu, K. S. and Sternberg, P. W. (1995) Sensory regulation of male mating behaviour in *Caenorhabditis elegans*. *Neuron*, **14**, 79.
59. Loer, C. M. and Kenyon, C. J. (1993) Serotonin-deficient mutants and male mating behaviour in the nematode *Caenorhabditis elegans*. *J. Neurosci.*, **13**, 5407.
60. Hodgkin, J. (1983) Male phenotypes and mating efficiency in *C. elegans*. *Genetics*, **103**, 43.
61. Jin, Y., Jorgensen, E., Hartwieg, E., and Horvitz, H. R. (1999) The *C. elegans* gene *unc-25* encodes glutamic acid decarboxylase and is required for synaptic transmission but not synaptic development. *J. Neurosci.*, **19**, 539.
62. Hart, A. C., Sims, S., and Kaplan, J. M. (1995) Synaptic code for sensory modalities revealed by *C. elegans* GLR-1 glutamate receptor. *Nature*, **378**, 82.

63. Lee, R. Y. N., Sawin, E. R., Chalfie, M., Horvitz, H. R., and Avery, L. (1999) EAT-4, a homolog of a mammalian sodium-dependent inorganic phosphate cotransporter, is necessary for glutamatergic neurotransmission in *Caenorhabditis elegans*. *J. Neurosci.*, **19**, 159.
64. Lewis, J. A., Wu, C. H., Levine, J. H., and Berg, H. (1980) Levamisole-resistant mutants of the nematode *Caenorhabditis elegans* appear to lack pharmacological acetylcholine receptors. *Neuroscience*, **5**, 967.
65. Fleming, J. T., Squire, M. D., Barnes, T. M., Tornoe, C., Matsuda, K., Ahnn, J., *et al.* (1997) *Caenorhabditis elegans* levamisole resistance genes *lev-1*, *unc-29*, and *unc-38* encode functional nicotinic acetylcholine receptor subunits. *J. Neurosci.*, **17**, 5843.
66. Schafer, W. R. and Kenyon, C. J. (1995) A calcium-channel homologue required for adaptation to dopamine and serotonin in *Caenorhabditis elegans*. *Nature*, **375**, 73.
67. Rand, J. B. and Johnson, C. D. (1995) Genetic pharmocology: Interactions between drugs and gene products in *C. elegans*. In *Caenorhabditis elegans: Modern biological analysis of an organism* (ed. H. F. Epstein and D. C. Shakes), Vol. 48, p. 187. Academic Press, San Diego.
68. Lewis, J. A. and Fleming, J. T. (1995) Basic culture methods. In *Caenorhabditis elegans: Modern biological analysis of an organism* (ed. H. F. Epstein and D. C. Shakes), Vol. 48, p. 3. Academic Press, San Diego.
69. Epstein, H. F. and Liu, F. (1995) Proteins and protein assemblies. In *Caenorhabditis elegans: Modern biological analysis of an organism* (ed. H. F. Epstein and D. C. Shakes), Vol. 48, p. 437. Academic Press, San Diego.
70. Harada, S., Hori, I., Yamamoto, H., and Hosono, R. (1994) Mutations in the unc-41 gene cause elevation of acetylcholine levels. *J. Neurochem.*, **63**, 439.
71. Kaech, S. M., Whitfield, C. W., and Kim, S. K. (1998) The LIN-2/LIN-7/LIN-10 complex mediates basolateral membrane localization of the *C. elegans* EGF receptor LET-23 in vulval epithelial cells. *Cell*, **94**, 761.
72. Maruyama, I. N. and Brenner, S. (1991) A phorbol ester/diacylglycerol-binding protein encoded by the *unc-13* gene of *Caenorhabditis elegans*. *Proc. Natl. Acad. Sci. USA*, **88**, 5729.
73. Ahmed, S., Maruyama, I. N., Kozma, R., Lee, J., Brenner, S., and Lim, L. (1992) The *Caenorhabditis elegans unc-13* gene product is a phospholipid-dependent high-affinity phorbol ester receptor. *Biochem. J.*, **287**, 995.
74. Ogawa, H., Harada, S., Sassa, T., Yamamoto, H., and Hosono, R. (1998) Functional properties of the *unc-64* gene encoding a *Caenorhabditis elegans* syntaxin. *J. Biol. Chem.*, **273**, 2192.
75. Chalfie, M., Tu, Y., Euskirchen, G., Ward, W. W., and Prasher, D. C. (1994) Green fluorescent protein as a marker for gene expression. *Science*, **263**, 802.
76. Dwyer, N. D., Troemel, E. R., Sengupta, P., and Bargmann, C. I. (1998) Odorant receptor localization to olfactory cilia is mediated by ODR-4, a novel membrane-associated protein. *Cell*, **93**, 455.
77. Jorgensen, E. M., Hartwieg, E., Schuske, K., Nonet, M. L., Jin, Y., and Horvitz, H. R. (1995) Defective recycling of synaptic vesicles in synaptotagmin mutants of *Caenorhabditis elegans*. *Nature*, **378**, 196.
78. Rongo, C., Whitfield, C. W., Rodal, A., Kim, S. K., and Kaplan, J. M. (1998) LIN-10 is a shared component of the polarized protein localization pathways in neurons and epithelia. *Cell*, **94**, 751.
79. Nonet, M. L. (1998) Visualization of presynaptic terminal specializations in live *C. elegans* using synaptic vesicle protein-GFP fusions. *J. Neurosci. Methods*, **89**, in press.
80. Nonet, M. L., Holgado, A. M., Brewer, F., Serpe, C. J., Norbeck, B. A., Holleran, J. *et al.*

(1999) UNC-11, a *C. elegans* AP180 homolog, regulates the size and protein composition of synaptic vesicles. *Mol. Biol. Cell*, **10**, 2343.

81. Hallam, S. J. and Jin, Y. (1998) *lin-14* regulates the timing of synaptic remodelling in *Caenorhabditis elegans*. *Nature*, **395**, 78.
82. White, J. G., Southgate, E., and Thomson, J. N. (1992) Mutations in the *Caenorhabditis elegans unc-4* gene alter the synaptic input to ventral cord motor neurons. *Nature*, **355**, 838.
83. Hall, D. H. and Hedgecock, E. M. (1991) Kinesin-related gene *unc-104* is required for axonal transport of synaptic vesicles in *C. elegans*. *Cell*, **65**, 837.
84. Hall, D. (1995) Electron microscopy and three-dimensional image reconstruction. In *Caenorhabditis elegans: Modern biological analysis of an organism* (ed. H. F. Epstein and D. C. Shakes), Vol. 48, p. 395. Academic Press, San Diego.
85. Walrond, J. P., Kass, I. S., Stretton, A. O., and Donmoyer, J. E. (1985) Identification of excitatory and inhibitory motoneurons in the nematode Ascaris by electrophysiological techniques. *J. Neurosci.*, **5**, 1.
86. Raizen, D. M. and Avery, L. (1994) Electrical activity and behaviour in the pharynx of *Caenorhabditis elegans*. *Neuron*, **12**, 483.
87. Raizen, D. M., Lee, R. Y. N., and Avery, L. (1995) Interacting genes required for pharyngeal excitation by motor neuron MC in *Caenorhabditis elegans*. *Genetics*, **141**, 1365.
88. Dent, J. A., Davis, M. W., and Avery, L. (1997) *avr-15* encodes a chloride channel subunit that mediates inhibitory glutamatergic neurotransmission and ivermectin sensitivity in *Caenorhabditis elegans*. *EMBO J.*, **16**, 5867.
89. Avery, L. (1993) Motor neuron M3 controls pharyngeal muscle relaxation timing in *Caenorhabditis elegans*. *J. Exp. Biol.*, **175**, 283.
90. Iwasaki, K., Staunton, J., Saifee, O., Nonet, M. L., and Thomas, J. (1997) *aex-3* encodes a novel regulator of presynaptic activity in *C. elegans*. *Neuron*, **18**, 613.
91. Fasshauer, D., Otto, H., Eliason, W. K., Jahn, R., and Brunger, A. (1997) Structural changes are associated with soluble *N*-ethylmaleimide-sensitive fusion protein attachment protein receptor complex formation. *J. Biol. Chem.*, **272**, 28036.
92. Sutton, R. B., Fasshauer, D., Jahn, R., and Brunger, A. T. (1998) Crystal structure of a SNARE complex involved in synaptic exocytosis at 2.4 Å resolution. *Nature*, **395**, 347.
93. Hanson, P. I., Roth, R., Morisaki, H., Jahn, R., and Heuser, J. E. (1997) Structure and conformational changes in NSF and its membrane receptor complexes visualized by quick-freeze/deep-etch electron microscopy. *Cell*, **90**, 523.
94. Bennett, M. K., Garcia-Arraras, J. E., Elferink, L. A., Peterson, K., Fleming, A. M., Hazuka, C. D., *et al.* (1993) The syntaxin family of vesicular transport receptors. *Cell*, **74**, 863.
95. Ibaraki, K., Horikawa, H. P., Morita, T., Mori, H., Sakimura, K., Mishina, M., *et al.* (1995) Identification of four different forms of syntaxin 3. *Biochem. Biophys. Res. Commun.*, **211**, 997.
96. Laage, R. and Langosch, D. (1997) Dimerization of the synaptic vesicle protein synaptobrevin (vesicle-associated membrane protein) II depends on specific residues within the transmembrane segment. *Eur. J. Biochem.*, **249**, 540.
97. van Swinderen, B., Saifee, O., Shebester, L. D., Robertson, R. S., Nonet, M. L., and Crowder, C. M. (1999) A neomorphic syntaxin mutation blocks volatile anesthetic action in *Caenorhabditis elegans*. *Proc. Natl. Acad. Sci. USA*, **96**, 2479.
98. Maclver, M. B., Mikulec, A. A., Amagasu, S. M., and Monroe, F. A. (1996) Volatile anesthetics depress glutamate transmission via presynaptic actions. *Anesthesiology*, **85**, 823.

99. Hosono, R., Hekimi, S., Kamiya, Y., Sassa, T., Murakami, S., Nishiwaki, K., *et al.* (1992) The *unc-18* gene encodes a novel protein affecting the kinetics of acetylcholine metabolism in the nematode *Caenorhabditis elegans*. *J. Neurochem.*, **58**, 1517.
100. Hata, Y., Slaughter, C. A., and Südhof, T. C. (1993) Synaptic vesicle fusion complex contains *unc-18* homologue bound to syntaxin. *Nature*, **366**, 347.
101. Pevsner, J., Hsu, S. C., and Scheller, R. H. (1994) n-Sec1: a neural-specific syntaxin-binding protein. *Proc. Natl. Acad. Sci. USA*, **91**, 1445.
102. Calakos, N., Bennett, M. K., Peterson, K. E., and Scheller, R. H. (1994) Protein-protein interactions contributing to the specificity of intracellular vesicular trafficking. *Science*, **263**, 1146.
103. Okamoto, M. and Südhof, T. (1997) Mints, Munc18-interacting proteins in synaptic vesicle exocytosis. *J. Biol. Chem.*, **272**, 31459.
104. Sulston, J. E. and Horvitz, H. R. (1981) Abnormal cell lineages in mutants of the nematode *Caenorhabditis elegans*. *Dev. Biol.*, **82**, 41.
105. Schiavo, G., Gmachi, M. J. S., Stenbeck, G., Sollner, T. H., and Rothman, J. E. (1995) A possible docking and fusion particle for synaptic transmission. *Nature*, **378**, 733.
106. Clary, D. O., Griff, I. C., and Rothman, J. E. (1990) SNAPs, a family of NSF attachment proteins involved in intracellular membrane fusion in animals and yeast. *Cell*, **61**, 709.
107. Song, I., Kamboj, S., Xia, J., Dong, H., Liao, D., and Huganir, R. L. (1998) Interaction of the *N*-ethylmaleimide-sensitive factor with AMPA receptors. *Neuron*, **21**, 393.
108. Osten, P., Srivastava, S., Inman, G. J., Vilim, F. S., Khatri, L., Lee, L. M., *et al.* (1998) The AMPA receptor GluR2 C terminus can mediate a reversible, ATP-dependent interaction with NSF and alpha- and beta-SNAPs. *Neuron*, **21**, 99.
109. Nishimune, A., Isaac, J. T., Molnar, E., Noel, J., Nash, S. R., Tagaya, M., *et al.* (1998) NSF binding to GluR2 regulates synaptic transmission. *Neuron*, **21**, 87.
110. Schafer, W. R., Sanchez, B. M., and Kenyon, C. J. (1996) Genes affecting sensitivity to serotonin in *Caenorhabditis elegans*. *Genetics*, **143**, 1219.
111. Sheng, Z. H., Rettig, J., Cook, T., and Catterall, W. A. (1996) Calcium-dependent interaction of N-type calcium channels with the synaptic core complex. *Nature*, **379**, 451.
112. Sheng, Z. H., Westenbroek, R. E., and Catterall, W. A. (1998) Physical link and functional coupling of presynaptic calcium channels and the synaptic vesicle docking/fusion machinery. *J. Bioenerg. Biomembr.*, **30**, 335.
113. Lee, R. Y., Lobel, L., Hengartner, M., Horvitz, H. R., and Avery, L. (1997) Mutations in the alpha1 subunit of an L-type voltage-activated Ca^{2+} channel cause myotonia in *Caenorhabditis elegans*. *EMBO J.*, **16**, 6066.
114. Perez-Reyes, E., Cribbs, L. L., Daud, A., Lacerda, A. E., Barclay, J., Williamson, M. P., *et al.* (1998) Molecular characterization of a neuronal low-voltage-activated T-type calcium channel. *Nature*, **391**, 896.
115. Perin, M. S., Fried, V. A., Mignery, G. A., Jahn, R., and Südhof, T. C. (1990) Phospholipid binding by a synaptic vesicle protein homologous to the regulatory region of protein kinase C. *Nature*, **345**, 260.
116. Verhage, M., de Vries, K. J., Roshol, H., Burbach, J. P., Gispen, W. H., and Südhof, T. C. (1997) DOC2 proteins in rat brain: complementary distribution and proposed function as vesicular adapter proteins in early stages of secretion. *Neuron*, **18**, 453.
117. Brose, N., Hofmann, K., Hata, Y., and Sudhof, T. C. (1995) Mammalian homologues of *Caenorhabditis elegans unc-13* gene define novel family of C2-domain proteins. *J. Biol. Chem.*, **270**, 25273.

118. Wang, Y., Okamoto, M., Schmitz, F., Hofmann, K., and Südhof, T. C. (1997) Rim is a putative Rab3 effector in regulating synaptic-vesicle fusion. *Nature*, **388**, 593.
119. Shirataki, H., Kaibuchi, K., Sakoda, T., Kishida, S., Yamaguchi, T., Wada, K., *et al.* (1993) Rabphilin-3A, a putative target protein for smg p25A/rab3A p25 small GTP-binding protein related to synaptotagmin. *Mol. Cell Biol.*, **13**, 2061.
120. Li, C., Ullrich, B., Zhang, J. Z., Anderson, R. G., Brose, N., and Südhof, T. C. (1995) Ca($^{2+}$)-dependent and -independent activities of neural and non-neural synaptotagmins. *Nature*, **375**, 594.
121. Nguyen, M., Alfonso, A., Johnson, C. D., and Rand, J. B. (1995) *Caenorhabditis elegans* mutants resistant to inhibitors of acetylcholinesterase. *Genetics*, **140**, 527.
122. Zhang, J. Z., Davletov, B. A., Südhof, T. C., and Anderson, R. G. W. (1994) Synaptotagmin I is a high affinity receptor for clathrin AP-2: implications for membrane recycling. *Cell*, **78**, 751.
123. Betz, A., Ashery, U., Rickmann, M., Augustin, I., Neher, E., Sudhof, T. C., *et al.* (1998) Munc13-1 is a presynaptic phorbol ester receptor that enhances neurotransmitter release. *Neuron*, **21**, 123.
124. Betz, A., Okamoto, M., Benseler, F., and Brose, N. (1997) Direct interaction of the rat *unc-13* homologue Munc-13-1 with the N-terminus of syntaxin. *J. Biol. Chem.*, **272**, 2520.
125. Ann, K., Kowalchyk, J. A., Loyet, K. M., and Martin, T. F. (1997) Novel Ca^{2+}-binding protein (CAPS) related to UNC-31 required for Ca^{2+}-activated exocytosis. *J. Biol. Chem.*, **272**, 19637.
126. Livingstone, D. (1991) Studies on the *unc-31* gene of *Caenorhabditis elegans*. Ph. D. dissertation, University of Cambridge.
127. Avery, L., Bargmann, C. I., and Horvitz, H. R. (1993) The *Caenorhabditis elegans unc-31* gene affects multiple nervous system-controlled functions. *Genetics*, **134**, 455.
128. Berwin, B., Floor, E., and Martin, T. F. (1998) CAPS (mammalian UNC-31) protein localizes to membranes involved in dense-core vesicle exocytosis. *Neuron*, **21**, 137.
129. Fischer von Mollard, G., Mignery, G. A., Baumert, M., Perin, M. S., Hanson, T. J., Burger, P. M., *et al.* (1990) rab3 is a small GTP-binding protein exclusively localized to synaptic vesicles. *Proc. Natl. Acad. Sci. USA*, **87**, 1988.
130. Ferro-Novick, S. and Novick, P. (1993) The role of GTP-binding proteins in transport along the exocytic pathway. *Annu. Rev. Cell Biol.*, **9**, 575.
131. Wada, M., Nakanishi, H., Satoh, A., Hirano, H., Obaishi, H., Matsuura, Y., *et al.* (1997) Isolation and characterization of a GDP/GTP exchange protein specific for the Rab3 subfamily small G proteins. *J. Biol. Chem.*, **272**, 3875.
132. Sasaki, T., Kikuchi, A., Araki, S., Hata, Y., Isomura, M., Kuroda, S., *et al.* (1990) Purification and characterization from bovine brain cytosol of a protein that inhibits the dissociation of GDP from and the subsequent binding of GTP to smg p25A, a ras p21-like GTP-binding protein. *J. Biol. Chem.*, **265**, 2333.
133. Fukui, K., Sasaki, T., Imazumi, K., Matsuura, Y., Nakanishi, H., and Takai, Y. (1997) Isolation and characterization of a GTPase activating protein specific for the Rab3 subfamily of small G proteins. *J. Biol. Chem.*, **272**, 4655.
134. Nagano, F., Sasaki, T., Fukui, K., Asakura, T., Imazumi, K., and Takai, Y. (1998) Molecular cloning and characterization of the noncatalytic subunit of the Rab3 subfamily-specific GTPase-activating protein. *J. Biol. Chem.*, **273**, 24781.
135. Brundage, L., Avery, L., Katz, A., Kim, U. J., Mendel, J. E., Sternberg, P. W., *et al.* (1996) Mutations in a *C. elegans* Gqalpha gene disrupt movement, egg laying, and viability. *Neuron*, **16**, 999.

136. Koelle, M. R. and Horvitz, H. R. (1996) EGL-10 regulates G protein signaling in the *C. elegans* nervous system and shares a conserved domain with many mammalian proteins. *Cell*, **84**, 115.
137. Bejanin, S., Cervini, R., Mallet, J., and Berrard, S. (1994) A unique gene organization for two cholinergic markers, choline acetyltransferase and a putative vesicular transporter of acetylcholine. *J. Biol. Chem.*, **269**, 21944.
138. Erickson, J. D., Varoqui, H., Schafer, M. K., Modi, W., Diebler, M. F., Weihe, E., *et al.* (1994) Functional identification of a vesicular acetylcholine transporter and its expression from a 'cholinergic' gene locus. *J. Biol. Chem.*, **269**, 21929.
139. McCaffery, C. A. and DeGennaro, L. J. (1986) Determination and analysis of the primary structure of the nerve terminal specific phosphoprotein, synapsin I. *EMBO J.*, **5**, 3167.
140. Duerr, J. S., Frisby, D. L., Gaskin, J. G., Duke, A., Asermely, K., Huddleston, D., *et al.* (1999) The *cat-1* gene of *C. elegans* encodes a vesicular monoamine transporter required for specific monoamine-dependent behaviours. *J. Neurosci.*, **19**, 72.
141. Perin, M. S., Fried, V. A., Stone, D. K., Xie, X. S., and Sudhof, T. C. (1991) Structure of the 116-kDa polypeptide of the clathrin-coated vesicle/synaptic vesicle proton pump. *J. Biol. Chem.*, **266**, 3877.
142. Ahle, S. and Ungewickell, E. (1986) Purification and properties of a new clathrin assembly protein. *EMBO J.*, **5**, 3143.
143. Morris, S. A., Schroder, S., Plessmann, U., Weber, K., and Ungewickell, E. (1993) Clathrin assembly protein AP180: primary structure, domain organization and identification of a clathrin binding site. *EMBO J.*, **12**, 667.
144. Ye, W. and Lafer, E. M. (1995) Clathrin binding and assembly activities of expressed domains of the synapse-specific clathrin assembly protein AP-3. *J. Biol. Chem.*, **270**, 10933.
145. Ye, W. and Lafer, E. M. (1995) Bacterially expressed F1-20/AP-3 assembles clathrin into cages with a narrow size distribution: implications for the regulation of quantal size during neurotransmission. *J. Neurosci. Res.*, **41**, 15.
146. McPherson, P. S., Garcia, E. P., Slepnev, V. I., David, C., Zhang, X., Grabs, D., *et al.* (1996) A presynaptic inositol-5-phosphatase. *Nature*, **379**, 353.
147. Hao, W., Tan, Z., Prasad, K., Reddy, K. K., Chen, J., Prestwich, G. D., *et al.* (1997) Regulation of AP-3 function by inositides. Identification of phosphatidylinositol 3,4,5-trisphosphate as a potent ligand. *J. Biol. Chem.*, **272**, 6393.
148. Gaidarov, I., Chen, Q., Falck, J. R., Reddy, K. K., and Keen, J. H. (1996) A functional phosphatidylinositol 3,4,5-trisphosphate/phosphoinositide binding domain in the clathrin adaptor AP-2 alpha subunit. Implications for the endocytic pathway. *J. Biol. Chem.*, **271**, 20922.
149. Andrews, J., Smith, M., Merakovsky, J., Coulson, M., Hannan, F., and Kelly, L. E. (1996) The stoned locus of *Drosophila melanogaster* produces a dicistronic transcript and encodes two distinct polypeptides. *Genetics*, **143**, 1699.
150. Stimson, D. T., Estes, P. S., Smith, M., Kelly, L. E., and Ramaswami, M. (1998) A product of the *Drosophila stoned* locus regulates neurotransmitter release. *J. Neurosci.*, **18**, 9638.
151. Ohno, H., Stewart, J., Fournier, M. C., Bosshart, H., Rhee, I., Miyatake, S., *et al.* (1995) Interaction of tyrosine-based sorting signals with clathrin-associated proteins. *Science*, **269**, 1872.
152. Lee, J., Jongeward, G. D., and Sternberg, P. W. (1994) *unc-101*, a gene required for many aspects of *Caenorhabditis elegans* development and behaviour, encodes a clathrin-associated protein. *Genes Dev.*, **8**, 60.

153. van der Bliek, A. M. and Meyerowitz, E. M. (1991) Dynamin-like protein encoded by the *Drosophila shibire* gene associated with vesicular traffic. *Nature*, **351**, 411.
154. Poodry, C. and Edgar, L. (1979) Reversible alteration in the neuromuscular junctions of *Drosophila melanogaster* bearing a temperature-sensitive mutation, *shibire*. *J. Cell Biol.*, **81**, 520.
155. Smirnova, E., Shurland, D. L., Ryazantsev, S. N., and van der Bliek, A. M. (1998) A human dynamin-related protein controls the distribution of mitochondria. *J. Cell Biol.*, **143**, 351.
156. Stenius, K., Janz, R., Südhof, T. C., and Jahn, R. (1995) Structure of synaptogyrin (p29) defines novel synaptic vesicle protein. *J. Cell Biol.*, **131**, 1801.
157. Leube, R. E., Kaiser, P., Seiter, A., Zimbelmann, R., Franke, W. W., Rehm, H., *et al.* (1987) Synaptophysin: molecular organization and mRNA expression as determined from cloned cDNA. *EMBO J.*, **6**, 3261.
158. De Camilli, P., Benfenati, F., Valtorta, F., and Greengrad, P. (1990) The synapsins. *Annu. Rev. Cell Biol.*, **6**, 433.
159. Brand, S. H. and Castle, J. D. (1993) SCAMP 37, a new marker within the general cell surface recycling system. *EMBO J.*, **12**, 3753.
160. Brand, S. H., Laurie, S. M., Mixon, M. B., and Castle, J. D. (1991) Secretory carrier membrane proteins 31–35 define a common protein composition among secretory carrier membranes. *J. Biol. Chem.*, **266**, 18949.
161. Zinsmaier, K. E., Hofbauer, A., Heimbeck, G., Pflugfelder, G. O., Buchner, S., and Buchner, E. (1990) A cysteine-string protein is expressed in retina and brain of *Drosophila*. *J. Neurogenet.*, **7**, 15.
162. Janz, R., Hofmann, K., and Sudhof, T. C. (1998) SVOP, an evolutionarily conserved synaptic vesicle protein, suggests novel transport functions of synaptic vesicles. *J. Neurosci.*, **18**, 9269.
163. Feany, M. B., Lee, S., Edwards, R. H., and Buckley, K. M. (1992) The synaptic vesicle protein SV2 is a novel type of transmembrane transporter. *Cell*, **70**, 861.
164. Bajjalieh, S. M., Peterson, K., Shinghal, R., and Scheller, R. H. (1992) SV2, a brain synaptic vesicle protein homologous to bacterial transporters. *Science*, **257**, 1271.
165. Fujita, Y., Shirataki, H., Sakisaka, T., Asakura, T., Ohya, T., Kotani, H., *et al.* (1998) Tomosyn: a syntaxin-1-binding protein that forms a novel complex in the neurotransmitter release process. *Neuron*, **20**, 905.
166. Maroteaux, L., Campanelli, J. T., and Scheller, R. H. (1988) Synuclein: a neuron-specific protein localized to the nucleus and presynaptic nerve terminal. *J. Neurosci.*, **8**, 2804.
167. Skehel, P. A., Martin, K. C., Kandel, E. R., and Bartsch, D. (1995) A VAMP-binding protein from *Aplysia* required for neurotransmitter release. *Science*, **269**, 1580.
168. Lelianova, V. G., Davletov, B. A., Sterling, A., Rahman, M. A., Grishin, E. V., Totty, N. F., *et al.* (1997) Alpha-latrotoxin receptor, latrophilin, is a novel member of the secretin family of G protein-coupled receptors. *J. Biol. Chem.*, **272**, 21504.
169. Bean, A. J., Seifert, R., Chen, Y. A., Sacks, R., and Scheller, R. H. (1997) Hrs-2 is an ATPase implicated in calcium-regulated secretion. *Nature*, **385**, 826.
170. Geppert, M., Bolshakov, V. Y., Siegelbaum, S. A., Takei, K., De Camilli, P., Hammer, R. E., *et al.* (1994) The role of Rab3A in neurotransmitter release. *Nature*, **369**, 493.
171. Geppert, M., Goda, Y., Hammer, R. E., Li, C., Rosahl, T. W., Stevens, C. F., *et al.* (1994) Synaptotagmin I: a major Ca^{2+} sensor for transmitter release at a central synapse. *Cell*, **79**, 717.
172. Reist, N. E., Buchanan, J., Li, J., DiAntonio, A., Buxton, E. M., and Schwarz, T. L. (1998)

Morphologically docked synaptic vesicles are reduced in *synaptotagmin* mutants of *Drosophila*. *J. Neurosci.*, **18**, 7662.

173. Bennett, M. K., Calakos, N., and Scheller, R. H. (1992) Syntaxin: a synaptic protein implicated in docking of synaptic vesicles at presynaptic active zones. *Science*, **257**, 255.
174. McMahon, H. T., Bolshakov, V. Y., Janz, R., Hammer, R. E., Siegelbaum, S. A., and Sudhof, T. C. (1996) Synaptophysin, a major synaptic vesicle protein, is not essential for neurotransmitter release. *Proc. Natl. Acad. Sci. USA*, **93**, 4760.
175. Johnstone, D. B., Wei, A., Butler, A., Salkoff, L., and Thomas, J. H. (1997) Behavioural defects in *C. elegans egl-36* mutants result from potassium channels shifted in voltage-dependence of activation. *Neuron*, **19**, 151.
176. Treinin, M. and Chalfie, M. (1995) A mutated acetylcholine receptor subunit causes neuronal degeneration in *C. elegans*. *Neuron*, **14**, 871.
177. Liu, J., Schrank, B., and Waterston, R. H. (1996) Interaction between a putative mechanosensory membrane channel and a collagen. *Science*, **273**, 361.
178. Collins, F. S., Patrinos, A., Jordan, E., Chakravarti, A., Gesteland, R., and Walters, L. (1998) New goals for the U.S. Human genome project: 1998–2003. *Science*, **282**, 682.
179. Jorgensen, E. M. and Nonet, M. L. (1995) Neuromuscular junctions in the nematode *C. elegans*. *Semin. Dev. Biol.*, **6**, 207.
180. Alfonso, A., Grundahl, K., Duerr, J. S., Han, H. P., and Rand, J. B. (1993) The *Caenorhabditis elegans unc-17* gene: a putative vesicular acetylcholine transporter. *Science*, **261**, 617.
181. Otsuka, A. J., Jeyaprakash, A., Garcia-Anoveros, J., Tang, L. Z., Fisk, G., Hartshorne, T., *et al.* (1991) The *C. elegans unc-104* gene encodes a putative kinesin heavy chain-like protein. *Neuron*, **6**, 113.

9 Dissecting the molecular mechanisms of neurotransmitter release in *Drosophila*

GIUSEPPA PENNETTA, MARK N. WU, and HUGO J. BELLEN

1. Introduction

Most intercellular communication in nervous systems relies on the release of neurotransmitters at chemical synapses. A detailed understanding of the presynaptic mechanisms underlying neurotransmitter release may not only provide basic insights into regulated secretion, but may also help us in the design of therapeutic agents to modulate this process. It has become apparent in the past five years that the vast majority of the proteins implicated in neurotransmitter release are extremely well conserved in the animal kingdom. Hence, model systems amenable to sophisticated genetic analyses such as *C. elegans* and *Drosophila*, are playing an important role in improving our understanding of the function of these proteins in vertebrates.

This chapter will focus on the molecular machinery required for synaptic vesicle (SV) trafficking and exocytosis, with an emphasis on proteins that are required for docking, priming, or fusion of SVs. Biochemical purification of synaptic proteins in vertebrate systems (Chapter 3), and genetic analyses in yeast, *C. elegans* (Chapter 8), and *Drosophila*, have identified a cast of molecular players that have been implicated in vesicle trafficking. To study the function of these proteins in neurotransmitter release *in vivo*, mutations in several of the genes encoding these proteins have been characterized in *Drosophila*. Here, we will discuss the advantages of using *Drosophila* as a model system to study neurotransmitter release, and the recent discoveries related to the function of specific proteins required for this process in *Drosophila*.

2. *Drosophila* as a model system

2.1 Introduction

Many attributes of *Drosophila* make it an attractive system to study the molecular mechanisms of neurotransmitter release. First, fruitflies undergo rapid development, which allows for large scale genetic screens. Embryos hatch to become 1st instar larvae in less than one day, and these larvae go through a series of moults to progress through 2nd and 3rd instar larval stages in about two to three days. After undergoing pupation, adult flies emerge about five days later (1). Secondly, elegant genetic manipulations can be used to generate many different point mutations in specific genes (2). The distinct phenotypes associated with different alleles often provide valuable information on the mechanisms of protein action. In addition, temperature-sensitive point mutations in genes required for neurotransmitter release allow for elegant experimental designs and protocols that are not possible in other species (3–6) (Chapter 11). Thirdly, several strategies have been developed to overexpress proteins at specific developmental times or in specific tissues or cells in the fruitfly. These strategies can be used to overexpress wild-type or mutant proteins in wild-type or mutant genetic backgrounds. These studies, in conjunction with loss-of-function studies, provide further information on the function of a protein *in vivo*. Fourthly, a variety of electrophysiological assays can be performed at embryonic and larval neuromuscular junctions (NMJs), in order to quantify the precise defects caused by the absence of or the presence of mutated proteins (7). Finally, the electrophysiological studies at the NMJ can be complemented with dye uptake experiments (Chapter 2) and transmission electron microscopy data analyses (8). The availability of this array of manipulations and assays at a single synapse provides a powerful combination of tools to study the specific function of proteins in neurotransmitter release *in vivo*.

2.2 Reverse genetics in *Drosophila*

The primary advantage of using *Drosophila* to study a biological process such as neurotransmitter release is the ability to carry out sophisticated genetic analyses. By studying the effects or 'phenotypes' of the absence or presence of an altered protein, one can draw conclusions about the normal function of this protein *in vivo*. One of the unique features about the neurotransmitter release field in the fruitfly is that it has thus far mainly relied on reverse genetic approaches, unlike most other fruitfly research areas. Most mutations in genes discussed in this chapter were not identified in genetic screens designed to isolate mutations which affect neurotransmitter release. Rather, they were isolated because the investigators were trying to obtain specific mutations in genes encoding proteins that have been implicated in neurotransmitter release in vertebrates or *C. elegans*. This approach is named 'reverse', as there is no known phenotype prior to the isolation of the mutations. It is likely that 'reverse genetics' will become more prevalent in all areas of *Drosophila* research in the near future, because the availability of the sequence of the entire *Drosophila* genome by the year 2000 will prompt many researchers to try to define the *in vivo* functions of evolutionarily conserved proteins.

The first step in the 'reverse genetics' process is to isolate and sequence the fruitfly gene encoding the protein of interest. Unfortunately, it is not yet possible to generate targeted mutations through homologous recombination in *Drosophila* (Chapter 10). Hence, the following strategies are typically followed. The gene is cytological mapped to polytene chromosome bands, which typically permits mapping to within a 50–100 kb interval. Upon searching the *Drosophila* database for available mutations (`http://flybase.bio.indiana.edu`), it may be possible to identify previously existing mutations in the gene of interest. There are two major ways to generate mutations: chemical mutagens and P-elements. Chemical mutagens such as ethyl methyl sulfonate (EMS) often create point mutations in genes, which can range in severity from mild partial loss-of-function mutations to null (complete absence of protein) mutations. As an example of identifying a previously generated EMS induced mutation, the gene encoding dNSF-1 was mapped to a cytological band where a previously characterized temperature-sensitive paralytic mutation, *comatose*, had been mapped by meiotic recombination (9–11). The other typical way that a mutation in the gene of interest can be obtained is by identifying a P-element in the gene (12). P-elements are transposable elements which often integrate into the 5′ regulatory region of a gene and can therefore interfere with the expression of that gene (13). The genomic fragments next to the P-element can then be cloned and sequenced to determine whether the P-element is integrated into the gene of interest. For example, mutations in *syntaxin* (14) and the AP180 *Drosophila* homologue, *lap* (15), were identified using this strategy. If a P-element causes a mutation in the gene, more mutations can be created by imprecise excision of the transposable element. If the P-element is inserted in a nearby gene (less than 50–100 kb), a local hopping strategy combined with a variety of molecular methods can be attempted to recover a P-element in the gene (16). This strategy permitted the isolation of a P-element insertion in *synaptotagmin* (17), *cysteine string protein* (18), *syntaxin* (19), and *synaptobrevin* (20). The advantage of the latter strategy is that no assumptions are made about the phenotype associated with mutations in the gene. Even if the loss of the gene does not cause a phenotype, mutants can still be isolated. The disadvantage of this approach is that local hopping often causes deletions instead of local hops, thereby removing adjacent genes (21).

If no P-elements are available, or if one wants to create new or additional alleles, the following strategy is usually pursued. More than 90% of the polytene chromosome bands are uncovered by one or more 'deficiencies' that are available from the *Drosophila* Stock Centers. These 'deficiencies' are fly strains that carry a chromosome that is missing a defined region. When this strategy to isolate mutations in the gene of interest is used, a basic assumption must be made, i.e. mutations will cause lethality, semi-lethality, or another easily identifiable phenotype. A chemical mutagenesis screen is then initiated to isolate mutations in all the essential/semi-lethal/identifiable genes that are uncovered by the deletion. Subsequently, all heterozygous flies (mutants/+) are crossed to each other, and mutations that fail to complement each other (e.g. are lethal in combination with one another) are assigned to the same 'complementation group' or gene. Once all the genes have been identified, rescue of the phenotype associated with a given complementation group is attempted with a transgene containing the cDNA or a genomic fragment of the gene for which mutants are desired. The different mutations or alleles of a complementation group that are rescued by the transgene then form the allelic series of the gene. This series provides the basis for a detailed phenotypic analysis. A caveat of this strategy is that removal of the gene may not cause a lethal phenotype or an easily observable phenotype. Nevertheless, this approach allowed isolation of mutations in *synaptotagmin* (17, 22, 23), *rop* (24), *synaptobrevin* (20), *syntaxin* (25), and *hrs* (M. W., T. Lloyd, and H. J. B., unpublished).

The ability to generate different kinds of mutations is important since different aspects of protein function can be revealed by different mutations. For example, generation of a 'null' mutation allows one to study the effects of the absence of the gene. On the other hand, such a phenotype may be so severe that it obscures some functions of the protein, which may be uncovered by studying a partial loss-of-function mutation that affects a specific domain of a protein. In addition, specific alleles may allow one to screen for mutations in other genes which interact with a specific gene. Hence, the geneticist always strives to collect a wealth of different mutations in the same gene.

2.3 Overexpression of proteins

A complementary approach to the analysis of loss-of-function mutations is the analysis of transgenic flies that overexpress a specific protein to determine its effect on neurosecretion. Overexpressed full-length protein may cause a positive, negative, or no effect on neurotransmitter release, as outlined in Chapter 7. However, without other biochemical or genetic information, the significance of an overexpression phenotype can be difficult to interpret (for caveats see Chapter 7). Alternatively, toxins that cleave specific proteins involved in neurotransmitter release can be expressed to mimic a loss-of-function phenotype, such as the clostridial neurotoxins (see Chapter 6 and the synaptobrevin section below).

Two strategies are typically used to overexpress a protein in *Drosophila*. In the first, a cDNA is cloned in a P-element vector under the transcriptional control of a promoter that is inducible by heat shock. This vector is integrated into the germline of a *Drosophila* embryo, and transgenic flies bearing this construct are obtained. To induce protein expression, a one hour heat shock at 37 °C, followed by a one hour rest period to allow protein synthesis and transport, typically leads to three- to tenfold induction above the wild-type protein level (26). This paradigm is followed by functional assays described in Sections 2.4 and 2.5. The major drawback of this strategy is that the protein is expressed in all tissues, not just at the synapse, and hence may affect the postsynaptic environment as well.

The second strategy uses a two component system. Many fly strains exist in the fly community that express the yeast GAL4 transcription factor in specific tissues, e.g. all neurons, or all muscles (27). By crossing these flies to flies that contain a transgene in which the binding sites of the GAL4 transcription factor (UAS sites) are upstream of the gene of interest, one can target the expression of the gene to specific tissues. The levels of overexpression vary greatly since they depend on the level, and/or activity of GAL4 produced by the driver construct. If the protein levels are too high, the overexpression can be lowered by keeping the flies at 18 °C as the GAL4 protein is more active at 29 °C than at 18 °C (2). One caveat of this approach is that the protein is often expressed for a protracted amount of time prior to the functional assays. This expression may cause lethality or allow for compensatory adaptations. However, many more GAL4 'driver' lines are becoming available, allowing not only improved spatial control but also better temporal control (12).

2.4 The morphology of the neuromuscular junction or NMJ

NMJs contain a robust ensemble of synapses that can be induced to secrete more than 10000 vesicles within a few minutes. Although the physiological properties of NMJs are similar in vertebrates and invertebrates, their morphological appearances and neurotransmitters are different (see Chapter 1 and below). In addition, fly NMJs develop at a spectacular rate when compared to vertebrate NMJs: there is a dramatic increase in the complexity, size, and number of boutons and active zones between the embryonic and the third instar larval stage. For example, the largest embryonic muscles (muscle 6 and 7) have about 17 boutons, each containing about two active zones. Two days later these muscles have approximately 180 boutons, many containing more than ten active zones (28). Hence, the number of active zones increases 50-fold in two days.

The nerve endings innervating the body wall muscles can be morphologically divided into two categories: type 1 endings, with short localized terminal branches and relatively large presynaptic release sites or boutons (averaging 3–4 μm in diameter); and, type 2 endings, with relatively long filaments and small boutons (1 μm in diameter). Large type I boutons contain about 40 active zones, whereas small type II boutons contain only 5–10 active zones (29). As shown in Fig. 1, active zones contain T-bars, which are composed of a stem (about 50 nm in height), and a bar-like roof separated by a very narrow gap; clear SVs (40 nm in diameter) are concentrated and docked at these sites. These T-bars are thought to be comparable with active zones of vertebrates NMJs. Although SVs are clustered at T-bars or active zones, each bouton also contains thousands of SVs that are thought to represent a 'reserve' pool (Fig. 1). These SVs are distributed over the entire volume of the bouton. Clear SVs are typically filled with L-glutamate, the excitatory neurotransmitter at the *Drosophila* NMJ (30). Most NMJs also contain large dense core vesicles that are filled with other neurotransmitters (8).

The pre- and postsynaptic membranes of a NMJ appear electron dense over a stretch of several hundred nanometres and are separated by a regular cleft of about 15 nm. The synaptic cleft contains electron-opaque material, which is strikingly

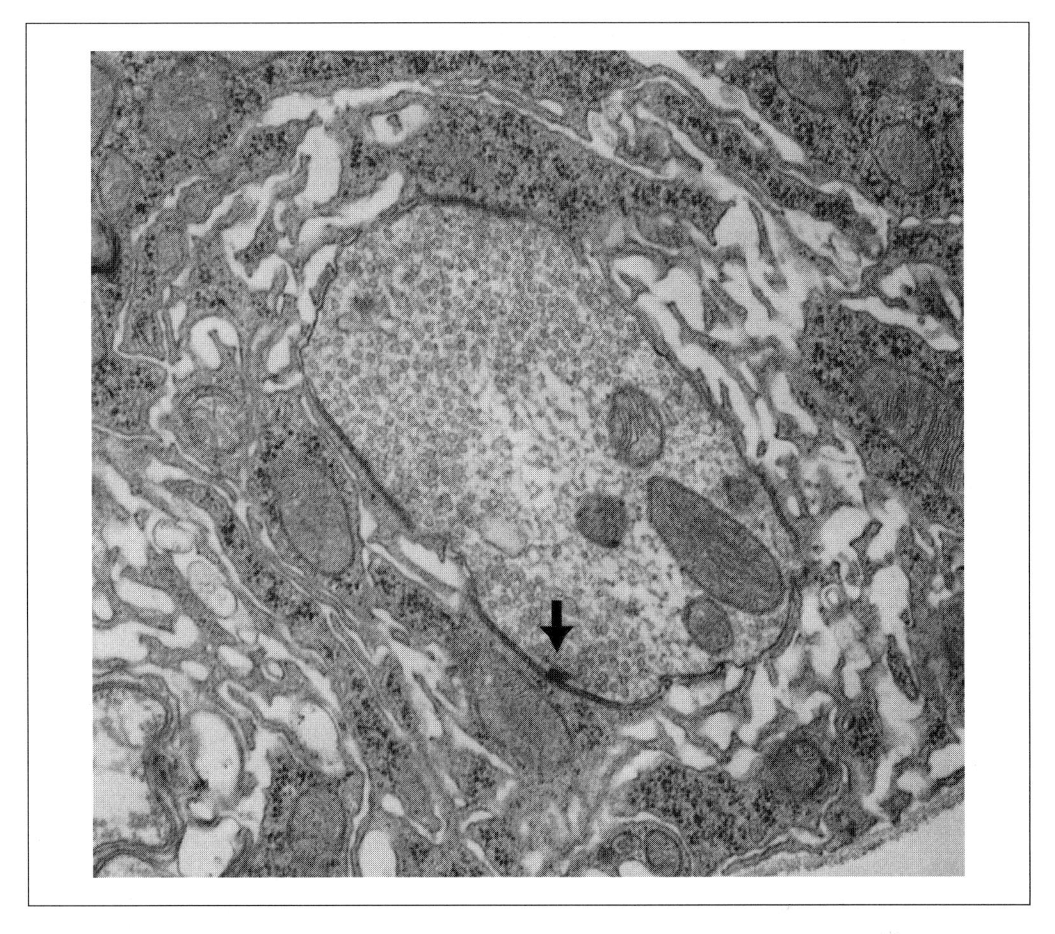

Fig. 1 Transmission electron micrograph of a wild-type synaptic bouton of a neuromuscular junction (NMJ) of a *Drosophila* third instar larva. Note the presynaptic density (arrow) and the associated cluster of SVs. (Provided by Vivian Budnik, University of Amherst.)

periodic in structure. Transmission electron microscopy studies of embryonic and third instar larval NMJs are relatively easy to perform and permit a detailed description of the number, size, and distribution of the clear SVs, as well as other ultrastructural aspects of NMJ synapses (8). As shown below, ultrastructural analyses have proven to provide some of the most valuable information in the analysis of mutants implicated in neurotransmitter release.

The NMJs most commonly chosen to characterize electrophysiological and morphological abnormalities in neurotransmitter release are those of muscles 6, 7, 12, and 13 because they are easily accessible, and are the simplest NMJs in the abdomen. Furthermore, they have been well characterized both physiologically and morphologically in the larva (30–33) and in the embryo (7, 34). Finally, their size and position make them amenable to a variety of experimental manipulations.

2.5 Electrophysiological paradigms

The most important tools used to study neurotransmitter release are electrophysiological assays. Electrophysiological recordings can be performed in embryos, larvae, pupae, and adults (for review see ref. 7). In adult flies, for example, electroretinograms (ERGs) can be performed by flashing light onto the fly eye and recording the depolarization of the photoreceptors as well as on/off transients. The on/off transients are typically used as a measure of synaptic transmission between the photoreceptors and the first-order interneurons (35). However, compared to NMJ preparations, ERGs are a relatively crude assay for studying synaptic transmission.

The loss of many synaptic proteins causes late embryonic lethality. Therefore, the embryonic NMJ preparation is extremely valuable as one can characterize the effects of the loss of essential genes at native synapses. However, given the technical difficulty of the embryonic NMJ recordings, most electrophysiologists prefer to carry out experiments at the third instar NMJ. This preparation is preferred when mutants survive until or beyond the third instar stage. Many of the mutants that affect neurotransmitter release survive as third instar larvae because even dramatic reductions of neurotransmitter release do not seem to significantly affect viability. A reduction of 60–80% in the size of the evoked response at the third instar NMJ has been documented for several viable mutants (15, 17, 36, 37). Yet few of these mutants are viable as adults, possibly because adult behaviour is much more complex than larval behaviour. *Drosophila* larvae therefore seem less sensitive to perturbations in neurotransmitter release than adult flies and in this respect may be comparable to *C. elegans*. However, fly larvae tend to be more sensitive to perturbations in proteins affecting neurotransmitter release than worms (Chapter 8).

The main physiological properties that are typically studied at the embryonic or larval NMJs are the size and shape of the evoked response (after nerve stimulation) and the size and frequency of spontaneous vesicle fusion events, also called 'minis' (Fig. 2). Both are measured postsynaptically as synaptic currents or as changes in membrane potential. Embryonic recordings are made using a whole-cell patch-clamp technique (7). The muscles are typically recorded from in a voltage-clamp configuration at –60 to –80 mV, to measure the synaptic current of the muscles. In the third instar larva, recordings are typically carried out by intracellular recordings using a single electrode to measure the de- and re-polarization of muscles rather than the current fluxes. Two electrode voltage-clamp muscle recordings must be used to measure synaptic currents at the third instar larva NMJ. The advantage of keeping the muscle membrane potential constant by voltage-clamping is that other factors such as voltage-dependent channels do not contribute to the evoked response. In addition, the linear current increase allows one to measure the kinetic properties of release as well.

In the absence of stimulation, one can record the depolarization or current fluxes caused by the fusion of a single vesicle with the presynaptic membrane (a 'mini'). The size of these fusion events is an important parameter for the interpretation of electrophysiological phenotypes. An alteration in 'mini' size can be due to a pre- or a

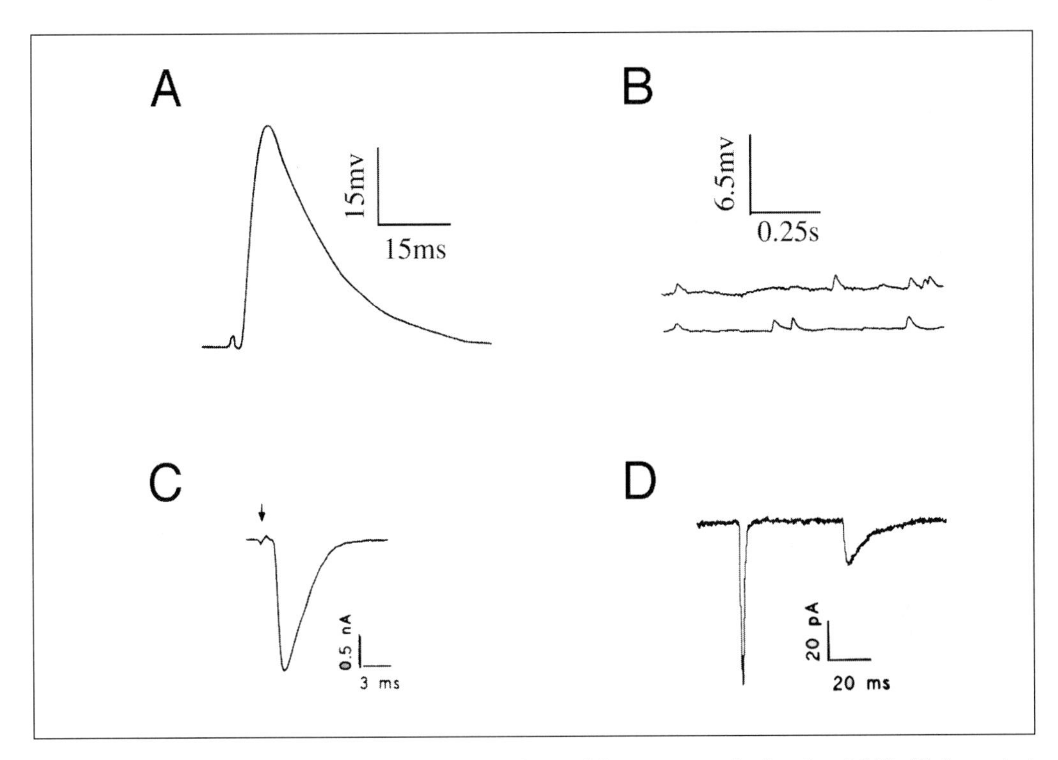

Fig. 2 Typical electrophysiological recordings at the *Drosophila* neuromuscular junction (NMJ). (A) An evoked response causes an excitatory junctional potential (EJP). (B) Spontaneous neurotransmitter release or MEJPs. (C) Excitatory junctional current (EJC) caused by an evoked response (voltage-clamp mode). (D) Minis or MEJCs. Note that at the wild-type (wt) NMJ there are two classes of MEJCs: class 1 with a larger amplitude and a faster time course (*left*), and class 2 with a smaller amplitude and slower time course (*right*). (From refs 107 and 120; courtesy of Cell Press.)

postsynaptic defect. The most likely cause for a presynaptic defect is that the amount of neurotransmitter in SVs differs from normal, e.g. the vesicles are larger or smaller, or more or less neurotransmitter is packaged in vesicles than normal (Chapters 5 and 11). Alternatively, it is possible that the fusion machinery is altered such that SVs are unable to fully release all of the packaged neurotransmitters, or that the kinetic properties of release have been altered. If the nature of the altered 'mini' size is postsynaptic in origin, a possible cause is the abnormal distribution or number of the postsynaptic neurotransmitter receptors, i.e. glutamate receptors. A postsynaptic defect can be confirmed using L-glutamate iontophoresis (7). By puffing neurotransmitter on the postsynaptic membrane and measuring the muscle response, a qualitative assessment of the ability of the muscles to respond to neurotransmitter release can be obtained. Given that a 'mini' is the basic unit used to measure the total amount of vesicles that are released, it is important to ensure that the size of the 'minis' is known in order to properly measure the quantal content or number of SVs that are released upon an evoked response.

The evoked response provides a measure of the amount of neurotransmitter release upon nerve stimulation. To elicit an evoked response, the segmental nerves are severed and drawn into a glass suction electrode. The nerve is then stimulated to measure excitatory junctional currents or EJCs (in voltage-clamp mode), or excitatory junctional potentials or EJPs. The total number of vesicles released, or quantal content, can then be calculated by dividing the size of the evoked or EJP response by the size of the 'minis' after correction for non-linearity using a simple equation (38). The number of vesicles that fuse at the NMJ to elicit a normal evoked response vary greatly during development of a muscle, i.e. from 5–10 late in embryogenesis, to about 60 in the mature third instar at an extracellular Ca^{2+} concentration of 1.8 mM (31, 39).

3. Determining the function of *Drosophila* proteins in neurotransmitter release

3.1 Introduction

Ca^{2+} entry through presynaptic voltage gated Ca^{2+} channels triggers evoked neurotransmitter release. Given the physiological significance of the Ca^{2+} signal, we will first summarize our knowledge of the *Drosophila* proteins that have been implicated in Ca^{2+} signalling. These proteins include Ca^{2+} channels, Synaptotagmin, and Cysteine String Protein or CSP (according to *Drosophila* nomenclature, protein names always start with a capital letter). These proteins closely interact with a set of proteins called the core complex that have been implicated in the fusion of SVs with the presynaptic membrane. These include Synaptobrevin, Syntaxin, and SNAP-25, and will be discussed in a second section. In the third section, proteins that are known to interact with proteins of the core complex and are thought to regulate their function, e.g. ROP and NSF, will be discussed. Finally, other proteins implicated in neurotransmission but whose role is less well understood, will be mentioned. A summary of the proteins discussed in this chapter is presented in Table 1.

3.2 The calcium signalling machinery

3.2.1 Introduction

Classical electrophysiological work has demonstrated the requirement for Ca^{2+} in neurotransmitter release (40) (Chapter 1). More recent biophysical and electrophysiological studies have emphasized the high concentrations ($> 100\ \mu M$) and fast kinetics ($< 200\ \mu sec$) of the Ca^{2+} signal (41, 42). These biophysical requirements imply that the molecular machinery required for synaptic vesicle exocytosis is tethered near Ca^{2+} channels. After the influx of Ca^{2+} through the channel, it is thought that a presynaptic protein or protein complex with a low affinity for Ca^{2+} acts as a Ca^{2+} sensor, initiating the fusion reaction of the SV with the presynaptic membrane, leading to transmitter release. Classical studies performed by Dodge and Rahaminoff (43) at the frog NMJ showed a non-linear relationship between Ca^{2+} and

Table 1 *Drosophila* proteins discussed in this chapter[a]

Protein	Lethality	Subcellular localization	Evoked response	Spontaneous response	Ultrastructural analysis	Functions in non-neuronal secretion	Functions in neural secretion	References
Calcium-channel alpha 1 subunit type A	Embryonic	?	↓	?	?	?	Required for depolarization of photoreceptors and for synaptic transmission	57, 58, 59
Calmodulin	Larval	?	↑ At low $[Ca^{2+}]$ in the presence of quinidine	Normal	?	Yes	Decreases synaptic depolarization	63, 64, 65
Frequenin	None	Synapse	↑ At high frequ.stimulat	Normal	?	?	Modulates synaptic efficacy	51, 68–69
Synaptotagmin	Embryonic and larval	SV	↓	↑	Decrease in docked SVs	No	Ca^{2+} sensor, implicated in docking and endocytosis	17, 36, 47 73, 74, 76–78, 90, 95, 96
Cysteine string proteins	Variable (escapers)	SV	↓	Normal	Decrease in SVs (?)	Yes (?)	Controls Ca^{2+} influx	18, 97, 104. 105, 107, 113–115
Syntaxin	Embryonic	Presynap membrane	Blocked	Blocked	Increase in docked SVs	Yes	Essential for fusion of SVs	4, 14, 19, 25, 120, 132
n-Synaptobrevin	Embryonic	SV	Blocked	Decrease in frequency	Increase in docked SVs	No	Essential for induced exocytosis	20, 22, 120, 147, 148
Rop	Embryonic	Cytoplasm	↑ or ↓	↓	?	Yes	Rate-limiting regulator of release with positive and negative roles	24, 37, 172, 173
dNSF1	ts paralytic	?	↓	?	Increase in docked SVs	?	Priming of SVs for release	4, 6, 9–11, 181–183
Stoned A	ts lethal	Presynap. terminals	↓	↑	Normal	?	Required for evoked response and excitation-secretion fidelity, decreases spontaneous release. May be involved in endocytosis	187–189, 191

[a] SV, synaptic vesicle; ts, temperature-sensitive.

transmitter release, namely, that transmitter release is dependent upon the fourth power of the Ca^{2+} concentration in the extracellular bath. This implies that four Ca^{2+} ions act in a co-operative manner to cause release of a single SV. Numerous electrophysiological experiments in many different synapses of different species (44, 45), including *Drosophila* NMJs (46) have confirmed, with few exceptions (47), this third- or fourth-order power relationship. In addition, Augustine and Charlton (44) have demonstrated that there is a fourth-order relationship between the presynaptic inward Ca^{2+} current and the evoked response, indicating that the Ca^{2+} sensor must act after Ca^{2+} influx in the presynaptic terminal. Given the speed of neurotransmitter release following Ca^{2+} entry (< 200 μsec) (48, 49) it is thought that very few or no intermediary steps can occur between Ca^{2+} entry and fusion of the SV with the presynaptic membrane. Hence, the most favoured models are those in which Ca^{2+} binding causes a conformational change of a protein or protein complex. The most obvious candidates for a Ca^{2+} sensor should thus be able to bind three or four Ca^{2+} ions in a co-operative manner, and be localized near Ca^{2+} channels and SVs at the active zone. At least seven potential Ca^{2+} binding proteins have been localized at synapses, including Calmodulin (50), Frequenin (51), Synaptotagmin (52), rabphilin (53), Doc2α (54), munc-13 (55), and Rim (56). Since the latter four proteins have not yet been described in *Drosophila*, we will focus on Calmodulin, Frequenin, and Synaptotagmin. We will also discuss other proteins involved in the generation of the Ca^{2+} signal: Ca^{2+} channels and cysteine string proteins.

3.2.2 Calcium channels

Little is known about neuronal Ca^{2+} channels in *Drosophila*. On the basis of the kinetic properties of Ca^{2+} channels in cultured *Drosophila* neurons, and their different sensitivities to a spider toxin, Leung and Byerly (57) have suggested that there are at least two types of neuronal calcium channels in *Drosophila* neurons. Recently, the cloning of the alpha1 subunit of Ca^{2+} channels (Dmca1A) has been reported (58) and shown to be encoded by a gene, which when defective causes abnormal courtship (*cacophony*), defects in vision (*night-blindA*), and lethality (*L13*) (59). For the sake of simplicity we will call this gene *cacophony*. The Dmca1A subunit is preferentially expressed in the nervous system. Electroretinograms (ERGs) showed a variety of defects in *cacophony* mutants, demonstrating a role for the channel subunit in the depolarization of photoreceptor cells. In addition, the ERGs also revealed defects in the on/off transients, suggesting a synaptic transmission defect when the channel is impaired (59). Recently, another Calcium channel alpha1 subunit has been cloned in *Drosophila* and it has been designated Dmca1D (60). Severe loss-of-function mutations in both *Dmca1* cause lethality at late embryonic stages, while weaker mutants die as pharate adults (61). Unfortunately, electrophysiological characterization of both mutants at the NMJ is not available yet.

3.2.3 Calmodulin

Calmodulin is a ubiquitous protein that binds four Ca^{2+} ions. It is highly conserved, and modulates the action of many cellular signalling proteins (for review see ref. 62), including Ca^{2+}/Calmodulin-dependent protein kinase II at *Drosophila* nerve terminals (63). Although null mutations in Calmodulin are first instar lethal, partial loss-of-function mutations have been generated (64), allowing a detailed electrophysiological characterization at the third instar larval NMJ of one allelic combination (65). These studies show no primary defects in mini amplitude, shape, and frequency, nor do they identify defects of evoked responses or quantal content at various Ca^{2+} concentrations under standard recording conditions. However, a noticeable increase in neurotransmitter release is observed at low extracellular Ca^{2+} concentrations in the presence of quinidine, a drug which blocks the delayed rectifier K^+ current in the larval muscle. Application of this drug enhances the effects of ion channel mutations such as Shaker and Hyperkinetic (66), and the authors therefore propose a role for Calmodulin in decreasing nerve terminal depolarization, or Ca^{2+} buffering, and not in Ca^{2+} sensing. Although the precise role of Calmodulin in neurotransmitter release is not well understood in *Drosophila*, a recent study by Peters and Mayer (67) examined the requirements for calmodulin in homotypic (between identical membranes) vacuolar fusion in yeast using an *in vitro* assay. They concluded that calmodulin is required as a Ca^{2+} sensor in this system. Whether Calmodulin acts as a Ca^{2+} sensor at synapses remains to be seen, although the evidence from *Drosophila* does not currently support a role in Ca^{2+} sensing.

3.2.4 Frequenin

The gene encoding Frequenin is one of the few examples where a 'forward' genetic approach led to the isolation of a gene involved in neurotransmitter release. A translocation between two chromosomes was originally recognized to cause a hyperexcitable phenotype, i.e. a single stimulus created more than one action potential (68–70). Similarly, larval NMJs showed increased neurotransmitter release when compared to wild-type under high frequency stimulation (68). Hence, the gene was named *Frequenin* as the mutation caused a frequency-dependent facilitation of neurotransmitter release at the NMJ (51).

The gene has been shown to encode a Ca^{2+} binding protein with four EF-hand motifs, exhibiting 41% homology with recoverin, a Ca^{2+}-dependent activator of guanylyl cyclase in bovine rod cells (71, 72). Frequenin has also been shown to maximally activate guanylyl cyclase in the absence of Ca^{2+}, and very low (nanomolar) amounts of Ca^{2+} inhibited Frequenin activity. The protein has been localized to the neuropil of the CNS and to boutons of NMJs, providing further evidence that it may play an important role in neurotransmitter release. Pongs *et al.* (51) also demonstrated that the phenotype induced by the *Frequenin* mutation was due to an overexpression of transcript and protein. The authors concluded that Frequenin may function as a Ca^{2+} sensor that facilitates neurotransmitter release, and may modulate excitation–secretion, enhancing the sensitivity of a synapse to an incoming stimulus. Unfortunately, the loss-of-function phenotype associated with the gene is unknown, and, as pointed out previously, drawing conclusions on the function of a protein *in vivo* without the loss-of-function phenotype is difficult.

3.2.5 Synaptotagmin

Drosophila Synaptotagmin (Syt) is an abundant and evolutionarily conserved integral membrane protein of SVs (73, 74). Biochemical studies have identified several functional domains shown in Fig. 3A:

- an intravesicular amino-terminal domain that is glycosylated
- a single transmembrane region
- two C2 domains (protein kinase C-homologous repeats that mediate Ca^{2+}-associated membrane interactions)
- a conserved carboxy-terminal sequence (52, 74).

Biochemical and molecular data suggest that Synaptotagmin might participate in different synaptic functions, including Ca^{2+} sensing, docking, endocytosis, and fusion (for review see refs 73 and 75).

RNA localization studies showed that syt message is localized to all or most *Drosophila* neurons (Fig. 3B). Immunohistochemical studies using an anti-Synaptotagmin antibody that specifically recognizes all known Synaptotagmin I proteins from *C. elegans* to rat, demonstrated that Synaptotagmin is specifically localized to NMJs and to the neuropil of the central nervous system of *Drosophila* embryos, larvae, and adults (Fig. 3C). *Drosophila* Synaptotagmin is only localized to synapses, consistent with a specific role in synaptic transmission (76).

To assess the *in vivo* role of Synaptotagmin, mutagenesis screens were carried out using both EMS and 'local hopping' of a P-element, allowing the isolation of more than 15 different alleles (17, 23, 77). A null allele of *synaptotagmin* was generated by imprecise excision of the P-element inserted in the gene, while the chemically-induced alleles yielded mutations of varying severity. Lack of Synaptotagmin protein causes a severe reduction in co-ordinated muscle contractions in embryos. These co-ordinated contractions are required for hatching of the embryo from the egg case, and hence most *synaptotagmin* mutant embryos are unable to hatch, although occasionally a first instar will escape (17, 77). Partial loss-of-function alleles cause severe sluggishness in third instar larvae, and a complete lack of co-ordination in surviving adults. These behavioural defects, combined with the synaptic localization of the protein, and the absence of morphological defects at NMJ synapses when viewed at the light microscopy level (78), strongly suggests that the protein is required for synaptic transmission.

Complementation tests between strong loss-of-function alleles showed intragenic complementation between several homozygous lethal alleles. In other words, specific mutations were lethal when homozygous or when tested over a deficiency, but were viable in combination with each other (36). This phenomenon is often observed in genes that encode proteins that contain different functional domains and/or form homomultimeric protein complexes (79). The defect associated with one particular allele in one functional domain is often partially rescued by a mutation in another functional domain, allowing partial restoration of function. Alternatively,

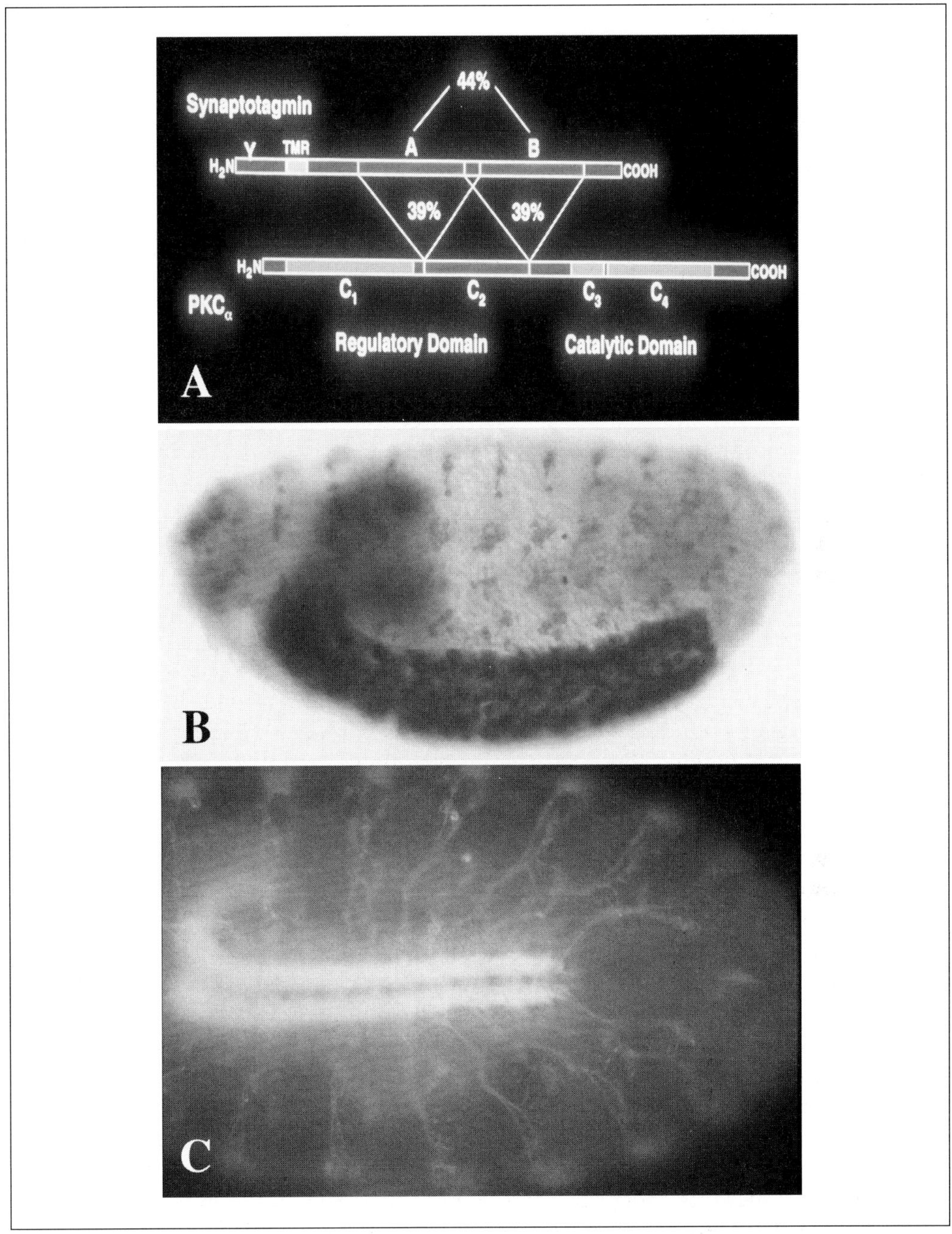

Fig. 3 Synaptotagmin structure and pattern of expression. (A) Domain structure of Synaptotagmin. TMR, transmembrane region; Y, glycosylation site; A and B, C2 domains. (B) *In situ* hybridization of whole mount embryos using a digoxygenin-labelled *sytnaptotagmin* cDNA as a probe. (C) Immunohistochemical staining of embryos with an anti-Synaptotagmin antibody (green and yellow, see colour plate section) and Monoclonal Ab. 22C10 (red and orange). Note the green dots in the periphery. They correspond to NMJs. (From ref. 17, courtesy of the company of Biologists Limited). See colour plate section between pages 46 and 47.

di- or multimerization of different defective proteins may form a partially functional complex. The genetic evidence therefore supported previous biochemical studies indicating that Synaptotagmin has at least two functionally important domains: the C2A and C2B domains, and/or acts as a homotetramer *in vivo* (74).

Based on biochemical studies, synaptotagmin has been proposed to function as a Ca^{2+} sensor. Indeed, synaptotagmin has a low affinity for Ca^{2+} and binds Ca^{2+} in the presence of phospholipids in the 100 micromolar range, consistent with the physiological Ca^{2+} concentrations present at the active zone when SV exocytosis occurs (80, 81). Moreover, the C2A domain of synaptotagmin has been shown to bind Ca^{2+} in a co-operative manner with a Hill coefficient of 3 (82), a property similar to that of the Ca^{2+} sensor (43). In addition, biochemical evidence shows that synaptotagmin interacts with the Ca^{2+} channel, placing the protein at the proper site in the presynaptic environment (83–85). Hence, *in vitro* data suggest that synaptotagmin has the essential properties to act as a Ca^{2+} sensor *in vivo*.

Although these biochemical studies provide *in vitro* data suggesting that synaptotagmin is a Ca^{2+} sensor, the data discussed below contribute *in vivo* support for this hypothesis. Given the importance of this protein in the field, we will discuss the experimental data gathered in *Drosophila* in some detail. On the basis of experimental evidence *in vivo*, synaptotagmin has been proposed to be:

- required for endocytosis
- required for docking of SVs
- a Ca^{2+} sensor.

In the following sections we will critically review the evidence supporting each of these functions.

A role for synaptotagmin in endocytosis was suggested on the basis that synaptotagmin binds to the Adaptor Protein 2 complex (AP2) (86), a protein complex required for endocytosis in *Drosophila* (87) (Chapter 11). Mutations in the gene encoding synaptotagmin in *C. elegans* cause a dramatic decrease in the number of SVs, and it was therefore argued that the protein plays a role in endocytosis *in vivo* (88) (Chapter 8). Similar observations were also reported in the squid with antibody blocking experiments (89) (Chapter 7). Finally, a reduction of approximately 50% of the total number of SVs was also observed in *Drosophila* mutants that lack Synaptotagmin (90). Hence, the biochemical, genetic, and squid data are consistent with a role for synaptotagmin in endocytosis. The simplest interpretation of these data is to propose that SV endocytosis is more efficient in the presence of synaptotagmin, but does not absolutely require its presence in the patch of membrane that is endocytosed. In the absence of the protein, the signal to initiate endocytosis via AP2 or another adaptor protein is impaired at different rates in different species (Chapter 8 and 11). It has been proposed that the second C2 domain, C2B, is required for this process to occur normally (89). In contrast to these data, it should be noted that no decrease in the number of vesicles were observed in mice that lack synaptotagmin I (91) (Chapter 10), possibly because of the presence of other synaptotagmins (92).

Although these data imply that Synaptotagmin is required for efficient SV endocytosis, direct evidence for such a role is still lacking.

A role for synaptotagmin in SV docking was suggested on the basis of biochemical data showing that this SV integral membrane protein binds to two members of the presynaptic core complex: syntaxin (93) and SNAP-25 (94). The only *in vivo* data to support a role for Synaptotagmin in docking comes from a careful TEM analysis of *Drosophila* mutants. Reist *et al.* (90) showed that, although the total number of SVs is decreased by 50%, the number of docked SVs is decreased by 75% in *synaptotagmin* mutants when compared to proper controls. Hence, mutant embryos that lack synaptotagmin have only 25% of the wild-type number of SVs docked at the presynaptic active zone. The authors therefore conclude that synaptotagmin stabilizes docked vesicles (90).

The main question that remains to be addressed is whether the above defects account for the electrophysiological defects observed in *synaptotagmin* mutants in *Drosophila*? We will first summarize the electrophysiological observations, and then interpret and discuss the data in the context of different models. Figure 4 shows a summary of the key electrophysiological observations made in a variety of *Drosophila synaptotagmin* mutants (17, 47, 95). First, as shown in Fig. 4A, intracellular recordings at the NMJ of third instar larvae that are transheterozygous for two different partial loss-of-function alleles exhibit no neurotransmitter release at 0.4 mM extracellular Ca^{2+}, showing that the protein is required for neurotransmitter release. At higher Ca^{2+} concentrations, some release can be observed in these partial loss-of-function mutants. Interestingly, repeated stimulation does not cause a decrease in evoked response at 1.8 mM Ca^{2+}, even when stimulated for minutes. As shown in Fig. 4B, 4C, and 4D, all *synaptotagmin* mutants tested require higher extracellular Ca^{2+} than wild-type animals to elicit an evoked response. Hence, all *synaptotagmin* mutants exhibit a strong reduction of evoked release. Secondly, at the larval NMJ, specific mutants exhibit a Ca^{2+} dependence of release of 3.6, whereas others show a Ca^{2+} dependence of release 1.5 or 1.6, showing that mutations in *synaptotagmin* can alter the Ca^{2+} co-operativity required for neurotransmitter release (Fig. 4B). In contrast, as shown in Fig. 4D, mutant embryos that lack Synaptotagmin protein exhibit a Ca^{2+} dependence similar to wild-type embryos. In both cases the co-operativity is 1.8, not 3 or 4, as is observed in most synapses (43). This implies that a different mechanism for sensing Ca^{2+} may be used at the embryonic NMJ vs. the larval NMJ. Thirdly, as shown in Fig. 4E and 4F, the release properties of *synaptotagmin* mutants that lack the protein are quite peculiar. Although an action potential fails to cause release 60% of the time, 10% of the time one vesicle is released, whereas in the remainder 30% anywhere from two to ten vesicles are released. This demonstrates that the absence of the protein causes a high failure rate and a dramatic reduction in the fidelity of release. This is nicely illustrated in Fig. 4F in which a series of traces are shown. Note that the scale bars for the recordings in the mutant are one-tenth of the wild-type controls. Fourthly, the recordings also show that there is a significant delay of the onset of release upon stimulation when compared to wild-type embryos. In addition, instead of an evoked response resulting from the synchronized release of neurotransmitter,

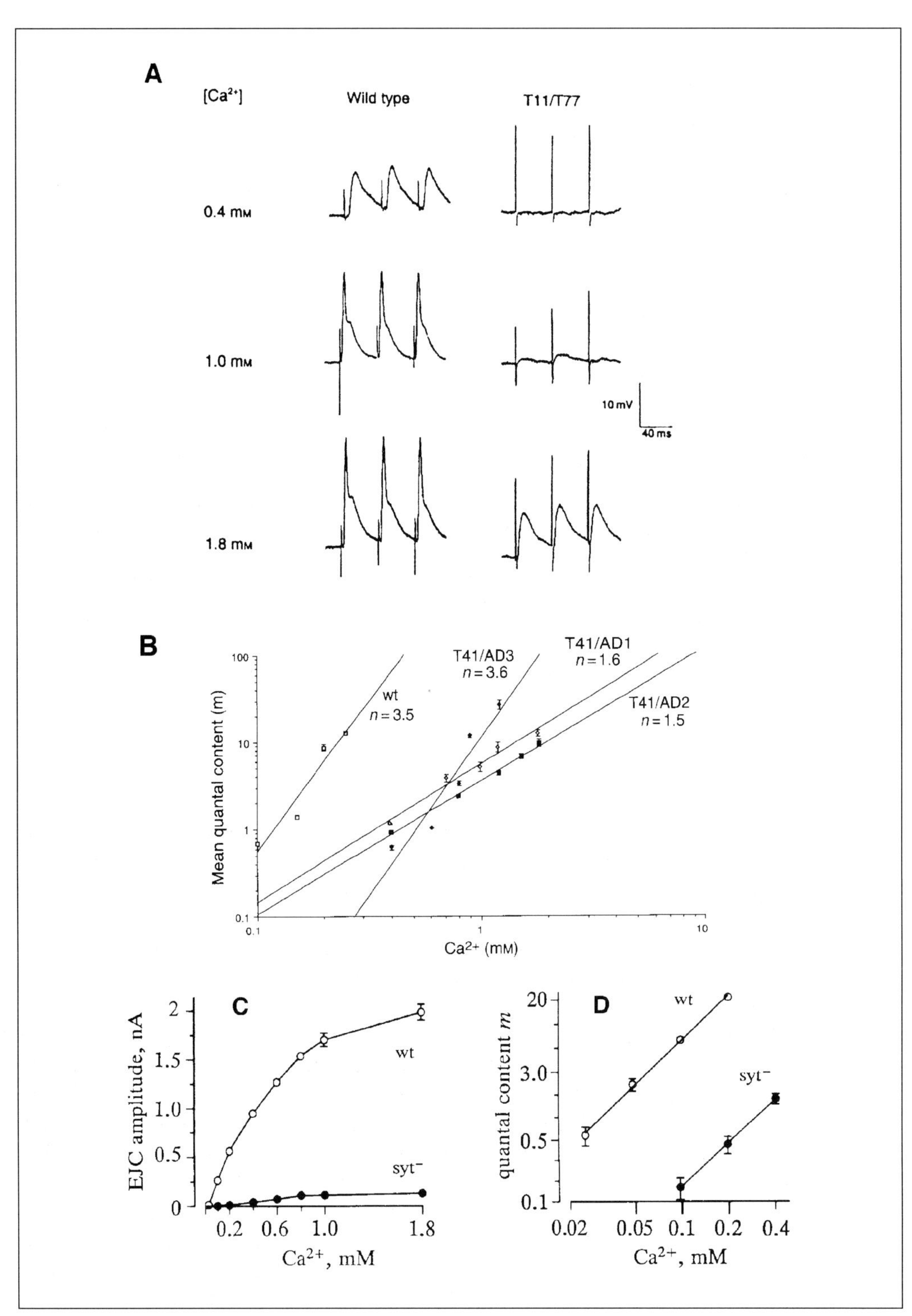
A
[Ca²⁺]
Wild type
T11/T77
0.4 mM
1.0 mM
10 mV
40 ms
1.8 mM
B
Mean quantal content (m)
wt
n = 3.5
T41/AD3
n = 3.6
T41/AD1
n = 1.6
T41/AD2
n = 1.5
Ca²⁺ (mM)
C
EJC amplitude, nA
wt
syt⁻
Ca²⁺, mM
D
quantal content m
wt
syt⁻
Ca²⁺, mM

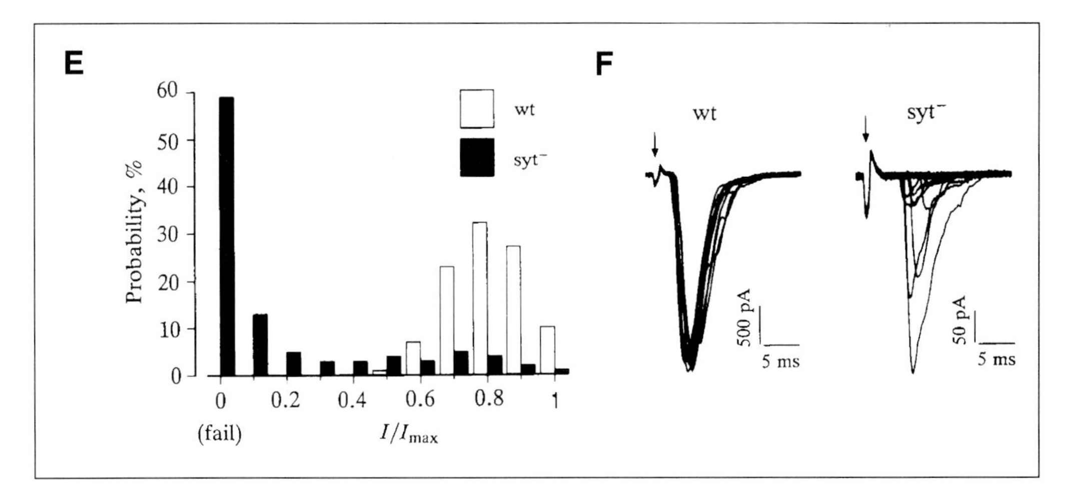

Fig. 4 Electrophysiological characterization of *synaptotagmin* mutants. (A) EJPs recorded at different Ca^{2+} concentrations in wild-type (*left*), and *synaptotagmin* mutants T11/T77 (*right*). (B) Ca^{2+} dependence of neurotransmitter release in *synaptotagmin* partial loss-of-function mutations. A double log plot of the mean quantal content at different Ca^{2+} concentrations is reported. (C) The mean amplitude of the evoked synaptic current is plotted versus the extracellular Ca^{2+} concentration in wild-type and mutants lacking *synaptotagmin* function. (D) The mean quantal content is reported relative to Ca^{2+} concentrations on a log-log plot for wild-type and null *synaptotagmin* mutations. (E) Current amplitude (I) is normalized against the maximal response (I_{max}) for wild-type and null *synaptotagmin* mutations, and plotted against the probability (%) of the response. Note the decrease in the amplitude and the greater variability of the evoked response in *synaptotagmin* mutants compared to the wild-type. (F) Superimposed current traces of evoked responses following repeated stimulation, from wild-type (wt) and *synaptotagmin* mutants. (From ref. 73 (A and B), courtesy of Elseviers; and ref. 47 (C–F), courtesy of Proceedings of the National Academy, USA.)

release is asynchronous, suggesting that the kinetics of release in response to a Ca^{2+} influx are altered. Finally, all *synaptotagmin* mutants tested have normal size 'minis', but the mini frequency is increased two- to fivefold. This finding suggests that Synaptotagmin may also act as a 'fusion clamp' to inhibit spontaneous SV fusion (Chapter 10).

The above data clearly show that in the absence of *synaptotagmin*, Ca^{2+} influx and vesicle releases are largely uncoupled. In other words, although neurotransmitter can be released, the amount and timing of neurotransmitter release is variable. These electrophysiological data, when combined with the localization and biochemical data showing that synaptotagmin binds Ca^{2+} in the 100 micromolar range (80) and with a co-operativity of 2–3 (82), argue that Synaptotagmin acts as the main Ca^{2+} sensor. However, it is likely that a less reliable and ineffective sensor is still present in *synaptotagmin* null mutant synapses. This sensor is presumably responsible for the remaining Ca^{2+} sensitivity in null mutant embryos. These hypotheses are further supported by the observation that the electrophysiological properties of an embryonic NMJ that lacks Synaptotagmin very much resembles that of a wild-type NMJ when recorded at the lowest possible extracellular Ca^{2+} concentration at which a response can still be elicited, i.e. 0.05 mM Ca^{2+} (47). These data, in conjunction with the previously mentioned biochemical data, demonstrate that Synaptotagmin is an excellent candidate for a key Ca^{2+} sensor at the synapse.

Reist *et al.* (90) have argued that a deficit in SV docking stability can account for the electrophysiological defects seen in *synaptotagmin* mutants. However, this appears unlikely for the following reasons. First, 25% of the SVs are still docked in null mutants, yet the amount of release that can be elicited is at best 5% of wild-type (Fig. 4C). Secondly, the presence of 25% of the normally docked vesicles does not explain the low fidelity and asynchronous release observed in the mutants unless other defects are invoked. Thirdly, the electrophysiological properties of at least five other mutants that play a role in neurotransmitter release have been described in *Drosophila* (see below release *syntaxin*, *rop*, *comatose*, *synaptobrevin*, and *cysteine string protein*), and none of these mutants resemble the phenotypes described for *synaptotagmin* mutants. In these five mutants as well as many others that were previously characterized, one does not observe a low fidelity of release and asynchronous release at 1.8 mM Ca^{2+}, as would be expected for a Ca^{2+} sensor. Hence, these features seem to be unique to *synaptotagmin* mutants. Fourthly, recently, *Drosophila lap* mutants have been isolated that show a TEM phenotype that is very similar to that described for *synaptotagmin* null mutants (15, 90) (Chapter 11), yet *lap* null mutants clearly display a much milder phenotype than *synaptotagmin* null mutants. Indeed, *lap* mutants do not exhibit the very low fidelity of release and highly asynchronous release observed in *synaptotagmin* mutants. Finally, some viable *synaptotagmin* mutants have been shown to exhibit an altered dependence on Ca^{2+}, further suggesting that Synaptotagmin is a Ca^{2+} sensor (for review see ref. 73). Some of these data have been disputed by Parfitt *et al.* (96), who argue that the Ca^{2+} concentrations were saturating and that the Ca^{2+} dependence is not altered in one of these mutants. However, other *synaptotagmin* mutants measured at the same 'saturating Ca^{2+} conditions' did exhibit a Ca^{2+} dependence of 3.5, showing that under these conditions, different *synaptotagmin* mutants display different Ca^{2+} dependencies. In summary, the evidence that Synaptotagmin is a Ca^{2+} sensor is compelling. However, the experimental data also support a role for this protein in other SV trafficking events as well, such as endocytosis and docking.

3.2.6 Cysteine string proteins

Cysteine String Proteins (CSPs) were initially identified by screening hybridoma libraries for *Drosophila* synapse-specific monoclonal antibodies (97). A single gene was shown to encode three isoforms (18). These proteins contained a contiguous string of 11 cysteine residues that gave the protein its name (97). A first glimpse of the potential function of CSP was revealed by Gundersen and Umbach (98) who attempted to identify subunits of a Ca^{2+} channel using suppression cloning in *Xenopus* oocytes. They identified a *Torpedo californica* homologue of the *Drosophila* CSPs and showed that the presence of CSP enhanced the level of the Ca^{2+} current of the ω conotoxin-sensitive Ca^{2+} channel when injected in *Xenopus* oocytes. They therefore proposed that CSP was either a subunit of a Ca^{2+} channel or a protein involved in targeting, turnover, or modulation of a Ca^{2+} channel. Homologues of

CSP have now been described in many vertebrates (99–102) but here we will mainly review the work in *Drosophila*. Although a variety of roles have been attributed to CSPs, including acting as a Ca^{2+} channel subunit or functioning in endocytosis (103), studies in the fruitfly suggest that CSP is specifically required for evoked release.

CSPs were found to be mostly associated with SVs in Torpedo (104) and *Drosophila* (105), not with the presynaptic membrane as expected for a subunit of a Ca^{2+} channel. Each SV has been estimated to harbour eight monomers with both the amino- and the carboxy-termini on the cytoplasmic face of the SV (106). These observations suggested that CSPs are not novel subunits of presynaptic calcium channels, but rather that they may participate in a regulatory interaction between synaptic vesicles and Ca^{2+} channels (104, 107). However, the nature and the function of this interaction remained elusive.

To determine the *in vivo* function of CSPs, Zinsmaier and colleagues (18) initiated a genetic analysis and created a series of P-element insertions and excisions allowing the removal of the gene. Interestingly, embryos that lack CSPs are semi-lethal and display a temperature-sensitive phenotype. Few adult escapers were observed at the permissive temperature (22 °C) while no survivors were obtained at the restrictive temperature (29 °C). Escapers that survive to adulthood showed a reduced life span (four to five days), became paralysed at 29 °C, and died. Prior to death, they exhibited several neurological symptoms, such as uncoordination, spasmic jumping, shaking, paralysis, and sluggishness (18). These behavioural data are consistent with a role for CSPs in neurotransmitter release.

Electrophysiological studies of *csp* mutants at the larval NMJ in *Drosophila* have provided the most convincing evidence that CSPs are involved in the regulation of the presynaptic release machinery (18, 107). As shown in Fig. 5A, at the permissive temperature, EJP amplitudes were reduced to about 50% of those of control. In addition, these NMJ synapses showed a predisposition to failures in evoked release, which was exacerbated by high frequency stimulation. At the restrictive temperature, a complete blockage of evoked neurotransmitter release is observed. However, as shown in Fig. 5A, spontaneous release persisted in *csp* mutants at all temperatures (107).

As shown in Fig. 5B, the Ca^{2+} dependence of neurotransmitter release is shifted to the right, indicating that more extracellular Ca^{2+} is required for a given stimulus. However, the slope, and hence the number of Ca^{2+} ions required to elicit release was not changed between mutants and wild-type controls (107). CSPs may therefore be involved in coupling the presynaptic membrane depolarization to neurotransmitter release. Since flies can survive the complete absence of CSP, the block of neurotransmission at higher temperatures indicates that CSP may also function in stabilizing the components of neurotransmission release machinery (107).

In 1993, CSPs were shown to contain another domain besides the cysteine-rich domain, namely a 'J' domain found in dnaJ proteins (108, 109). This domain mediates a co-operative interaction with heat shock protein hsp70, which function as molecular chaperones (108–110). The idea that CSP either alone or in concert with other proteins may act as a molecular chaperone was consistent with the proposal

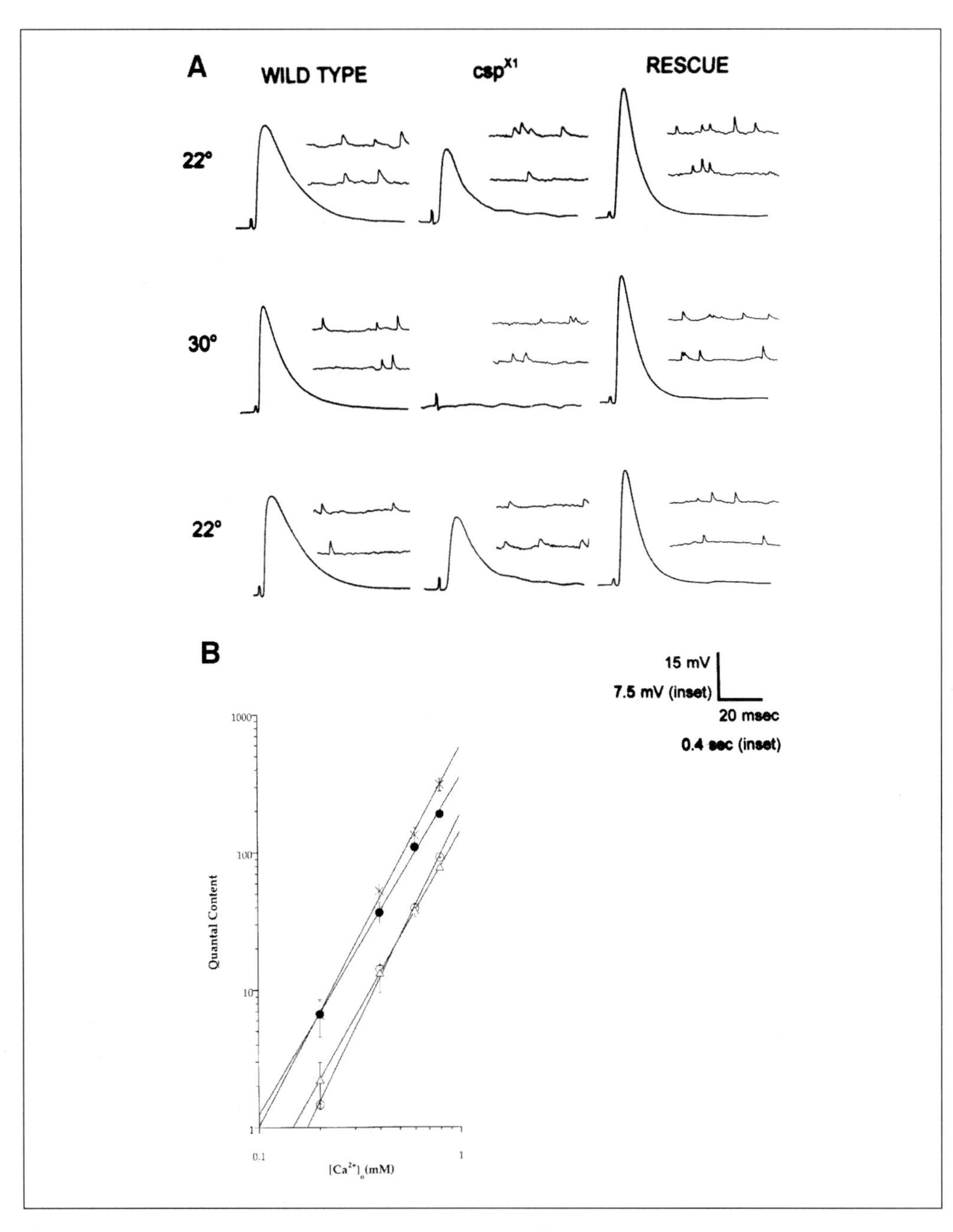

Fig. 5 Electrophysiological properties of *csp* mutants. (A) Each column shows evoked and spontaneous potentials (insets) recorded at 22 °C, after 15 min at 30 °C and 20 min after returning the temperature to 22 °C for wild-type (wt), *csp* mutants (csp^{x1}), and for mutant larvae 'rescued' by the expression of a wt copy of the gene. (B) Ca^{2+} dependence of quantal content: quantal content is plotted versus extracellular Ca^{2+} concentration for wt (crosses), for *csp* mutants: csp^{x1} (open circles), csp^{R1} (open triangles), and rescued larvae (closed circles). (From ref. 107; courtesy of Cell Press.)

that it may stabilize protein complexes involved in synaptic vesicle exocytosis. Interestingly, one member of the family of hsp70 chaperones is involved in uncoating clathrin-coated vesicles (111). Moreover, mammalian CSP has been shown to stimulate the ATPase activity of the mammalian HSC70 protein (99, 112). It was therefore proposed that CSPs might participate in the removal of the clathrin coat after endocytosis (Chapter 11). This hypothesis was strengthened by the observation that adult *csp* mutants exhibit a decreased number of synaptic vesicles at the presynaptic terminals of first-order interneurons of the adult visual system (18). The decreased number of synaptic vesicles at nerve endings may therefore result from a defect in vesicle membrane recycling.

To test the role of CSPs in endocytosis, Ranjan *et al.* (115) designed a series of elegant experiments using the FM1-43 tracer dye to show that endocytosis is not blocked at the restrictive temperature in *csp* mutants. In addition, they and others (113) showed that black widow spider venom (BWSV) or calcium ionophores can bypass the requirement for CSP and induce normal neurotransmitter release at the restrictive temperature in the mutants (114). Both of these treatments induce synaptic vesicle exocytosis in wild-type neurons, but are unable to do so in *syntaxin* and *synaptobrevin* mutants (see below). These data suggest that CSP is required at the level of Ca^{2+} ion entry and is therefore required upstream of the core complex (113, 115).

Recently, a biochemical interaction between CSP and the Ca^{2+} channel has been demonstrated (116), further suggesting a role for CSP in Ca^{2+}-coupled exocytosis. Finally, Umbach *et al.* (114) demonstrated a correlation of the temperature-sensitive block of neurotransmitter release in *csp* mutants with a reduction in Ca^{2+} entry using a Ca^{2+} dye indicator to monitor changes in cytosolic Ca^{2+} in the synaptic boutons of *csp* mutants. In summary, the current data indicate that the CSPs function at the level of the Ca^{2+} channel, i.e. Ca^{2+} entry prior to fusion.

3.3 The core complex

3.3.1 Introduction

The core complex consists of three proteins that form a stable, SDS-resistant *in vitro* complex (117) (Chapter 3). This complex consists of the SV integral membrane protein Synaptobrevin, and two presynaptic membrane-associated proteins, SNAP-25 and Syntaxin-1A. Vesicle-associated proteins like Synaptobrevin have also been named v-SNAREs, whereas the membrane-associated proteins have been named t-SNAREs (Chapter 3). The core complex (or SNARE complex) was originally proposed to mediate the docking of vesicles with their target membranes. The proposed model suggested that the specificity of the interaction between a vesicle and its target membrane might be imparted by the specific interaction between these proteins located on the vesicle and target membrane (118). However, recent studies from a variety of systems have provided evidence for a role of this complex after docking and perhaps in the fusion event itself (119–121).

The *Drosophila SNAP-25* gene has been identified and shown to exhibit 61% identity with the mouse SNAP-25 gene (122, 123). The protein is specifically localized to synapses (T. Lloyd and H. J. B., personal communication). However, since no mutant has yet been described in *Drosophila*, it will not be discussed any further. In contrast, mutations have been identified in *syntaxin* and *synaptobrevin* in *Drosophila*, and we will discuss the data from these studies that support a post-docking role for these proteins.

3.3.2 Syntaxin-1A

Syntaxin-1A is a 35 kDa plasma membrane protein which has been proposed to participate in a variety of events in synaptic vesicle exocytosis, including the targeting, docking, priming, or fusing of synaptic vesicles, as well as the tethering of the fusion machinery to Ca^{2+} channels (Chapter 3). The wide range of functions attributed to this protein can be explained in part by the observation that amongst all the synaptic proteins implicated in neurotransmission, syntaxin appears to have the most binding partners. For instance, syntaxin homologues have been shown to bind at least 11 other proteins, not including non-synaptic partners, and this list continues to grow (81, 93, 118, 124–131). Thus far, genetic analysis in *Drosophila* has mainly provided evidence for a function of this protein in the fusion of synaptic vesicles. In contrast to the large number of syntaxin genes found in vertebrates (17 thus far), only two syntaxin homologues have been identified in *Drosophila*. The plasma membrane homologue in *Drosophila* is Syntaxin-1A (Syx-1A) (14, 132), which shows 70% identity to its vertebrate counterpart. The other is dSED5, which is localized to the Golgi compartment (133). In this section, we will only discuss Syx-1A.

Immunostaining with anti-Syntaxin antibodies reveals that Syx-1A protein is not confined to neurons, but is also found in epithelial cells, suggesting that it may function in secretion of both neuronal and non-neuronal tissues (14, 120) (Chapter3). This expression pattern is distinct from Synaptotagmin, which is expressed specifically in neurons. In order to study the function of Syntaxin *in vivo*, a strong loss-of-function mutation was identified by obtaining a fly strain that contained a P-element 500 bp upstream of the *syx-1A* gene (*P{syx}*) (14). This mutation reduces protein expression to about 20% of wild-type levels. *P{syx}* mutants are embryonic lethal and are unable to hatch from the egg case since they are paralysed. To assess the effects of the complete absence of Syx-1A protein, a null mutant was generated by imprecisely excising the transposable P-element removing the majority of the *syx-1A* ORF. These mutants display a more severe phenotype compared to *P{syx}* mutants. First, null mutants show developmental defects, such as a failure to secrete cuticle, which is essentially an exoskeleton that is apically secreted by epidermal cells. Secondly, the gut does not fully develop. Thirdly, further studies have shown that Syx-1A is required for cellularization of the *Drosophila* embryo (19, 25). Together these data suggest that Syx-1A is normally required not only for neurosecretion, but also a variety of non-neuronal secretory events. Recently, a temperature-sensitive paralytic mutation in Syntaxin has been identified (4). Flies bearing this mutation survive as adults, but show behavioural defects at room temperature, such as a reduced ability to fly. Exposure of these mutants to 38 °C causes paralysis within seconds suggesting an impairment of synaptic transmission. The point mutation causing this temperature-sensitive paralysis inhibits the interaction of Syntaxin with Synaptobrevin and SNAP-25 at the restrictive temperature, supporting an important role for Syntaxin and the core complex in synaptic transmission (4).

Based largely on biochemical studies, Syntaxin has been suggested to function in the targeting and/or docking of synaptic vesicles. However, work in *Drosophila* does not support such a role for Syx-1A. Unlike proteins such as Synaptotagmin, which are found exclusively at synapses (76), Syntaxin-1A protein is also found along axons (14). This observation has also been made in vertebrates (134, 135) and argues against a primary role for this protein in defining specific sites of targeting and docking of synaptic vesicles. More importantly, ultrastructural analysis of the synaptic boutons of *syntaxin* null mutants that completely lack Syntaxin protein reveals an increase, not a decrease, in the number of docked vesicles (120). These data show that vesicles can properly target and dock by morphological criteria in the absence of Syntaxin. In addition, these vesicles are functionally docked, since application of hypertonic saline, which has been shown to induce the fusion of the readily releasable pool of docked vesicles (136), results in vesicle release (see below) (Fig. 8). These observations are consistent with genetic and toxin-mediated removal of Synaptobrevin (see below) and together suggest that the core complex is not required for docking.

If Syntaxin is not required for docking, at what step might it function? Electrophysiological studies of partial loss-of-function and null mutations in *syntaxin* suggest a role for this protein in fusion. Whole-cell patch-clamp studies at the embryonic NMJ of partial loss-of-function *P{syx}* mutants reveal an absence of the normal bursts of synaptic activity, which in wild-type embryos correlates with hatching activity. Furthermore, evoked release in these mutants is reduced to about 20% of controls, indicating that Syx-1A protein is required for neurotransmitter release *in vivo*. The synaptic transmission phenotype of *syx* null mutants is even more severe (Fig. 6, syx). Although null mutants show a variety of developmental defects, NMJ synapses form properly, because focal application of glutamate at these synapses results in a robust excitatory junctional current (Fig. 6) (14). The reason that these synapses develop properly is because of maternal contribution of *syx-1A* message and protein. Although the postsynaptic muscle can respond to glutamate, *syx* null mutant animals display a complete failure of evoked (Fig. 6) and spontaneous neurotransmitter release (Fig. 7), which demonstrates that Syx-1A protein is absolutely required for both evoked and spontaneous neurotransmission. This phenotype suggests an essential role for this protein in the actual fusion event itself. Furthermore, it was shown that latroinsectotoxin which can elicit neurotransmitter release in *csp* mutants, is unable to do so in *syx* null mutants (120), indicating that this protein functions at a late step in synaptic vesicle exocytosis. Finally, as described earlier, the only treatment thus far that can elicit any release from *syx* null mutants is the use of hypertonic saline, which induces the release of docked vesicles (Fig. 8). At lower concentrations, hypertonic saline induces an increased frequency of minis. In *syntaxin* null mutants, these minis are altered in amplitude and are variable. This is in

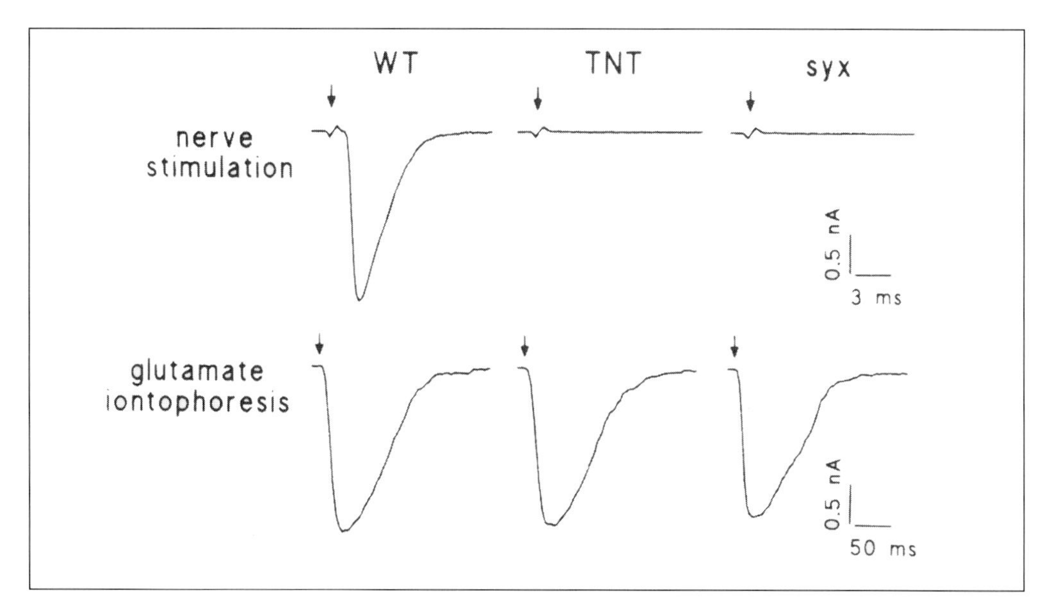

Fig. 6 Evoked response at NMJs lacking either functional Synaptobrevin or Syntaxin. The upper traces show evoked synaptic currents recorded after stimulation in wild-type (wt), tetanus toxin expressing embryos (TNT), and *syx* null embryos. Glutamate iontophoresis is shown in the bottom panel for all three genotypes. (From ref. 120; courtesy of Cell Press.)

contrast to 'minis' elicited by hypertonic saline in the absence of n-Synaptobrevin (see below and Fig. 8), where mini frequency is dramatically enhanced by hyperosmotic treatment, and the mini amplitude is unchanged from controls. One interpretation for the reduced and variable amplitude of minis in *syx* null mutants is that Syntaxin is required for the fusion event itself and that, in its absence, full fusion of the vesicle and subsequent release of all the contents of that vesicle is impaired.

Interestingly, subcellular fractionation studies have revealed that *Drosophila* Syx-1A protein can be found not only in the presynaptic membrane but also in synaptic vesicles (14). Vertebrate syntaxin has also been observed in synaptic vesicles (137, 138). The role of this vesicular Syx-1A is at present unclear, although it is tempting to speculate that vesicular Syntaxin together with plasma membrane Syntaxin may function in the fusion machinery. Along these lines, Syntaxin homologues have been shown to bind themselves (133, 139) (M. Wu, T. Lloyd, H. J. B., unpublished results) and in yeast a t-SNARE homologue, Ufe1p, has been suggested to function as a homomultimer in the absence of a v-SNARE to mediate a homotypic fusion event (140).

Recent biochemical and biophysical studies have supported a role for Syntaxin and the core complex in fusion. Structural studies of the core complex indicate that it forms a parallel four-helical bundle that could potentially pull the vesicle and the membrane together (141, 142). The formation of this bundle is associated with the release of energy, which could assist in the energetically unfavourable act of bringing

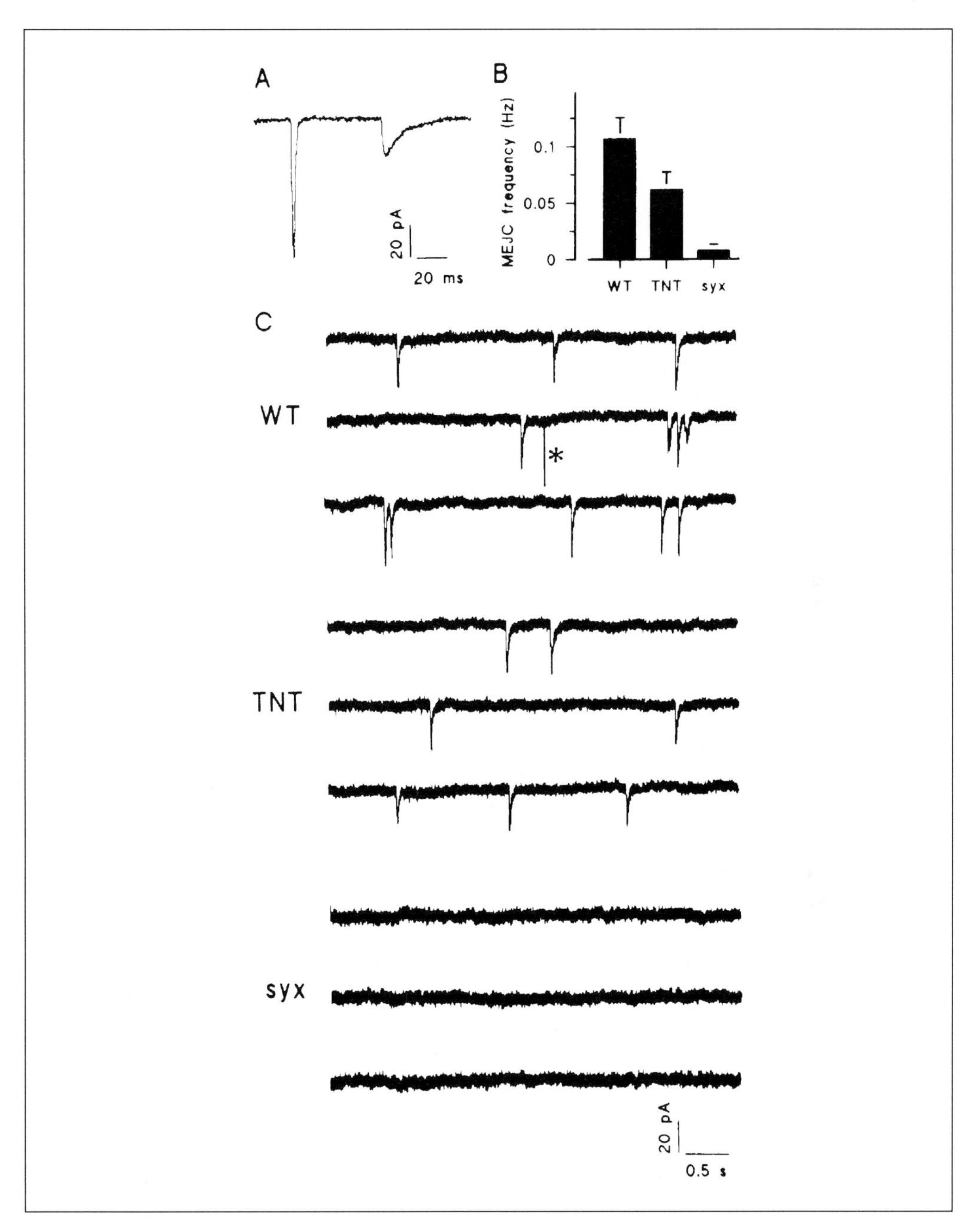

Fig. 7 Spontaneous neurotransmitter release in neuromuscular junctions (NMJs) lacking either functional Synaptobrevin or Syntaxin. (A) Minis recorded at wild-type (wt) NMJs. (B) The frequency of minis is shown for wt, TNT expressing embryos, and *syx* null mutants. (C) MEJCS recorded in wt, in the absence of functional Synaptobrevin, and in *syx* null mutants. (From ref. 120; courtesy of Cell Press.)

vesicle and target membrane close together (143). Furthermore, recent studies have shown that the core complex reconstituted in artificial lipid bilayers can mediate fusion, albeit inefficiently (144). These studies, taken together with the genetic and electrophysiological data from *Drosophila*, suggest that Syx-1A functions after docking and likely in the fusion event of a synaptic vesicle with the presynaptic membrane.

3.3.3 Synaptobrevin

Synaptobrevin (Syb) is a small 18 kDa integral SV protein that is part of the core complex together with syntaxin and SNAP-25 (145, 146) (Chapter 3). Two different genes encoding Synaptobrevins have been isolated in *Drosophila*: an ubiquitous Syb (cellular Syb) which is enriched in the gut and only weakly expressed in neuronal cells (147); and, a neuronal isoform, n-Syb, that is specifically expressed in the nervous system (22). The two isoforms share similar structural features. They both have a proline-rich amino-terminus, a highly conserved central cytoplasmic region, and a single transmembrane domain. Here we will focus on the neural isoform or n-Syb since no mutants have been described for Syb.

In the embryo, n-Syb is localized to the ventral nerve cord and NMJ synapses (20). As mentioned in the previous section, Synaptobrevin, like Syntaxin, has been hypothesized to play a role in vesicle targeting or docking at the active zone. If this hypothesis is correct, removal of n-Syb should lead to a defect in vesicle targeting or docking. This hypothesis was tested by overexpressing the light chain of tetanus toxin (which cleaves n-Syb but not cellular Syb) using the UAS-GAL4 system to target expression to neurons (148). The experiments using the toxin-induced cleavage of n-Syb have been discussed in Chapter 6 (148), and we will therefore only highlight the main points. Electron microscopy analysis of NMJs from embryos expressing tetanus toxin revealed that synaptic vesicles are correctly targeted at the active zones, indicating that n-Syb is not required for vesicle docking or targeting to the active zone (120). Hunt *et al.* (121) made similar observations by examining squid synapses treated with tetanus toxin (Chapter 7). Furthermore, using hyperosmotic saline solutions to induce the release of docked vesicles, Broadie *et al.* (120) showed that the morphologically docked vesicles of tetanus poisoned NMJs are functionally docked, since they can be induced to release (Fig. 8). In contrast, BWSV treatment of NMJs lacking n-Syb does not cause neurotransmitter to be released. In general, these phenotypes are very similar to those observed for *syntaxin* mutants and led to the conclusion that n-Syb, like Syntaxin, is required after targeting/docking. Electrophysiological recordings of tetanus toxin-treated synapses revealed that while evoked response was abolished (Fig. 6), minis persisted (Fig. 7), although they were reduced in frequency (120). Therefore, n-Synaptobrevin is absolutely required for evoked neurotransmitter release, but may not be essential for all synaptic vesicle fusion events.

Although clostridial neurotoxins have been widely used, some caution is appropriate when this kind of approach is adopted. First, the toxin may not completely cleave all the target protein at the synapse. Furthermore, tetanus toxin cleaves only

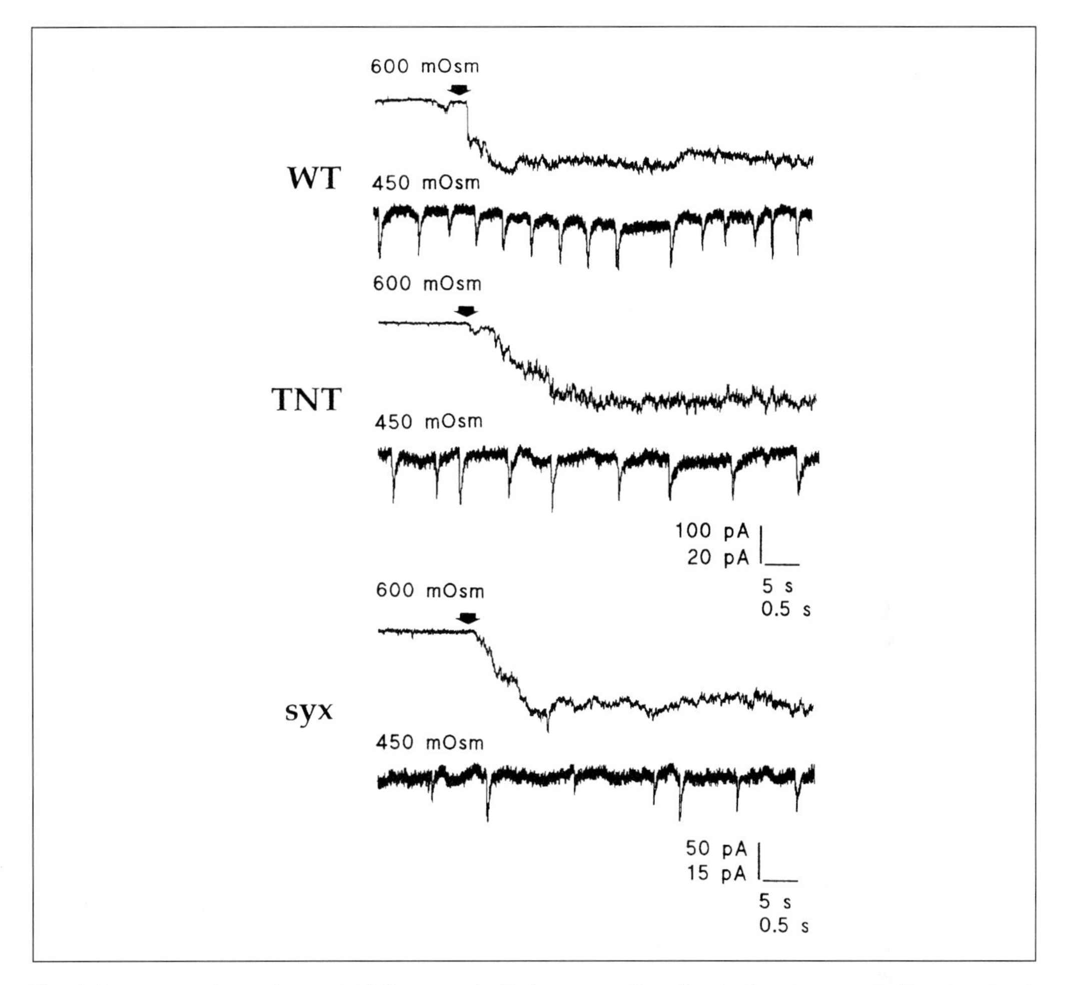

Fig. 8 Neurotransmitter release at NMJs treated with hyperosmotic saline in the absence of either functional Synaptobrevin or *syx*. Two concentrations (600 and 450 mOsm) of hyperosmotic salines were used. (From ref. 120; courtesy of Cell Press.)

the C-terminal part such that the truncated protein could still retain some residual function. Finally, each toxin may have additional undefined targets or additional enzymatic activities (Chapter 6) (149, 150). Hence, toxin-induced cleavage of synaptic proteins does not substitute for genetic approaches. To address these concerns, Deitcher and colleagues (20) created null mutations in *n-syb* using a P-element hopping strategy. Unlike *syntaxin* null mutants, the complete absence of n-Syb does not cause any obvious morphological defects when viewed with light microscopy, possibly because non-neuronal secretory functions are mediated by cellular Syb. However, mutant embryos that lack n-Syb are paralysed and fail to hatch from the egg case. Electrophysiological measurements from the NMJs of these mutants revealed that the excitatory synaptic current is completely abolished whereas the frequency of minis is reduced by 75%. Hence, tetanus-treated synapses and synapses that lack n-Syb exhibit very similar properties (20).

The phenotype of the loss of n-Syb is similar to the loss of Syntaxin in that evoked release is abolished, but differs in that spontaneous fusion persists. The persistence of spontaneous transmitter release in *n-syb* mutants raises two possibilities. One possibility is that these spontaneous fusion events use an alternative Synaptobrevin, such as the ubiquitous Syb isoform. However, the ubiquitous form is present at very low levels in synaptic regions (20). Alternatively, minis may occur in the absence of Synaptobrevins, a hypothesis that is supported by data obtained on synaptobrevins or v-SNAREs in other species (151). In every organism studied to date (including *Drosophila*), genetic removal of synaptobrevin is always less severe than removal of syntaxin. For instance, in *C. elegans*, *syb* mutants can perform some uncoordinated movements, whereas *syx* mutants are completely paralysed (152, 153). In a yeast vacuolar fusion assay, deleting the synaptobrevin homologue causes a less severe phenotype than deleting the syntaxin homologue (154). In yeast Golgi to plasma membrane secretion, there are two synaptobrevin homologues (SNC1 and SNC2) and two syntaxin homologues (SSO1 and SSO2). SNC1/SNC2 double mutants are capable of limited secretion and are conditionally lethal, while SSO1/SSO2 double mutants are unconditionally lethal (155, 156). Since the *Drosophila* genome has not been entirely sequenced, it remains possible that the milder phenotype of *n-syb* mutants compared to *syx* mutants is caused by redundancy of another Synaptobrevin, but for yeast and *C. elegans* (whose genomes are fully sequenced), this cannot be the case. Therefore, although Synaptobrevin plays a similar role in neurotransmitter release to Syntaxin, the persistence of 'minis', combined with data obtained in other species on synaptobrevin homologues, indicate that Synaptobrevin plays a distinct role from Syntaxin in the release process. Alternatively, syntaxin may play multiple roles. However, the data suggest that synaptobrevins are not essential for the fusion event itself, but rather are required for SVs that participate in the evoked response just prior to fusion. This in turn suggests that while the core complex may be required for fusion, all three proteins do not necessarily participate in the fusion itself, as has been proposed elsewhere (Chapter 3).

3.4 Proteins that regulate the core complex

3.4.1 Introduction

In the previous sections we discussed how the core complex is essential for evoked neurotransmitter release. Clearly, at the synapse the fusion machinery must be controlled by a protein or protein complex that can respond to the Ca^{2+} influx, such as Synaptotagmin. In addition to this regulation by a Ca^{2+} sensor at a late step, there are additional earlier regulatory steps that modulate the release process. For example, electrophysiological data have suggested that SVs undergo a maturation or 'priming' step after docking, since not all docked vesicles are capable of being released (157, 158). These and other regulatory steps appear to ensure that the proper sequence of protein interactions occurs to allow productive exocytosis, but may also provide a means to modify the release process, which could contribute to the synaptic plasticity required for learning and memory (159, 160).

The core complex is an important target for regulation. Proteins which can biochemically interact with Syntaxin and have been proposed to regulate its function include ROP/n-Sec1/Munc-18 (126), complexins (161), tomosyn (125), α-SNAP, and NSF (162). Synaptobrevin has been shown to bind VAP-33 (163) and synaptophysin (164), whereas SNAP-25 was shown to interact with Hrs (165). These proteins are likely to regulate the availability or conformation of the proteins of the core complex. In *Drosophila*, mutations in ROP and NSF have been studied in detail and will be discussed below.

3.4.2 Ras Opposite or ROP

The n-Sec1 family of proteins has been proposed to function in regulation of secretory events. Biochemical analyses using vertebrate Sec1 family members (Munc-18/n-Sec1) show that Sec1 family members can interact with syntaxin (126, 166, 167). This interaction is very strong (nanomolar affinity) and can inhibit the ability of syntaxin to bind to synaptobrevin and SNAP-25 and hence to participate in the core complex (168), suggesting that the Sec1 family may inhibit neurotransmitter release. On the other hand, mutations in Sec1 homologues in yeast and *C. elegans* block or reduce secretion (169–171), indicating that these proteins perform a positive role in this process. In *Drosophila*, the Sec1 family member implicated in neurotransmission is called ROP (for Ras Opposite). It was named so because the gene shares a bidirectional promoter with the *ras* gene (172). ROP shows 65% identity to its vertebrate homologue Munc-18a/n-Sec1. In *Drosophila*, studies of ROP have shown that it can perform both positive and negative roles in neurosecretion *in vivo*. Importantly, ROP is a rate-limiting component for SV exocytosis at the synapse.

To study the role of ROP, an EMS mutagenesis screen was performed and 14 different mutations in ROP were isolated, including partial loss-of-function alleles and null alleles (24). Null mutants are embryonic lethal, while partial loss-of-function mutants die at stages ranging from embryos to adults. Analysis of a null *rop* mutant reveals an important role for ROP in non-neuronal secretion. Similar to Syntaxin, but unlike Synaptotagmin or n-Synaptobrevin, ROP is present not only in neurons, but also in non-neuronal tissues such as garland cells, gut, and salivary glands. As in the case of *syx* null mutants, complete absence of zygotic ROP expression causes defects in non-neuronal secretion, such as a failure to secrete cuticle. However, *rop* mutants also display defects not observed in *syx* mutants, such as an inability to secrete uric acid in the Malpighian tubules or to secrete glue protein in the salivary glands (24). Thus, ROP and Syntaxin seem to be components of a shared machinery used for different kinds of secretion.

In addition to expression in non-neuronal tissues, ROP is also strongly enriched in the nervous system, and like Syntaxin, is localized not only to the synapse, but also along axons (173). Unfortunately, the electrophysiological phenotype of the complete loss of ROP function could not be assessed at the embryonic NMJ, since the synapse does not develop properly. In this case, maternal contribution of ROP is apparently not sufficient to allow the NMJ to develop. However, 12 of the 14 mutants identified in the EMS screen were partial loss-of-function mutants which developed sufficiently to allow electrophysiological analysis. As in the case for *synaptotagmin*, studying the phenotypes caused by different point mutations revealed important insights into different aspects of protein function. First, electroretinograms show a loss of on/off transients in adult *rop* mutants (24), indicating a loss of synaptic transmission between photoreceptors and the first-order interneurons. These findings are consistent with studies in yeast and *C. elegans*, which demonstrated that Sec1 family members perform a positive function that is required for secretion.

These studies were extended by overexpression experiments using heat shock-inducible constructs. Overexpression of ROP causes a 30–40% reduction of evoked neurotransmitter release as well as a 50% reduction in the frequency of spontaneous events, compared to controls. These data suggest that ROP can inhibit neurotransmission (173) and are consistent with biochemical data suggesting that ROP may inhibit core complex formation (126, 168). This inhibition of neurotransmitter release caused by overexpressing ROP can be suppressed by simultaneously overexpressing Syntaxin. This indicates that ROP and Syntaxin interact *in vivo* (37), and that the Syntaxin–ROP complex is an important modulatory complex in synaptic transmission.

The question of whether ROP is an activator or an inhibitor of synaptic vesicle exocytosis has been addressed by electrophysiological analyses of partial loss-of-function mutations which revealed that partial loss-of-function *rop* mutants exhibit two distinct phenotypes at the larval NMJ (37). As shown in Fig. 9A, certain mutations caused a significant reduction in evoked release, while others caused an increase in evoked release (Fig. 9B). These data show that ROP has two different functions in synaptic vesicle exocytosis, a positive role and a negative role.

Further analysis of heteroallelic combinations of *rop* mutants revealed that ROP is a rate-limiting component at the synapse (37). Heterozygous combinations of different partial loss-of-function mutations showed that evoked transmission is very sensitive to the levels of ROP function. In other words, significantly different amounts of evoked release were observed for each allelic combination (Fig. 9C). More importantly, a 50% reduction in the amount of ROP protein resulted in a 50% reduction of evoked response (Fig. 9D). This phenotype has not been observed in any other *Drosophila* mutants implicated in neurotransmitter release and indicates that ROP is a rate-limiting component of synaptic vesicle exocytosis.

In summary, the studies of *rop* mutants in *Drosophila* have demonstrated that this protein is required for both non-neuronal and neuronal secretion. Furthermore, ROP interacts with Syntaxin *in vivo* and performs both positive and inhibitory functions in synaptic vesicle exocytosis. ROP appears to be an important modulator of neurotransmitter release, and to regulate Syntaxin function. In addition, the finding that ROP is a rate-limiting component of release makes it itself an attractive target for regulation, since regulation of its activity or availability by phosphorylation (174, 175) or other means would have a key impact on neurosecretion.

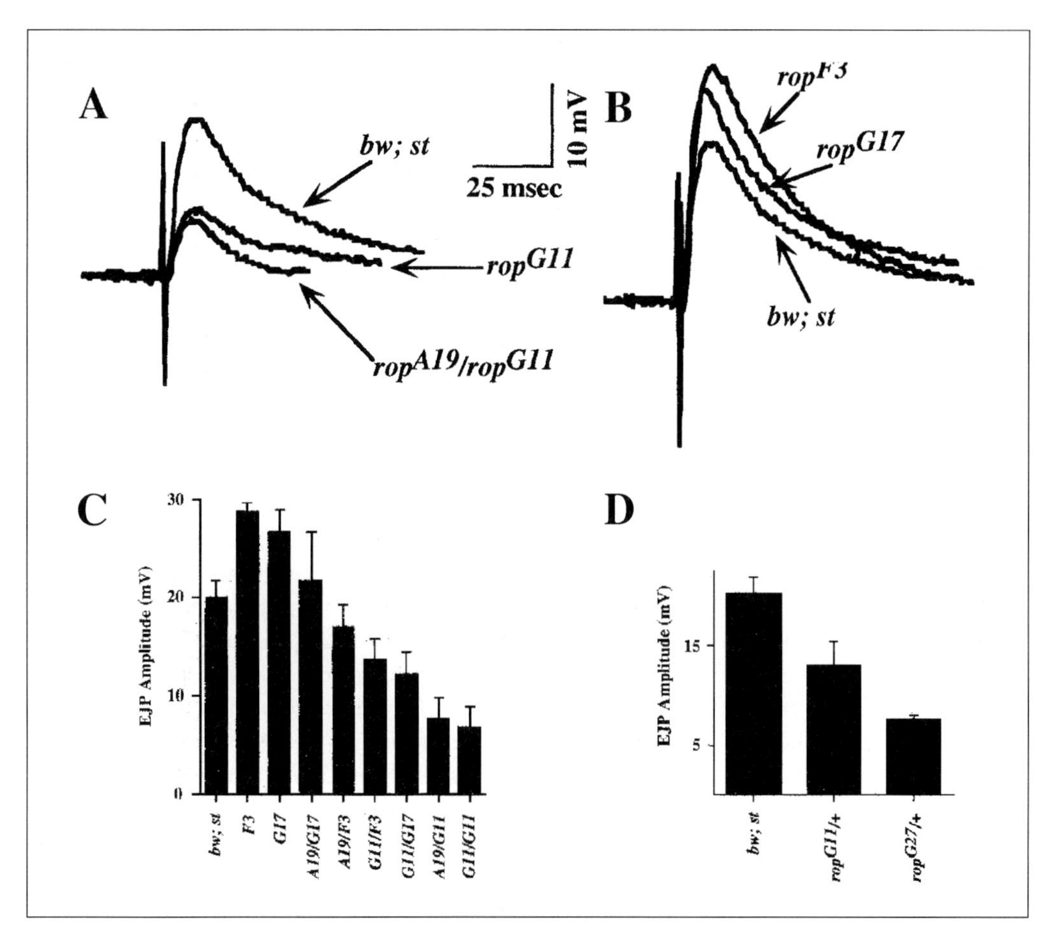

Fig. 9 Electrophysiological analysis of *rop* mutants. (A) Representative EJPs are shown from controls (*bw; st*), homozygous (*rop*^{G11}), and heterozygous (*rop*^{A19}/*rop*^{G11}) mutant larvae. (B) EJPs recorded from another class of *rop* alleles *rop*^{F3} and *rop*^{G17}. (C) EJPs amplitudes from controls (*bw; st*) and several homo and heterozygous combinations of *rop* mutant alleles are plotted. (D) EJPS amplitudes from controls (*bw; st*) and larvae carrying a wild-type copy of *rop* and either the *rop*^{G11} and the *rop*^{G27} alleles. In (C) and (D) error bars represent standard error of mean, SEM. (From ref. 37; courtesy of Oxford University Press.)

3.4.3 *N*-ethylmaleimide-sensitive fusion protein or NSF

Another important regulator of neurosecretion is NSF, an ATPase that was first identified as a protein important for Golgi transport (Chapter 3). Using a cell-free transport assay, Rothman and colleagues demonstrated that intra-Golgi transport vesicle fusion requires cytosolic factors including ATP, GTP, and several cytosolic proteins (176, 177). Transport was blocked by the action of *N*-ethylmaleimide, a sulfydryl alkylating agent. A single protein, NSF, was shown to be sufficient to rescue this blockage (178).

NSF is the mammalian counterpart of the yeast SEC18 protein (179). The rapid inactivation of a temperature-sensitive mutant allele of the SEC18 gene allowed

Graham and Emr (1991) to show that NSF/SEC18 is required for transport from the endoplasmic reticulum to Golgi complex, within the Golgi complex, and from the Golgi complex to the cell surface (180). According to the SNARE hypothesis (Chapter 3), SV targeting and docking at release sites involves the assembly of the core complex. The complex then recruits SNAPs (soluble NSF attachment proteins) and NSF to form a pre-fusion complex. Hydrolysis of ATP by NSF disassembles the complex in a step postulated to initiate membrane fusion (118). As shown below, mutational analyses in *Drosophila* point towards a role for this protein in disassembly of the core complex, either before or after, but not during fusion. These data are in agreement with a number of other *in vivo* manipulations discussed in Chapters 3 and 7.

To study the function of NSF in neurotransmitter release in *Drosophila*, a homologue of NSF (dNSF-1) has been isolated. Mapping showed that the gene encoding NSF and a series of temperature-sensitive paralytic mutations of the *comatose* (*comt*) locus mapped to the same region. It was then confirmed that *comt* mutants carry point mutations in dNSF (9–11). More recently, a second homologue of NSF, dNSF-2, has been reported in *Drosophila* (181, 182). dNSF-2 is 84% identical to dNSF-1, but no mutants have been described thus far. Here, we will discuss the electrophysiological, ultrastructural, and biochemical properties of the core complexes associated with mutations in dNSF-1 (4, 6, 11, 183). As shown below, these observations suggest that dNSF-1 is required for the disassembly of the core complex, either prior to or after fusion of the SVs with the presynaptic membrane.

One of the hallmarks of the paralysis associated with *comt* larvae or adults is that the paralysis is slow when compared to that observed in a temperature-sensitive *syntaxin* mutant (4). Similarly, recovery from paralysis is slow in *comt* mutants (11). The slower paralysis kinetics correlate with a slower time course for the blockade of synaptic transmission. Analysis of ERGs of temperature-sensitive *syntaxin* and *comatose* mutants has shown that loss of on/off transients occurs in a few seconds in *syntaxin* mutants, whereas *comt* mutants exhibit a loss after three minutes (4). These data suggest different kinetic requirements for these two proteins *in vivo*. This may be explained if Syntaxin is required for fusion *per se* (see above), whereas *comt* is required for a process that is much slower than fusion. Alternatively, *comt* mutations may cause a slow unfolding of the protein at the restrictive temperature. However, this is unlikely as the kinetics of paralysis are very similar for different *comt* mutations.

Electrophysiological analysis of adult NMJs in paralysed dNSF-1 flies showed no obvious defects in the absence of repetitive stimulation. However, as shown in Fig. 10A and B, during repetitive stimulation, a progressive, activity-dependent reduction in neurotransmitter release has been observed (183). It is important to note that these flies are null mutants, i.e. they lack functional protein (183). This suggests that dNSF-1 is required for the maintenance of neurotransmitter release, but not for vesicle fusion itself. Failure to maintain a normal level of neurotransmitter release following repetitive stimulation also implies that the readily releasable pool of SVs cannot be maintained at the restrictive temperature in these mutants. Possible causes

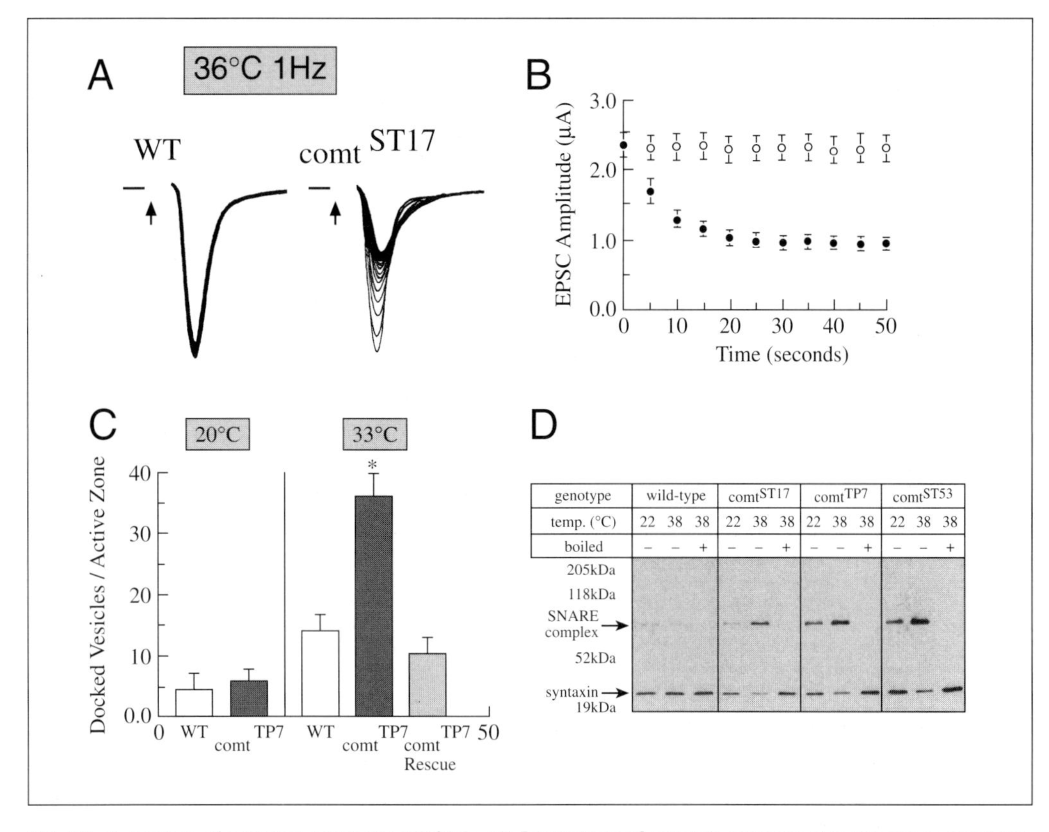

Fig. 10 Phenotype of *comatose* mutants (dNSF-1). (A) Recordings of synaptic currents at restrictive temperature for wild-type (wt) and ST17 *comt* mutants following a 1 Hz stimulation train. (B) Peak amplitude measurements of synaptic currents at the restrictive temperature with a 1 Hz stimulation train for wt and *comt* mutants. (C) Number of docked vesicles in wt and *comt* flies at the permissive and restrictive temperature. 'Rescue' flies are *comt* mutants where the mutant phenotype has been rescued by transgenic expression of a wt copy of the gene. (D) Western blot analysis of the SNARE complexes extracted from *comt* mutants and wt either at permissive or restrictive temperature. Samples marked + were boiled before gel loading. (From ref. 183 (A–C), and ref. 6 (D); courtesy of Journal of Neuroscience.)

include a partial block in endocytosis, a defect in targeting or docking of vesicles, or a defect in SV 'priming'. As shown in Fig. 10C, ultrastructural analysis of *comt* synaptic terminals at restrictive temperatures revealed an increase in the number of docked vesicles at the active zones, suggesting that dNSF-1 does not affect endocytosis, targeting, or docking, but rather the 'priming' of docked SVs (4, 183).

What might be the nature of such a 'priming' step carried out by NSF? Given that NSF has been shown to disassemble the core complex *in vitro* (Chapter 3), one obvious possibility is that NSF primes SVs by disassembling unproductive core complexes. Inactive or unproductive core complexes could arise from formation of such complexes in the same membrane compartment (such as SVs) (184) or by the residual presence of core complexes after fusion. In order to measure the *in vivo* amounts of this complex in *comt* mutants at the restrictive temperature, Tolar and Pallanck (6) designed a simple assay based on the SDS resistance of the proteins forming the core complex. As expected, they observed a marked increase of the core complexes in *comt* mutant heads at the restrictive temperature, showing that disassembly *in vivo* requires dNSF-1 (Fig. 10D). Similar observations were also made by Littleton *et al.* (4). They reported that the core complexes mainly accumulate with SVs suggesting that dNSF may act after SV fusion. However, a thorough examination of this question by Tolar and Pallanck (6) determined that more than 97% of the core complexes in the mutants are associated with the synaptic vesicles and plasma membranes. These data can be most easily interpreted by proposing that dNSF-1 acts after SV docking in the disassembly of the core complex to maintain the readily releasable pool of SVs prior to fusion. In summary, the slow time course of paralysis, the electrophysiological phenotype of *comt* mutants, the accumulation of docked SVs, and the accumulation of core complexes *in vivo*, all point to a role for dNSF-1, not in fusion of SVs, but rather in 'priming' of SVs, or in recycling of SVs by disassembling unproductive core complexes after fusion (4, 6, 183).

3.5 Other proteins required for proper neurotransmitter release

3.5.1 Introduction

A number of other players have been implicated in neurotransmitter release, but their precise roles *in vivo* or *in vitro* are less defined in *Drosophila*. These include the gene products of the *stoned* locus, as well as *Drosophila* Synapsin. In addition, a variety of genes have been implicated in learning and memory in *Drosophila,* and mutations in some of these genes clearly affect neurotransmitter release and behaviour. These mutants have been reviewed extensively (185, 186). The observation that genes which may be required for neurotransmitter release at the NMJ are also involved in learning and memory strengthens the notion that understanding the mechanistic processes underlying synaptic transmission at a simple model synapse may provide insights into higher order nervous system functions.

3.5.2 Stoned

In Chapter 11, Zhang and Ramaswami discuss the role of most of the proteins involved in endocytosis. However, some of the proteins required for neurotransmitter release may play a role in exo- and endocytosis. This is precisely what has been proposed by Stimson *et al.* (191) for the proteins encoded by the *stoned* locus. Mutations in the *stoned* locus were first identified in mutagenesis screens designed to isolate temperature-sensitive paralytic mutants (187). Independent alleles were isolated in screens to identify stress-sensitive behavioural mutants (188). Molecular analysis revealed that the locus produces a dicistronic transcript that encodes two proteins: Stoned A and Stoned B (183). Stoned A is a novel protein, whereas Stoned B has a region of 42% homology with AP50, a subunit of the AP2 adaptor protein complex implicated in endocytosis of SVs (Chapter 11). However, Stoned B is not the *Drosophila* AP50 orthologue, which has been identified, and shown to display 86% homology to human AP50 [Flybase]. Stoned B also exhibits strong homology with the *C. elegans* gene *unc-41*, a gene that is believed to play a role in synaptic vesicle recycling (190). Mutations that affect Stoned B have not been characterized, and we will therefore confine our discussion to Stoned A.

The Stoned A protein was shown to be localized to presynatic terminals (191). Interestingly, the phenotypes associated with *stoned A* mutations are quite reminiscent of the phenotypes previously described for *synaptotagmin* mutants (see above). EJC recordings at third instar larval NMJs showed that the EJC amplitude of one of the *stoned A* alleles was only 15% of controls. In addition, *stoned A* mutants have a threefold increase in mini frequency when compared to wild-type larvae. Stimson *et al.* (191) also reported that the failure rate upon stimulation is increased in mutants, and that this increase is accompanied with a reduction in the fidelity of release. These electrophysiological findings are reminiscent of *synaptotagmin* mutants, and indeed, Synaptotagmin protein is mislocalized in *stoned A* NMJ terminals (191). Therefore, Stoned A does not likely serve as a Ca^{2+} sensor itself, but may be required for the proper localization of Synaptotagmin in SVs. Given that no morphological defects are observed at the ultrastructural level using TEM in *stoned A* mutants, Stimson *et al.* (131) were unable to conclude if the observed defects in *stoned* mutants are due to a defect in endo- or exocytosis.

3.5.3 Synapsin

Relatively little is known about invertebrate synapsins, especially in contrast to the wealth of information gathered in vertebrates (Chapter 3 and 10). Synapsins are a small family of SV-associated phosphoproteins which may regulate targeting of vesicles by interacting with the actin cytoskeleton. Klagges *et al.* (1996) (192) identified a *synapsin* gene in *Drosophila,* which encodes at least two potential proteins, a small and a large isoform, that may be generated by the read-through of a stop codon. The proteins contain a highly conserved domain, which has been shown to bind vesicles and actin in vertebrates. The *Drosophila* synapsins are localized to the nervous system and, more specifically, to synaptic terminals. Unfortunately, the mutational analysis of *synapsin* has revealed no phenotype so far (Klagges, personal communication), and it is not known if other synapsin homologues can substitute for its function in flies.

4. Summary and conclusions

In this chapter, we have summarized the salient features of *Drosophila* biology specifically related to neurotransmitter release. The fruitfly NMJ is one of the best systems to define the *in vivo* functions of proteins involved in neurotransmitter release because of the ability to genetically manipulate the organism, to efficiently carry out TEM studies, and to perform electrophysiological assays. As will become apparent in the next chapter, one of the main advantages of genetic analysis in the fruitfly is that its genome is smaller and more compact than the vertebrate genome, i.e. for each fly gene there are typically four vertebrate homologues (12). Hence, the issue of redundancy, which is a genuine concern of reverse genetic strategies, has been much less of a problem than in vertebrates (Chapter 10). In any case, the question of redundancy will be even less relevant when the complete sequence of the *Drosophila* genome is determined.

The major contributions of the fly neurotransmitter release field have, and will continue to be the ability to define the phenotype associated with the loss or alteration of proteins *in vivo*, providing vital data about function. For example, mutational analysis of *Drosophila synaptotagmin* provided the first compelling *in vivo* evidence that Synaptotagmin is a Ca^{2+} sensor (73). In addition, almost all functional data about CSP has come from studying *csp* mutants in *Drosophila*, which indicate that CSP is required in SV exocytosis by regulating the Ca^{2+} signal. Studies of Syntaxin, Synaptobrevin, and NSF have helped to revise the original SNARE hypothesis. In the original hypothesis (Chapter 3), the core complex had been proposed to be required for targeting/docking of SVs, while NSF catalysed the fusion event itself. However, work in *Drosophila*, as well as other systems, has indicated that both Syntaxin and Synaptobrevin are required after docking (14, 120). In addition, while Syntaxin appears to be essential for fusion, work from several species suggests that Synaptobrevin has a distinct post-docking role (151). Similarly, analysis of *comatose* mutants, together with a number of studies from yeast to vertebrates, has shown that NSF is not required for fusion, but rather for priming or disassembling inactive core complexes (4, 6, 183). Finally, analysis of *rop* mutations in *Drosophila* has provided *in vivo* evidence for both positive and inhibitory roles of the Sec1 family in SV exocytosis and indicates that this protein is rate-limiting in this process (37).

The last five years have seen an explosion of the number of proteins implicated in neurotransmitter release. More than 50 players are thought to be required for exo- and endocytosis of SVs. The full genome sequences have become available for *C. elegans* and yeast, and the fruitfly genome will likely be finished in the next two years. This information will be a treasure trove for *Drosophila* biologists, as all the homologues of proteins implicated in neurotransmitter release will automatically be known and mapped. It will thus soon become possible to define the *in vivo* function of most of the proteins described in this book using the approaches outlined in this chapter. Although reverse genetics will become much easier with the completion of the genomic sequence, one largely untapped frontier is the possibility of harnessing the power of 'forward' genetics to study neurotransmitter release in *Drosophila*. In other words, mutagenic screens can be carried out to functionally identify new components of this process. This combination of reverse and forward genetics should continue to provide important advances in unravelling the mechanisms underlying SV exocytosis.

Acknowledgements

We would like to thank Konrad Zinsmaier for providing very useful comments. We would also like to thank Jay Bhave, Bing Zhang, and Thomas Lloyd for comments on the manuscript. G. P. is supported by the Howard Hughes Medical Institute (H. H. M. I.), and M. W. is supported by a National Research Service Award from the National Institute of Mental Health. H. J. B. is an Associate Investigator of the H. H. M. I.

References

1. Lawrence, P. A. (1992) *The making of a fly: the genetics of animal design*, p. 1. Blackwell Scientific Publications, Great Britain.
2. Greenspan, R. J. (1997) *Fly pushing: the theory and practice of Drosophila genetics*, p. 1. CSHL Press, Cold Spring Harbor.
3. Kuromi, H. and Kidokoro, Y. (1998) Two distinct pools of synaptic vesicles in single presynaptic boutons in a temperature-sensitive *Drosophila* mutant, *shibire*. *Neuron*, **20**, 917.
4. Littleton, J. T., Chapman, E. R., Kreber, R., Garment, M. B., Carlson, S. D., and Ganetzky, B. (1998) Temperature-sensitive paralytic mutations demonstrate that synaptic exocytosis requires SNARE complex assembly and disassembly. *Neuron*, **21**, 401.
5. Ramaswami, M., Krishnan, K. S., and Kelly, R. B. (1994) Intermediates in synaptic vesicle recycling revealed by optical imaging of *Drosophila* neuromuscular junctions. *Neuron*, **13**, 363.
6. Tolar, L. A. and Pallanck, L. (1998) NSF function in neurotransmitter release involves rearrangement of the SNARE complex downstream of synaptic vesicle docking. *J. Neurosci.*, **18**, 10250.
7. Broadie, K. S. *Drosophila* electrophysiology. In *Drosophila methods* (ed. Ashburner, Hawley, and Sullivan). CSHL Press, Cold Spring Harbor, New York, (in press).
8. Bellen, H. J. and Budnik, V. The neuromuscular junction. In *Drosophila methods* (ed. Ashburner, Hawley, and Sullivan). CSHL Press, Cold Spring Harbor, New York, (in press).
9. Ordway, R. W., Pallanck, L., and Ganetzky, B. (1994) Neurally expressed *Drosophila* genes encoding homologs of the NSF and SNAP secretory proteins. *Proc. Natl. Acad. Sci. USA*, **91**, 5715.
10. Pallanck, L., Ordway, R. W., and Ganetzky, B. (1995) A *Drosophila* NSF mutant. *Nature*, **376**, 25.
11. Siddiqi, O. and Benzer, S. (1976) Neurophysiological defects in temperature-sensitive paralytic mutants of *Drosophila melanogaster*. *Proc. Natl. Acad. Sci. USA*, **73**, 3253.
12. Bellen, H. J. Ten years of enhancer detection: a fly viewpoint. *Plant Cell*, (in press).
13. Wilson, C., Pearson, R. K., Bellen, H. J., O'Kane, C. J., Grossniklaus, U., and Gehring, W. J. (1989) P-element-mediated enhancer detection: an efficient method for isolating and characterizing developmentally regulated genes in *Drosophila*. *Genes Dev.*, **3**, 1301.
14. Schulze, K. L., Broadie, K., Perin, M. S., and Bellen, H. J. (1995) Genetic and electrophysiological studies of *Drosophila* syntaxin-1A demonstrate its role in nonneuronal secretion and neurotransmission. *Cell*, **80**, 3113.
15. Zhang, B., Koh, Y. H., Beckstead, R. B., Budnik, B., Ganetzky, B., and Bellen, H. J. (1998) Synaptic vesicle size and number are regulated by a clathrin adaptor protein required for endocytosis. *Neuron*, **21**, 1465.
16. Zhang, P. and Spradling, A. C. (1993) Efficient and dispersed local P element transposition from *Drosophila* females. *Genetics*, **133**, 361.

17. Littleton, J. T., Stern, M., Schulze, K., Perin, M., and Bellen, H. J. (1993) Mutational analysis of *Drosophila synaptotagmin* demonstrates its essential role in Ca^{2+}-activated neurotransmitter release. *Cell*, **74**, 1125.
18. Zinsmaier, K. E., Eberle, K. K., Buchner, E., Walter, N., and Benzer, S. (1994) Paralysis and early death in cysteine string protein mutants of *Drosophila*. *Science*, **263**, 977.
19. Burgess, R. W., Deitcher, D. L., and Schwarz, T. L. (1997) The synaptic protein syntaxin1 is required for cellularization of *Drosophila* embryos. *J. Cell Biol.*, **138**, 861.
20. Deitcher, D. L., Ueda, A., Stewart, B. A., Burgess, R. W., Kidokoro, Y., and Schwarz, T. L. (1998) Distinct requirements for evoked and spontaneous release of neurotransmitter are revealed by mutations in the *Drosophila* gene neuronal-synaptobrevin. *J. Neurosci.*, **18**, 2028.
21. Bhat, M. A., Izaddoost, S., Lu, Y., Cho, K.-O., Choi, K.-W., and Bellen, H. J. (1999) Discs lost, a multi PDZ domain protein, interacts with Neurexin IV and Crumbs, and is required for epithelial polarity. *Cell*, **96**, 833.
22. DiAntonio, A., Burgess, R. W., Chin, A. C., Deitcher, D. L., Scheller, R. H., and Schwarz, T. L. (1993) Identification and characterization of *Drosophila* genes for synaptic vesicle proteins. *J. Neurosci.*, **13**, 4924.
23. Littleton, J. T. and Bellen, H. J. (1994) Genetic and phenotypic analysis of thirteen essential genes in cytological interval 22F1–2; 23B1–2 reveals novel genes required for neural development in *Drosophila*. *Genetics*, **138**, 111.
24. Harrison, S. D., Broadie, K., van de Goor, J., and Rubin, G. M. (1994) Mutations in the *Drosophila Rop* gene suggest a function in general secretion and synaptic transmission. *Neuron*, **13**, 555.
25. Schulze, K. L. and Bellen, H. J. (1996) *Drosophila syntaxin* is required for cell viability and may function in membrane formation and stabilization. *Genetics*, **144**, 1713.
26. Pirrotta, V. (1988) Vectors for P-element transformation in *Drosophila*. In *Series vectors for P-element transformation in Drosophila* (ed. R. L. Rodriguez and D. T. Denhardt), p. 437. Butterworths, Boston, Massachusetts.
27. Brand, A. H. and Perrimon, N. (1993) Targeted gene expression as a means of altering cell fates and generating dominant phenotypes. *Development*, **118**, 401.
28. Schuster, C. M., Davis, G. W., Fetter, R. D., and Goodman, C. S. (1996) Genetic dissection of structural and functional components of synaptic plasticity. II. Fasciclin II controls presynaptic structural plasticity. *Neuron*, **17**, 641.
29. Atwood, H. L., Govind, C. K., and Wu, C. F. (1993) Differential ultrastructure of synaptic terminals on ventral longitudinal abdominal muscles in *Drosophila* larvae. *J. Neurobiol.*, **24**, 1008.
30. Jan, L. Y. and Jan, Y. N. (1976) L-glutamate as an excitatory transmitter at the *Drosophila* larval neuromuscular junction. *J. Physiol. (London)*, **262**, 215.
31. Jan, L. Y. and Jan, Y. N. (1976) Properties of the larval neuromuscular junction in *Drosophila melanogaster*. *J. Physiol.*, **262**, 189.
32. Johansen, J., Halpern, M. E., Johansen, K. M., and Keshishian, H. (1989) Stereotypic morphology of glutamatergic synapses on identified muscle cells of *Drosophila* larvae. *J. Neurosci.*, **9**, 710.
33. Keshishian, H., Broadie, K., Chiba, A., and Bate, M. (1996) The *Drosophila* neuromuscular junction: A model system for studying synaptic development and function. *Annu. Rev. Neurosci.*, **19**, 545.
34. Halpern, M. E., Chipa, A., Johansen, J., and Keshishian, H. (1991) Growth cone behavior underlying the development of stereotypic synaptic connections in *Drosophila* embryos. *J. Neurosci.*, **11**, 3227.

35. Coombe, P. E. (1986) The large monopolar cells L1 and L2 are responsible for ERG transients in *Drosophila*. *J. Comp. Physiol.*, **159**, 655.
36. Littleton, J. T., Stern, M., Perin, M., and Bellen, H. J. (1994) Calcium dependence of neurotransmitter release and rate of spontaneous vesicle fusions are altered in *Drosophila synaptotagmin* mutants. *Proc. Natl. Acad. Sci. USA*, **91**, 10888.
37. Wu, M. N., Littleton, J. T., Bhat, M. A., Prokop, A., and Bellen, H. J. (1998) ROP, the *Drosophila* Sec1 homolog, interacts with syntaxin and regulates neurotransmitter release in a dosage-dependent manner. *EMBO J.*, **17**, 127.
38. Martin, A. R. (1955) A further study of the statistical composition of the end-plate potential. *J. Physiol.*, **130**, 114.
39. Broadie, K. S. and Bate, M. (1993) Development of the embryonic neuromuscular synapse of *Drosophila melanogaster*. *J. Neurosci.*, **13**, 144.
40. Katz, B. (1969) *The release of neural transmitter substances*. Liverpool: Liverpool University Press.
41. Matthews, G. (1996) Neurotransmitter release. *Annu. Rev. Neurosci.*, **19**, 219.
42. Zucker, R. S. (1996) Exocytosis: a molecular and physiological perspective. *Neuron*, **17**, 1049.
43. Dodge, F. A. and Rahaminoff, R. (1967) Cooperative action of Ca^{2+} ions in transmitter release at the neuromuscular junction. *J. Physiol. (London)*, **193**, 419.
44. Augustine, G. J. and Charlton, M. P. (1986) Calcium dependence of presynaptic calcium current and post-synaptic response at the squid giant synapse. *J. Physiol. (London)*, **381**, 619.
45. Lando, L. and Zucker, R. S. (1994) Ca^{2+} cooperativity in neurosecretion measured using photolabile Ca^{2+} chelators. *J. Neurophysiol.*, **72**, 825.
46. Jan, Y. N. and Jan, L. Y. (1978) Genetic dissection of short-term and long-term facilitation at the *Drosophila* neuromuscular junction. *Proc. Natl. Acad. Sci. USA*, **75**, 515.
47. Broadie, K., Bellen, H. J., DiAntonio, A., Littleton, J. T., and Schwarz, T. L. (1994) The absence of synaptotagmin disrupts excitation-secretion coupling during synaptic transmission. *Proc. Natl. Acad. Sci. USA*, **91**, 10727.
48. Cope, T. C. and Mendell, L. M. (1982) Distributions of EPSP latency at different group Ia-fiber alpha-motoneuron connections. *J. Neurosci.*, **47**, 469.
49. Llinas, R., Steinberg, I. Z., and Walton, K. (1981) Presynaptic Ca^{2+} currents in squid giant synapse. *Biophys. J.*, **33**, 289.
50. DeLorenzo, R. J. (1981) The calmodulin hypothesis of neurotransmission. *Cell Calcium*, **2**, 365.
51. Pongs, O., Lindemeire, J., Zhu, X. R., Theil, T., Engelkamp, D., Krah-Jentgens, I., *et al.* (1993) Frequenin, a novel calcium-binding protein that modulates synaptic efficacy in the *Drosophila* nervous system. *Neuron*, **11**, 15.
52. Perin, M. S., Fried, V. A., Mignery, G. A., Jahn, R., and Südhof, T. C. (1990) Phospholipid binding by a synaptic vesicle protein homologous to the regulatory region of protein kinase C. *Nature*, **345**, 260.
53. Wada, K., Mizoguchi, A., Kaibuchi, K., Shirataki, H., Ide, C., and Takai, Y. (1994) Localization of rabphilin-3A, a putative target protein for Rab3A, at the sites of $Ca(^{2+})$-dependent exocytosis in PC12 cells. *Biochem. Biophys. Res. Commun.*, **198**, 158.
54. Orita, S., Sasaki, T., Naito, A., Komuro, R., Ohtsuka, T., Maeda, M., *et al.* (1995) Doc2: a novel brain protein having two repeated C2-like domains. *Biochem. Biophys. Res. Commun.*, **206**, 439.
55. Brose, N., Hofmann, K., Hata, Y., and Sudhof, T. C. (1995) Mammalian homologues of

Caenorhabditis elegans unc-13 gene define novel family of C2-domain proteins. *J. Biol. Chem.*, **270**, 25273.

56. Wang, Y., Okamoto, M., Schmitz, F., Hofmann, K., and Sudhof, T. C. (1997) Rim is a putative Rab3 effector in regulating synaptic-vesicle fusion. *Nature*, **388**, 593.
57. Leung, H. T. and Byerly, L. (1991) Characterization of single calcium channels in *Drosophila* embryonic nerve and muscle cells. *J. Neurosci.*, **11**, 3047.
58. Smith, L. A., Wang, X., Peixoto, A. A., Neumann, E. K., Hall, L. M., and Hall, J. C. (1996) A *Drosophila* calcium channel alpha1 subunit gene maps to a genetic locus associated with behavioral and visual defects. *J. Neurosci.*, **16**, 7868.
59. Smith, L. A., Peizoto, A. A., Kramer, E. M., Villella, A., and Hall, J. C. (1998) Courtship and visual defects of cacophony mutants reveal functional complexity of a calcium-channel alpha1 subunit in *Drosophila*. *Genetics*, **149**, 1407.
60. Zheng, W., Feng, G., Ren, D., Eberl, D. F., Hannan, F., Dubald, M., *et al.* (1995) Cloning and characterization of a calcium channel alpha 1 subunit from *Drosophila melanogaster* with similarity to the rat brain type D isoform. *J. Neurosci.*, **15**, 1132.
61. Eberl, D. F., Ren, D., Feng, G., Lorenz, L. J., Van Vactor, D., and Hall, L. M. (1998) Genetic and developmental characterization of Dmca1D, a calcium channel alpha1 subunit gene in *Drosophila melanogaster*. *Genetics*, **148**, 1159.
62. Cohen, P. and Klee, C. B. (1988) *Calmodulin: molecular aspects of cellular regulation*, Vol. 5, p. 1. Elsevier Press, Amsterdam.
63. Griffith, L. C. (1997) *Drosophila melanogaster* as a model system for the study of the function of calcium/calmodulin-dependent protein kinase II in synaptic plasticity. *Inv. Neurosci.*, **3**, 93.
64. Nelson, H. B., Heiman, R. G., Bolduc, C., Kovalick, G. E., Whitley, P., Stern, M., *et al.* (1997) Calmodulin point mutations affect *Drosophila* development and behavior. *Genetics*, **147**, 1783.
65. Arredondo, L., Nelson, H. B., Beckingham, K., and Stern, M. (1998) Increased transmitter release and aberrant synapse morphology in a *Drosophila* calmodulin mutant. *Genetics*, **150**, 265.
66. Stern, M. and Ganetzky, B. (1989) Altered synaptic transmission in *Drosophila* hyperkinetic mutants. *J. Neurogenet.*, **5**, 215.
67. Peters, C. and Mayer, A. (1998) Ca^{2+}/calmodulin signals the completion of docking and triggers a late step of vacuole fusion. *Nature*, **396**, 575.
68. Mallart, A., Angaut-Petit, D., Bourret-Poulain, C., and Ferrus, A. (1991) Nerve terminal excitability and neuromuscular transmission in T(X;Y)V7 and Shaker mutants of *Drosophila melanogaster*. *J. Neurogenet.*, **7**, 75.
69. Rivosecchi, R., Pongs, O., and Mallart, A. (1994) Implication of frequenin in the facilitation of transmitter release in *Drosophila*. *J. Physiol.*, **474**, 223.
70. Tanouye, M. A., Ferrus, A., and Fujita, S. C. (1981) Abnormal action potentials associated with the *Shaker* complex locus of *Drosophila*. *Proc. Natl. Acad. Sci. USA*, **78**, 6548.
71. Dizhoor, A. M., Ray, S., Kumar, S., Niemi, G., Spencer, M., Brolley, D., *et al.* (1991) Recoverin: a calcium sensitive activator of retinal rod guanylate cyclase. *Science*, **251**, 915.
72. Lambrecht, H. G. and Koch, K. W. (1991) A 26 kd calcium binding protein from bovine rod outer segments as modulator of photoreceptor guanylate cyclase. *EMBO J.*, **10**, 793.
73. Littleton, J. T. and Bellen, H. J. (1995) Synaptotagmin controls and modulates synaptic-vesicle fusion in a Ca^{2+}-dependent manner. *Trends Neurosci.*, **18**, 177.
74. Perin, M. S., Brose, N., Jahn, R., and Südhof, T. C. (1991) Domain structure of synaptotagmin (p65). *J. Biol. Chem.*, **266**, 623.

75. Sudhof, T. C. and Rizo, J. (1996) Synaptotagmins: C2-domain proteins that regulate membrane traffic. *Neuron*, **17**, 379.
76. Littleton, J. T., Bellen, H. J., and Perin, M. S. (1993) Expression of synaptotagmin in *Drosophila* reveals transport and localization of synaptic vesicles to the synapse. *Development*, **118**, 1077.
77. DiAntonio, A., Parfitt, K., and Schwarz, T. L. (1993) Synaptic transmission persists in synaptotagmin mutants of *Drosophila*. *Cell*, **73**, 1281.
78. Littleton, J. T., Upton, L., and Kania, A. (1995) Immunocytochemical analysis of axonal outgrowth in synaptotagmin mutations. *J. Neurochem.*, **65**, 32.
79. Fincham, J. R. S. (1966) *Genetic complementation.* W. A. Benjamin, Inc., Amsterdam and New York.
80. Brose, N., Petrenko, A. G., Sudhof, T. C., and Jahn, R. (1992) Synaptotagmin: a Ca^{2+} sensor on the synaptic vesicle surface. *Science*, **256**, 1021.
81. Chapman, E. R., Hanson, P. I., An, S., and Jahn, R. (1995) Ca^{2+} regulates the interaction between synaptotagmin and syntaxin 1. *J. Biol. Chem.*, **270**, 23667.
82. Davletov, B. A. and Sudhof, T. C. (1993) A single C2 domain from synaptotagmin I is sufficient for high affinity Ca^{2+}/phospholipid binding. *J. Biol. Chem.*, **268**, 26386.
83. Charvin, N., L'eveque, C., Walker, D., Berton, F., Raymond, C., Kataoka, M., *et al.* (1997) Direct interaction of the calcium sensor protein synaptotagmin I with a cytoplasmic domain of the alpha1A subunit of the P/Q-type calcium channel. *EMBO J.*, **16**, 4591.
84. Sheng, Z. H., Yokoyama, C. T., and Catterall, W. A. (1997) Interaction of the synprint site of N-type Ca^{2+} channels with the C2B domain of synaptotagmin I. *Proc. Natl. Acad. Sci. USA*, **94**, 5405.
85. Yoshida, A., Oho, C., Akira, O., Kuwahara, R., Ito, T., and Takahashi, M. (1992) HPC-1 is associated with synaptotagmin and the ω-conotoxin receptor. *J. Biol. Chem.*, **267**, 24925.
86. Zhang, J. Z., Davletov, B. A., Sudhof, T. C., and Anderson, R. G. (1994) Synaptotagmin I is a high affinity receptor for clathrin AP-2: implications for membrane recycling. *Cell*, **78**, 751.
87. Gonzales-Gaitan, M. and Jackle, H. (1997) Role of *Drosophila* α-adaptin in presynaptic vesicle recycling. *Cell*, **88**, 767.
88. Jorgensen, E. M., Hartwieg, E., Schuske, K., Nonet, M. L., Jin, Y., and Horvitz, H. R. (1995) Defective recycling of synaptic vesicles in synaptotagmin mutants of *Caenorhabditis elegans*. *Nature*, **378**, 196.
89. Fukuda, M., Moreira, J. E., Lewis, F. M., Sugimori, M., Niinobe, M., Mikoshiba, K., *et al.* (1995) Role of the C2B domain of synaptotagmin in vesicular release and recycling as determined by specific antibody injection into the squid giant synapse preterminal. *Proc. Natl. Acad. Sci. USA*, **92**, 10708.
90. Reist, N. E., Buchanan, J., Li, J., DiAntonio, A., Buxton, E. M., and Schwarz, T. L. (1998) Morphologically docked synaptic vesicles are reduced in synaptotagmin mutants of *Drosophila*. *J. Neurosci.*, **18**, 7662.
91. Geppert, M., Goda, Y., Hammer, R. E., Li, C., Roshal, T. W., Stevens, C. F., *et al.* (1994) Synaptotagmin I: a major Ca^{2+} sensor for transmitter release at a central synapse. *Cell*, **79**, 717.
92. Li, C., Ullrich, B., Zhang, J. Z., Anderson, R. G., Brose, N., and Sudhof, T. C. (1995) $Ca(^{2+})$-dependent and -independent activities of neural and non-neural synaptotagmins. *Nature*, **375**, 594.
93. Bennett, M. K., Calakos, N., and Scheller, R. H. (1992) Syntaxin: a synaptic protein implicated in docking of synaptic vesicles at presynaptic active zones. *Science*, **257**, 255.

94. Schiavo, G., Stenbeck, G., Rothman, J. E., and Söllner, T. H. (1997) Binding of the synaptic vesicle v-SNARE, synaptotagmin, to the plasma membrane t-SNARE, SNAP-25, can explain docked vesicles at neurotoxin-treated synapses. *Proc. Natl. Acad. Sci. USA*, **94**, 997.
95. DiAntonio, A. and Schwarz, T. L. (1994) The effect on synaptic physiology of synaptotagmin mutations in *Drosophila*. *Neuron*, **12**, 909.
96. Parfitt, K., Reist, J. L., Burgess, R., Deitcher, D., DiAntonio, A., and Schwarz, T. L. (1995) *Drosophila* genetics and the functions of synaptic proteins. *Cold Spring Harbor Symp. Quant. Biol.*, **60**, 371.
97. Zinsmaier, K. E., Hofbauer, A., Heimbeck, G., Pflugfelder, G. O., Buchner, S., and Buchner, E. (1990) A cysteine-string protein is expressed in retina and brain of *Drosophila*. *J. Neurogenet.*, **7**, 15.
98. Gundersen, C. B. and Umbach, J. A. (1992) Suppression cloning of the cDNA encoding a candidate subunit of a presynaptic calcium channel. *Neuron*, **9**, 527.
99. Braun, J., Wilbanks, S. M., and Scheller, R. H. (1996) The cysteine string secretory vesicle protein activates Hsc70 ATPase. *J. Biol. Chem.*, **271**, 25989.
100. Buchner, E. and Gundersen, C. B. (1997) The DnaJ-like cysteine string protein and exocytotic neurotransmitter release. *Trends Neurosci.*, **20**, 223.
101. Coppola, T. and Gundersen, C. (1996) Widespread expression of human cysteine string proteins. *FEBS Lett.*, **391**, 269.
102. Mastrogiacomo, A. and Gundersen, C. B. (1995) The nucleotide and deduced amino acid sequence of a rat cysteine string protein. *Brain Res. Mol. Brain Res.*, **28**, 12.
103. Sudhof, T. C. (1995) The synaptic vesicle cycle: a cascade of protein-protein interactions. *Nature*, **375**, 645.
104. Mastrogiacomo, A., Parsons, S. M., Zampighi, G. A., Jenden, D. J., Umbach, J. A., and Gunderson, C. B. (1994) Cysteine string proteins: a potential link between synaptic vesicles and presynaptic calcium channels. *Science*, **263**, 981.
105. van de Goor, J., Ramaswami, M., and Kelly, R. (1995) Redistribution of synaptic vesicles and their proteins in temperature-sensitive *shibire(ts1)* mutant *Drosophila*. *Proc. Natl. Acad. Sci. USA*, **92**, 5739.
106. Gundersen, C. B., Mastrogiacomo, A., Faull, K., and Umbach, J. A. (1994) Extensive lipidation of a Torpedo cysteine string protein. *J. Biol. Chem.*, **269**, 19197.
107. Umbach, J. A., Zinsmaier, K. E., Eberle, K. K., Buchner, E., Benzer, S., and Gundersen, C. B. (1994) Presynaptic dysfunction in *Drosophila csp* mutants. *Neuron*, **13**, 899.
108. Caplan, A. J., Cyr, D. M., and Douglas, M. G. (1993) Eukaryotic homologues of *Escherichia coli* dnaJ: a diverse protein family that functions with hsp70 stress proteins. *Mol. Biol. Cell*, **4**, 555.
109. Cyr, D. M. and Douglas, M. G. (1994) Differential regulation of Hsp70 subfamilies by the eukaryotic DnaJ homologue YDJ1. *J. Biol. Chem.*, **269**, 9798.
110. Glover, J. R. and Lindquist, S. (1998) Hsp104, Hsp70, and Hsp40: a novel chaperone system that rescues previously aggregated proteins. *Cell*, **94**, 73.
111. Chappell, T. G., Welch, W. J., Schlossman, D. M., Palter, K. B., Schlesinger, M. J., and Rothman, J. E. (1986) Uncoating ATPase is a member of the 70 kilodalton family of stress proteins. *Cell*, **45**, 3.
112. Chamberlain, L. H. and Burgoyne, R. D. (1997) The molecular chaperone function of the secretory vesicle cysteine string proteins. *J. Biol. Chem.*, **272**, 31420.
113. Umbach, J. A. and Gundersen, C. B. (1997) Evidence that cysteine string proteins regulate an early step in the Ca^{2+}-dependent secretion of neurotransmitter at *Drosophila* neuromuscular junctions. *J. Neurosci.*, **17**, 7203.

114. Umbach, J. A., Saitoe, M., Kidokoro, Y., and Gundersen, C. B. (1998) Attenuated influx of calcium ions at nerve endings of *csp* and *shibire* mutant *Drosophila*. *J. Neurosci.*, **18**, 3233.
115. Ranjan, R., Bronk, P., and Zinsmaier, K. (1998) Cysteine string protein is required for calcium-secretion coupling of evoked neurotransmission in *Drosophila* but not for vesicle recycling. *J. Neurosci.*, **18**, 956.
116. Leveque, C., Pupier, S., Marqueze, B., Geslin, L., Kataoka, M., Takahashi, M., *et al.* (1998) Interaction of cysteine string proteins with the alpha1A subunit of the P/Q-type calcium channel. *J. Biol. Chem.*, **273**, 13488.
117. Hayashi, T., McMahon, H., Yamasaki, S., Binz, T., Hata, Y., Südhof, T. C., *et al.* (1994) Synaptic vesicle membrane fusion complex: action of clostridial neurotoxins on assembly. *EMBO J.*, **13**, 5051.
118. Sollner, T., Whiteheart, S. W., Brunner, M., Erdjument-Bromage, H., Geromanos, S., Tempst, P., *et al.* (1993) SNAP receptors implicated in vesicle targeting and fusion. *Nature*, **362**, 318.
119. Banerjee, A., Kowalchyk, J. A., DasGupta, B. R., and Martin, T. F. J. (1996) SNAP-25 is required for a late post-docking step in Ca^{2+}-dependent exocytosis. *J. Biol. Chem.*, **271**, 20227.
120. Broadie, K., Prokop, A., Bellen, H. J., O'Kane, C. J., Schulze, K. L., and Sweeney, S. T. (1995) Syntaxin and synaptobrevin function downstream of vesicle docking in *Drosophila*. *Neuron*, **15**, 663.
121. Hunt, J. M., Charlton, M. P., Kistner, A., Habermann, E., Augustine, G. J., and Betz, H. (1994) A post-docking role for synaptobrevin in synaptic vesicle fusion. *Neuron*, **12**, 1269.
122. Risinger, C., Blomqvist, A. G., Lundell, I., Lambertsson, A., Nassel, D., Pieribone, V., *et al.* (1993) Evolutionary conservation of synaptosome-associated protein 25 kDa (SNAP-25) shown by *Drosophila* and *Torpedo* cDNA clones. *J. Biol. Chem.*, **268**, 24408.
123. Risinger, C., Deitcher, D. L., Lundell, I., Schwarz, T. L., and Larhammar, D. (1997) Complex gene organization of synaptic protein SNAP-25 in *Drosophila melanogaster*. *Gene*, **194**, 169.
124. Betz, A., Okamoto, M., Benseler, F., and Brose, N. (1997) Direct interaction of the rat unc-13 homologue Munc13–1 with the N terminus of syntaxin. *J. Biol. Chem.*, **272**, 2520.
125. Fujita, Y., Shirataki, H., Sakisaka, T., Asakura, T., Ohya, T., Kotani, H., *et al.* (1998) Tomosyn: a syntaxin-1-binding protein that forms a novel complex in the neurotransmitter release process. *Neuron*, **20**, 905.
126. Hata, Y., Slaughter, C. A., and Südhof, T. C. (1993) Synaptic vesicle fusion complex contains *unc-18* homologue bound to syntaxin. *Nature*, **366**, 347.
127. Hsu, S. C., Ting, A. E., Hazuka, C. D., Davanger, S., Kenny, J. W., Kee, Y., *et al.* (1996) The mammalian brain rsec6/8 complex. *Neuron*, **17**, 1209.
128. Kee, Y., Lin, R. C., Hsu, S.-C., and Scheller, R. H. (1995) Distinct domains of syntaxin are required for synaptic vesicle fusion complex formation and dissociation. *Neuron*, **14**, 991.
129. Krasnoperov, V. G., Bittner, M. A., Beavis, R., Kuang, Y., Salnikow, K. V., Chepurny, O. G., *et al.* (1997) α-latrotoxin stimulates exocytosis by the interaction with a neuronal G-protein-coupled receptor. *Neuron*, **18**, 925.
130. McMahon, H. T. and Sudhof, T. C. (1995) Synaptic core complex of synaptobrevin, syntaxin, and SNAP25 forms high affinity alpha-SNAP binding site. *J. Biol. Chem.*, **270**, 2123.
131. Sheng, Z.-H., Rettig, J., Takahashi, M., and Catterall, W. A. (1994) Identification of a syntaxin-binding site on N-type calcium channels. *Neuron*, **13**, 1303.

132. Cerezo, J. R., Jimenez, F., and Moya, F. (1995) Characterization and gene cloning of *Drosophila* syntaxin 1 (Dsynt1): the fruit fly homologue of rat syntaxin 1. *Brain Res. Mol. Brain Res.*, **29**, 245.
133. Banfield, D. K., Lewis, M. J., Rabouille, C., Warren, G., and Pelham, H. R. (1994) Localization of Sed5, a putative vesicle targeting molecule, to the cis-Golgi network involves both its transmembrane and cytoplasmic domains. *J. Cell Biol.*, **127**, 357.
134. Garcia, E. P., McPherson, P. S., Chilcote, T. J., Takei, K., and De Camilli, P. (1995) rbSec1A and B colocalize with syntaxin 1 and SNAP-25 throughout the axon, but are not in a stable complex with syntaxin. *J. Cell Biol.*, **129**, 105.
135. Sesack, S. R. and Snyder, C. L. (1995) Cellular and subcellular localization of syntaxin-like immunoreactivity in the rat striatum and cortex. *Neuroscience*, **67**, 993.
136. Stevens, C. F. and Tsujimoto, T. (1995) Estimates for the pool size of releasable quanta at a single central synapse and for the time required to refill the pool. *Proc. Natl. Acad. Sci. USA*, **92**, 846.
137. Kretzschmar, S., Volknandt, W., and Zimmermann, H. (1996) Colocalization on the same synaptic vesicles of syntaxin and SNAP-25 with synaptic vesicle proteins: a re-evaluation of functional models required? *Neurosci. Res.*, **26**, 141.
138. Walch-Solimena, C., Blasi, J., Edelmann, L., Chapman, E. R., von Mollard, G. F., and Jahn, R. (1995) The t-SNAREs syntaxin 1 and SNAP-25 are present on organelles that participate in synaptic vesicle recycling. *J. Cell Biol.*, **128**, 637.
139. Nicholson, K. L., Munson, M., Miller, R. B., Filip, T. J., Fairman, R., and Hughson, F. M. (1998) Regulation of SNARE complex assembly by an N-terminal domain of the t-SNARE Sso1p. *Nat. Struct. Biol.*, **5**, 793.
140. Patel, S. K., Indig, F. E., Olivieri, N., Levine, N. D., and Latterich, M. (1998) Organelle membrane fusion: a novel function for the syntaxin homolog Ufe1p in ER membrane fusion. *Cell*, **92**, 611.
141. Poirier, M. A., Xiao, W., Macosko, J. C., Chan, C., Shin, Y. K., and Bennett, M. K. (1998) The synaptic SNARE complex is a parallel four-stranded helical bundle. *Nat. Struct. Biol.*, **5**, 765.
142. Sutton, R. B., Fasshauer, D., Jahn, R., and Brunger, A. T. (1998) Crystal structure of a SNARE complex involved in synaptic exocytosis at 2.4 Å resolution. *Nature*, **395**, 347.
143. Fasshauer, D., Bruns, D., Shen, B., Jahn, R., and Brunger, A. T. (1997) A structural change occurs upon binding of syntaxin to SNAP-25. *J. Biol. Chem.*, **272**, 4582.
144. Weber, T., Zemelman, B. V., McNew, J. A., Westermann, B., Gmachl, M., Parlati, F., *et al.* (1998) SNAREpins: minimal machinery for membrane fusion. *Cell*, **92**, 759.
145. Baumert, M., Maycox, P. R., Navone, F., DeCamilli, P., and Jahn, R. (1989) Synaptobrevin: an integral membrane protein of 18,000 daltons present in small synaptic vesicles of rat brain. *EMBO J.*, **8**, 379.
146. Trimble, W. S., Cowan, D. M., and Scheller, R. H. (1988) VAMP-1: a synaptic vesicle-associated integral membrane protein. *Proc. Natl. Acad. Sci. USA*, **85**, 4538.
147. Chin, A. C., Burgess, R. W., Wong, B. R., Schwarz, T. L., and Scheller, R. H. (1993) Differential expression of transcripts from *syb*, a *Drosophila melanogaster* gene encoding a VAMP (synaptobrevin) that is abundant in non-neuronal cells. *Gene*, **131**, 175.
148. Sweeney, S. T., Broadie, K., Keane, J., Niemann, H., and O'Kane, C. J. (1995) Targeted expression of tetanus toxin light chain in *Drosophila* specifically eliminates synaptic transmission and causes behavioral defects. *Neuron*, **14**, 341.
149. Ashton, A. C., Li, Y., Doussau, F., Weller, U., Dougan, G., Poulain, B., *et al.* (1995) Tetanus

toxin inhibits neuroexocytosis even when its Zn($^{2+}$)-dependent protease activity is removed. *J. Biol. Chem.*, **270**, 31386.
150. Foran, P., Lawrence, G. W., Shone, C. C., Foster, K. A., and Dolly, J. O. (1996) Botulinum neurotoxin C1 cleaves both syntaxin and SNAP-25 in intact and permeabilized chromaffin cells: correlation with its blockade of catecholamine release. *Biochemistry*, **35**, 2630.
151. Wu, M. N. and Bellen, H. J. (1997) The dissection of synaptic transmission in *Drosophila*. *Curr. Opin. Neurobiol.*, **7**, 624.
152. Nonet, M. L., Saifee, O., Zhao, H., Rand, J. B., and Wei, L. (1998) Synaptic transmission deficits in *Caenorhabditis elegans* synaptobrevin mutants. *J. Neurosci.*, **18**, 70.
153. Saifee, O., Wei, L., and Nonet, M. L. (1998) The *Caenorhabditis elegans* unc-64 locus encodes a syntaxin that interacts genetically with synaptobrevin. *Mol. Biol. Cell*, **9**, 1235.
154. Nichols, B. J., Ungermann, C., Pelham, H. R. B., Wickner, W. T., and Haas, A. (1997) Homotypic vacuolar fusion mediated by t- and v-SNAREs. *Nature*, **387**, 199.
155. Aalto, M. K., Ronne, H., and Keranen, S. (1993) Yeast syntaxins Sso1p and Sso2p belong to a family of related membrane proteins that function in vesicular transport. *EMBO J.*, **12**, 4095.
156. Protopopov, V., Govindan, B., Novick, P., and Gerst, J. E. (1993) Homologs of the Synaptobrevin/VAMP family of synaptic vesicle proteins function on the late secretory pathway in *S. cerevisiae*. *Cell*, **74**, 855.
157. Hessler, N. A., Shirke, A. M., and Malinow, R. (1993) The probability of transmitter release at a mammalian central synapse. *Nature*, **366**, 569.
158. Wickelgren, W. O., Leonard, J. P., Grimes, M. J., and Clark, R. D. (1985) Ultrastructural correlates of transmitter release in presynaptic areas of lamprey reticulospinal axons. *J. Neurosci.*, **5**, 1188.
159. Hawkins, R. D., Kandel, E. R., and Siegelbaum, S. (1993) Learning to modulate transmitter release: themes and variations in synaptic plasticity. *Annu. Rev. Neurosci.*, **16**, 625.
160. Kandel, E. R. (1981) Calcium and the control of synaptic strength by learning. *Nature*, **293**, 697.
161. McMahon, H. T., Missler, M., Li, C., and Südhof, T. C. (1995) Complexins: cytosolic proteins that regulate SNAP receptor function. *Cell*, **83**, 111.
162. Wilson, D. W., Whiteheart, S. W., Wiedman, M., Brunner, M., and Rothman, J. E. (1992) A multisubunit particle implicated in membrane fusion. *J. Cell Biol.*, **117**, 531.
163. Skehel, P. A., Martin, K. C., Kandel, E. R., and Bartsch, D. (1995) A VAMP-binding protein from Aplysia required for neurotransmitter release. *Science*, **236**, 1580.
164. Edelmann, L., Hanson, P. I., Chapman, E. R., and Jahn, R. (1995) Synaptobrevin binding to synaptophysin: a potential mechanism for controlling the exocytotic fusion machine. *EMBO J.*, **14**, 224.
165. Bean, A. J., Seifert, R., Chen, Y. A., Sacks, R., and Scheller, R. H. (1997) Hrs-2 is an ATPase implicated in calcium-regulated secretion. *Nature*, **385**, 826.
166. Garcia, E. P., Gatti, E., Butler, M., Burton, J., and De Camilli, P. (1994) A rat brain Sec1 homologue related to Rop and UNC18 interacts with syntaxin. *Proc. Natl. Acad. Sci. USA*, **91**, 2003.
167. Pevsner, J., Hsu, S.-C., and Scheller, R. (1994) n-Sec1: a neural-specific syntaxin-binding protein. *Proc. Natl. Acad. Sci. USA*, **91**, 1445.
168. Pevsner, J., Hsu, S.-C., Braun, J. E. A., Calakos, N., Ting, A. E., Bennett, M. K., *et al.* (1994) Specificity and regulation of a synaptic vesicle docking complex. *Neuron*, **13**, 353.
169. Brenner, S. (1974) The genetics of *Caenorhabditis elegans*. *Genetics*, **77**, 71.

170. Hosono, R., Hekimi, S., Kamiya, Y., Sassa, T., Murakami, S., Nishiwaki, S., *et al.* (1992) The *unc-18* gene encodes a novel protein affecting the kinetics of acetylcholine metabolism in the nematode *Caenorhabditis elegans*. *J. Neurochem.*, **58**, 1517.
171. Novick, P., Field, C., and Schekman, R. (1980) Identification of 23 complementation groups required for post-translational events in the yeast secretory pathway. *Cell*, **21**, 205.
172. Salzberg, A., Cohen, N., Halachmi, N., Kimchie, Z., and Lev, Z. (1993) The *Drosophila Ras 2* and *Rop* gene pair: a dual homology with a yeast Ras-like gene and a suppressor of its loss-of-function phenotype. *Development*, **117**, 1309.
173. Schulze, K. L., Littleton, J. T., Salzberg, A., Halachmi, N., Stern, M., Lev, Z., *et al.* (1994) *rop*, a *Drosophila* homolog of yeast Sec1 and vertebrate n-Sec1/Munc-18 proteins, is a negative regulator of neurotransmitter release *in vivo*. *Neuron*, **13**, 1099.
174. Fujita, Y., Sasaki, T., Fukui, K., Kotani, H., Kimura, T., Hata, Y., *et al.* (1996) Phosphorylation of Munc-18/n-Sec1/rbSec1 by protein kinase C. *J. Biol. Chem.*, **271**, 7265.
175. Sassa, T., Ogawa, H., Kimoto, M., and Hosono, R. (1996) The synaptic protein UNC-18 is phosphorylated by protein kinase C. *Neurochem. Int.*, **29**, 543.
176. Rothman, J. E. (1994) Mechanisms of intracellular protein transport. *Nature*, **372**, 55.
177. Wilson, D. W., Wilcox, C. A., Flynn, G. C., Chen, E., Kuang, W.-J., Henzel, W. J., *et al.* (1989) A fusion protein required for vesicle-mediated transport in both mammalian cells and yeast. *Nature*, **339**, 355.
178. Block, M. R., Glick, B. S., Wilcox, D. A., Wieland, F. T., and Rothman, J. E. (1988) Purification of an *N*-ethylamaleimide-sensitive protein catalyzing vesicular transport. *Proc. Natl. Acad. Sci. USA*, **85**, 7852.
179. Eakle, K. A., Bernstein, M., and Emr, S. D. (1988) Characterization of a component of the yeast secretion machinery: identification of the SEC18 gene product. *Mol. Cell Biol.*, **8**, 4098.
180. Graham, T. R. and Emr, S. D. (1991) Compartmental organization of Golgi-specific protein modification and vacuolar protein sorting events defined in a yeast sec18 (NSF) mutant. *J. Cell Biol.*, **114**, 207.
181. Boulianne, G. L. and Trimble, W. S. (1995) Identification of a second homolog of *N*-ethylmaleimide-sensitive fusion protein that is expressed in the nervous system and secretory tissues of *Drosophila*. *Proc. Natl. Acad. Sci. USA*, **92**, 7095.
182. Pallanck, L., Ordway, R. W., Ramaswami, M., Chi, W. Y., Krishnan, K. S., and Ganetzky, B. (1995) Distinct roles for *N*-ethylmaleimide-sensitive fusion protein (NSF) suggested by the identification of a second *Drosophila* NSF homolog. *J. Biol. Chem.*, **270**, 18742.
183. Kawasaki, F., Mattiuz, A. M., and Ordway, R. W. (1998) Synaptic physiology and ultrastructure in comatose mutants define an *in vivo* role for NSF in neurotransmitter release. *J. Neurosci.*, **18**, 10241.
184. Otto, H., Hanson, P. I., and Jahn, R. (1997) Assembly and disassembly of a ternary complex of synaptobrevin, syntaxin, and SNAP-25 in the membrane of synaptic vesicles. *Proc. Natl. Acad. Sci. USA*, **94**, 6197.
185. Bellen, H. J. (1998) The fruit fly: a model organism to study the genetics of alcohol abuse and addiction. *Cell*, **93**, 909.
186. Davis, R. L. (1996) Physiology and biochemistry of *Drosophila* learning mutants. *Physiol. Rev.*, **76**, 299.
187. Grigliatti, T. A., Hall, L., Rosenbluth, R., and Suzuki, D. T. (1973) Temperature-sensitive mutations in *Drosophila melanogaster*. XIV. A selection of immobile adults. *Mol. Gen. Genet.*, **120**, 107.

188. Homyk, T., Jr. (1977) Behavioral mutants of *Drosophila melanogaster*. II. Behavioral analysis and focus mapping. *Genetics*, **87**, 105.
189. Andrews, J., Smith, M., Merakovsky, J., Coulson, M., Hannan, F., and Kelly, L. E. (1996) The *stoned* locus of *Drosophila melanogaster* produces a dicistronic transcript and encodes two distinct polypeptides. *Genetics*, **143**, 1699.
190. Cremona, O. and De Camilli, P. (1997) Synaptic vesicle endocytosis. *Curr. Opin. Neurobiol.*, **7**, 323.
191. Stimson, D. T., Estes, P. S., Smith, M., Kelly, L. E., and Ramaswami, M. (1998) A product of the *Drosophila* stoned locus regulates neurotransmitter release. *J. Neurosci.*, **18**, 9638.
192. Klagges, B. R., Heimbeck, G., Godenschwege, T. A., Hofbauer, A., Pflugfelder, G. O., Reifegerste, R., Reisch, D., Schaupp, M., Buchner, S. and Buchner, E. (1996) Invertebrate synapsins: a single gene codes for several isoforms in *Drosophila*. *J. Neurosci.* **16**, 3154.

10 Genetic analysis of neurotransmitter release in mice and humans

THOMAS E. LLOYD and HUGO J. BELLEN

1. Introduction

While organisms such as *C. elegans*, *Drosophila*, and squid are excellent model systems for studying the fundamental principles of neurotransmitter release *in vivo*, one of the ultimate goals of these studies is to understand the mechanisms underlying neurotransmitter release in humans. In this respect, the use of invertebrate model systems is validated by the observation that most proteins and processes implicated in synaptic transmission are highly conserved between invertebrates and vertebrates. However, the simplicity of invertebrate model systems does not substitute for the complexity of vertebrate systems, as many sophisticated physiological paradigms and behavioural analyses can only be performed in complex systems such as mice. These studies may provide clues into not only the normal function of genes, but also the role these genes may play in human behaviour and disease. Thus, given the similarities between the mouse and human nervous system, a reverse genetic approach using knock-out technology in mice is at present, and will continue in the foreseeable future, to be one of the most valuable ways to define the function of proteins that are putatively involved in neurotransmitter release in humans.

We will first outline the techniques used to generate and study mutant mice. Next, the phenotype of each knock-out mouse will be described and compared with the phenotypes of mutations obtained in other systems in order to help elucidate the function of the mutated gene. Particular attention will be placed on studies of synapsin, synaptotagmin, and rab3A knock-out mice. In addition, non-targeted mutations in both mouse and human genes implicated in neurotransmitter release will also be discussed. Finally, the advantages and disadvantages of using mouse as a model system will be compared with other systems, and directions for future studies will be discussed.

2. Generating mutant mice and phenotypic analyses

2.1 'Knock-out' technology

Generating homozygous null mutations in mice using homologous recombination is now performed routinely (for review, see refs 1 and 2). Briefly, a targeting vector is made in which an exon or exons of a particular gene is replaced by a selectable marker. The vector containing the altered genomic DNA is then linearized and electroporated into embryonic stem (ES) cells. The cells that successfully integrate the DNA are then injected into mouse embryos at the blastocyst stage and transplanted into pseudopregnant female mice. Because ES cells are totipotent, mice can be made which contain germline cells derived from ES cells containing the altered gene. These 'chimeric' mice can then be bred to generate homozygous mutant mice.

The first step in generating a knock-out mouse is the isolation of a genomic fragment of the gene of interest. As shown in Fig. 1, a targeting vector is then created by replacing one or more exons of the genomic fragment with a neomycin-resistance gene or other selectable marker. Ideally the first exon, or an exon encoding an important domain, is replaced in an attempt to ensure that the construct will produce a null mutation. A second selectable marker such as thymidylate kinase is added beyond the end of the distal genomic fragment to negatively select ES cells with a random integration rather than the desired targeted integration (2). When the

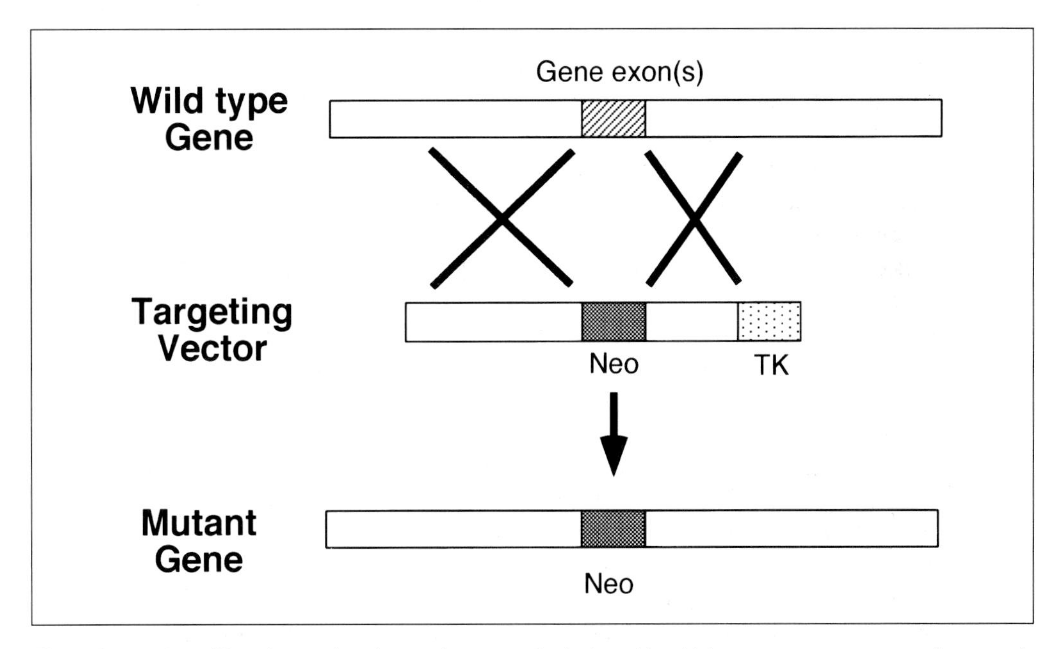

Fig. 1 Generation of knock-out mice. A targeting vector is designed in which a gene exon or exons of a genomic fragment is replaced with the neomycin-resistance gene. The thymidylate kinase (TK) gene at the end of the vector allows negative selection of random integration events. When electroporated into ES cells, the vector should undergo homologous recombination (Xs) with the cell's genomic DNA, replacing the exon(s) of the wild-type gene with the neo gene.

targeting construct is linearized and electroporated into ES stem cells, it should undergo homologous recombination to replace the wild-type exon(s) with the neomycin-resistance gene (Fig. 1). After allowing ES cells to integrate the construct by growing them on non-selectable media, cells are transferred to media containing neomycin. This positively selects for cells that have successfully integrated the construct into the genome. Because homologous recombination will cause loss of the thymidylate kinase, a second selection is performed to negatively select constructs which have undergone random integration and still contain the TK gene. ES cells are then screened using PCR or Southern blot for cells which have successfully 'knocked-out' the gene.

Once the proper ES cells have been generated, they are injected into mouse embryos at the blastocyst stage when the cells are still totipotent. These embryos are then transplanted into the uterus of pseudopregnant female mice. Newborn mice will appear chimeric, as ES cells and embryos are derived from different mouse strains with different phenotypic markers such as hair colour. If chimeric males contain germ cells derived from the altered ES cell, they should produce progeny heterozygous for the knocked-out gene when bred to wild-type females. Homozygous mutant mice are then generated by inbreeding heterozygotes and identified by performing PCR on genomic DNA extracted from their tails.

2.2 Preliminary phenotypic analysis

Once mouse lines with homozygous mutations in the gene of interest are established, an array of assays are performed to ascertain the biochemical and morphological nature of the mutant defect. Because almost all mutations in mouse proteins implicated in neurotransmitter release thus far published develop to term, techniques used to study embryonic lethal mutations will not be discussed. First and foremost, immunoblotting is performed on newborn mouse brains to determine if the protein of interest is absent in homozygous mutant mice (referred to as 'mutants'). The antibodies used in these experiments must be able to specifically detect the protein made by the mutated gene, as there are often multiple isoforms of genes which may cross-react with the antibody. Even though the targeted homologous insertion may remove all of an exon, it is possible that the remaining exons can be properly spliced together and be translated into a functional protein. For example, in the synaptotagmin I knock-out, the deleted exon corresponds to a 38 amino acid stretch in the C-terminal half of the syntaptotagmin protein (3). Fortunately, levels of the truncated protein in the mutant mice were less than 5% of wild-type level, suggesting that the truncated protein is unstable and degraded.

Once the biochemical nature of the lesion is established, phenotypic characterization is performed. First, the viability of homozygous mutant mice is assessed. If newborn mutant pups are healthy, one-quarter of progeny from heterozygous matings should be homozygous for the mutation. Mutations that cause failure to thrive, on the other hand, will often show a bias in the Mendelian ratio. For example, matings between mice heterozygous for synaptotagmin I mutations produce half the expected number of mutant pups, possibly because the mutant pups are unhealthy and are cannibalized by their mothers (3). Once identified, newborn mutant mice are followed to determine if they have increased mortality or morbidity when compared with heterozygous or wild-type mice from the same litter. Mutant mice are analysed for their ability to breathe and suckle milk from their mothers, two fundamental behaviours needed for early survival. Next, the strength, coordination, response to stimulation, and reflexes are assayed to determine if mice have a functional nervous system. Because mice and humans with altered brain function are often prone to developing seizures, the frequency of both spontaneous and stimulation-induced seizures can be measured in mutant mice. Finally, behavioural abnormalities such as hyperactivity are also beginning to be analysed in some mutants, and in the future, complex behavioural assays may allow mutations in different genes implicated in neurotransmitter release to be associated with specific behavioural abnormalities.

After determination of the gross mutant phenotype, the brains of mutant mice are analysed for morphological or biochemical alterations. Immunohistochemistry is performed on brain slices from mutant mice using a variety of antibodies to synaptic proteins to analyse the overall architecture of the brain and the synapses. Thus far, mutations in proteins implicated in neurotransmitter release have not shown obvious morphological defects, suggesting that synaptogenesis may occur normally even with alterations in neurotransmitter release. Finally, transmission electron microscopy can be performed on mutant brains to determine if synapses contain normal numbers and distribution of synaptic vesicles.

2.3 Electrophysiological analysis of knock-out mice

The most sensitive assays for identifying alterations in neurotransmitter release are electrophysiological analyses of mutant neurons. One advantage of using mice is that electrophysiology can be performed on central synapses of the brain, rather than the neuromuscular junction (NMJ) which is typically used in *Drosophila* and other systems (see Chapter 9). Central synapses are very different from NMJs, as the latter have many more boutons and release sites. Synapses in the central nervous system, on the other hand, are finely-tuned, and normally release only one SV per synapse with each stimulation.

Two electrophysiological preparations commonly used to study central synapses are the hippocampal slice and cultured hippocampal neurons. Hippocampal slices are used to study synaptic function for several reasons:

(a) The anatomic structure of the hippocampus allows it to be easily removed from the brain and sectioned in ways that preserve much of the normal cellular circuitry.

(b) Fibre tracts and soma in different layers and pathways can be distinguished at low magnification.

(c) A single slice contains several axonal projections of different neurons due to their lamellar organization.

(d) Most of the physiological parameters recorded in hippocampal slices *in vitro* mimic those recorded in the hippocampus *in vivo* (reviewed in ref. 4).

Extracellular electrophysiological recordings of hippocampal slices reliably measure the changes in postsynaptic potential (EPSP) after stimulation of presynaptic inputs. Figure 2A diagrams the synapses of the hippocampus that may be analysed using the slice preparation. Typically, the Schaffer collateral fibres are stimulated, and the response is recorded in CA1 pyramidal neurons that receive these inputs. Other synapses such as the Mossy fibre/CA3 and perforant fibre/dentate gyrus granule cell synapse can also be analysed using this preparation.

Cultured hippocampal neurons can be used to study the electrophysiology of mutant mice which do not survive long enough for the hippocampus to mature, as is the case in synaptotagmin mutants (3). Additionally, cultured neurons can be used to measure more sophisticated properties of synaptic transmission such as the probability of release (5, 6). In this preparation, intracellular recordings are performed on isolated pairs of neurons using the whole-cell patch-clamping technique. However, because electrophysiological recordings of hippocampal slices are thought to be very similar to *in vivo* recordings, slices are used whenever possible.

The two fundamental parameters of synaptic function are the evoked release of neurotransmitter due to nerve stimulation, and spontaneous neurotransmitter release in the absence of stimulation (see Chapter 9). The amplitude of evoked release (EPSP) is measured following a stimulus of the presynaptic fibres or neuron. The frequency and amplitude of spontaneous release is measured by recording postsynaptic responses (mEPSPs) in absence of stimulation or after applying hypertonic sucrose to increase the frequency of spontaneous release (7, 8).

Synapses undergo activity-dependent changes in synaptic transmission referred to as short- and long-term plasticity. Figure 2B demonstrates several forms of short-term plasticity, including facilitation, depression, and potentiation (reviewed in ref. 9). Synaptic facilitation is the phenomenon whereby the amplitude of the EPSP increases with each successive stimulation (10). Facilitation is often analysed by measuring paired-pulse facilitation (PPF), the enhancement of EPSP amplitude following the second of a pair of stimuli. Facilitation is generally thought to occur due to residual calcium build up with increasing stimulation (11). When the intracellular calcium level is not restored to resting levels prior to a subsequent stimulus, the increase in intracellular calcium will enhance the amount of neurotransmitter release. Furthermore, because the amount of neurotransmitter release is dependent upon the fourth power of intracellular calcium, increasing stimulation frequency or extracellular calcium concentration exponentially increases the amount of facilitation (12). Finally, some synapses undergo depression (PPD) rather than facilitation (PPF) in response to paired stimuli, but the mechanism is unknown (13, 14).

As shown in Fig. 2B, during repetitive high frequency stimulation, the initial facilitation eventually becomes a depression as the supply of readily-releasable SVs becomes exhausted (15, 16). For this reason, this property of activity-dependent depression is often referred to as fatigue. The time that it takes the EPSP amplitude to recover from this depression is thought to reflect the ability of the synapse to replenish its readily-releasable vesicle pool (17). Finally, after a synapse undergoes tetanic stimulation (a short period of high frequency stimulation) and is allowed to recover from the activity-dependent depression, it may undergo a post-tetanic potentiation (PTP) (see Fig. 2B) (9). Like facilitation, PTP is also thought to be at least

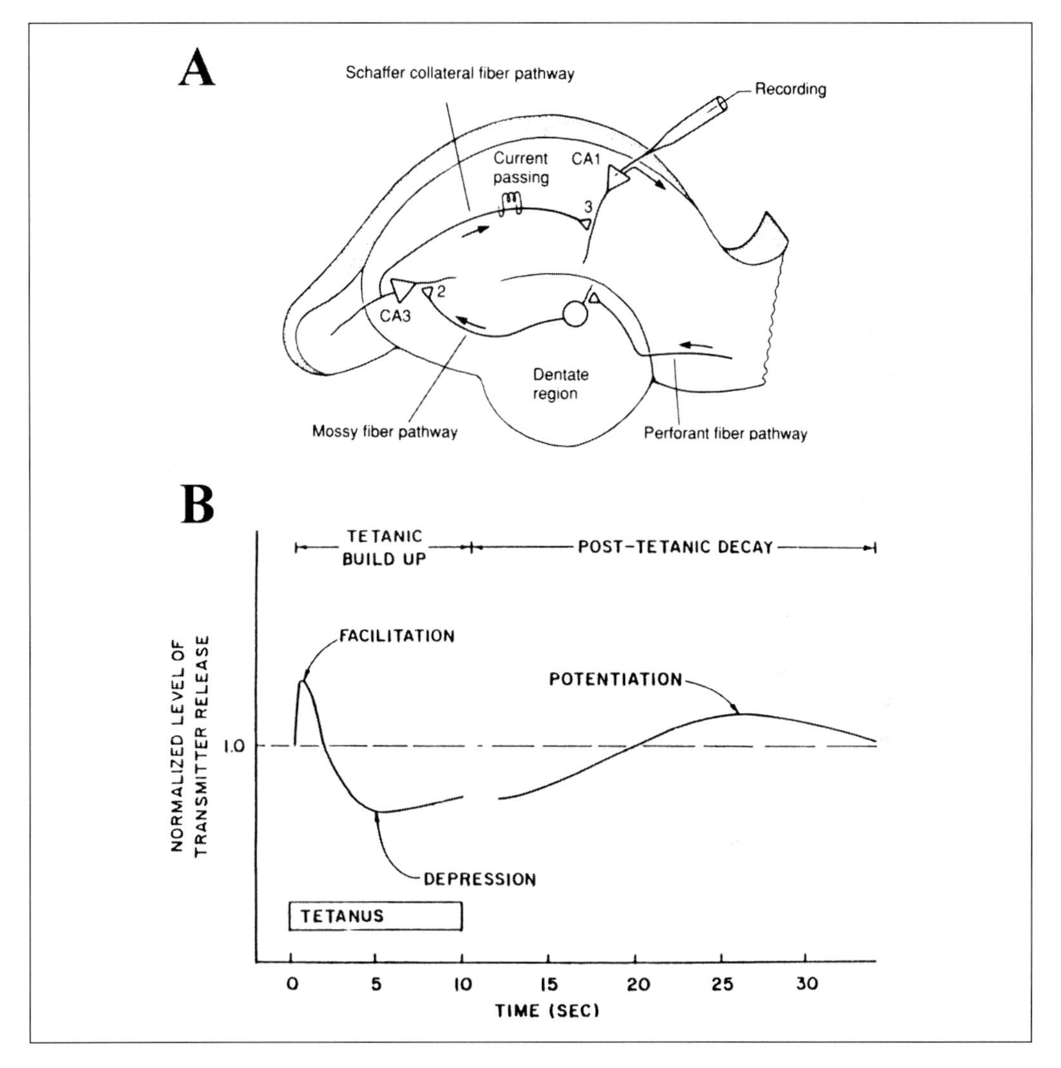

Fig. 2 Hippocampal electrophysiology. (A) Diagram of the major hippocampal pathways commonly studied using the hippocampal slice preparation. Synaptic transmission is analysed by stimulating the pathways (or fibres) and measuring the response in the postsynaptic neurons (arrows denote direction of impulse flow). (B) Parameters of short-term synaptic plasticity. During repetitive, high frequency (tetanic) stimulation, there is an initial enhancement of neurotransmitter release (facilitation), followed by a period of depression (or fatigue). After the tetanus, a post-tetanic potentiation of neurotransmitter release results with slower kinetics of induction and decay than facilitation. (Adapted from ref. 158 (A), courtesy of Appleton & Lange; ref. 9 (B), courtesy of Annual Reviews, Inc.)

partially due to residual calcium; however, other mechanisms such as increased intracellular sodium due to the opening of voltage gated sodium channels also appear to play a role (18, 19).

In addition to short-term plasticity, synapses may also undergo long-term changes in synaptic strength. The best characterized form of long-term plasticity is long-term potentiation (LTP) (reviewed in ref. 20), which is thought to be a cellular mechanism for memory (21). Like PTP, LTP also occurs following a tetanus of stimulation, but unlike PTP which lasts on the order of minutes, LTP lasts for hours or even days. Furthermore, LTP requires a stringent stimulation protocol for induction. The mechanisms for LTP are complex and may have both presynaptic and postsynaptic components depending on the type of synapse studied (22).

At least one form of hippocampal LTP induction requires the activation of NMDA-type glutamate receptors on the postsynaptic membrane (reviewed in ref. 20). NMDA receptors are special glutamate receptors which require a depolarization of the postsynaptic membrane in addition to binding glutamate for the channel to open. When activated, these channels conduct both calcium and sodium, unlike other glutamate receptors which only conduct sodium. The entry of calcium into the postsynaptic cell then triggers a series of events including protein phosphorylation and gene transcription which ultimately leads to increased synaptic strength. The postsynaptic cell also sends a retrograde signal to the presynaptic cell, possibly mediated through nitric oxide (NO), which allows the pre- and postsynaptic neurons to communicate with each other, thereby increasing their synaptic connectivity (20, 23, 24).

A second form of LTP is thought to be predominantly presynaptic in nature. Unlike NMDA-dependent LTP which occurs in the CA1 region of the hippocampus, presynaptic LTP occurs predominantly in Mossy fibres of the CA3 region (25). This presynaptic LTP appears to require the cAMP-dependent protein kinase (PKA) and is at least partially due to a potentiation in neurotransmitter release; however, the mechanism of this form of LTP is unknown (26–28).

Together, short- and long-term mechanisms of plasticity allow the brain to perform many of its complex functions. Both short- and long-term plasticity are at least partially mediated through modulations of neurotransmitter release. As will become apparent in later sections, many mouse mutations in genes implicated in neurotransmitter release specifically alter different forms of plasticity. Analysis of knock-out mice may allow an elucidation of the complex mechanisms of synaptic plasticity and correlation of changes in synaptic function with alterations in the animal's behaviour.

2.4 Are mouse knock-outs informative?

Critics of knock-out approaches often raise three major reservations concerning the approach. First, the lack of a protein during development may alter the developmental program and/or induce compensatory developmental mechanisms which may lead to false interpretations about the function of the protein in the adult animal.

Secondly, removal of a protein may not show a phenotype because other, related or unrelated, proteins compensate for its loss. Thirdly, the lack of a protein may cause a phenotype because it disrupts the transcription, synthesis, stability, or localization of other proteins required for synaptic transmission. Although these criticisms are occasionally borne out to be correct, they apply to all genetic approaches in all systems amenable to genetic analysis and certainly do not preclude the use of genetic approaches.

We will briefly address each of these criticisms. First, loss of a gene early in development may trigger compensatory changes in mice. This may be especially true in the brain, as various mechanisms of synaptic plasticity discussed in the previous section may play a role in activity-dependent developmental mechanisms. For this reason, in some cases studying the chronic loss of a protein using knock-out mice may be less informative than studying the acute loss by injecting antisense mRNA or antibodies or using toxins (see Chapter 6). However, injection experiments may lack the sensitivity and/or specificity of the knock-out approach. While the loss of a gene may alter development and thereby confound the analysis of the adult phenotype, it does allow one to study the role of the gene in normal development by studying mice at earlier time points. Also, in many cases, developmental defects should be apparent by morphological analysis, and such defects have not been observed in most mouse mutations of genes implicated in synaptic transmission. Finally, if the lack of a protein is shown or thought to cause developmental defects, this can be overcome with conditional knock-outs using the Cre-Lox recombination technology (29). In this system, bigenic mice are engineered which contain:

- the Cre-recombinase gene under control of a tissue-specific or inducible promoter
- an exon or exons of a gene contained (or 'floxed') within loxP sites.

When the Cre-recombinase protein is turned on, it enzymatically excises the floxed gene. This approach provides temporal and/or spatial specificity to the knock-out approach to overcome developmental defects, and will likely become a powerful technique for analysing genes implicated in neurotransmitter release in the future.

Secondly, mice and humans typically have approximately four homologues for each *Drosophila* or *C. elegans* gene, the two other model systems most amenable to genetic analysis of synaptic transmission (see Chapters 8 and 9). The fact that these different protein products or isoforms may play redundant roles may complicate the interpretation of the phenotype associated with the removal or lack of a gene in mice. However, this caveat plagues every geneticist in every organism to different extents. In addition, redundancy can also be a windfall, especially if redundant proteins allow the animal to survive, thereby permitting the study of phenotypes that are otherwise difficult to tackle. This pitfall can be addressed by making double and triple knock-outs to remove homologous genes.

Finally, it has been argued that the phenotype of some knock-out mice is due to alterations in the expression of another gene. For example, in rab3A mutant mice, there is a 70% reduction in rabphillin protein, presumably because the interaction of rabphillin with rab3A stabilizes the protein (30). The question then arises, is the knock-out phenotype due to the loss of rab3A or the loss of rabphillin? However, this criticism is not valid from a geneticist's perspective, as the phenotype associated with a knock-out mouse still provides information about the role of the protein *in vivo*. In summary, while there are limitations, no experimental approach surpasses the removal of a gene *in vivo*, and the knock-out of a gene in mice must be considered one of the key experiments in our attempts to define the function of a gene or gene family *in vivo*.

3. Mutations in synaptotagmin and rab3A: implications for a role in the final steps of exocytosis

Recent studies on synaptotagmin and rab3A in mice suggest that these two proteins play opposite roles in the final steps of exocytosis (reviewed in ref. 31). Synaptotagmin has been proposed to be the primary calcium sensor, allowing synaptic vesicles to fuse with the presynaptic membrane in response to calcium influx. Rab3A, on the other hand, has been proposed to play an inhibitory role in allowing only one synaptic vesicle to fuse at each active zone. This section will discuss the phenotypes associated with mutations in these proteins in mice, and how these results compare with studies in other systems.

3.1 Synaptotagmin

A role for synaptotagmin I in sensing the calcium signal for exocytosis was first proposed based on biochemical studies. Synaptotagmin I is specifically expressed in the brain as a synaptic vesicle integral membrane protein (32, 33). Synaptotagmin I binds to both syntaxin and SNAP-25, suggesting that its interaction with the core complex may regulate the docking and/or fusion of synaptic vesicles with the presynaptic membrane (34, 35). Members of the synaptotagmin family contain two C2 domains (C2A and C2B), and C2A binds to calcium and phospholipids in concentrations similar to that found in the active zone during exocytosis (33, 36, 37). The binding of calcium to synaptotagmin displays cooperativity with a Hill coefficient of $n = 2–3$, consistent with measurements of the calcium dependence of neurotransmitter release (38, 39). Moreover, synaptotagmin binds calcium with higher affinity than strontium and barium and co-immunoprecipitates with N-type calcium channels, two other proposed properties of the calcium sensor (40, 41). Thus, biochemical data suggest that synaptotagmin may be a calcium sensor for neurotransmitter release.

In order to determine the function of synaptotagmin I *in vivo*, a deletion mutation in the synaptotagmin I gene was generated using the strategy discussed earlier. Because extensive screening revealed only one genomic DNA clone containing the exon corresponding to amino acids 270–308 (within the C2B domain), this construct was used to make the synaptotagmin deletion (3). Although it is possible that the remaining protein exhibits residual activity, the expression level of the truncated protein in mutant mice was only 5% of the wild-type protein. While mice homozygous for the synaptotagmin mutation are indistinguishable from wild-type litter mates at birth, within two to three hours after birth, they become increasingly weak, have impaired suckling ability, and die within 48 hours. The morphology of the brain is normal in mutant mice, suggesting that synaptotagmin is not required for normal early brain development. Furthermore, the mutation in synaptotagmin I did not alter the protein levels of other synaptic proteins, including synaptotagmin II and rab3A.

Because synaptotagmin I mutant mice died shortly after birth, cultured hippocampal neurons rather than hippocampal slices were used to determine the electrophysiological consequences of the loss of synaptotagmin in these mice. While the overall morphology and number of synapses formed in cultured neurons from mutant mice was indistinguishable from wild-type cultures, electrophysiological analysis demonstrated significant differences between the two neuronal cultures. As shown in Fig. 3A and 3B, whole-cell patch-clamp electrophysiology demonstrated a marked decrease in the amplitude of evoked postsynaptic current (EPSCs) from mutant cultures, whereas the frequency of miniature EPSCs (mEPSCs) was unchanged. This suggests that synaptotagmin I is required for calcium-regulated neurotransmitter release but does not play a role in spontaneous release.

Furthermore, three sets of experiments demonstrated that vesicles were docked normally in synaptotagmin mutants and capable of undergoing fusion with the presynaptic membrane. First, transmission electron microscopy (TEM) did not show qualitative differences in the number or distribution of docked vesicles between wild-type and mutant mice. Secondly, triggering release with hypertonic sucrose or α-latrotoxin gave close to wild-type responses in mutant cultures, suggesting that the core exocytosis machinery was not altered in these mutants. Finally, analysis of the rate of release in mutant and wild-type cultures suggested that the fast (synchronous) component of release was largely absent, whereas the slow (asynchronous) component was unchanged (Fig. 3C and 3D). These findings are consistent with synaptotagmin playing a positive role in the late stages of fast, calcium-regulated exocytosis.

These observations in mice, combined with biochemical data and other *in vivo* and *in vitro* manipulations suggest that synaptotagmin is a major calcium sensor *in vivo* (reviewed in ref. 42). Possibly the best evidence that synaptotagmin is a calcium sensor is the finding that synaptotagmin partial loss-of-function alleles in *Drosophila* show altered calcium dependence for neurotransmitter release (43). The finding that synaptotagmin knock-out mice completely lack the fast evoked response yet have normal slow evoked response further supports the hypothesis that synaptotagmin plays an active role in allowing fusion in response to a calcium signal (3). However, in order to test this hypothesis directly, the critical experiment would be to determine if mutating the synaptotagmin residues responsible for calcium binding also leads to a significant reduction in evoked release or an altered calcium dependence of release. This would provide direct evidence *in vivo* that the calcium binding function of synaptotagmin is essential for normal evoked neurotransmission.

In addition to sensing the calcium signal, synaptotagmin has also been proposed to play other roles as well. For example, synaptotagmin mutants in *Drosophila* have an increased frequency of spontaneous neurotransmitter release, suggesting that synaptotagmin may play an inhibitory role in preventing vesicle release in the

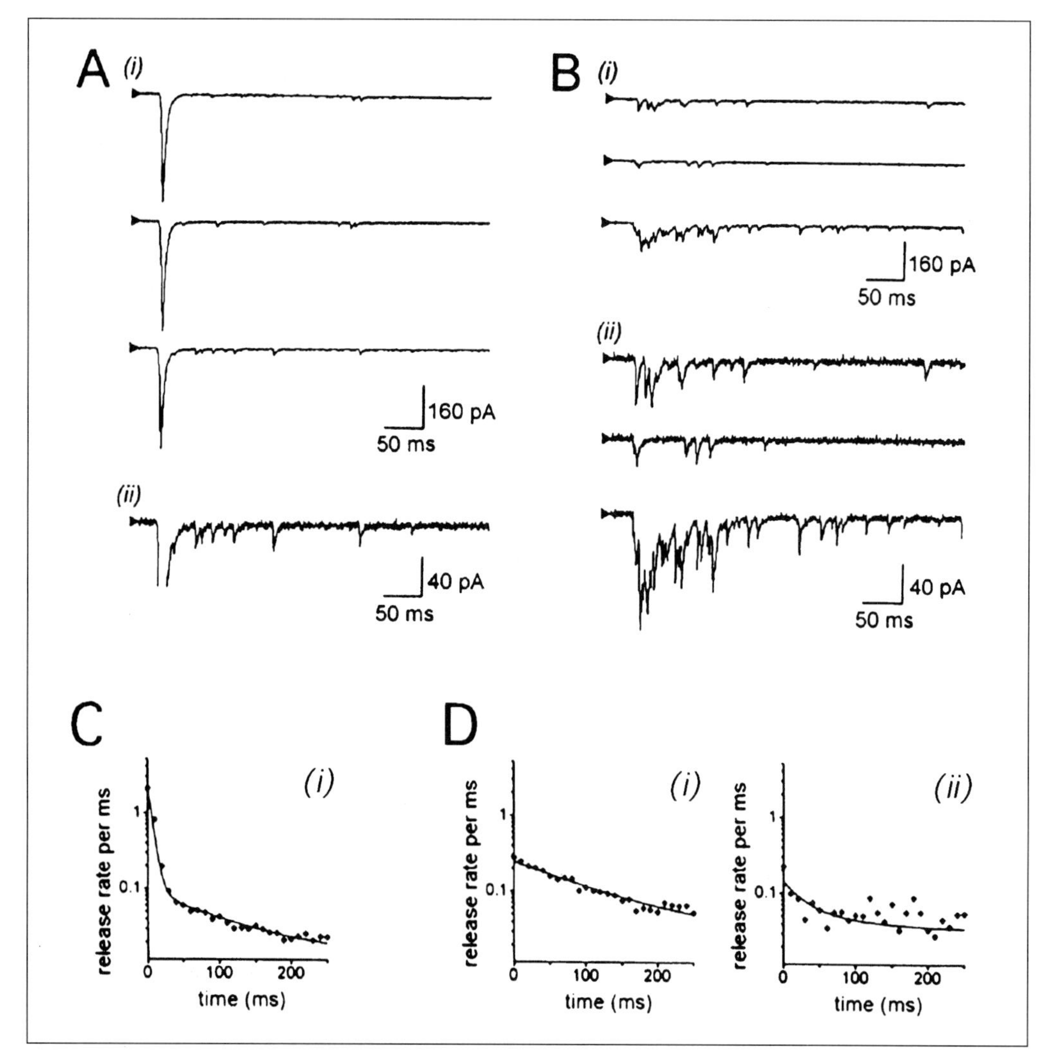

Fig. 3 Phenotype of synaptotagmin knock-out mice. (A, B) (*i*) Whole-cell patch-clamp electrophysiological measurements of EPSCs from three different pairs of wild-type (A) and mutant (B) hippocampal neurons. (*ii*) Fourfold increase in vertical scale to show mEPSCs. Although somewhat variable, the amplitude of EPSPs is severely reduced in mutant neurons. (C, D) Analysis of the rate of release per msec after stimulation in wild-type (C) and mutant (D) neurons, demonstrating the severe reduction of the fast component of release in mutant neurons. Recordings were carried out in 0.5 mM Mg^{2+} and 10 mM Ca^{2+}. (From ref. 3, courtesy of Cell Press.)

absence of calcium (see Chapter 9). This 'fusion clamp' model is also suggested by studies performed in *Aplysia* neurons which displayed a decrease in EPSP amplitude with synaptotagmin overexpression and an increase in amplitude with antisense mRNA (44). CHO fibroblast cells transfected with synaptotagmin, on the other hand, had increased calcium-evoked release but reduced spontaneous release of endocytosed acetylcholine, suggesting that synaptotagmin enhances the efficiency of excitation–coupling (45). How do we reconcile these observations with the phenotype of synaptotagmin knock-out mice? It is possible that an increase in mEPSC frequency is not observed in mutant mice because there is a small amount of remaining synaptotagmin or that other synaptotagmin isoforms perform this inhibitory role.

Furthermore, recent EM studies demonstrate a reduction in the number of morphologically docked vesicles in *Drosophila* synaptotagmin mutants (see Chapter 9) (46), supporting the hypothesis that synaptotagmin may play a role in the docking of synaptic vesicles (34, 47, 48). In addition, studies in *C. elegans* suggest that synaptotagmin is also critical for endocytosis of synaptic vesicles, possibly through its interaction with the clathrin adapter protein AP2 (49, 50). Finally, while synaptotagmin is likely a major calcium sensor in the fast component of release, other calcium sensors must also exist to explain the remaining asynchronous component of calcium-evoked release. In summary, synaptotagmin likely functions in multiple steps of neurotransmitter release, including docking, fusion in response to calcium, and endocytosis.

3.2 Rab3A

While synaptotagmin plays an active role in allowing synaptic vesicles to fuse in response to calcium, rab3A has recently been proposed to play an inhibitory role in allowing only one synaptic vesicle to fuse at each active zone. This data obtained in rab3A mutant mice is in contrast to data from yeast and *C. elegans* which suggest that rabs play an active role in vesicle docking. This section will discuss four studies of the rab3A knock-out mouse and will compare these results with data obtained in other systems.

Proteins of the rab family are small GTPases implicated in multiple stages of vesicular transport (reviewed in ref. 51). Similar to members of the syntaxin and synaptobrevin family, different members of the rab family are essential for different steps of vesicular transport. Rabs associate with vesicular membranes in the GTP-bound form, but dissociate upon hydrolysis of GTP. However, the mechanism of rab action in vesicular transport is unknown. Rab3A has been implicated as the rab family member involved in neurotransmitter release due to its neural-specific localization (52–54).

Rab3A knock-out mice were generated through deletion of a portion of the promoter along with the first two exons, and resulted in undetectable levels of rab3A protein (30). Nevertheless, homozygous mutant mice were viable, and no morphological or gross behavioural defects were found using the standard assays previously described. Electrophysiogical analysis of CA1 hippocampal neurons after stimulation of Schaffer collaterals in hippocampal slices suggested that the amplitude of excitatory postsynaptic currents (EPSCs) was normal, as were most of the standard assays for synaptic plasticity: paired-pulse facilitation (PPF), post-tetanic potentiation (PTP), and long-term potentiation (LTP). As shown in Fig. 4A, the only phenotypic abnormality initially found in these mice was an increase in synaptic

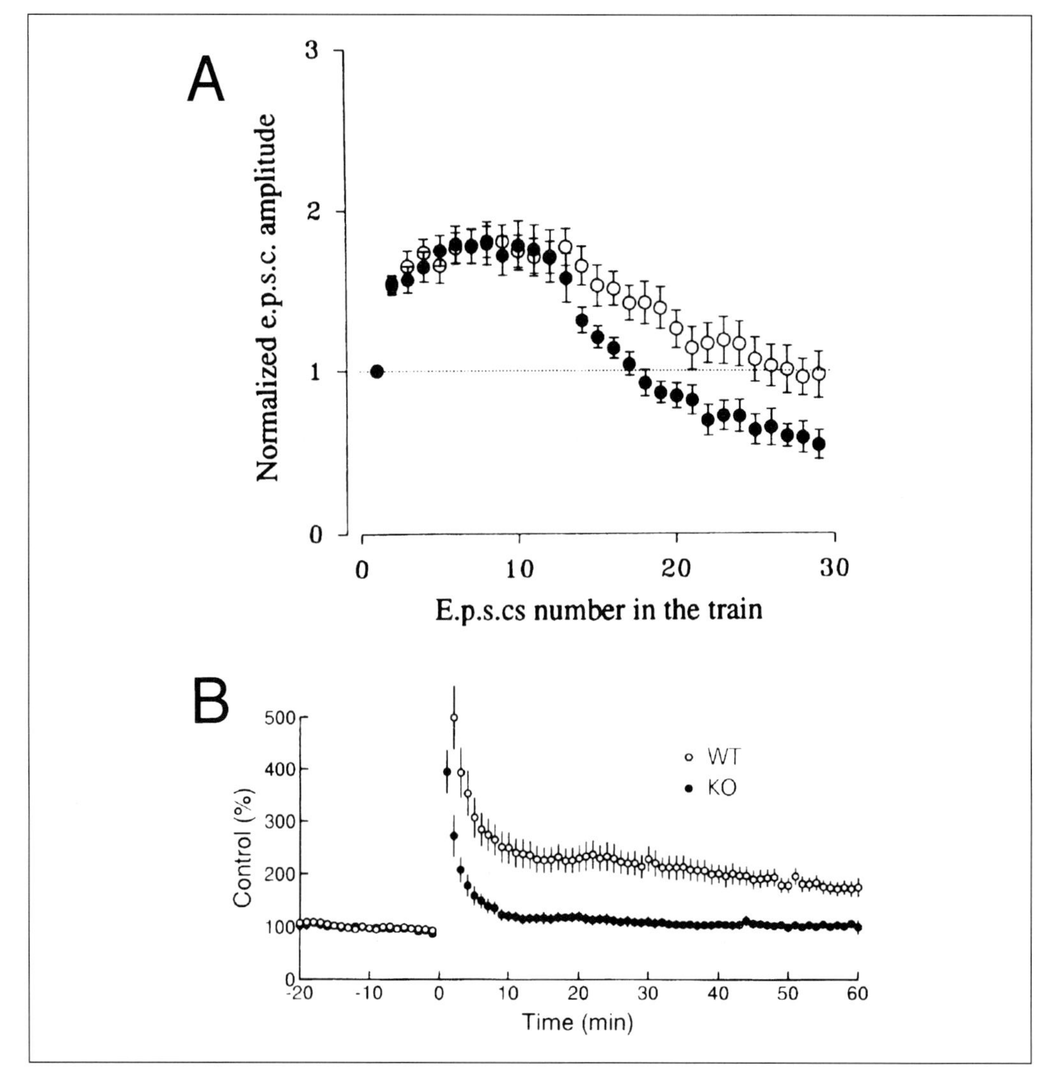

Fig. 4 Phenotype of rab3A knock-out mice. (A) Summary of recordings during 14 Hz stimulation of hippocampal slices from wild-type (open circles) and knock-out (closed circles) mice. Synaptic fatigue is enhanced in rab3A mutant mice. (B) Summary of recordings of Mossy fibre LTP in wild-type (open circles) and knock-out (closed circles) mice. Mossy fibre LTP is essentially absent in mutant mice. LTP was induced by one train lasting 5 sec at 25 Hz in the presence of 100 μM D-AP5. (From ref. 30 (A) and ref. 60 (B), courtesy of Nature Press.)

depression following a train of 15 to 30 stimuli. Thus, these results suggest that rab3A is not essential for neurotransmitter release, but that it may play a role in recruiting SVs from the reserve pool and/or in SV docking, because when rab3A is missing, the synapse fatigues more rapidly than a wild-type synapse with repetitive stimulation.

One possible explanation for the subtle phenotype observed in rab3A knock-out mice is that other proteins compensated for the loss of rab3A. However, rab3C, the other brain-specific rab3 isoform, was weakly expressed in both wild-type and mutant brains. Furthermore, analysis of GTP binding proteins in synaptosomes from wild-type and mutant mice suggested that rab3A is the major GTP binding protein at the synapse (30). Interestingly, the protein expression, but not the message, of rabphillin, a proposed effector of rab3A, was reduced almost 70%, suggesting that its interaction with rab3A is required for stabilization of the protein. Furthermore, the remaining rabphillin in rab3A-deficient mice was present in neural perikaryon, suggesting that rab3A is also necessary for the synaptic localization of rabphillin (55).

In order to test the hypothesis that rab3A is involved in docking, further electrophysiological studies were performed using cultured hippocampal neurons to determine the quantal properties of neurotransmitter release in the absence of rab3A (6). Since the amplitude and frequency of mini EPSCs and the number of readily releasable vesicles were normal in mutant cultures, this suggested that the number of docked vesicles was not altered in rab3A-deficient mice. Furthermore, measurements of the rate of recovery of the readily releasable pool following stimulation suggested that the synaptic vesicle refilling rate was also normal in mutant mice. These studies suggest that docking occurs normally in knock-out mice, and thus, the defect in rab3A knock-out mice occurs after docking. However, EM analysis of the number of docked SVs in rab3A mutant mice has not been performed.

Because the enhanced fatigue in mutant mice did not appear to be due to a defect in SV docking, the authors next explored the possibility that the fatigue was due to an increase in fusion. Central synapses normally release only one SV per bouton with a certain probability, P (5, 56). The probability that each bouton will release one SV is greatly increased with each stimulation, but the number of SVs released at each synapse ordinarily does not change. Interestingly, there was a twofold increase in the amount of quantal release per bouton with each stimulation of mutant neurons. Furthermore, when paired-pulse facilitation of hippocampal slices was reanalysed using shorter interval times (< 50 msec), a significant increase in facilitation was observed in mutant slices. The increase in quantal release suggests that either the probability of the release of one SV is increased or the number of SVs released when a release event does occur is increased in mutant neurons. To distinguish between these possibilities, the probability of release was determined by measuring the evoked response in the presence of an open NMDA channel blocker, MK-801. Because MK-801 irreversibly blocks open NMDA channels, it should have a greater effect on synapses releasing increased amounts of neurotransmitter (57, 58). The blocking rate of MK-801 was not altered in rab3A mutant mice, suggesting that the overall probability of neurotransmitter release is not affected in knock-out mice. However, because the release of one SV has been shown to saturate postsynaptic receptors of central synapses, these results would not allow the detection of an increased number of SVs released per release event (56, 59). Therefore, to increase the sensitivity of the assay, recordings were also done in the presence of a competitive NMDA receptor blocker, and an increase in the blocking rate of NMDA receptors was found in mutant mice. These data suggested that the overall probability of release was not affected in rab3A-deficient mice, but rather the number of quanta released per release event was increased. In slice recordings, the EPSP amplitude was normal because increased evoked quantal release does not lead to an increase in the number of activated postsynaptic receptors, as the release of one SV is saturating. The authors proposed a model whereby after one vesicle is released, rab3A or one of its effectors inhibits further release by destabilizing the membrane.

Because other rab3 isoforms are also expressed in the hippocampus, it is possible that their expression may compensate for loss of rab3A. Interestingly, the stratum lucidum region of the hippocampus which contain Mossy fibres only expresses rab3A and not other rab3 isoforms, making the Mossy fibre synapse an ideal location to study the effect of the loss of rab3A (60). This synapse was found to have normal paired-pulse facilitation and post-tetanic potentiation in mutant mice, but was essentially devoid of long-term potentiation (LTP) (Fig. 4B). This marked, but specific defect in LTP is consistent with earlier studies which suggest that LTP is predominantly presynaptic in the Mossy fibre synapse (see Section 2.3). Furthermore, the absence of a defect in LTP in the Schaffer collateral/CA1 synapse may be due to the fact that in these synapses, either LTP is predominantly postsynaptic or rab3C is also expressed. Physiological consequences of this lack of LTP such as an alteration in learning and memory have not yet been examined in these mice.

As mentioned in Section 2.3, the presynaptic component of LTP is thought to involve the activation of PKA in response to accumulation of cAMP (26–28). Therefore, the response to activators of the cAMP pathway were examined in wild-type and knock-out mice. The authors hypothesized that the lack of LTP at Mossy fibre synapses was due to an inability to potentiate SV release, because the loss of rab3A already allowed increased numbers of SVs to be released. However, the potentiation following application of forskolin, an activator of adenylate cyclase, and a membrane permeable cAMP analogue was normal in knock-out mice. If Mossy fibre LTP requires both rab3A and the cAMP pathway, how does increasing cAMP levels allow LTP to occur in rab3A knock-out mice? The answer to this apparent paradox came after analysing the mechanisms of forskolin-mediated enhancement of glutamate release from synaptosomes (61). Forskolin potentiates neurotransmitter release through three pathways:

- increasing the readily releasable pool, as tested by administration of hypertonic sucrose
- increasing calcium influx, as tested by depolarizing cells with KCl to open voltage gated calcium channels

- increasing SV fusion in response to calcium, as tested by addition of the calcium ionophore, ionomycin.

Only the third mechanism of forskolin potentiation was altered in synaptosomes from rab3A knock-out mice. Thus, increased cAMP levels allow potentiation of SV release through multiple mechanisms, and only the calcium-mediated fusion of SVs is altered in rab3A knock-out mice. This provides further evidence that rab3A is involved in regulating SV fusion, and suggests that the cAMP pathway modulates Mossy fibre LTP through an interaction with rab3A.

In summary, rab3A appears to play a regulatory role in neurotransmitter release, allowing only one synaptic vesicle to fuse at each synapse. This role for rab3A in the fusion process may appear to be inconsistent with the initial electrophysiological analysis of the rab3A knock-out mice, which suggested a defect in SV docking or recruitment. However, the increased synaptic fatigue at high stimulation frequencies can be explained by enhanced depletion of SVs due to the loss of rab3A's inhibitory action on SV fusion. This inhibitory function of rab3A is consistent with *in vitro* studies in which overexpression of GTPase-deficient rab3 mutants in *Aplysia* neurons, bovine adrenal chromaffin cells, and PC12 cells inhibited neurotransmitter release (62–64). Similarly, inhibition of rab3A with antisense oligonucleotides increased evoked release of adrenal chromaffin cells after repetitive stimulation (64). However, in pituitary cells, rab3A antisense oligonucleotides had no effect, while rab3B antisense inhibited neurotransmitter release (65, 66).

Despite evidence for an inhibitory role of rab3 action, *in vivo* data from *C. elegans* suggests that rab3 plays an active role in exocytosis. In *C. elegans* in which only one rab3 gene has been identified, *rab3* mutants had subtle behavioural abnormalities, but nevertheless had impaired synaptic transmission when electropharyngeograms were recorded (67). Furthermore, *rab3* mutants were resistant to the acetylcholinesterase inhibitor aldicarb, suggesting that release of acetylcholine was decreased. Finally, EM analysis demonstrated an increase in the number of SVs in axons and a decreased number in nerve terminals, suggesting a defect in SV recruitment or docking with the plasma membrane. This proposed function is consistent with the initial study of rab3A knock-out mice, showing an increase in synaptic fatigue. While a subsequent study used complex electrophysiological analysis of the size and recycling ability of the SV pool to determine that SVs are docked normally in knock-out mice, ultrastructural studies were not performed.

In conclusion, there is evidence that rab3A may regulate the fusion of vesicles with the plasma membrane. While a need for inhibiting multivesicle fusion in neurons of central synapses seems like a logical mechanism for modulating neurotransmitter release, it seems unlikely that rabs in other cell types and involved in different membrane trafficking events require such an inhibitory restraint on membrane fusion. Furthermore, while it has been proposed that rab3 acts through an interaction with effector molecules such as rabphillin and rim, there are no such homologues in yeast, which clearly use rab homologues for vesicle trafficking. Thus, the precise mechanism of rab action still awaits future studies.

3.3 Guanine nucleotide dissociation inhibitor 1

As discussed in Chapter 3, the cycling between the GTP- and GDP-bound forms of rab family members is crucial to their function and is regulated by several different families of proteins (reviewed in ref. 68). GTPase accelerating proteins (GAPs) increase the GTPase activity of rabs, thereby favouring the GDP-bound conformation. Furthermore, guanine nucleotide exchange factors (GEFs) accelerate the exchange of GDP for GTP, whereas guanine nucleotide dissociation inhibitors (GDI) stabilize the GDP-bound form. Members of the GDI family are thought to be particularly important in their regulation of rab function, as they retrieve rabs from membranes, thereby restoring the cytoplasmic pool.

Recently, the human αGDI1 gene was found to be mutated in two families with X-linked non-specific mental retardation (69). Interestingly, one of the mutations was a point mutation (L92P), and this protein product had decreased binding to rab3A. The other mutation identified was a null mutation. The authors hypothesized that the mental retardation in these patients is due to an inability of synapses to replenish their cytoplasmic pool of rab3A following stimulation. However, these types of questions are difficult to answer in humans, and the mechanism of GDI action *in vivo* will most likely await the generation of GDI knock-out mice.

4. Mutations in synaptic vesicle phosphoproteins

Several synaptic vesicle proteins have been discovered which are thought to regulate synaptic vesicle docking and/or fusion as a function of their phosphorylation state. Synapsins, the predominant phosphoproteins of synaptic vesicles, are phosphorylated in response to both calcium and cAMP. Synaptophysin and synaptogyrin are weakly homologous synaptic vesicle proteins that are both tyrosine phosphorylated on their C-terminus. Synapsins and synaptophysin have been knocked-out in mice; however, the knock-out phenotype is very subtle.

4.1 Synapsins

Synapsins are abundant SV phosphoproteins that are present in almost all nerve terminals and comprise almost 10% of total SV protein (70, 71). Synapsin I is peripherally associated with synaptic vesicles and binds to actin microfilaments at its carboxy-terminal end (72–74). Upon nerve stimulation, the carboxy-terminus of synapsin I is phosphorylated, leading to a decrease in affinity for both SVs and actin. Furthermore, synapsin I is also phosphorylated on its amino-terminus *in vivo* in response to both cAMP and calcium (75). Synapsin II has similar biochemical properties, but is expressed weakly in the brain and is not phosphorylated on its carboxy-terminus (71, 76). These biochemical properties have led to the hypothesis that synapsins play a role in recruiting synaptic vesicles from the reserve pool through their interactions with cytoskeletal components in a phosphorylation-dependent manner.

The *in vivo* function of synapsins was examined by creating synapsin I, synapsin II, and double knock-out (DKO) mice (77–79). Homozygous synapsin I and synapsin II knock-out mice expressed none of the mutated synapsin protein, but expressed wild-type amounts of the other synapsin. The two mice were bred together to produce synapsin I/synapsin II double knock-out (DKO) mice (78). All synapsin mutant mice were viable and healthy and had no gross morphological or behavioural defects (77–79). However, when subjected to either vestibular or electrical stimulation, these mutant mice developed grand mal (throughout the brain) seizures with an incidence proportional to the number of mutant alleles (78, 80). Thus, as in the case of rab3A, synapsins also appear to not be essential genes, but may play a modulatory role in brain function.

Ultrastructural analysis of mutant synapses was performed to test the hypothesis that synapsins function to recruit SVs from the reserve pool. Electron microscopic analysis of synaptic terminals revealed that although the overall synaptic morphology was normal in synapsin I and DKO mice, the number of synaptic vesicles at all synapses investigated was reduced by almost 50% (78, 79). Furthermore, all synaptic vesicle proteins examined were slightly reduced in single knock-out mice and much more reduced in DKO mice, consistent with the overall decrease in the number of synaptic vesicles (78). The cause of this SV depletion is unknown, but it has been proposed that loss of synapsin prevents the attachment of SVs with the presynaptic matrix, thereby causing either increased SV turnover or spillage of SVs into the axon (81).

Analysis of SV distribution revealed conflicting results in synapsin I and DKO mice. Although there was no alteration in the distribution of SVs in DKO mice (78), detailed analysis of axon terminals from synapsin I mutant mice revealed that only SVs between 150–500 nm from the active zone were depleted, whereas there were a normal number of SVs within 150 nm (79, 80). This suggests that docking of SVs at the active zone occurs normally in synapsin I-deficient mice, but there is a decrease in the SV reserve pool. It is unclear why the distribution of SVs was not altered in DKO mice, but as will be apparent later, the phenotype of the synapsin I and synapsin II knock-outs are different, and the phenotype of the DKO more closely resembles that of the synapsin II knock-out.

Synaptic transmission was examined in knock-out mice by performing extracellular recordings at the CA1/Schaffer collateral synapse of hippocampal slices. Although the amplitude of EPSPs was normal in synapsin knock-out mice (77), different properties of short-term plasticity were altered in synapsin I and synapsin II knock-out mice (Fig. 5). Paired-pulse facilitation (PPF) was normal in synapsin II and DKO mice, but was significantly increased in synapsin I knock-out mice at all inter-stimulus frequencies tested (77, 78) (Fig. 5a, 5b). One possible cause for increased PPF is a decrease in the amount of neurotransmitter released following the first stimulus, thereby allowing more to be released during subsequent stimuli (9). However, it was argued that the increase in PPF observed in synapsin I knock-out mice is not likely

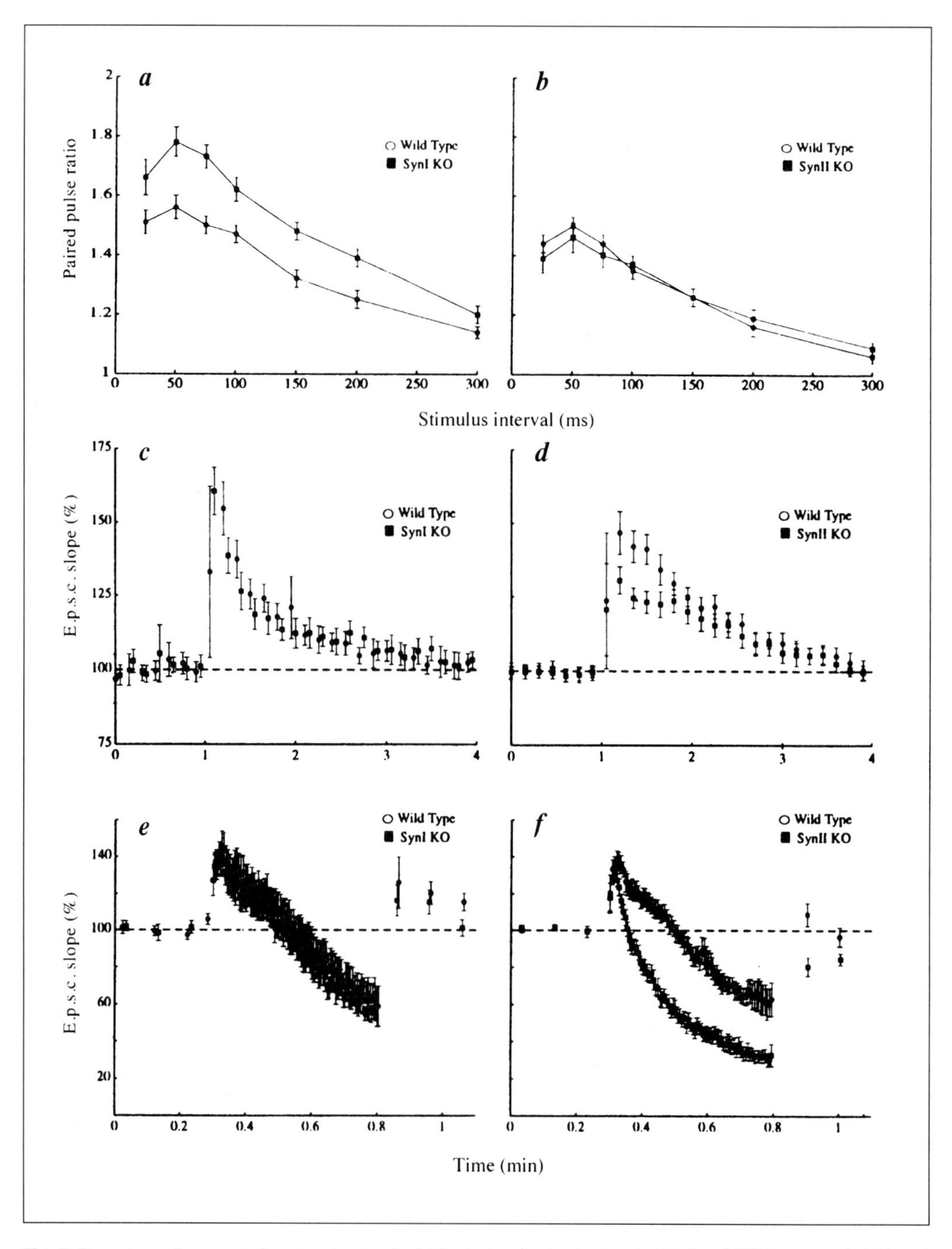

Fig. 5 Phenotype of synapsin knock-out mice. (a, b) Analysis of paired-pulse facilitation (PPF) in synapsin I (a) and synapsin II (b) knock-outs. (c, d) Analysis of post-tetanic potentiation (PTP), induced by a 1 sec 100 Hz stimulus at 1 min. (e, f) Synaptic response to repetitive stimulation. At 0.3 min, synapses were stimulated at 10 Hz for 0.5 min. (From ref. 78, courtesy of Nature Press.)

due to a general decrease in the probability of release, because the amplitude of EPSPs is normal, and synaptosomes from knock-out mice released normal amounts of [^{3}H]noradrenaline when stimulated *in vitro* (77). This is in contrast to subsequent studies which revealed decreased neurotransmitter release from synaptosomes when stimulated under similar conditions (80). Similarly, studies using the lipophilic dye FM1-43 demonstrated that the number of vesicles exocytosed during brief action potential trains was significantly decreased in synapsin I mutants (82). Thus, the increase in PPF observed in synapsin I mutants may either be due to a decrease in neurotransmitter release, or it may be due to the loss of synapsin I which normally functions to specifically inhibit facilitation.

Interestingly, the synapse between the medial perforant path and the dentate gyrus granule cells normally undergoes depression (PPD) rather than facilitation (PPF) in response to paired stimuli (83). PPD at this synapse was enhanced to a similar degree as PPF at the CA1/Schaffer collateral synapse in synapsin I knock-out mice (77), suggesting that both PPF and PPD rely on similar mechanisms.

In contrast to PPF, post-tetantic potentiation (PTP) was normal in synapsin I knock-out mice, but was significantly decreased in both synapsin II and DKO mice (Fig. 5c, 5d). Furthermore, neurotransmitter release recorded in synapsin II and DKO mice fatigued rapidly in response to high frequency stimulation, similar to that found in the original study of rab3A knock-out mice (78) (Fig. 5e, 5f). Although synapsin I knock-out mice did not exhibit a similar increase in fatigue, the recovery time following tetanic stimulation was significantly increased (80). Finally, LTP was not altered in synapsin knock-out mice (77–79). Both the decrease in PTP and the increased synaptic fatigue during high frequency stimulation observed in synapsin II and DKO mice can be explained by the overall decrease in SVs observed with EM. While it is not clear why synapsin I knock-out mice did not exhibit decreased PTP or enhanced fatigue with high frequency stimulation, the increased recovery time suggests that synapsin I knock-outs have a decreased ability to replenish their readily releasable pool.

Taken together, these studies suggest that synapsins play a role in the formation or maintenance of a SV reserve pool at the synapse. However, electrophysiological data suggests that synapsin I and synapsin II may play different roles in regulating neurotransmitter release, as mutations in synapsin I and synapsin II have different phenotypes. In synapsin II and DKO mice, there is a decrease in the overall number of SVs, consistent with the enhanced fatigue and decreased PTP at high stimulation frequencies. In synapsin I knock-out mice, on the other hand, there is a specific loss of distal SVs, consistent with studies of the lamprey ribbon synapse which show a depletion of the SV reserve pool when synapsin is blocked by specific antibodies (84). Furthermore, synapsin I knock-out mice also have increased PPF. One possibility is that both synapsins function in synaptic vesicle recruitment, whereas synapsin I plays an additional role in regulating the fusion process. Despite these differences, it is clear that synapsins serve to maintain a reserve pool of SVs which can rapidly be recruited to the active zone upon SV depletion, thereby potentially regulating the plasticity required for complex brain function.

4.2 Synaptophysin

Another abundant synaptic vesicle phosphoprotein is synaptophysin (85–88). Synaptophysin is a four-pass transmembrane protein of synaptic vesicles which forms a multimeric complex with synaptobrevin (89–91). Although synaptophysin is phosphorylated by serine, threonine, and tyrosine kinases, it represents one of the most abundant tyrosine-phosphorylated proteins at the synapse (92–94). In addition to synaptophysin, a highly homologous protein named synaptoporin (also called synaptophysin II) is also a multipass transmembrane protein of synaptic vesicles (95). Synaptophysin is expressed in almost all synapses while synaptoporin has a more selected distribution, suggesting that synaptoporin may have a more specialized function (96, 97). *In vitro* studies in *Xenopus* oocytes blocking synaptophysin with antibodies or antisense mRNA and overexpressing synaptophysin suggested that synaptophysin may play an active and essential role in neurotransmitter release (98, 99). Other studies suggested that synaptophysin may function as a calcium sensor, a transporter, or even play a role in synaptic vesicle biogenesis (100–103).

Because the proposed roles for synaptophysin were very diverse, knock-out mice were generated to study its role *in vivo* (104). Homozygous mutant mice expressed no synaptophysin detectable by Western blot, but nevertheless, were healthy and fertile. Levels of over 20 other proteins including synaptoporin were unchanged, with the exception of a mild (20%) decrease in the level of synaptobrevin II. Similar to the other knock-outs previously described, there was no detectable abnormality in the brain morphology of mutant mice. Electron microscopic analysis of synaptic regions demonstrated a normal size and number of SVs (105). These studies suggest that synaptophysin is not essential to normal mouse development or gross function.

Electrophysiological recordings at the Schaffer collateral/CA1 synapse of hippocampal slices showed no alterations in amplitude of EPSCs or in any of the previously discussed parameters of synaptic plasticity (104). Studies of the probability and amplitude of quantal release suggest that calcium-dependent release is unaffected in synaptophysin null mice. Furthermore, no changes were found in the mEPSC frequency or in the response to stimulation by α-latrotoxin or hypertonic sucrose, suggesting that the calcium-independent release is also unaffected by the loss of synaptophysin.

Thus, the absence of a phenotype suggests that either the lack of synaptophysin can be compensated for by other synaptic vesicle phosphoproteins or the protein does not play a role which can be detected by the above assays. Both synaptoporin and synaptogyrin have significant homology to synaptophysin and may be able to compensate for the loss of synaptophysin. Thus, studies of synaptophysin knock-out mice suggest that synaptophysin does not play an essential role in neurotransmitter release, in contrast to the studies performed in *Xenopus* oocytes using antisense mRNA or antibodies to block synaptophysin function. This may be explained by either non-specific effects of antisense mRNA or antibodies, or because knock-out mice develop compensatory mechanisms which oocytes lack either the capacity or time frame to develop. As discussed in Section 2.4, the difference in the two studies may be due to the difference between the acute and chronic loss of a protein. However, in order to study the role of the multipass transmembrane proteins in mice, double, triple, or conditional knock-outs should be made to address these issues.

5. Other mutations in proteins implicated in neurotransmitter release

5.1 Neurexin Iα

α-Latrotoxin is a potent neurotoxin derived from black widow spider venom which causes a rapid, calcium-independent release of neurotransmitters when applied to nerve terminals (see Chapter 6). A search to find the receptor for this potent toxin originally led to the identification of neurexin Iα (47). Members of the neurexin family are neuron-specific transmembrane proteins with multiple isoforms and spliced variants (106, 107). Neurexins have been shown to interact with synaptotagmin, suggesting that the binding of α-latrotoxin to neurexin may disrupt synaptotagmin's function, leading to massive exocytosis (47, 108). However, synaptotagmin mutant mice respond normally to α-latrotoxin (3). Moreover, neurexin Iα only binds to α-latrotoxin in the presence of calcium, whereas α-latrotoxin still stimulates neurotransmitter release in the absence of calcium (109). Interestingly, another high affinity α-latrotoxin receptor has recently been discovered named latrophilin or CIRL (calcium-independent receptor for latrotoxin) which binds to α-latrotoxin in the absence of calcium (110, 111). Thus, despite the well characterized biochemical interaction between neurexin Iα and α-latrotoxin, the *in vivo* relevance of this interaction remains questionable (see Chapter 6).

To determine the *in vivo* role of neurexin Iα, in particular its role as the α-latrotoxin receptor, neurexin Iα knock-out mice were generated (112). Homozygous mutants were healthy and fertile, despite the complete absence of neurexin Iα protein. The only abnormality observed in these mice was a decreased ability for mutant mothers to take care of their offspring, suggesting that the lack of neurexin Iα may alter the behaviour of these mice. Electrophysiological analysis of hippocampal slices was not reported for these mice. Thus, neurexin Iα is not an essential gene, suggesting that it is either functionally redundant in its role of neurotransmitter release, or it does not play an important role in this process.

To test whether neurexin Iα is an α-latrotoxin receptor *in vivo*, the ability for α-latrotoxin to stimulate neurotransmitter release in neurexin Iα knock-out mice was determined. Interestingly, while no change in calcium-independent binding of α-latrotoxin to brain membranes was observed in mutant mice, the calcium-dependent binding was reduced by approximately half. Furthermore, synaptosomes from mutant mice loaded with radiolabelled norepinephrine released significantly less neurotransmitter when stimulated by α-latrotoxin in the presence but not in the absence of calcium. These results suggest that while not essential for α-latrotoxin action, neurexin Iα does indeed mediate some of the effects of α-latrotoxin in the presence but not in the absence of calcium. However, the role that neurexin Iα plays in neurotransmitter release, if any, remains to be established.

5.2 Sec 8

One of the key players in neurotransmitter release yet to be identified is the protein(s) required for targeting synaptic vesicles to active zones. In budding yeast, spatial organization of secretion is mediated through an 'exocyst' complex containing many proteins, including sec6, sec8, and sec15 (113, 114). Recently, the mammalian homologues of sec6 and sec8 have been shown to form a complex which specifies vesicle transport to the basolateral membrane in epithelial cells (115) (see Chapter 3). A similar complex is also present in brains and co-immunoprecipitates with septin filaments (116, 117), suggesting that this complex may play a role in targeting synaptic vesicles to the plasma membrane.

An insertional mutation into an intron in the 3′ end of the sec8 gene was identified using a gene trapping technique for embryonic lethal mutations (118). Interestingly, homozygous mutant mice are lethal early in embryonic life. E7.5 embryos have normal endoderm and epiblast formation, but lack formation of the mesoderm. Analysis of genes normally expressed in mesoderm such as sonic hedgehog, brachyury, and Mox-1 suggested that mutant embryos were able to initiate gastrulation, but were unable to form paraxial mesoderm, leading to the arrest of embryonic development. Because growth factor secretion is known to be critical at this stage of development, and a mutation in the mammalian homologue of PEP8, a yeast secretion mutant, is also lethal at this stage, the authors proposed that these two genes are required for early secretion of growth factors and early embryogenesis (119, 120). Although this data suggests that these genes play a role in general secretion, the role they play in neurotransmitter release, if any, is unknown.

6. Mutations in the core complex of proteins implicated in neurotransmitter release

As discussed in previous chapters, fusion of synaptic vesicles with the presynaptic membrane requires the formation of a core complex comprised of syntaxin and SNAP-25 on the presynaptic membrane and synaptobrevin on the synaptic vesicle. Targeted mutations in these genes have not yet been generated in mice, presumably because the essential role of these proteins has already been demonstrated *in vivo* using clostridial neurotoxins (see Chapter 6). However, spontaneous deletions have been discovered which remove SNAP-25 in mice and syntaxin 1A in humans. Mice heterozygous for a deficiency removing the SNAP-25 gene have spontaneous locomotor activity which is rescued by introducing a SNAP-25 transgene (121, 122). Similarly, Williams syndrome occurs in human patients with a hemizygous deletion of approximately 13 genes, one of which is syntaxin 1A. Although the features of Williams syndrome have not yet been shown to be due to loss of syntaxin, the neurological and gastrointestinal symptoms of these patients are consistent with the localized expression of syntaxin 1A in the brain and gut, respectively. These hemizygous deletion syndromes may suggest that the levels of SNAP-25, and possibly syntaxin, are critical to normal synapse function *in vivo*.

6.1 Coloboma mice are deficient in SNAP-25

SNAP-25 (synaptosomal-associated protein of 25 kDa) is a neural-specific protein which peripherally associates with the cytoplasmic surface of the presynaptic membrane (123) (see Chapter 3). SNAP-25 contributes two of the four coiled-coil domains of the core complex thought to be essential for the fusion of synaptic vesicles with the presynaptic membrane (124). Studies using botulinum neurotoxin A and E, which selectively cleaves SNAP-25 at its C-terminus, have established SNAP-25 as an essential member of the secretion apparatus (125, 126), yet little is known about the function of SNAP-25 *in vivo*.

As a first step in studying SNAP-25 in mice, the SNAP-25 gene was mapped to a ~18 cM region on the second chromosome using genetic recombination mapping and somatic cell hybrid analysis (122). Once the approximate location of the SNAP-25 gene was known, previously isolated mouse mutations within this region were screened for restriction fragment length polymorphisms (RFLPs) using Southern blot. Interestingly, the intensity of all restriction fragments was reduced in coloboma mutant mice, suggesting that they carry a deficiency of SNAP-25. Coloboma mutant mice are heterozygous (*Cm*/+), and homozygous mutants (*Cm*/*Cm*) are early embryonic lethal (127). The reduction of SNAP-25 protein, message, and genomic DNA in coloboma mice was estimated to be approximately 50%, suggesting that most, if not all, of one copy of the SNAP-25 gene was deleted in these mice. Later studies demonstrated the coloboma deletion encompassed a 1–2 cM region including the phospholipase Cb-1 gene as well as other genes (128).

Heterozygous mice (*Cm*/+) are easily identified by postnatal day 14 due to characteristic ophthalmic deformation, head bobbing, and severe spontaneous hyperactivity (122, 127, 129). The spontaneous hyperactivity was quantified to be three to ten times the normal level using automated open field activity cages, which measure the number of photocell beam breaks over a 24 hour period (122). Because hyperactivity is a component of several childhood onset behaviour disorders such as attention deficit-hyperactivity disorder (ADHD) and Tourette's syndrome (130, 131), it was proposed that the coloboma mice may be a good model to study the neurobiological basis of hyperactivity (122). To test this model, the response of these mice to various psychostimulants was tested (121). Similar to children with ADHD, amphetamines reduced the hyperactivity in coloboma mice, while amphetamines increase activity of normal children and wild-type mice (132, 133). In contrast, methylphenidate, the drug most commonly used to treat patients with ADHD, increased the activity of both wild-type and coloboma mice, suggesting that the pathophysiological mechanisms of hyperactivity may be different between children with ADHD and coloboma mice.

To determine if the phenotype of coloboma mutants was due to haploinsufficiency of the SNAP-25 gene, a SNAP-25 transgene was introduced into mutant mice (121). The SNAP-25 transgene was made by fusing a 10 kb genomic region 5′ of the SNAP-25 start site with the SNAP-25 open reading frame. When the transgene was bred onto the coloboma mutant strain, SNAP-25 expression was increased by 30% in the striatum and cerebellum, two regions of the brain which control motor activity. As shown in Fig. 6, this localized increase in expression was able to completely rescue the hyperactivity observed in coloboma mutant mice. Furthermore, the transgene also reversed the effects of amphetamines on coloboma mutant mice, making their response similar to wild-type mice. The involvement of loss of SNAP-25 in the hyperactivity of these mice is consistent with the fact that both SNAP-25 and amphetamines act presynaptically. Finally, although the transgene rescued the spontaneous locomotor hyperactivity of the coloboma mice, it did not rescue the head bobbing or eye deformity, suggesting that either these features are due to the deletion of a gene other than SNAP-25, or that the expression of the SNAP-25 transgene was not sufficient to reverse these phenotypes.

Electroencephlography revealed that coloboma mice had markedly reduced

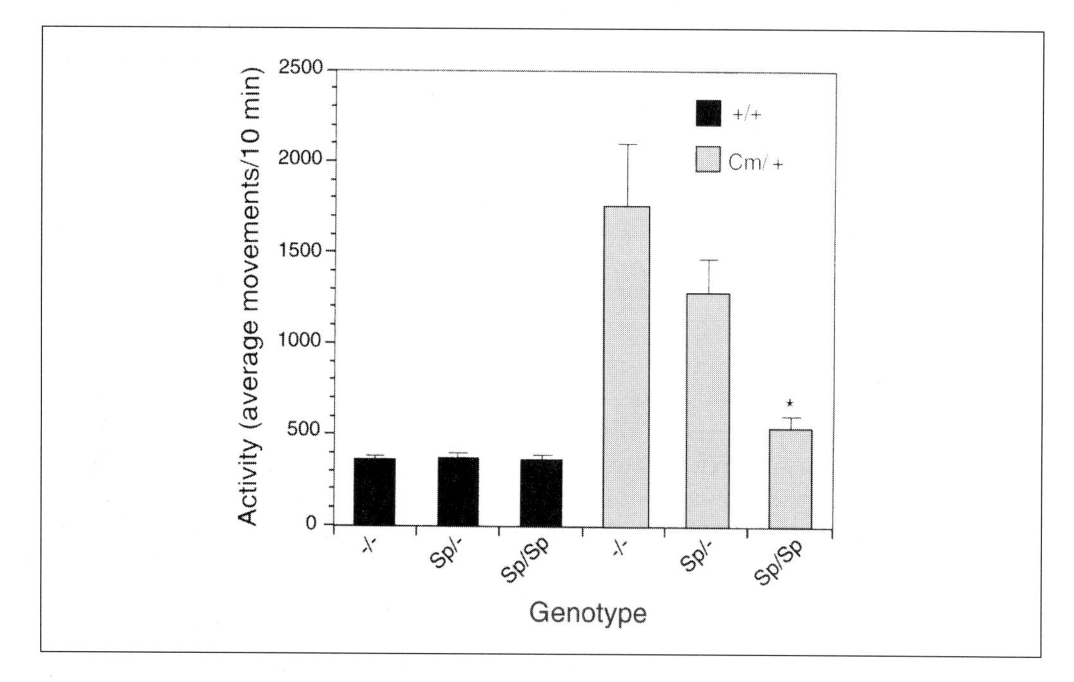

Fig. 6 Rescue of coloboma hyperactivity with SNAP-25 transgene. One (Sp/–) or two (Sp/Sp) copies of the SNAP-25 transgene were bred onto the wild-type (filled) or coloboma (*Cm*) (open) background. The activity of mice was measured by counting the number of photocell beam breaks every minute. Hyperactivity of coloboma mutant mice was almost completely reversed by addition of two copies of the SNAP-25 transgene (From ref. 121, courtesy of Society for Neuroscience.)

theta rhythmic activity in response to sensory activation (134). Furthermore, electrophysiological recordings of the dentate gyrus of the hippocampus showed normal evoked release, but increased paired-pulse depression (PPD) and decreased LTP when compared with wild-type litter mates. Because dopamine release has been shown to be essential for theta activity and may modulate LTP, it was suggested that coloboma mice may have defects in dopamine release. Indeed, dopamine secretion from dorsal striatum but not ventral striatum is completely blocked in coloboma mice, suggesting that this neurochemical alteration may be responsible for the hyperactivity observed in these mice (135). Other neurochemical alterations in coloboma mice include decreased glutamate release from cortical synaptosomes, decreased serotonin release from the ventral striatum, and decreased corticotrophin releasing factor secretion in response to acetylcholine. However, it should be emphasized that coloboma mice containing the SNAP-25 transgene were not used as controls in these studies, and thus, it can not be ruled out that genes in the coloboma deficiency other than SNAP-25 may be responsible for these electrophysiological or neurochemical defects.

These studies clearly implicate SNAP-25 as essential for normal behaviour in the mouse. The hyperactivity in coloboma mice is likely due to decreased neurotransmitter release in certain inhibitory synapses of the brain where high levels of SNAP-25 protein are critical. Furthermore, it is possible that mutations in the human SNAP-25 gene are responsible for some human behavioural disorders such as ADHD, although in some families with apparently single gene-mediated ADHD, there was no genetic linkage with the human region syntenic to the coloboma deficiency (136). While these studies give insight into the role of SNAP-25 in behaviour, they do not shed light on the specific role that SNAP-25 plays in neurotransmitter release. Thus, future studies will likely focus on making targeted mutations in SNAP-25 and examining the electrophysiological and behavioural phenotype of mutant mice.

6.2 Williams syndrome patients are haploinsufficient for syntaxin 1A

A deletion of a gene encoding a member of the core complex was also discovered in humans. Patients with Williams syndrome (WS) are hemizygous for syntaxin 1A (STX1A) as well as at least 13 other genes within a small region of chromosome 17 (137–139). These patients have several distinguishing characteristics including dysmorphic facial anomalies, mild to moderate mental retardation, specific behavioural abnormalities, and cardiovascular defects, especially supravalvular aortic stenosis (SVAS) (reviewed in ref. 140). Although WS patients have low IQs similar to other mental retardation syndromes, these patients perform exceptionally well on certain 'right-sided' brain tasks which measure creativity and language expression (141). In contrast, Williams patients perform poorly on tasks which require logical reasoning or visuo-spatial recognition.

Recently, key insights into the genetic basis of the WS phenotype have been made by studying patients which only have certain features of the disease. For example, families of patients with SVAS but lacking other characteristics of WS were found to have small, hemizygous deletions or translocations in the elastin gene (142–144). Because one copy of the elastin gene is deleted in 96% of WS patients, this highly suggests that the SVAS and likely other cardiovascular abnormalities of this syndrome are due to a 50% reduction in the amount of elastin (145, 146). Furthermore, non-classical WS patients have been identified which have most of the features of the syndrome, but perform well on tasks of visuo-spatial recognition. Interestingly, these patients maintain two copies of the LIM-kinase 1 (LIMK1) gene, a gene which is ordinarily deleted in classical WS patients (147). This provides evidence to suggest that LIMK1 may play a role in mediating visuo-spatial recognition, and in fact, this is one of the first genes implicated in a specific behaviour.

Thus, studies are currently being performed to determine which of the aspects of the WS phenotype are due to which of the 13 or so genes deleted in these patients. Although neither non-classical WS patients which lack a deletion of syntaxin, nor patients with a mutation in syntaxin only have thus far been identified, several pieces of evidence suggest that syntaxin may play a role in the WS phenotype. First, over 70% of WS patients also suffer from attention-deficit and hyperactivity disorder (ADHD). As discussed previously, a hemizygous deletion of SNAP-25 results in hyperactive mice, making it likely that loss of other members of the core complex may have a similar phenotype. Secondly, WS patients also have a high incidence of gastrointestinal disorders, and syntaxin has been shown to modulate the activity of the CFTR ion channel in epithelial cells of the gut (148, 149). Finally, the STX1A gene is specifically expressed in neurons and gastrointestinal epithelial cells (148, 150), thus allowing the gastrointestinal and neurological disorders of WS patients to be explained by haploinsufficiency of STX1A. However, further analysis of both classical and non-classical WS patients must be performed in order to determine the association, if any, of WS characteristics with loss of STX1A.

7. Conclusions

Despite the complexities and shortcomings of the knock-out approach, knock-out mice have significantly contributed to our understanding of genes and proteins involved in neurotransmitter release. Table 1 summarizes the key aspects of the phenotype of each of the mutant mice discussed in this chapter. With the exception of the mutation in synaptotagmin and sec8, all the other mutations thus far published have resulted in healthy mice, with phenotypes apparent only after using sophisticated electrophysiological techniques. This suggests that these genes are not essential, most likely because homologous genes are also present which have overlapping functions, even though homologous genes are not found to have increased expression in mutant mice. Although in some cases it may appear that a gene is redundant, it is unlikely that evolution would maintain multiple genes with identical functions. It is more likely that these genes are evolutionarily maintained because

Table 1 Summary of the phenotypes of mouse mutations discussed in this chapter[a]

	Synaptotagmin I	**Rab3A**	**Synapsins**	**Synaptophysin**	**Neurexin Iα**	**Coloboma mice (SNAP-25)**
Gross phenotype	Lethal by 48 hours. Newborn mice are weak and rejected by mothers	Normal	↑ evoked seizures w/ ↑ no. of mutant alleles	Normal	Normal	Spontaneous hyperactivity, eye defects, head bobbing
Ultrastructural changes	No qualitative change in SV number or distribution	ND	↓ SVs, impaired distal SV clustering in syn I K.O.	Normal	ND	ND
Biochemical alterations	None found	70% ↓ in rabphillin	20–40% ↓ in levels of SV proteins	None found	↓ latrotoxin binding to synaptosomes	↓ in specific neurotransmitters
Evoked release	Fast (synchronous) component essentially eliminated, but slow component normal	Normal EPSPs, 2x ↑ in quantal release	Normal	Normal	ND	Normal
Spontaneous release	Normal	Normal	Normal	Normal	ND	ND
Synaptic plasticity	ND	↑ PPF ↑ fatigue Lack of Mossy fibre LTP	↑ PPF in syn I K.O. ↓ PTP, ↑ fatigue in syn II, DKO	Normal	ND	↑ PPD ↓ LTP
Role in neurotransmitter release	Calcium sensor, active role in evoked release	Allows only one SV to be released at each synapse	Docking or recruiting of SVs at the active zone	No essential role	No essential role	Member of core complex for fusion

[a] ND, not determined; SV, synaptic vesicle; syn, synapsin; K.O., knock-out; DKO, double knock-out; ↑ , increase; ↓ , decrease.

they impart a selective advantage to the organism, even though this advantage may not be apparent by the assays used to study mutant mice. Thus, although mice lacking rab3A, synaptophysin, synapsin, and neurexin Iα appear to be normal, it is likely that they have subtle behavioural changes which would decrease their fitness in the wild.

While behavioural assays have not yet been performed on mice deficient in genes implicated in neurotransmitter release, these assays have been widely used in analysing knock-out mice with defects in other aspects of synaptic transmission (for review, see refs 151 and 152). In many cases, these analyses have allowed specific genetic defects to be correlated with both behavioural abnormalities and electrophysiological defects. For example, mice lacking alpha-calcium-calmodulin-dependent kinase II (αCaMKII) have an abnormal fear response, aggressive behaviour, and deficient spatial learning (153, 154). The spatial learning defect has been correlated with deficient hippocampal LTP, implicating both αCaMKII and LTP in spatial learning (155). Similarly, mice deficient in the gamma isoform of protein kinase C (PKC) have modest impairments in spatial and contextual learning in addition to alterations in LTP (156, 157). In the future, similar behavioural assays applied to the knock-out mice discussed in this chapter may allow a correlation of specific alterations in neurotransmitter release with abnormalities in behaviour.

In summary, there are both advantages and disadvantages to using mice as a model system for studying neurotransmitter release. Mutations can be introduced in mouse genes using homologous recombination; however, this process is time-consuming and expensive in comparison with more simple organisms. Furthermore, although more sophisticated electrophysiological techniques can be performed in mice than in *Drosophila* or *C. elegans*, these complex assays are also required in order to find a defect in neurotransmitter release in mutant mice. Nevertheless, the complexity of the mechanisms of neurotransmitter release must be understood in mice in order to eventually understand these processes in humans. Thus, although deciphering the roles of synaptic proteins may be more challenging in mice, information obtained is likely to be very relevant to humans, and should lead to enhanced understanding of human behaviour and disease.

Acknowledgements

We would like to thank David Sweat, Harvey McMahon, Mark Wu, Bing Zhang, and Guissy Pennetta for their careful reading of this chapter, and Salpy Sarikhanian for help in the preparation of the manuscript.

References

1. Capecchi, M. R. (1989) Altering the genome by homologous recombination. *Science*, **244**, 1288.
2. Joyner, A. L. (1993) In *Gene Targeting: a practical approach*, p. 1. Oxford University Press, Oxford.

3. Geppert, M., Goda, Y., Hammer, R. E., Li, C., Roshal, T. W., Stevens, C. F., *et al.* (1994) Synaptotagmin I: a major Ca^{2+} sensor for transmitter release at a central synapse. *Cell*, **79**, 717.
4. Lynch, G. and Schubert, P. (1980) The use of *in vitro* brain slices for multidisciplinary studies of synaptic function. *Annu. Rev. Neurosci.*, **3**, 1.
5. Bekkers, J. M. and Stevens, C. F. (1995) Quantal analysis of EPSCs recorded from small numbers of synapses in hippocampal cultures. *J. Neurophysiol.*, **73**, 1145.
6. Geppert, M., Goda, Y., Stevens, C., and Südhof, T. C. (1997) The small GTP-binding protein rab3A regulates a late step in synaptic vesicle fusion. *Nature*, **387**, 810.
7. Bekkers, J. M. and Stevens, C. F. (1989) NMDA and non-NMDA receptors are co-localized at individual excitatory synapses in cultured rat hippocampus. *Nature*, **341**, 230.
8. Fatt, P. and Katz, B. (1952) Spontaneous subthreshold activity at motor nerve endings. *J. Physiol.*, **117**, 109.
9. Zucker, R. S. (1989) Short-term synaptic plasticity. *Annu. Rev. Neurosci.*, **12**, 13.
10. Katz, B. and Miledi, R. (1968) The role of calcium in neuromuscular facilitation. *J. Physiol. (Lond)*, **195**, 481.
11. Manabe, T., Wyllie, D. J., Perkel, D. J., and Nicoll, R. A. (1993) Modulation of synaptic transmission and long-term potentiation: effects on paired pulse facilitation and EPSC variance in the CA1 region of the hippocampus. *J. Neurophysiol.*, **70**, 1451.
12. Creager, R., Dunwiddie, T., and Lynch, G. (1980) Paired-pulse and frequency facilitation in the CA1 region of the *in vitro* rat hippocampus. *J. Physiol. (Lond)*, **299**, 409.
13. Debanne, D., Guerineau, N. C., Gahwiler, B. H., and Thompson, S. M. (1996) Paired-pulse facilitation and depression at unitary synapses in rat hippocampus: quantal fluctuation affects subsequent release. *J. Physiol.*, **491**, 163.
14. Wilcox, K. S. and Dichter, M. A. (1994) Paired pulse depression in cultured hippocampal neurons is due to a presynaptic mechanism independent of GABAB autoreceptor activation. *J. Neurosci.*, **14**, 1775.
15. Hubbard, J. I. (1963) Repetitive stimulation at the neuromuscular junction, and the mobilization of transmitter. *J. Physiol.*, **169**, 641.
16. Ryan, T. A. and Smith, S. J. (1995) Vesicle pool mobilization during action potential firing at hippocampal synapses. *Neuron*, **14**, 983.
17. Rosemund, C. and Stevens, C. F. (1996) Definition of the readily releasable pool of vesicles at hippocampal synapses. *Neuron*, **16**, 1197.
18. Swandulla, D., Hans, M., Zipser, K., and Augustine, G. J. (1991) Role of residual calcium in synaptic depression and posttetanic potentiation: fast and slow calcium signaling in nerve terminals. *Neuron*, **7**, 915.
19. Zucker, R. S. (1993) Calcium and transmitter release. *J. Physiol. (Paris)*, **87**, 25.
20. Bliss, T. V. and Collingridge, G. L. (1993) A synaptic model of memory: long-term potentiation in the hippocampus. *Nature*, **361**, 31.
21. Stevens, C. F. (1998) A million dollar question: does LTP = memory? *Neuron*, **20**, 1.
22. Lopez-Garcia, J. C. (1998) Two different forms of long-term potentiation in the hippocampus. *Neurobiology*, **6**, 75.
23. Constantine-Paton, M. and Cline, H. T. (1998) LTP and activity-dependent synaptogenesis: the more alike they are, the more different they become. *Curr. Opin. Neurobiol.*, **8**, 139.
24. Holscher, C. (1997) Nitric oxide, the enigmatic neuronal messenger: its role in synaptic plasticity. *Trends Neurosci.*, **20**, 298.
25. Nicoll, R. A. and Malenka, R. C. (1995) Contrasting properties of two forms of long-term potentiation in the hippocampus. *Nature*, **377**, 115.

26. Huang, Y.-Y., Kandel, E. R., Varshavsky, L., Brandon, E. P., Qi, M., Idzerda, R. L., *et al.* (1995) A genetic test of the effects of mutations in PKA on mossy fiber LTP and its relation to spatial and contextual learning. *Cell*, **83**, 1211.
27. Huang, Y.-Y., Li, X.-C., and Kandel, E. (1994) cAMP contributes to mossy fiber LTP by initiating both a covalently mediated early phase and macromolecular synthesis-dependent late phase. *Cell*, **79**, 69.
28. Weisskopf, M. G., Castillo, P. E., Zalutsky, R. A., and Nicoll, R. A. (1994) Mediation of hippocampal mossy fiber long-term potentiation by cyclic AMP. *Science*, **265**, 1878.
29. Rajewsky, K., Gu, H., Kuhn, R., Betz, U. A. K., Muller, W., Roes, J., *et al.* (1996) Conditional gene targeting. *J. Clin. Invest.*, **98**, 600.
30. Geppert, M., Bolshakov, V. Y., Siegelbaum, S. A., Takei, K., De Camilli, P., Hammer, R. E., *et al.* (1994) The role of rab3a in neurotransmitter release. *Nature*, **369**, 493.
31. Geppert, M. and Südhof, T. C. (1998) RAB3 and synaptotagmin: the yin and yang of synaptic membrane fusion. *Annu. Rev. Neurosci.*, **21**, 75.
32. Matthew, W. D., Tsavaler, L., and Reichardt, L. F. (1981) Identification of a synaptic vesicle-specific membrane protein with a wide distribution in neuronal and neurosecretory tissue. *J. Cell Biol.*, **91**, 257.
33. Perin, M. S., Fried, V. A., Mignery, G. A., Jahn, R., and Südhof, T. C. (1990) Phospholipid binding by a synaptic vesicle protein homologous to the regulatory region of protein kinase C. *Nature*, **345**, 260.
34. Bennett, M. K., Calakos, N., and Scheller, R. H. (1992) Syntaxin: a synaptic protein implicated in docking of synaptic vesicles at presynaptic active zones. *Science*, **257**, 255.
35. Schiavo, G., Stenbeck, G., Rothman, J. E., and Söllner, T. H. (1997) Binding of the synaptic vesicle v-SNARE, synaptotagmin, to the plasma membrane t-SNARE, SNAP-25, can explain docked vesicles at neurotoxin-treated synapses. *Proc. Natl. Acad. Sci. USA*, **94**, 997.
36. Brose, N., Petrenko, A. G., Südhof, T. C., and Jahn, R. (1992) Synaptotagmin: a Ca^{2+} sensor on the synaptic vesicle surface. *Science*, **256**, 1021.
37. Llinas, R., Sugimori, M., and Silver, R. B. (1992) Microdomains of high Ca^{2+} concentration in a presynaptic terminal. *Science*, **256**, 677.
38. Davletov, B. A. and Südhof, T. C. (1993) A single C2 domain from synaptotagmin I is sufficient for high affinity Ca^{2+}/phospolipid binding. *J. Biol. Chem.*, **268**, 26386.
39. Dodge, F. A. and Rahaminoff, R. (1967) Cooperative action of Ca^{2+} ions in transmitter release at the neuromuscular junction. *J. Physiol. (Lond)*, **193**, 419.
40. Leveque, C., Hoshino, T., David, P., Shoji-Kasai, Y., Leys, K., Omori, A., *et al.* (1992) The synaptic vesicle protein synaptotagmin associates with calcium channels and is a putative Lambert-Eaton myasthenic syndrome antigen. *Proc. Natl. Acad. Sci. USA*, **89**, 3625.
41. Li, C., Davletov, B. A., and Südhof, T. C. (1995) Distinct Ca^{2+} and Sr^{2+} binding properties of synaptotagmins. *J. Biol. Chem.*, **270**, 24898.
42. Littleton, J. T. and Bellen, H. J. (1995) Synaptotagmin controls and modulates synaptic-vesicle fusion in a Ca^{2+}-dependent manner. *Trends Neurosci.*, **18**, 177.
43. Littleton, J. T., Stern, M., Perin, M., and Bellen, H. J. (1994) Calcium dependence of neurotransmitter release and rate of spontaneous vesicle fusions are altered in *Drosophila synaptotagmin* mutants. *Proc. Natl. Acad. Sci. USA*, **91**, 10888.
44. Martin, K. C., Hu, Y., Armitage, B. A., Siegelbaum, S. A., Kandel, E. R., and Kaang, B. K. (1995) Evidence for synaptotagmin as an inhibitory clamp on synaptic vesicle release in Aplysia neurons. *Proc. Natl. Acad. Sci. USA*, **92**, 11307.
45. Morimoto, T., Popov, S., Buckley, L. M., and Poo, M.-M. (1995) Calcium-dependent transmitter secretion from fibroblasts: modulation by synaptotagmin I. *Neuron*, **15**, 689.

46. Reist, N. E., Buchanan, J., Li, J., DiAntonio, A., Buxton, E. M., and Schwarz, T. L. (1998) Morphologically docked synaptic vesicles are reduced in synaptotagmin mutants of *Drosophila*. *J. Neurosci.*, **18**, 7662.
47. Petrenko, A. G., Perin, M. S., Bazbek, A., Davletov, B. A., Ushkaryov, Y. A., Geppert, M., *et al.* (1991) Binding of synaptotagmin to the α-latrotoxin receptor implicates both in synaptic vesicle exocytosis. *Nature*, **353**, 65.
48. Sollner, T., Bennett, M. K., Whiteheart, S. W., Scheller, R. H., and Rothman, J. E. (1993) A protein assembly-disassembly pathway *in vitro* that may correspond to sequential steps of synaptic vesicle docking, activation, and fusion. *Cell*, **75**, 409.
49. Jorgensen, E. M., Hartwieg, E., Schuske, K., Nonet, M. L., Jin, Y., and Horvitz, H. R. (1995) Defective recycling of synaptic vesicles in synaptotagmin mutants of *Caenorhabditis elegans*. *Nature*, **378**, 196.
50. Zhang, J. Z., Davletov, B. A., Südhof, T. C., and Anderson, R. G. (1994) Synaptotagmin I is a high affinity receptor for clathrin AP-2: implications for membrane recycling. *Cell*, **78**, 751.
51. Novick, P. and Zerial, M. (1997) The diversity of Rab proteins in vesicle transport. *Curr. Opin. Cell Biol.*, **9**, 496.
52. Fischer von Mollard, G., Mignery, G. A., Baumert, M., Perin, M. S., Hanson, T. J., Burger, P. M., *et al.* (1990) Rab3 is a small GTP-binding protein exclusively localized to synaptic vesicles. *Proc. Natl. Acad. Sci. USA*, **87**, 1988.
53. Fischer von Mollard, G., Südhof, T. C., and Jahn, R. (1991) A small GTP-binding protein dissociates from synaptic vesicles during exocytosis. *Nature*, **349**, 79.
54. Simons, K. and Zerial, M. (1993) Rab proteins and the road maps for intracellular transport. *Neuron*, **11**, 789.
55. Li, C., Takei, K., Geppert, M., Daniell, L., Stenius, K., Chapman, E. R., *et al.* (1994) Synaptic targeting of rabphilin-3A, a synaptic vesicle Ca^{2+}/phospholipid-binding protein, depends on rab3A/3C. *Neuron*, **13**, 885.
56. Perkel, D. J. and Nicoll, R. A. (1993) Evidence for all-or-none regulation of neurotransmitter release: implications for long-term potentiation. *J. Physiol. (Lond)*, **471**, 481.
57. Hessler, N. A., Shirke, A. M., and Malinow, R. (1993) The probability of transmitter release at a mammalian central synapse. *Nature*, **366**, 569.
58. Rosenmund, C., Clements, J. D., and Westbrook, G. L. (1993) Nonuniform probability of glutamate release at a hippocampal synapse. *Science*, **262**, 754.
59. Clements, J. D., Lester, R. A., Tong, G., Jahr, C. E., and Westbrook, G. L. (1992) The time course of glutamate in the synaptic cleft. *Science*, **258**, 1498.
60. Castillo, P. E., Janz, R., Südhof, T. C., Tzounopoulos, T., Makenka, R. C., and Nicoll, R. A. (1997) Rab3A is essential for mossy fibre long-term potentiation in the hippocampus. *Nature*, **388**, 590.
61. Lonart, G., Janz, R., Johnson, K. M., and Südhof, T. C. (1998) Mechanism of action of rab3A in mossy fiber LTP. *Neuron*, **21**, 1141.
62. Doussau, F., Clabecq, A., Henry, J. P., Darchen, F., and Poulain, B. (1998) Calcium-dependent regulation of rab3 in short-term plasticity. *J. Neurosci.*, **18**, 3147.
63. Holz, R. W., Brondyk, W. H., Senter, R. A., Kuizon, L., and Macara, I. G. (1994) Evidence for the involvement of Rab3A in Ca^{2+}-dependent exocytosis from adrenal chromaffin cells. *J. Biol. Chem.*, **269**, 10229.
64. Johannes, L., Lledo, P. M., Roa, M., Vincent, J. D., Henry, J. P., and Darchen, F. (1994) The GTPase Rab3a negatively controls calcium-dependent exocytosis in neuroendocrine cells. *EMBO J.*, **13**, 2029.

65. Lledo, P. M., Vernier, P., Vincent, J. D., Mason, W. T., and Zorec, R. (1993) Inhibition of Rab3B expression attenuates Ca^{2+}-dependent exocytosis in rat anterior pituitary cells. *Nature*, **364**, 540.
66. Tasaka, K., Masumoto, N., Mizuki, J., Ikebuchi, Y., Ohmichi, M., Kurachi, H., *et al.* (1998) Rab3B is essential for GnRH-induced gonadotrophin release from anterior pituitary cells. *J. Endocrinol.*, **157**, 267.
67. Nonet, M. L., Staunton, J. E., Kilgard, M. P., Fergestad, T., Hartwieg, E., Horvitz, H. R., *et al.* (1997) *Caenorhabditis elegans* rab-3 mutant synapses exhibit impaired function and are partially depleted of vesicles. *J. Neurosci.*, **17**, 8061.
68. Südhof, T. C. (1997) Function of Rab3 GDP-GTP exchange. *Neuron*, **18**, 519.
69. D'Adamo, P., Menegon, A., Lo Nigro, C., Grasso, M., Gulisano, M., Tamanini, F., *et al.* (1998) Mutations in GDI1 are responsible for X-linked non-specific mental retardation. *Nat. Genet.*, **19**, 134.
70. Huttner, W. B., Schlieber, W., Greengard, P., and DeCamilli, P. (1983) Synapsin I, a nerve terminal-specific phosphoprotein: III. Its association with synaptic vesicles studied in a highly-purified synaptic vesicle preparation. *J. Cell Biol.*, **96**, 1374.
71. Südhof, T. C., Czernik, A. J., Kao, H., Takei, K., Johnston, P. A., Horiuchi, A., *et al.* (1989) Synapsins: mosaics of shared and individual domains in a family of synaptic vesicle phosphoproteins. *Science*, **238**, 1142.
72. Bahler, M. and Greengard, P. (1987) Synapsin I bundles F-actin in a phosphorylation-dependent manner. *Nature*, **326**, 704.
73. Petrucci, T. C. and Morrow, J. (1987) Synapsin I: an actin-bundling protein under phosphorylation control. *J. Cell Biol.*, **105**, 1355.
74. Schiebler, W., Jahn, R., Doucet, J.-P., Rothlein, J., and Greengard, P. (1986) Characterization of synapsin I binding to small synaptic vesicles. *J. Biol. Chem.*, **261**, 8383.
75. Greengard, P., Valtorta, F., Czernik, A. J., and Benfenati, F. (1993) Synaptic vesicle phosphoproteins and regulation of synaptic function. *Science*, **259**, 780.
76. Südhof, T. C. and Jahn, R. (1991) Proteins of synaptic vesicles involved in exocytosis and membrane recycling. *Neuron*, **6**, 665.
77. Rosahl, T. W., Geppert, M., Spillane, D., Herz, J., Hammer, R. E., Malenka, R. C., *et al.* (1993) Short-term synaptic plasticity is altered in mice lacking synapsin I. *Cell*, **75**, 661.
78. Rosahl, T. W., Splllane, D., Missler, M., Herz, J., Selig, D. K., Wolff, J. R., *et al.* (1995) Essential functions of synapsins I and II in synaptic vesicle regulation. *Nature*, **375**, 488.
79. Takei, Y., Harada, A., Takeda, S., Kobayashi, K., Terada, S., Noda, T., *et al.* (1995) Synapsin I deficiency results in the structural change in the presynaptic terminals in the murine nervous system. *J. Cell Biol.*, **131**, 1789.
80. Li, L., Chin, L.-S., Shupliakov, O., Brodin, L., Sihra, T. S., Hvalby, O., *et al.* (1995) Impairment of synaptic vesicle clustering and of synaptic transmission, and increased seizure propensity, in synapsin I-deficient mice. *Proc. Natl. Acad. Sci. USA*, **92**, 9235.
81. De Camilli, P. (1995) Keeping synapses up to speed. *Nature*, **375**, 450.
82. Ryan, T. A., Li, L., Chin, L. S., Greengard, P., and Smith, S. J. (1996) Synaptic vesicle recycling in synapsin I knock-out mice. *J. Cell Biol.*, **134**, 1219.
83. McNaughton, B. (1980) Evidence for two physiologically distinct perforant pathways in the fascia dentata. *Brain Res.*, **199**, 1.
84. Pieribone, V. A., Shupliakov, O., Brodin, L., Hilfiker-Rothenfluh, S., Czernik, A. J., and Greengard, P. (1995) Distinct pools of synaptic vesicles in neurotransmitter release. *Nature*, **375**, 493.

85. Jahn, R., Schiebler, W., Ouimet, C., and Greengard, P. (1985) A 38 kilo dalton membrane protein (p38) present in synaptic vesicles. *Proc. Natl. Acad. Sci. USA*, **82**, 4137.
86. Leube, R. E., Kaiser, P., Seiter, A., Zimbelmann, R., Franke, W. W., Rehm, H., *et al.* (1987) Synaptophysin: molecular organization and mRNA expression as determined from cloned cDNA. *EMBO J.*, **6**, 3261.
87. Südhof, T. C., Lottspeich, F., Greengard, P., Mehl, E., and Jahn, R. (1987) A synaptic vesicle protein with a novel cytoplasmic domain and four transmembrane regions. *Science*, **238**, 1142.
88. Wiedenmann, B. and Franke, W. W. (1985) Identification and localization of synaptophysin, an integral membrane glycoprotein of Mr 38kD characteristic of presynaptic vesicles. *Cell*, **41**, 1017.
89. Calakos, N., Bennett, M. K., Peterson, K. E., and Scheller, R. H. (1994) Protein-protein interactions contributing to the specificity of intracellular vesicular trafficking. *Science*, **263**, 1146.
90. Edelmann, L., Hanson, P. I., Chapman, E. R., and Jahn, R. (1995) Synaptobrevin binding to synaptophysin: a potential mechanism for controlling the exocytotic fusion machine. *EMBO J.*, **14**, 224.
91. Johnston, P. A. and Südhof, T. C. (1990) The multisubunit structure of synaptophysin: relationship between disulfide bonding and homo-oligomerization. *J. Biol. Chem.*, **265**, 7849.
92. Barnekow, A., Jahn, R., and Schartl, M. (1990) Synaptophysin: a substrate for the protein tyrosine kinase pp60c-src in intact synaptic vesicles. *Oncogene*, **5**, 1019.
93. Pang, D. T., Wang, J. K., Valtorta, F., Benfenati, F., and Greengard, P. (1988) Protein tyrosine phosphorylation in synaptic vesicles. *Proc. Natl. Acad. Sci. USA*, **85**, 762.
94. Rubenstein, J. L., Greengard, P., and Czernik, A. J. (1993) Calcium-dependent serine phosphorylation of synaptophysin. *Synapse*, **13**, 161.
95. Knaus, P., Marqueze-Pouey, B., Scherer, H., and Betz, H. (1990) Synaptoporin, a novel putative channel protein of synaptic vesicles. *Neuron*, **5**, 453.
96. Fykse, E. M., Takei, K., Walch-Solimena, C., Geppert, M., Jahn, R., De Camilli, P., *et al.* (1993) Relative properties and localizations of synaptic vesicle protein isoforms: the case of the synaptophysins. *J. Neurosci.*, **13**, 4997.
97. Marqueze-Pouey, B., Wisden, W., Malosio, M. L., and Betz, H. (1991) Differential expression of synaptophysin and synaptoporin mRNAs in the postnatal rat central nervous system. *J. Neurosci.*, **11**, 3388.
98. Alder, J., Kanki, H., Valtorta, F., Greengard, P., and Poo, M. M. (1995) Overexpression of synaptophysin enhances neurotransmitter secretion at *Xenopus* neuromuscular synapses. *J. Neurosci.*, **15**, 511.
99. Alder, J., Xie, Z. P., Valtorta, F., Greengard, P., and Poo, M. (1992) Antibodies to synaptophysin interfere with transmitter secretion at neuromuscular synapses. *Neuron*, **9**, 759.
100. Rehm, H., Wiedenmann, B., and Betz, H. (1986) Molecular characterization of synaptophysin, a major calcium-binding protein of the synaptic vesicle membrane. *EMBO J.*, **5**, 535.
101. Cameron, P. L., Südhof, T. C., Jahn, R., and De Camilli, P. (1991) Colocalization of synaptophysin with transferrin receptors: implications for synaptic vesicle biogenesis. *J. Cell Biol.*, **115**, 151.
102. Linstedt, A. D. and Kelly, R. B. (1991) Synaptophysin is sorted from endocytotic markers in neuroendocrine PC12 cells but not transfected fibroblasts. *Neuron*, **7**, 309.

103. Thomas, L., Hartung, K., Langosch, D., Rehm, H., Bamberg, E., Franke, W. W., *et al.* (1988) Identification of synaptophysin as a hexameric channel protein of the synaptic vesicle membrane. *Science*, **242**, 1050.
104. McMahon, H. T., Bolshakov, V. Y., Janz, R., Hammer, R. E., Siegelbaum, S. A., and Südhof, T. C. (1996) Synaptophysin, a major synaptic vesicle protein, is not essential for neurotransmitter release. *Proc. Natl. Acad. Sci. USA*, **93**, 4760.
105. Eshkind, L. G. and Leube, R. E. (1995) Mice lacking synaptophysin reproduce and form typical synaptic vesicles. *Cell Tissue Res.*, **282**, 423.
106. Ullrich, B., Ushkaryov, Y. A., and Südhof, T. C. (1995) Cartography of neurexins: more than 1000 isoforms generated by alternative splicing and expressed in distinct subsets of neurons. *Neuron*, **14**, 497.
107. Ushkaryov, Y. A., Hata, Y., Ichtchenko, K., Moomaw, C., Afendis, S., Slaughter, C. A., *et al.* (1994) Conserved domain structure of β-neurexins. *J. Biol. Chem.*, **269**, 11987.
108. Hata, Y., Davletov, B., Petrenko, A. G., Jahn, R., and Südhof, T. C. (1993) Interaction of synaptotagmin with the cytoplasmic domains of neurexins. *Neuron*, **10**, 307.
109. Davletov, B. A., Krasnoperov, V., Hat, Y., Petrenko, A. G., and Südhof, T. C. (1995) High affinity binding of α-latrotoxin to recombinant neurexin Iα. *J. Biol. Chem.*, **270**, 23903.
110. Krasnoperov, V. G., Bittner, M. A., Beavis, R., Kuang, Y., Salnikow, K. V., Chepurny, O. G., *et al.* (1997) alpha-latrotoxin stimulates exocytosis by the interaction with a neuronal G-protein-coupled receptor. *Neuron*, **18**, 925.
111. Lelianova, V. G., Davletov, B. A., Sterling, A., Rahman, M. A., Grishin, E. V., Totty, N. F., *et al.* (1997) Alpha-latrotoxin receptor, latrophilin, is a novel member of the secretin family of G protein-coupled receptors. *J. Biol. Chem.*, **272**, 21504.
112. Geppert, M., Khvotchev, M., Krasnoperov, V., Goda, Y., Missler, M., Hammer, R. E., *et al.* (1998) Neurexin Iα is a major alpha-latrotoxin receptor that cooperates in alpha-latrotoxin action. *J. Biol. Chem.*, **273**, 1705.
113. Bowser, R., Muller, H., Govinan, B., and Novick, P. (1992) Sec8p and Sec15p are components of a plasma membrane-associated 19.5S particle that may function downstream of Sec4p to control exocytosis. *J. Cell Biol.*, **118**, 1041.
114. TerBush, D. R., Maurice, T., Roth, D., and Novick, P. (1996) The Exocyst is a multiprotein complex required for exocytosis in *Saccharomyces cerevisiae*. *EMBO J.*, **15**, 6483.
115. Grindstaff, K. K., Yeaman, C., Anandasabapathy, N., Hsu, S. C., Rodriguez-Boulan, E., Scheller, R. H., *et al.* (1998) Sec6/8 complex is recruited to cell-cell contacts and specifies transport vesicle delivery to the basal-lateral membrane in epithelial cells. *Cell*, **93**, 731.
116. Hsu, S. C., Hazuka, C. D., Roth, R., Foletti, D. L., Heuser, J., and Scheller, R. H. (1998) Subunit composition, protein interactions, and structures of the mammalian brain sec6/8 complex and septin filaments. *Neuron*, **20**, 1111.
117. Hsu, S. C., Ting, A. E., Hazuka, C. D., Davanger, S., Kenny, J. W., Kee, Y., *et al.* (1996) The mammalian brain rsec6/8 complex. *Neuron*, **17**, 1209.
118. Friedrich, G. A., Hildebrand, J. D., and Soriano, P. (1997) The secretory protein Sec8 is required for paraxial mesoderm formation in the mouse. *Dev. Biol.*, **192**, 364.
119. Bachhawat, A. K., Suhan, J., and Jones, E. W. (1994) The yeast homolog of Hb58, a mouse gene essential for embryogenesis, performs a role in the delivery of proteins to the vacuole. *Genes Dev.*, **8**, 1379.
120. Lee, J. J., von Kessler, D. P., Parks, S., and Beachy, P. A. (1992) Secretion and localized transcription suggest a role in positional signaling for products of the segmentation gene *hedgehog*. *Cell*, **71**, 33.

121. Hess, E. J., Collins, K. A., and Wilson, M. C. (1996) Mouse model of hyperkinesis implicates SNAP-25 in behavioral regulation. *J. Neurosci.*, **16**, 3104.
122. Hess, E. J., Jinnah, H. A., Kozak, C. A., and Wilson, M. C. (1992) Spontaneous locomotor hyperactivity in a mouse mutant with a deletion including the *Snap* gene on chromosome 2. *J. Neurosci.*, **12**, 2865.
123. Oyler, G. A., Higgins, G. A., Hart, R. A., Battenberg, E., Billingsley, M., Bloom, F. E., *et al.* (1989) The identification of a novel synaptosomal-associated protein, SNAP-25, differentially expressed by neuronal subpopulations. *J. Cell Biol.*, **109**, 3039.
124. Sutton, R. B., Fasshauer, D., Jahn, R., and Brunger, A. T. (1998) Crystal structure of a SNARE complex involved in synaptic exocytosis at 2.4 Å resolution. *Nature*, **395**, 347.
125. Blasi, J., Chapman, E. R., Link, E., Binz, T., Yamasaki, S., De Camilli, P., *et al.* (1993) Botulinum neurotoxin A selectively cleaves the synaptic protein SNAP-25. *Nature*, **365**, 160.
126. Schiavo, G., Rossetto, O., Catsicas, S., Polverino de Laureto, P., DasGupta, B. R., Benfenati, F., *et al.* (1993) Identification of the nerve terminal targets of botulinum neurotoxin serotypes A, D, and E. *J. Biol. Chem.*, **268**, 23784.
127. Theiler, K. and Varnum, D. S. (1981) Development of coloboma (Cm/+) mutation with anterior lens adhesion. *Anat. Embryol.*, **161**, 121.
128. Hess, E. J., Collins, K. A., Copeland, N. G., Jenkins, N. A., and Wilson, M. C. (1994) Deletion map of the coloboma (Cm) locus on mouse chromosome 2. *Genomics*, **21**, 257.
129. Searle, A. G. (1966) New mutants. II. Coloboma. *Mouse News Lett.*, **35**, 27.
130. Greenhill, L., Puig-Antich, J., Goetz, R., Hanlon, C., and Davies, M. (1983) Sleep architecture and REM sleep measures in prepubertal children with attention deficit disorder with hyperactivity. *Sleep*, **6**, 91.
131. Pauls, D. L., Hurst, C. R., Kruger, S. D., Leckman, J. F., Kidd, K. K., and Cohen, D. J. (1986) Gilles de la Tourette's syndrome and attention deficit disorder with hyperactivity. Evidence against a genetic relationship. *Arch. Gen. Psychiatry*, **43**, 1177.
132. Barkley, R. A. (1977) A review of stimulant drug research with hyperactive children. *J. Child Psychol. Psychiatry*, **18**, 137.
133. Shaywitz, S. E. and Shaywitz, B. A. (1984) Diagnosis and management of attention deficit disorder: a pediatric perspective. *Pediatr. Clin. North Am.*, **31**, 429.
134. Steffensen, S. C., Wilson, M. C., and Henriksen, S. J. (1996) Coloboma contiguous gene deletion encompassing Snap alters hippocampal plasticity. *Synapse*, **22**, 281.
135. Raber, J., Mehta, P. P., Kreifeldt, M., Parsons, L. H., Weiss, F., Bloom, F. E., *et al.* (1997) Coloboma hyperactive mutant mice exhibit regional and transmitter-specific deficit in neurotransmission. *J. Neurochem.*, **68**, 176.
136. Hess, E. J., Rogan, P. K., Domoto, M., Tinker, D. E., Ladda, R. L., and Ramer, J. C. (1995) Absence of linkage of apparently single gene medicated ADHD with the human syntenic region of the mouse mutant Coloboma. *Am. J. Med. Genet.*, **60**, 573.
137. Ashkenas, J. (1996) Williams syndrome starts making sense. *Am. J. Hum. Genet.*, **59**, 756.
138. Osborne, L. R., Soder, S., Shi, X. M., Pober, B., Costa, T., Scherer, S. W., *et al.* (1997) Hemizygous deletion of the syntaxin 1A gene in individuals with Williams syndrome. *Am. J. Hum. Genet.*, **61**, 449.
139. Wu, Y.-Q., Sutton, V. R., Nickerson, E., Lupski, J. R., Potocki, L., Korenberg, J. R., *et al.* (1998) Delineation of the common critical region in Williams syndrome and clinical correlation of growth, heart defects, ethnicity, and parental origin. *Am. J. Med. Genet.*, **78**, 82.

140. Burn, J. (1986) Williams syndrome. *J. Med. Genet.*, **23**, 389.
141. Wang, P. P. and Bellugi, U. (1993) Williams syndrome, Down syndrome, and cognitive neuroscience. *Am. J. Dis. Child.*, **147**, 1246.
142. Curran, M. E., Atkinson, D. L., Ewart, A. K., Morris, C. A., Leppert, M. F., and Keating, M. T. (1993) The elastin gene is disrupted by a translocation associated with supravalvular aortic stenosis. *Cell*, **73**, 159.
143. Ewart, A. K., Jin, W., Atkinson, D., Morris, C. A., and Keating, M. T. (1994) Supravalvular aortic stenosis associated with a deletion disrupting the *elastin* gene. *J. Clin. Invest.*, **93**, 1071.
144. Olson, T. M., Michels, V. V., Urban, Z., Csiszar, K., Christiano, A. M., Driscoll, D. J., *et al.* (1995) A 30 kb deletion within the elastin gene results in familial supravalvular aortic stenosis. *Hum. Mol. Genet.*, **4**, 1677.
145. Ewart, A. K., Morris, C. A., Atkinson, D., Jin, W., Sternes, K., Spallone, P., *et al.* (1993) Hemizygosity at the elastin locus in a developmental disorder, Williams syndrome. *Nat. Genet.*, **5**, 11.
146. Lowery, M. C., Morris, C. A., Ewart, A., Brothman, L. J., Zhu, X. L., Leonard, C. O., *et al.* (1995) Strong correlation of elastin deletions, detected by FISH, with Williams syndrome: evaluation of 235 patients. *Am. J. Hum. Genet.*, **57**, 49.
147. Frangiskakis, J. M., Ewart, A. K., Morris, C. A., Mervis, C. B., Bertrand, J., Robinson, B. F., *et al.* (1996) LIM-kinase1 hemizygosity implicated in impaired visuospatial constructive cognition. *Cell*, **86**, 59.
148. Naren, A. P., Nelson, D. J., Xie, W., Jovov, B., Pevsner, J., Bennett, M. K., *et al.* (1997) Regulation of CFTR chloride channels by syntaxin and Munc18 isoforms. *Nature*, **390**, 302.
149. Naren, A. P., Quick, M. W., Collawn, J. F., Nelson, D. J., and Kirk, K. L. (1998) Syntaxin 1A inhibits CFTR chloride channels by means of domain-specific protein-protein interactions. *Proc. Natl. Acad. Sci. USA*, **95**, 10972.
150. Bennett, M. K., Garcia-Arraras, J. E., Elferink, L. A., Peterson, K., Fleming, A. M., Hazuka, C. D., *et al.* (1993) The syntaxin family of vesicular transport receptors. *Cell*, **74**, 863.
151. Steele, P. M., Medina, J. F., Nores, W. L., and Mauk, M. D. (1998) Using genetic mutations to study the neural basis of behavior. *Cell*, **95**, 879.
152. Takahashi, J. S., Pinto, L. H., and Vitaterna, M. H. (1994) Forward and reverse genetic approaches to behavior in the mouse. *Science*, **264**, 1724.
153. Chen, C., Rainnie, D. G., Greene, R. W., and Tonegawa, S. (1994) Abnormal fear response and aggressive behavior in mutant mice deficient for alpha-calcium-calmodulin kinase II. *Science*, **266**, 291.
154. Silva, A. J., Paylor, R., Wehner, J. M., and Tonegawa, S. (1992) Impaired spatial learning in alpha-calcium-calmodulin kinase II mutant mice. *Science*, **257**, 206.
155. Silva, A. J., Stevens, C. F., Tonegawa, S., and Wang, Y. (1992) Deficient hippocampal long-term potentiation in alpha-calcium-calmodulin kinase II mutant mice. *Science*, **257**, 201.
156. Abeliovich, A., Chen, C., Goda, Y., Silva, A. J., Stevens, C. F., and Tonegawa, S. (1993) Modified hippocampal long-term potentiation in PKC gamma-mutant mice. *Cell*, **75**, 1253.
157. Abeliovich, A., Paylor, R., Chen, C., Kim, J. J., Wehner, J. M., and Tonegawa, S. (1993) PKC gamma mutant mice exhibit mild deficits in spatial and contextual learning. *Cell*, **75**, 1263.
158. Kandel, E. R., Schartz, J. H., and Jessel, T. (1991) *Principles of neural science*. Appleton & Lange Publishing Co., New York, NY.

11 Synaptic vesicle endocytosis and recycling

BING ZHANG and MANI RAMASWAMI

'The *yang* having reached its climax retreats in favor of the *yin*;
the *yin* having reached its climax retreats in favor of the *yang*.'
- Wang Chu'ung (AD 80) on *Tai Chi Tu*

1. Introduction

Exocytosis and endocytosis, the *yin* and *yang* elements of the vesicle cycle, must be well balanced to ensure continuous synaptic transmission. At nerve terminals, the number of synaptic vesicles is insufficient to account for the number of quanta released during a short burst of intense nerve activity. Axonal transport of newly synthesized vesicle proteins occurs too slowly to refill the vesicle pool at the required rates. Under these circumstances, the only way to rapidly replenish synaptic vesicles is to recycle them locally through endocytosis. Thus, a major challenge is to understand how synaptic vesicles are faithfully and efficiently recovered.

The nervous system employs at least two distinct pathways for vesicle recycling, both of which may coexist at the same nerve terminal. One pathway occurs at the active zone for vesicle fusion, with rapid kinetics, and may be calcium (Ca^{2+})-dependent. A second pathway occurs slowly and away from the sites of transmitter release and may be relatively insensitive to Ca^{2+}. Both pathways require cytoplasmic proteins such as dynamin, a GTPase required for membrane fission and, possibly, vesicle coat proteins such as clathrin and adaptor proteins required for vesicle formation. In addition to these demonstrated pathways, a seemingly efficient mechanism for vesicle fusion and recycling, termed 'kiss-and-run' has been postulated to exist in granule cells and at nerve terminals (1, 2). An extreme view of the 'kiss-and-run' model postulates that vesicles form a transient fusion pore with the plasma membrane to permit transmitter release. Because vesicle membrane does not mix with the plasma membrane, closing of the fusion pore stops transmitter release and accomplishes vesicle recycling with no obvious need for any accessory proteins. Even though it may operate in dense core granule cells and remains conceptually appealing, such a pathway has not yet been demonstrated for synaptic vesicles (1).

The available data argue that synaptic vesicles fuse completely with the plasma membrane during exocytosis and recycle by an endocytic mechanism that requires reassembly of synaptic vesicle components.

The goal of this chapter is to review our current understanding of the cellular and molecular mechanisms of synaptic vesicle endocytosis. We focus almost exclusively on small synaptic vesicles that release fast neurotransmitters. The endocytosis of other classes of secretory vesicles, although pertinent, is beyond the scope of this chapter.

2. Cell biology of synaptic vesicle recycling

2.1 Synaptic vesicle recycling: an overview and an alternative view

Our overview of synaptic vesicle recycling posits:

(a) Synaptic vesicles following exocytosis fuse completely with the plasma membrane and expose their lumenal space to the extracellular fluid.

(b) Vesicle proteins retrieved via endocytosis will be incorporated into vesicles along with a small portion of extracellular fluid.

(c) A specific molecular machinery exists to mediate vesicle recycling, but not vesicle exocytosis.

Evidence for all of these have been obtained at nerve terminals in several different organisms.

2.1.1 Synaptic vesicles collapse into plasma membrane

Bittner and Kennedy (1970) observed that the total number of synaptic vesicles at a crayfish nerve terminal was far less than the total quanta released during an experiment, and first suggested that synaptic vesicles are replenished locally (3). A series of pioneering studies at the frog neuromuscular synapse, by Heuser, Reese, and Ceccarelli and their associates, demonstrated that sustained high frequency stimulation or treatment with black widow spider venom (BWSV) results in depletion of synaptic vesicles together with a corresponding expansion of the plasma membrane (4–8). Freeze-fracture techniques used to image the presynaptic membrane milliseconds after nerve stimulation in the presence of 4-aminopyridine (a potassium channel blocker used to maximize exocytosis) revealed vesicle-sized pockets, indicating that synaptic vesicles collapse into the plasmalemma during transmitter release (4, 5, 9, 10). Finally, several studies have established that lumenal domains of synaptic vesicle membrane proteins are exposed to extracellularly applied antibodies following intense nerve stimulation. In addition, these vesicle proteins are found on the presynaptic plasmalemma (11–15). These and other studies establish that synaptic vesicles collapse into the plasma membrane during exocytosis, at least under some conditions.

2.1.2 Synaptic vesicle re-formation occurs via an endocytic pathway

If sufficient time is given for vesicle recycling, the expanded plasma membrane surface following the depletion of synaptic vesicles can be restored to its original state. This suggests that synaptic vesicles are likely to be regenerated from the plasma membrane itself. This notion is supported by the observation that newly endocytosed vesicles can take up extracellular tracers. If a fluid phase tracer, horse-radish peroxidase (HRP), is present during the recovery of vesicles, then HRP may be observed, first in large intraterminal cisternae and subsequently in newly formed synaptic vesicles (8). The observation that newly formed synaptic vesicles included a marker for extracellular fluid strongly implicates that these recycled vesicles originate from plasma membrane and are internalized via a conventional endocytic pathway. More recently, a fluorescent, lipophilic, endocytic tracer, FM1-43, has been used to observe recycling at living frog synapses. Synaptic vesicles recycled in the presence of extracellular FM1-43 are shown to contain the dye, further demonstrating that between fusion and re-formation, vesicles reside long enough on the plasma-lemma to take up the plasma membrane-associated molecules (16, 17) (also see Chapter 2).

As predicted by this model for vesicle cycling, both HRP and FM1-43 in recycled vesicles disappear after prolonged nerve stimulation in the absence of extracellular tracer due to exocytotic release of vesicle contents. Observations similar to those described above for the frog neuromuscular synapse have been made in a variety of other preparations including the crayfish neuromuscular junction (18), mammalian hippocampal cultures (19, 20), and *Drosophila* retinular (30) or motor synapses (22, 23). These studies demonstrate that synaptic vesicles after fusion are re-formed from the plasma membrane via an endocytic recycling mechanism.

2.1.3 Specific molecules are involved in vesicle recycling, but not in exocytosis

The most dramatic demonstration of endocytosis-dependent synaptic vesicle recycling comes from studies of the *Drosophila shibirets* (*shi*^{ts1}) mutant conditionally defective in the function of a cytoplasmic GTPase called dynamin (21, 24–28). In these mutants, restrictive temperatures block the GTPase activity of dynamin. Consequently, this leads to a use-dependent depletion of synaptic vesicles at the nerve terminal and a failure of evoked transmitter release (Fig. 1). Exocytosis is not affected in the mutants. Therefore, dynamin is a molecule specifically required for the endocytic but not exocytotic part of the synaptic vesicle cycle. The effect of high temperatures on *shi*^{ts1} mutant flies demonstrates the importance of synaptic vesicle recycling for an organism. On shifting from permissive to restrictive temperatures, the flies are paralysed in less than a minute due to vesicle depletion in critical synapses. Studies of *shi*^{ts1} mutants, some of which are discussed subsequently, have contributed significantly to our understanding of the cell biology of vesicle recycling. As in amphibian and mammalian synapses, the transfer of vesicle proteins to plasma membrane during exocytosis, and their subsequent recovery via endocytosis have been documented in shi^{ts1} synapses by a variety of different techniques (21, 23, 26, 29–31).

In addition to dynamin, a rapidly expanding collection of molecules involved in the endocytic recycling of synaptic vesicle has been identified. Particularly strong evidence exists for the involvement of the vesicle coat protein clathrin and its associated proteins, the adaptor AP2 complex and AP180. Clathrin, clathrin adaptor proteins, and dynamin, have been localized by electron microscopy to endocytic synaptic vesicles arrested on the plasma membrane at a stage requiring GTP

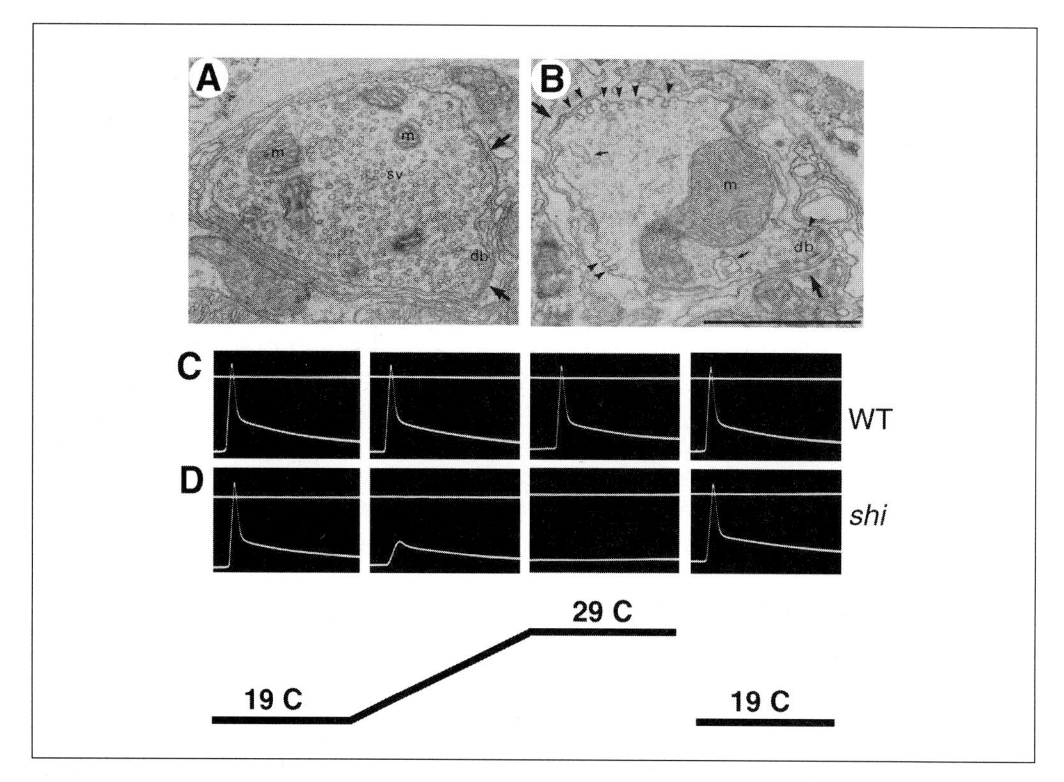

Fig. 1 Temperature-sensitive block of synaptic vesicle recycling in *shibire*. (A). Synaptic vesicles distribute evenly in a coxal synapse of a shi^{ts1} fly at 19°C. (B) Synaptic vesicles are depleted from a coxal synapse after a shi^{ts1} fly is exposed to 29°C for 8 min. Note that the accumulation of several small invaginations on the plasma membrane, which are collared with electron-dense substances around their necks (arrowheads). Large invaginations, whose necks are out of the plane of the section, are identified by small arrows. The release sites including pre- and postsynaptic densities and a presynaptic dense body (db) are indicated by large arrows. SV, synaptic vesicles; m, mitochondria. Scale bar, 1 μm; × 40000. (Reprinted with permission from ref. 21.Copyright 1989. Society for Neuroscience.) (C and D) Action potential recorded in adult dorsal longitudinal muscles in response to electrical stimulation of the giant fibre neuron in the head. (C) Action potentials in wild-type (WT) flies are not affected by temperatures ranging from 19°C to 29°C. (D) In shi^{ts1}, action potential is normal compared with wild-type at 19°C. However, as temperature increases, action potential disappears, leaving only a small subthreshold excitatory junctional potential (EJP). At 29°C, EJP is also absent, suggesting that synaptic transmission is completely blocked due to depletion of vesicles. This temperature block of synaptic transmission is reversible upon returning to 19°C. The white flat line indicates zero voltage level. (Reprinted with permission from Koenig *et al.* (1983). *J. Cell Biol.*, **96**,1517. Copyright 1983. The Rockefeller University Press.)

hydrolysis by dynamin (32, 33) (Fig. 2). However, none of these molecules are associated with vesicles during exocytosis. Thus, a distinct class of molecules exists that is required for synaptic vesicle recycling but not for vesicle exocytosis. This would be expected if transmitter release were to occur by a mechanism involving complete collapse of synaptic vesicles into plasma membrane. On the other hand, it should not be a complete surprise that synaptic vesicle proteins such as synaptotagmin may play a dual role in both exocytosis and endocytosis, because they may interact with endocytic proteins to facilitate vesicle reformation during endocytosis (34, 35).

2.1.4 An alternative view: the 'kiss-and-run' pathway for synaptic vesicle recycling

All discussions of a 'kiss-and-run' mechanism for vesicle cycling, must begin with the authors' personal definition of this mechanism (2). Betz and Angleson (1998) have discussed with excellent clarity the wide range of mechanistically different vesicle recycling mechanisms which could fall within a generous interpretation of 'kiss-and-run' (1). As we discuss further in this chapter, it should appear very clear that slow and fast pathways for synaptic vesicle recycling exist at nerve terminals. While we will argue that dynamin appears essential for all known pathways of synaptic vesicle recycling, it is less clear whether other accessory proteins like clathrin and clathrin-associated proteins are essential for the faster pathway.

We choose an unambiguous definition for a 'kiss-and-run' pathway for vesicle cycling. In this definition:

(a) Individual synaptic vesicles retain their lipid and protein components during transmitter release.

(b) The vesicles do not ever fuse with the plasma membrane.

(c) Transmitter release occurs via transient pores which span the two lipid bilayers between vesicle and plasma membrane.

Evidence for such a mode of vesicular release has been obtained by amperometric and capacitance analysis of dense core granules (see Chapter 2). Two phases for release of catecholamine from bovine adrenal chromaffin cell granules may be observed by amperometry: a fast activated dribble, referred to as a 'foot', which precedes a large 'spike' of release (36–38). The foot is likely to reflect a dribble of catecholamine release through transient fusion pores previously described by capacitance measurements (39, 40) while the spike reflects complete fusion of the vesicle. Instances where a foot is not followed by a spike occur at a reasonable frequency, and these may represent true instances of 'kiss-and-run' demonstrated by chromaffin granules. In support of this idea, docked granules appear to have less content than cytoplasmic granules as though the contents were reduced by several transient fusion events (41), 'a sort of granular incontinence', to borrow a picturesque description from Betz and Angleson (1998).

The idea of the 'kiss-and-run' mechanism for synaptic vesicle cycling is attractive,

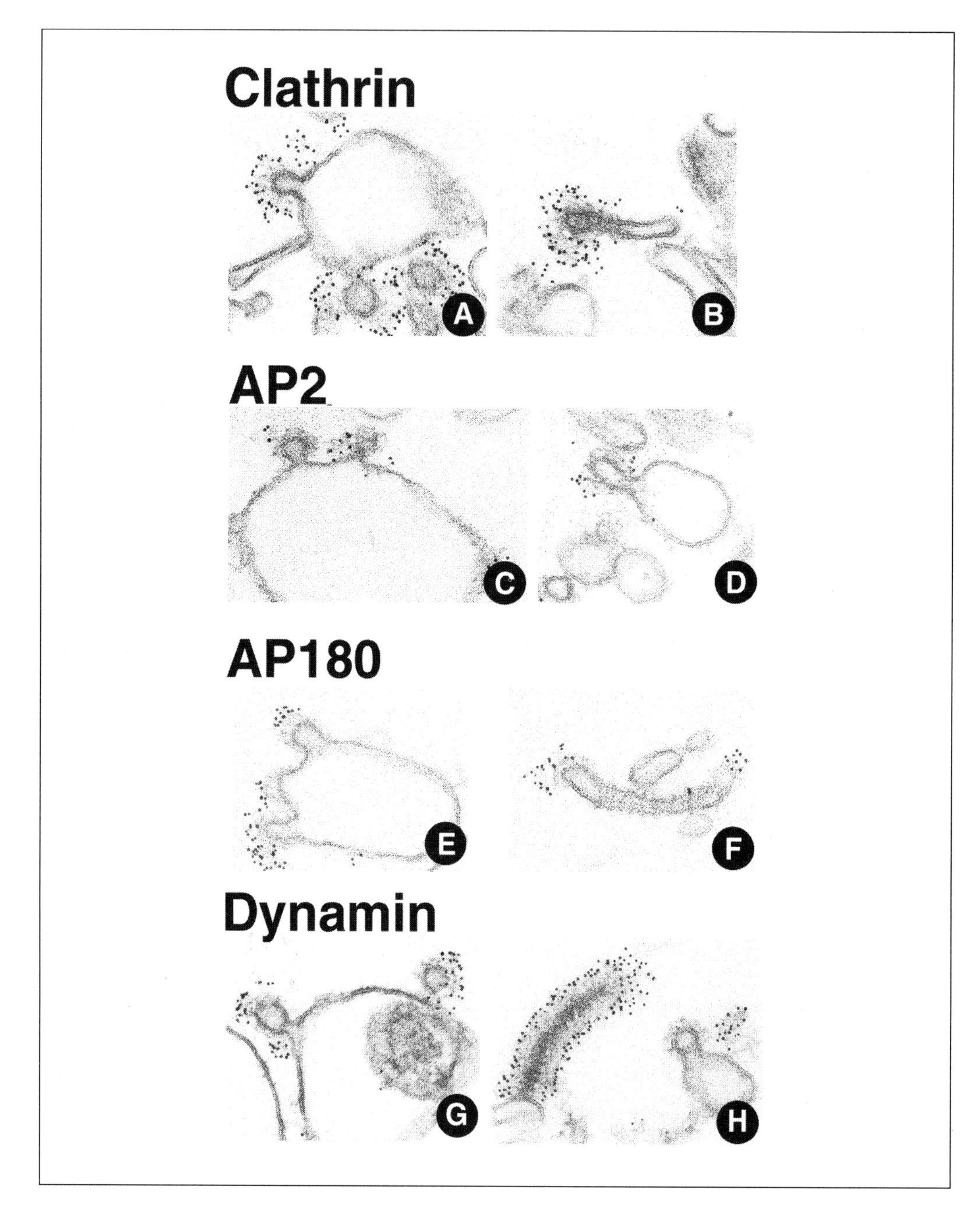

Fig. 2 Immunogold localization of clathrin, α-adaptin, AP180, and dynamin on endocytic vesicles of brain synaptosomes treated with ATP and GTPγS. (A and B) Clathrin homogeneously coats vesicular buds. (C and D) α-adaptin and (E and F) AP180 partially co-localize with clathrin on endocytic vesicles. Their distribution is not even around the clathrin coat. (G and H) Dynamin heavily surrounds the GTPγS-induced membrane tubules and unevenly localizes to clathrin coats. (Reprinted with permission from ref. 33. Copyright 1996. The Rockefeller University Press).

primarily because it offers high speed and efficiency for synaptic vesicle recycling. However, evidence for 'kiss-and-run', as we define it here, has not yet been obtained for clear synaptic vesicles involved in fast neurotransmitter release. None-the-less, substantial evidence exists for a fast recycling pathway different from 'kiss-and-run' and distinct from that described by Heuser and Reese (1973) at the frog neuromuscular synapse (8). Since fast and slow recycling pathways have been distinguished largely on the basis of kinetic measurements, we will first discuss the kinetics of vesicle recycling and the techniques used for these measurements.

2.2 Kinetics of synaptic vesicle recycling

To ensure continuous transmitter release, synaptic vesicles must be recycled at a reasonable rate to replenish the vesicle pool. The kinetics of vesicle recycling can be studied by tracing vesicle cycling using either HRP or FM 1-43 dye. Heuser and Reese (1973) used electron microscopy to observe recycling synaptic vesicles at nerve terminals recovering from intense nerve stimulation. In frog synapses recovering after prolonged stimulation, large cisternae containing the endocytic tracer HRP appear first. Subsequently, these cisternae resolve themselves into newly recycled vesicles. Careful experiments using HRP activity to visualize the endocytic compartment showed that the half-time for membrane internalization after brief stimulation is on the order of tens of seconds (8). More incisive studies of vesicle recycling were prevented by the difficulty of quantitative EM experiments and the relative inconvenience of HRP as an endocytic tracer.

Unlike HRP, FM1-43 dye is a fluorescent styryl dye that is used to label synaptic vesicle membrane. Its unique properties allow it to be internalized into recycling vesicles and subsequently released upon exocytosis. Thus, it allows direct observation and quantification of synaptic vesicle recycling at living nerve terminals (16, 17) (Chapter 2). Several innovative strategies using FM1-43 have offered insight into the kinetics of synaptic vesicle recycling (19, 42). Two of these procedures measure the time taken for a complete vesicle cycle (dead time) or the time for vesicle internalization after exocytosis (Fig. 3). Using these and other various procedures, the time constants for membrane internalization and total vesicle recycling have been determined at the frog neuromuscular junction (NMJ) and in synapses formed between hippocampal neurons in culture. Internalization times from these preparations are about 30–60 seconds, and the complete recycling times are about 45–75 seconds during a short period of stimulation (20, 42). The time course obtained from these FM 1-43 studies is in good agreement with that obtained earlier.

Despite the consistency among time constants for internalization and recycling obtained from these initial FM1-43 studies, capacitance measurements of presynaptic terminals have revealed the existence of a more rapid mechanism for vesicle recycling. von Gersdorff and Matthews (1994) showed that presynaptic capacitance increases rapidly following exocytosis of about 6000 vesicles and drops with a time constant of less than two seconds on cessation of the stimulus (43). Although it has yet to be demonstrated that this capacitance drop results from specific retrieval of

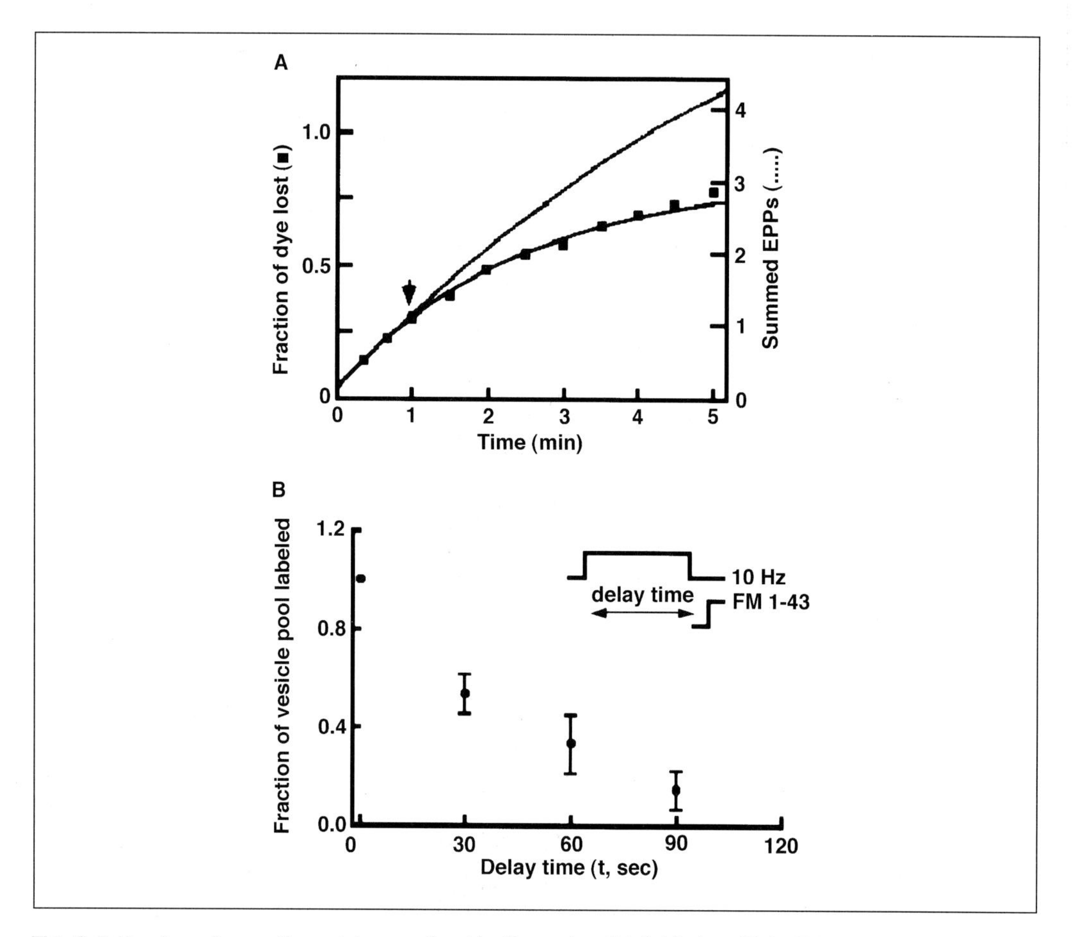

Fig. 3 Estimation of synaptic vesicle recycling kinetics using FM 1-43 dye. (A) In FM 1-43 loaded frog NMJ synapses, the rate of dye loss during continuous electrical stimulation (20 Hz) is reduced after newly internalized but unlabelled vesicles randomly enter the releasable pool and release transmitter. The time (~ 1 min) taken for new vesicles to decouple rates of dye loss and transmitter release (summed EPPs) reflects the complete cycling time for synaptic vesicles, including endocytosis, recycling, re-priming, and re-release. (Reprinted with permission from ref. 16. Copyright 1992. American Association for the Advancement of Science.) (B) Synaptic vesicle pool of hippocampal synapses in culture is labelled with FM1-43 at a variable delay following a 30 sec train (10 Hz) of stimulation. During the 30 sec stimulation period, half of synaptic vesicles are estimated to release transmitter within 10 sec. At the end of the train, approximately one-half of these vesicles fused with the plasma membrane are endocytosed and labelled with the dye. Therefore, the half-time for vesicle endocytosis is estimated to be ~ 20 sec. (Reprinted with permission from ref. 20. Copyright 1995. Cell Press.)

synaptic vesicle membrane, this observation suggests that the rate of rapid vesicle fission in goldfish bipolar neurons is more than tenfold faster than that previously estimated in frog NMJ and cultured central synapses.

A pathway for rapid internalization of synaptic vesicles may indeed exist in frog NMJs as well as in central synapses in mammals as demonstrated by the ingenious use of FM1-43 and related styryl dyes (20, 44). The dye FM1-43 has a hydrophobic (aliphatic hydrocarbon) tail, which allows the dye to departition from lipid bilayers of hippocampal neuron membrane with a time constant of about 2.5 seconds. Thus, dye loss during destaining is determined by the duration of vesicle fusion and by the rate of departition. When vesicle internalization occurs on a similar or faster time scale of the rate of departition, transmitter release from a dye-stained vesicle would occur, but the dye would be partially released. This has been observed in frog NMJ synapses treated with the protein kinase blocker staurosporine. In staurosporine-treated motor synapses, transmitter release is not affected, but FM1-43 destaining is selectively blocked. This suggests that synaptic vesicles fuse into the plasma membrane for a time period sufficient to release transmitter but insufficient to allow FM 1–43 to departition into the plasma membrane. As predicted in such a scenario, HRP uptake into recycling vesicles is not observed, and vesicle depletion normally observed at the frog motor synapses following prolonged stimulation is also completely blocked (45). It is not clear whether staurosporine simply accelerates the speed of vesicle recycling or switches the mode of exocytosis from a slow pathway to a fast one.

Another demonstration for a rapid recycling mechanism is in cultured hippocampal synapses (20, 44). Klingauf *et al.* (1998) showed that when stimulated with high K^+ solution, the initial dye release from cultured hippocampal synapses is incomplete, but it occurs rapidly with a time constant of 2.5 seconds for FM 1-43. The rate of dye loss also depends on the type of FM dyes used and varies inversely with the time constant for dye departitioning from lipid membranes (44) (Fig. 4). This study further suggests that the 'real' rate of vesicle recycling could even be faster were it not limited by the rate of departition of FM dyes available. It is intriguing to note that differential rates of destaining for FM2-10, FM1-43, FM1-84 are not observed when the hippocampal synapses are stimulated at 1 Hz stimulation (20). A possible explanation for this dichotomy is that this rapid pathway is more prominently dependent upon Ca^{2+} or nerve activity. As at frog motor synapses, the rate of destaining for a given dye is significantly diminished by treatment with staurosporine. Thus, a fast endocytic pathway operating within a time scale of a few seconds (i.e. the departitioning time constant for FM1-43) exists at both the peripheral and central synapses. We will argue in the next section that most existing data for synaptic vesicle recycling may be rationalized by postulating the coexistence of two kinetically, spatially, and mechanistically distinct pathways.

2.3 Distinct pathways of synaptic vesicle endocytosis

2.3.1 Two spatially and kinetically distinct pathways

Heuser and Reese (1981) used freeze-fracture analysis to observe synaptic vesicle collapse into plasma membrane during exocytosis. Following fusion, they observed vesicle membrane proteins diffuse to distinct sites away from the active zone where they were eventually internalized (46). However, in a similar analysis of frog neuromuscular synapses at lower rates of exocytosis, Ceccarelli, Hurlbut, and co-workers observed that vesicle internalization occurs at sites close to active zones (4, 47). These

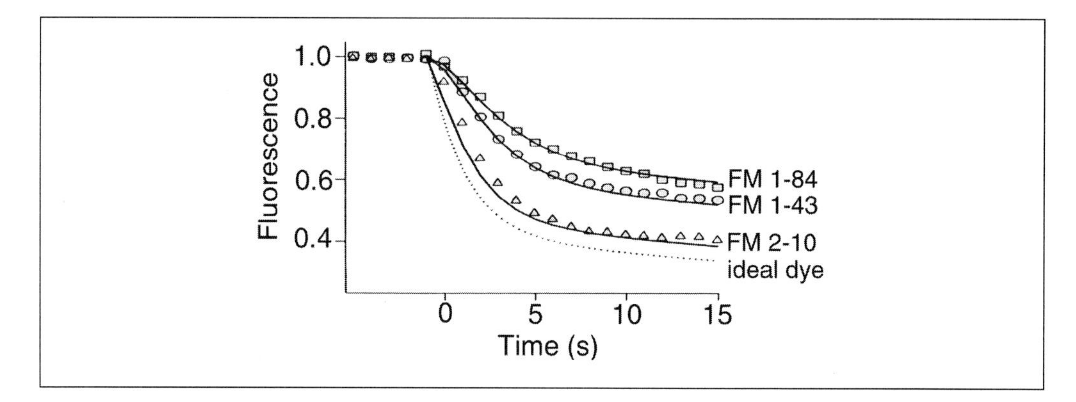

Fig. 4 Rapid endocytosis revealed by differences in departitioning kinetics of FM dyes. Presynaptic terminals in cultured hippocampal neurons are loaded with FM dye and destained with 90 mM K^+ and 2 mM Ca^{2+} solution. The time constant for rapid endocytosis is determined by the rate of membrane dissociation of each dye. The faster the rate a dye departitions, the smaller the time constants (FM 2-10, 0.6 sec > FM 1-43, 2.5 sec > FM 1-84, 4.7 sec). An even faster endocytosis rate is expected if the departitioning rate of a dye (ideal dye) can be accelerated. (Reprinted with permission from ref. 44. Copyright 1998. Macmillan Magazines Ltd.)

early experiments suggest two spatially distinct pathways for synaptic vesicle internalization: an active zone pathway, predominant under moderate stimulation conditions, and a non-active zone pathway predominant at high rates of stimulation (5, 46).

The most compelling evidence for two spatially distinct pathways of synaptic vesicle recycling has been recently obtained by Koenig and Ikeda through analysis of vesicle recovery at *shi*^{ts1} retinula synapses (30). If these synapses are first depleted by stimulation at 29 °C and then shifted to 26 °C, a temperature where dynamin activity is presumably only partially restored, long tubular invaginations are seen emanating from almost 100% of active zones within one minute. At this time point, relatively few endocytic pits or invaginations are visible away from the active zone. However, at later time points, accumulations of branched tubular invaginations of plasma membrane become apparent at sites away from the active zone (Fig. 5). These temporally and spatially separated endocytic pathways suggest distinct pathways for synaptic vesicle recycling: a relatively rapid pathway at the active zone for vesicle fusion, and a slower pathway away from the active zone (30).

Specialized regions for synaptic vesicle endocytosis are also suggested by immunofluorescence studies at the *Drosophila* larval NMJ (29). In *shi*^{ts1} NMJ terminals depleted of synaptic vesicles, synaptic vesicle membrane proteins may be seen relatively evenly dispersed on the presynaptic plasma membrane. In contrast, the GTPase dynamin, which should mark sites of endocytosis, is found sharply concentrated in spots on the plasma membrane (29). Other molecules likely to be involved in membrane internalization have also been detected in such hotspots at frog and fly neuromuscular synapses (45, 48, 49). These observations suggest, albeit indirectly, that vesicle internalization during synaptic vesicle recycling occurs in distinct regions of the plasma membrane. Examining the location of these endocytic hotspots

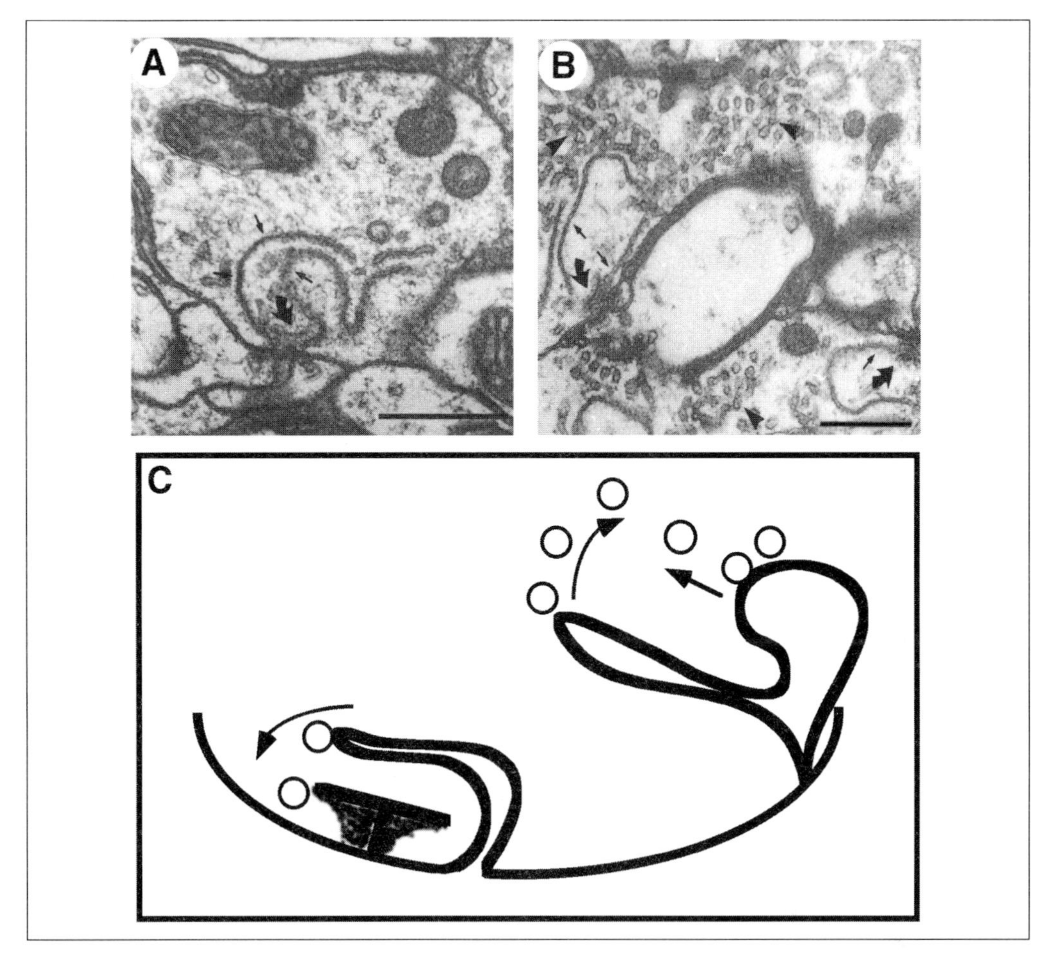

Fig. 5 Two distinct recycling pathways revealed within a single nerve terminal in *shibire* retinula cells. (A and B) Retinula terminals are depleted of synaptic vesicles after 30 sec exposure to 29 °C. Upon returning to 26 °C, two kinetically and spatially distinct pathways for vesicle membrane internalization are observed. One pathway occurs first, near the active zone under the dense body (curved arrows), and involves the formation of long, thin, and unbranched tubules (small arrows) emanating from the plasma membrane. (B) The other pathway occurs subsequently, away from the active zone, and involves the formation of branching tubular structures (arrow heads). Scale bars: (A) 0.5 μm, × 44000; (B) 0.5 μm, × 37000. (Reprinted with permission from ref. 30. Copyright 1996. The Rockefeller University Press.) (C) These two pathways are schematically illustrated. The 'T'-shaped area represents the dense body commonly found in the active zone of *Drosophila* synapses.

relative to active zones may reveal the predominance of each pathway under specific conditions.

2.3.2 Mechanistically distinct pathways

It is likely that fast and slow internalization of clear synaptic vesicles involve different cytosolic proteins. Fast and slow internalization of dense core granule membrane require different molecules: fast internalization appears clathrin-independent and calmodulin-dependent, while slow endocytosis is clathrin-dependent and calmodulin-independent (50, 51). However, different molecular requirements for fast and slow internalization of synaptic vesicle membrane have not yet been demonstrated. At nerve terminals, dynamin, the product of the *shibire* gene, is required for both the fast active zone pathway as well as for the slower non-active pathway for membrane internalization (30). The requirement for clathrin in at least one of these pathways is strongly suggested by a wealth of independent circumstantial data; however, differential requirements for clathrin in active zone and non-active zone internalization have not yet been analysed.

Heuser and Reese (1973) proposed that clathrin-coated vesicles after internalization uncoat and fuse to form cisternae-like structures (termed endosomes) from which recycled synaptic vesicles emerge (8). A major attraction of this model was parsimony: no new biological processes need to be invoked if synaptic vesicles recycle by a pathway similar to that used by receptors involved in the uptake of yolk proteins, transferrin, or low-density lipoprotein (LDL). However, several aspects of this model for clathrin-mediated recycling of synaptic vesicles have come into question from observations of *Drosophila* mutants (21, 30, 52), isolated central synapses in rat brain (33), and lamprey reticulospinal synapses (53). These studies suggest cisternae observed at nerve terminals after prolonged stimulation originate from an unusual bulk uptake pathway, rather than via fusion of endocytic vesicles. Nerve terminal cisternae appear to be large invaginations of the plasma membrane, often connected to the plasma membrane via small tubules. Because the bulk uptake occurs away from the active zone, a location corresponding to sites of slow internalization in shi^{ts1} mutants, this pathway is generally considered to have slower kinetics than the pathway at the active zone.

The two spatially and kinetically distinct pathways observed in *Drosophila* retinular synapses differ in their requirements for Ca^{2+} ions. When preparations are incubated in saline containing high Mg^{2+} and low Ca^{2+} ions, the fast active zone pathway is specifically inhibited. A likely explanation for this effect is the successful competition of Mg^{2+} for Ca^{2+} binding elements involved in fast synaptic vesicle recycling (30). Although Ca^{2+} and different kinases (including those sensitive to staurosporine) appear to play a role in modulating rates of vesicle recycling in many preparations, effects of these modulators specific to one or other recycling pathway have not been demonstrated.

2.4 Distinct pools of synaptic vesicles

It is not apparent why different pathways for recycling coexist at synapses. However, we propose one of several possible explanations for this phenomenon. A Ca^{2+}-dependent pathway at active zones would allow tight coupling between exocytic and endocytic pathways. However, under conditions of intense exocytosis, vesicle proteins accumulated on the plasma membrane may diffuse away from sites of transmitter release. Recycling from these distant locations must occur some distance from the cloud of high Ca^{2+} at active zones and well after exocytosis has ceased. For this reason, a second Ca^{2+}-independent mechanism may be utilized.

The fast and slow internalization pathways are thought to maintain the vesicle population in distinct pools at the nerve terminal. At *Drosophila* retinular synapses, the fast, active zone pathway contributes selectively to an active pool of synaptic vesicles close to the plasma membrane, while the slower pathway contributes predominantly to a pool of vesicles distant from sites of fusion (30). The active zone and non-active pool have been considered the active pool and the reserve pool, respectively. The active pool is believed to supply vesicles for exocytosis at low rates of stimulation, and the reserve pool becomes available only when the active pool is depleted by sustained exocytosis at higher rates than vesicle recycling (54–56). Thus, at low rates of stimulation, most exocytotic vesicles not only derive from the active pool but also recycle into this pool. At sustained high rates of stimulation, vesicles are mobilized from reserve to active pool and recycled to both pools. Significantly, in this scenario, the reserve pool of vesicles and the non-active zone recycling pathway are both recruited under identical circumstances (54).

Depletion of the active pool has been long considered to be responsible for synaptic depression in many synapses (57). In the recent past, mechanisms regulating the dynamic translocation of synaptic vesicles from the reserve pool to the active pool have begun to be analysed. Ca^{2+}, serotonin, and protein kinase C have been shown to increase the active pool of vesicles, and calcineurin (a synaptic phosphatase) may function to inhibit vesicle mobilization from the reserve pool (58–60). Regulation of active and reserve pool sizes may play important roles in synaptic plasticity.

2.5 The role of Ca^{2+} in endocytosis

A precise role for Ca^{2+} in synaptic vesicle recycling has been difficult to demonstrate, because endocytosis usually closely follows exocytosis, which itself requires Ca^{2+}. To resolve this issue, endocytosis has to be temporally isolated from exocytosis. Several different approaches have revealed considerable information on the Ca^{2+}-dependence of vesicle recycling.

The first direct demonstration for the role of extracellular Ca^{2+} in synaptic vesicle endocytosis comes from studies at the frog NMJ. Black widow spider venom (BWSV) induces Ca^{2+}-independent exocytosis at the frog NMJ (also see Chapter 6). Thus, prolonged treatment with BWSV leads to synaptic vesicle exocytosis in the presence or absence of Ca^{2+}. Under these circumstances, vesicle depletion occurs in the absence of extracellular Ca^{2+}, not in its presence, suggesting that Ca^{2+} is required for synaptic vesicle recycling (6, 7). The interpretation of these data is limited by observations that exposure of synapses to BWSV treatment in Ca^{2+}-free saline, or even prolonged incubation in Ca^{2+}-free saline alone, could reduce intracellular Ca^{2+} to non-physiologically low concentrations. Thus, the block in vesicle endocytosis observed in BWSV-treated synapses may simply indicate a requirement for Ca^{2+} at normal cytoplasmic concentrations, and not at high concentrations resulting from action potential evoked Ca^{2+} entry.

Recent studies have also suggested a requirement for Ca^{2+} in vesicle recycling in mammalian synaptosome and in recycling of large dense core granules in adrenal chromaffin cells. Exocytosis may be separated from endocytosis by substituting Ca^{2+} with different divalent cations. Ba^{2+} triggers synaptic vesicle exocytosis in both preparations. However, the exocytosed membrane is not retrieved unless Ca^{2+} is present (51, 61, 62). These studies demonstrate an essential role for Ca^{2+} in synaptic vesicle endocytosis, but do not answer several critical questions about the involvement of Ca^{2+} in synaptic vesicle endocytosis:

- When is Ca^{2+} required?
- At what concentration is Ca^{2+} required?
- How does Ca^{2+} exert its effects on endocytosis?

Current data suggest that low concentrations of Ca^{2+} are required at an early stage of membrane internalization prior to the formation of endocytic pits. At *Drosophila* larval NMJ synapses, steps of synaptic vesicle endocytosis following the stage arrested in *shi^{ts1}* mutants do not require extracellular or significantly elevated intracellular Ca^{2+} (23) (Fig. 6). At lamprey reticulospinal synapses, removal of extracellular Ca^{2+} after vesicle depletion by prolonged tetanic stimulation blocks vesicle recycling completely (53). After being maintained in Ca^{2+}-free saline for 90 minutes, shifting the preparation to saline containing 10 μM Ca^{2+} results in a high density of presynaptic clathrin-coated pits within minutes, and subsequent synaptic vesicle recycling. The effect of the 10 μM Ca^{2+} saline added extracellularly could reflect a requirement for even lower concentrations of intracellular Ca^{2+} for this early formation of clathrin-coated pits, if Ca^{2+} is required at all inside the cell for endocytosis. These data suggest a requirement of very low Ca^{2+} concentration (53). The rather minimal Ca^{2+} requirement probably accounts for a previous report that extracellular Ca^{2+} is not required for vesicle recycling in synapses among hippocampal cultured neurons (19).

Ca^{2+} plays an important role in regulating the rate of endocytosis in several cellular contexts including recycling of clear synaptic vesicles. However, the reported effects of Ca^{2+} manipulations differ among different nerve terminals. Increasing extracellular Ca^{2+} concentrations during stimulation, or increased rates of stimulation at constant extracellular Ca^{2+} speeds up the rate of endocytosis in synapses between cultured hippocampal neurons (44). However, increased intracellular Ca^{2+} inhibits endocytosis in terminals of goldfish retina bipolar cells (63) and high frequency stimulation reduces the rate of endocytosis at the frog neuromuscular synapse (64, 65). While further studies are needed to resolve these discrepancies, it appears clear that Ca^{2+} concentrations do regulate the rate of synaptic vesicle recycling. Potential molecular targets of Ca^{2+}- and kinase-dependent regulation of synaptic vesicle recycling are discussed below.

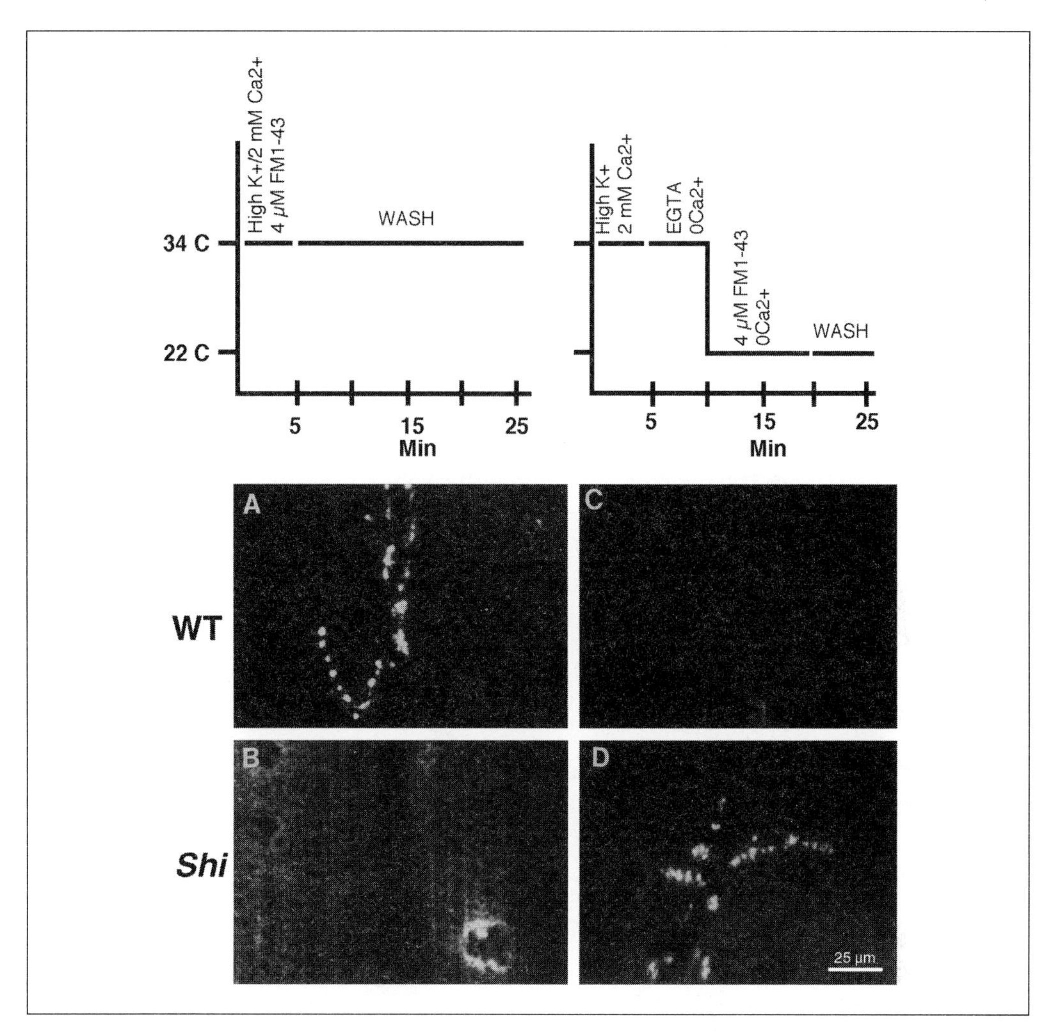

Fig. 6 The requirement of Ca^{2+} for synaptic vesicle recycling in *Drosophila* NMJ synapses. Experimental protocols are shown above images. High K^+ saline is used to stimulate exocytosis and related endocytosis in third instar larval NMJ synapses. (A and B) FM 1-43 is bath applied to wild-type and shi^{ts1} synapses at 34 °C in the presence of high K^+ saline. Only the wild-type synaptic boutons are labelled with the dye. (C and D) The nerve terminals of wild-type and shi^{ts1} are stimulated for 5 min at 34 °C and bathed in 0 Ca^{2+} saline containing EGTA. Upon returning to 22 °C, shi^{ts1} synapses, but not wild-type's, are now labelled with FM 1-43. This suggests that synaptic vesicle recycling subsequent to dynamin block does not depend on Ca^{2+}. (Reprinted with permission from ref. 23. Copyright 1994. Cell Press.)

3. Molecular mechanisms of synaptic vesicle recycling

In the preceding sections, we have developed an outline of the cellular processes involved in recycling synaptic vesicle membrane. These processes may be conceptually divided into three steps:

(a) Recovery of synaptic vesicle proteins from the plasma membrane and reassembly of synaptic vesicles.

(b) Vesicle fission from the plasma membrane.

(c) Subsequent recycling events.

We now proceed to review molecular mechanisms involved in each of these events.

To study molecular mechanisms involved in any process, three different tools must be available: candidate molecules, techniques to specifically and rapidly perturb these molecules, and functional assays for the process. While ideally, all three tools must be simultaneously employed to study vesicle recycling in one preparation *in vivo*, such ideal studies are often prohibitively difficult and, hence, quite rare. Until recently, candidate proteins involved in endocytosis at synapses were quite limited. However, currently, over ten proteins have been implicated to participate in the process, and this will likely continue to grow rapidly (Table 1, Fig. 7). With one exception (i.e. dynamin), the candidate presynaptic proteins involved in endocytosis have been identified due to either their known functions in other cellular contexts (clathrin and associated proteins) or their physical interactions *in vitro* with other components of the endocytic pathway (66).

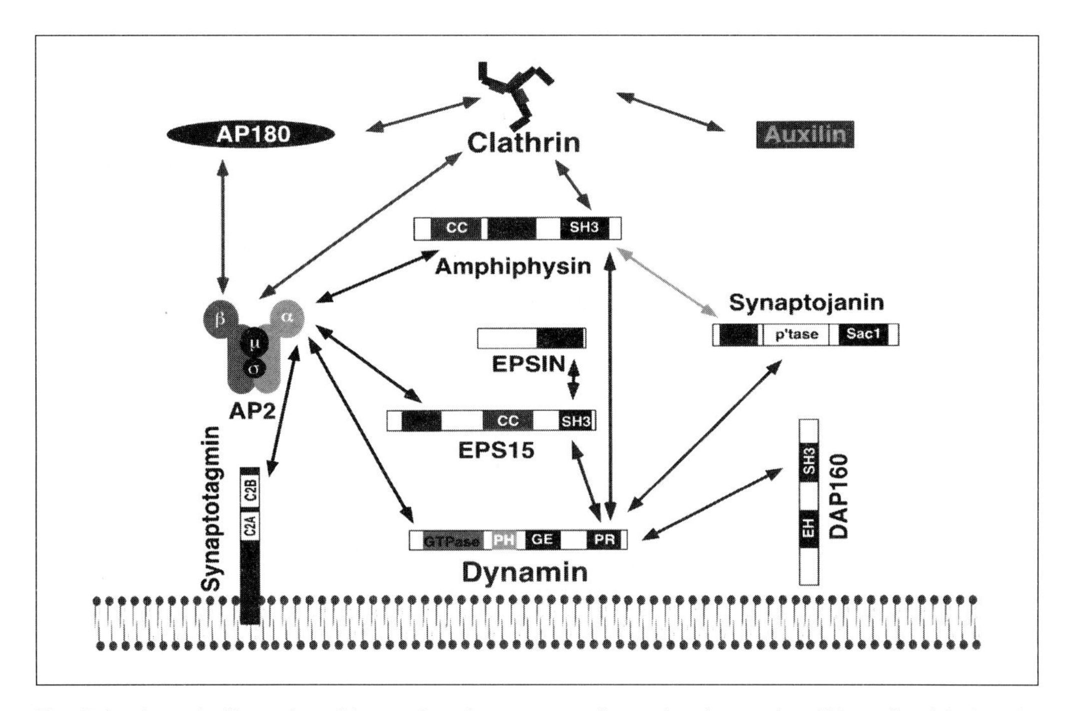

Fig. 7 A schematic illustration of interactions between putative endocytic proteins. PR, proline-rich domain; p'tase, phosphatase domain; Sac1, yeast Sac1p domain; SH3, Src homology 3 domain; CC, coil-coiled domain; PH, pleckstrin homology domain; C2A, C2B, protein kinase C domain repeat A and B, respectively; α, α-adaptin; β, β-adaptin; μ, the medium subunit, σ, the small subunit of AP2. See text for details of description of these proteins.

Table 1 Molecules involved in synaptic vesicle recycling

Gene product	Subcellular localization	Mutant or cell line phenotype	Proposed functions
Dynamin I	Vesicle neck	*Drosophila*: Reversible ts block in synaptic vesicle endocytosis results in active terminals depleted of vesicles (26), leading to synaptic depression and failure (22). Terminals exhibit collared pits on presynaptic PM (21) and do not take up FM1-43 (23). *C. elegans*: Reversible ts uncoordinated locomotion (130). HeLa cells: Expression of mutant dynamin I blocks coated pit formation (121).	May form constricting rings around neck of budding vesicle to promote fission from PM (69, 122). Acts as a mechanochemical enzyme *in vitro* and causes vesiculation of phospholipid tubes (123).
Clathrin heavy chain (CHC)	Vesicle coat	*Drosophila*: 4 alleles (3 lethal, 1 semi-lethal). Male sterility in some contexts (100). HeLa cells: Overexpression of dominant negative CHC blocks coated pit formation (181).	Trimeric triskelions form coats on vesicle and may provide mechanical force for budding (76).
Clathrin light chain	Vesicle coat		May regulate coat assembly and disassembly (169).
Clathrin-associated proteins			
Hsc70 ATPase	?	—	Strips clathrin from coated vesicles (165, 166).
Auxilin	On clathrin-coated vesicles	—	Recruits hsc 70c to coated vesicles via chaperone-like DNAJ domain (167).
Clathrin adaptor proteins			
α-adaptin	Coat	*Drosophila*: Reduced FM1-43 uptake by active terminals, synaptic vesicle depletion, and PM deformation in (99).	Facilitates formation of coated vesicles from PM by linking clathrin to target proteins in PM and by promoting clathrin cage assembly (87).
β_2-adaptin	Coat		Fly β-adaptin may be hybrid of β_1- and β_2-adaptin.
μ_2-adaptin (AP50)	Coat	—	
σ_2-adaptin (AP17)	Coat	—	
AP180	Vesicle coat	*Drosophila*: Mislocalizes clathrin and synaptobrevin at third larval NMJ; reduces the number of SVs and evoked EJP; increases the number of cisternae; enlarges vesicle size and quantal size (52). *C. elegans*: Mislocalizes synaptobrevin; enlarges SV size, but does not affect SV number (113).	A synapse-enriched clathrin, alpha adaptin, and phosphoinositide-binding protein (104, 108, 111, 174), assembles clathrin into homogeneously sized cages (110, 112). Regulates SV size and defines membrane to be retrieved by clathrin cage (52, 113). However, non-neuronal isoforms are found in yeast (182) and human (183). Unknown functions.

Table 1 (*Continued*)

Gene product	Subcellular localization	Mutant or cell line phenotype	Proposed functions
EPS15	On rims of coated pits, suggesting a role in vesicle fission.		Binds to alpha adaptin and may recruit several endocytic proteins to PM, including clathrin, AP2, and synaptojanin.
		HeLa cells: Overexpression of EPS15 C-terminus blocks transferrin uptake (163).	Interaction with AP2 is essential for receptor-mediated endocytosis (163). But does not remain in coated vesicles (164).
EPSIN		Blocks transferrin uptake in CHO cells (172).	Contains NPF repeats that bind to the EH domains of EPS15. Also interacts with α-adaptin via DPW repeats (172).
Amphiphysin	Enriched at nerve terminals	When SH3 peptide is microinjected into lamprey reticulospinal axons, it blocks SV at the fission step (68).	Binds AP2, dynamin, synaptojanin, endophilin, and clathrin (124, 126, 149, 150). Recruits dynamin to coated pits via SH3 domain homologous to yeast Rvs167 implicated in actin function (148).
		In fibroblasts (156) and 3T3-L1 adipocte cells (184), overexpression of amp/SH3 domains blocks receptor-mediated endocytosis. The phenotype can be 'rescued' by co-expression of dynamin (156).	
Dap160	Presynaptic; localized to membrane spots similar to dynamin in *shi* synapses depleted of vesicles.	—	Dynamin-binding protein purified from *Drosophila* brain. Contains an EH domain and multiple SH3 domains — may serve functions analogous to EH proteins and amphiphysin (49).
Synaptojanin I	Presynaptic	Yeast: Mutants lacking SJL1 and SJL2 are defective in receptor-mediated and fluid phase endocytosis (185).	An inositol-5-phosphatase that interacts with amphiphysin (124, 127) and endophilin (124). May regulate local concentration of PPIs to promote membrane fission. Contains a domain homologous to yeast SAC1 implicated in actin functions (186).
Endophilin		—	Binds to dynamin and synaptojanin via SH3 domain (187), but is excluded from amphiphysin–dynamin–synaptojanin complexes (188).
Regulators			
PKC	Cytosol	*Drosophila*: Induction of pseudosubstrate results in abnormal learning acquisition (189).	Phosphorylates dynamin I in resting nerve terminals (139).

Table 1 *(Continued)*

Gene product	Subcellular localization	Mutant or cell line phenotype	Proposed functions
Calcineurin	Cytosol	No mutants, but calcineurin inhibitors affect FM1-43 uptake into fly larval motor terminals (190). In rat brain synaptosomes, calcineurin blockers inhibit endocytosis (62). In adrenal chromaffin cells, blockers prevent Ca^{2+}-mediated inhibition of compensatory endocytosis (191).	Dephosphorylates dynamin I and reduces its GTPase activity (179); may act as Ca^{2+}-activated switch to mobilize dynamin–GTP to PM following exocytosis (178).
Synaptotagmin	Synaptic vesicle protein	*C. elegans* (34) and *Drosophila* (101): Reduces the number of SVs.	May act as high affinity receptor for AP2 to promote SV retrieval from PM (35).

While functional assays for synaptic vesicle recycling are most convenient in cultured neurons, frog NMJs, and goldfish retinal bipolar cells, biochemical perturbation methods are more feasible in squid and lamprey giant synapses, which allows microinjection of interfering peptides or antibodies (53, 67, 68). Synaptosomes, on the other hand, are ideal for EM studies (33, 69) and for introducing interfering peptides as conjugates to lipid or other carrier molecules (62). Another recently devised preparation for molecular perturbation in the context of synaptic transmission are synapses formed in culture from frog eggs microinjected with antisense RNA (70, 71). However, the most generally employed perturbation approach is genetic manipulations in *C. elegans*, *Drosophila*, and mouse, where precise assays for recycling synaptic vesicles are yet to reach their full potential (see Chapters 8, 9, and 10).

Following this brief introduction to the strategies used to analyse molecular mechanisms of vesicle recycling, we will discuss in turn the current knowledge of molecular mechanisms involved in synaptic vesicle reassembly, vesicle fission, and other events in vesicle recycling. Remarkably, almost all of the putative endocytic proteins known are implicated in a clathrin-mediated pathway of recycling. If clathrin is not required for fast synaptic vesicle recycling, it is conceivable that a different molecular assembly specific for fast recycling remains to be identified.

3.1 Recovery of vesicle proteins and reassembly of endocytic vesicles

3.1.1 Self-assembly

Upon exocytosis, synaptic vesicles in most nerve terminals fuse with the plasma membrane. Therefore, the major challenge for endocytosis is to recover synaptic vesicle components (lipids and proteins) and reassemble these components into new vesicles. This task is further complicated due to the fact that vesicular lipids and proteins are distinct from those in the plasma membrane (see Chapters 4 and 5).

The simplest method for synaptic vesicle proteins to sort themselves from plasma membrane proteins is to form stoichiometric lateral associations among themselves. In such a model synaptic vesicle proteins and perhaps even lipids could stay together as unit rafts which may be subsequently be internalized. Some evidence exists for lateral associations among synaptic vesicle proteins (72). This supports the idea that synaptic vesicle proteins may be held together as units, and that lateral interactions among vesicle proteins plays an important role in synaptic vesicle recycling. However, such interactions are almost certainly not sufficient: strong evidence indicates that clathrin and associated proteins, known to be involved in sorting membrane receptors into endocytic vesicles in non-neuronal cells, play an important role in synaptic vesicle recycling.

3.1.2 Clathrin-mediated coated pit formation

At several different stages of the secretory pathway, eukaryotic cells need to sort a subset of membrane proteins into transport vesicles. In general, such sorting involves coat proteins, a class of cytosolic molecules that first bind to and then aggregate around the cytoplasmic tails of membrane proteins. These coat proteins may concentrate synaptic vesicle proteins to a small region within the plasma membrane from where eventually a transport vesicle forms. Of several known coat protein complexes, a specific vesicle coat, composed of clathrin and associated proteins, is most likely to participate in synaptic vesicle recycling.

Electron-dense clathrin coats around endocytic vesicles were first described by electron microscopy of yolk protein endocytosis into developing mosquito oocytes (73) and soon found to be associated with synaptic vesicles (74, 75). Clathrin, the major component of this coat, was first purified along with a set of accessory proteins from bovine brain (76). Clathrin molecules are organized as a trimeric complex, called a triskelion, by three heavy chains (180 kDa) and three light chains (25–30 kDa). Heavy chains are highly conserved across species and ubiquitously expressed. There are two light isoforms, and each also has alternatively spliced brain isoforms with slightly heavier molecular weight (77–80). Purified clathrin forms triskelia *in vitro* and further polymerizes into polyhedral lattices composed of pentagonal and hexagonal faces with a natural curvature under appropriate conditions. Because clathrin polymerization results in the formation of curved baskets, it has properties expected for a molecule that drives vesicle formation. However, clathrin alone does not bind to membrane or membrane proteins. Membrane association of clathrin is mediated by one of three different tetrameric protein complexes termed 'adaptors', AP1 (81), AP2 (82), and AP3 (83–85), and a monomeric protein AP180 (104–112). AP1, AP2 (Fig. 7), and AP3 are heterotetrameric complexes consisting of four non-covalently bound subunits: two large subunits ($\gamma/\beta1$, $\alpha/\beta2$, and $\delta/\beta3$, respectively), one medium subunit ($\mu1$, $\mu2$, $\mu3$), and one small subunit ($\sigma1$, $\sigma2$, and $\sigma3$). These large subunits are also called adaptins. The β-class adaptins are very similar to each other, but α-, γ-, and δ-adaptins, as well as the μ- and σ-subunits are unique to each adaptor and appear to play an important role in targeting each adaptor to the right compartment within a cell (86). AP1 is involved primarily in protein sorting and transport from the *trans*-Golgi networks (TGN) to endosomes (87). AP2 and AP180 are involved in receptor and synaptic vesicle endocytosis at the plasma membrane (87, 88), while AP3 acts between the TGN and lysosomes as well as in endosomes (89–93). These adaptor proteins are thought to be sandwiched between empty clathrin cages and cargo membranes during clathrin-coated vesicle formation (87, 88).

The key adaptor involved in synaptic vesicle reassembly at the plasma membrane is the AP2 complex. AP2 may be targeted to the plasma membrane by an unknown mechanism (94). During receptor-mediated endocytosis, AP2 binds to the cytoplasmic tail of cargo proteins, probably through its μ-subunit (which binds tyrosine-based endocytosis signals) or the β-chain (known to recognize a di-leucine repeat, an alternative internalization signal) (95, 96), and to clathrin through α- or β-subunits (about 110 kDa each) (87, 97). The sequence of the binding events is not clear. Some evidence suggests that flat lattices containing clathrin and AP2 may show enhanced affinity for membrane proteins. Therefore, clathrin binding to AP2 may occur prior to AP2 association with cargo proteins (98).

Several lines of evidence support a role for clathrin and AP2 in recovering synaptic vesicle proteins during vesicle recycling. Clathrin and AP2 are enriched at presynaptic terminals, and are found on nascent vesicles derived from presynaptic plasma membrane (33, 52, 99). Furthermore, clathrin-coated vesicles purified from brain are highly enriched in synaptic vesicle membrane proteins (32). Finally, molecular interactions have been observed between the synaptic vesicle protein synaptotagmin and the AP2 adaptor (35). Other synaptic vesicle membrane proteins may be sorted into AP2-containing regions either by low affinity binding to AP2, or by weak lateral interactions among themselves. Such sorting is suggested by electron micrographs which show that synaptic vesicle proteins such as synaptotagmin are concentrated in clathrin-coated pits (33).

These morphological and biochemical data are supported by recent genetic perturbation studies. *Drosophila* embryos deficient in clathrin are lethal. Although these mutants have not been examined for vesicle cycling defects, the failure to hatch in mutant embryos is likely caused by functional defects in the nervous system (100). *Drosophila* mutants completely lacking the α-subunit of AP2 show altered presynaptic morphology consistent with a defect in synaptic vesicle endocytosis prior to the formation of clathrin-coated pits (99). An excess of presynaptic plasma membrane is seen in these mutants at the apparent expense of synaptic vesicle membrane. These observations lend indirect support to the hypothesis that clathrin-mediated endocytosis is involved in synaptic vesicle endocytosis. Furthermore, no endocytic pits are present in null mutants in *Drosophila* α-adaptin, indicating that clathrin–AP2-mediated endocytosis may be the major pathway for synaptic vesicle recycling at these embryonic CNS synapses. A requirement for synaptotagmin as an adaptor receptor in vesicle recycling is supported by analyses of *C. elegans* and

Drosophila synaptotagmin mutants (34, 101). Vesicle depletion at these nerve terminals is similar to, but less severe than, that observed in α-adaptin mutants of *Drosophila*.

3.1.3 Fine-tuning of synaptic vesicle size through clathrin-dependent reassembly

While clathrin-coated vesicles show a wide range of sizes in most cellular contexts, a distinctive feature of clear synaptic vesicles in most NMJs (52, 102) and mammalian brain (103) is their uniform small size. Particularly striking is that synaptic vesicles undergoing cycles of fusion and fission still maintain a rather constant size. This indicates that synaptic vesicle reassembly via clathrin-mediated endocytosis is a highly regulated process. Such regulation may be critically important to ensure high speed and precision in neuronal signalling.

A good candidate for such a regulatory protein is the synapse-specific clathrin-associated protein AP180. AP180 (formerly called NP185, F1-20, and AP3) was the third clathrin-binding protein to be identified and thus has been widely called AP3 (81, 104–109). Because AP180 is a monomeric protein distinct from the hetero-tetrameric complexes of AP1 and AP2, its name was taken over by the recently identified heterotetrameric AP3 complex (83) and is now called AP180. In addition to binding to clathrin, AP180 promotes clathrin cage formation (110–112). Ye and Lafer (1995) showed that in a cell-free system, AP180 restricts the size and size distribution of clathrin cages assembled by purified clathrin (112). This leads to the hypothesis that AP180 may regulate vesicle size *in vivo*.

Recent studies of mutants in highly conserved AP180 homologues named LAP (Like -AP180) in *Drosophila* (52) and UNC-11 in *C. elegans* (113) strongly support this hypothesis. Synaptic vesicles in both *lap* and *unc-11* mutant nerve terminals exhibit significantly increased size (Fig. 8). The altered vesicle size results in a concurrent increase in the size of quantal events measured electrophysiologically in post-synaptic muscles of *lap* mutants. If *lap* mutant synapses are examined after recovery from depletion of synaptic vesicles, quantal size remains large and variable as before. This indicates that altered synaptic vesicle size in the mutants reflects a specific synaptic recycling impairment, rather than defects in vesicle biosynthesis in the soma. The observation that altering the size of a clathrin-coated vesicle leads directly to changes in synaptic vesicle size has important implications for the cell biology of synaptic vesicle recycling. If a distinct clathrin-independent traffic station (an endosome) were to exist downstream of clathrin-coated vesicles, then alterations in coated vesicle size would not be expected to change synaptic vesicle diameter. Thus, studies on *lap* and *unc-11* mutants not only supports a role for AP180 in regulating coated vesicle size, but also supports the notion that synaptic vesicle recycling occurs via a single vesicle budding step involving clathrin, AP2, and AP180 (33). This study also implicates that vesicle refilling following endocytosis is dynamically regulated, as the amount of transmitter (i.e. glutamate) is proportionally increased in *lap* mutants.

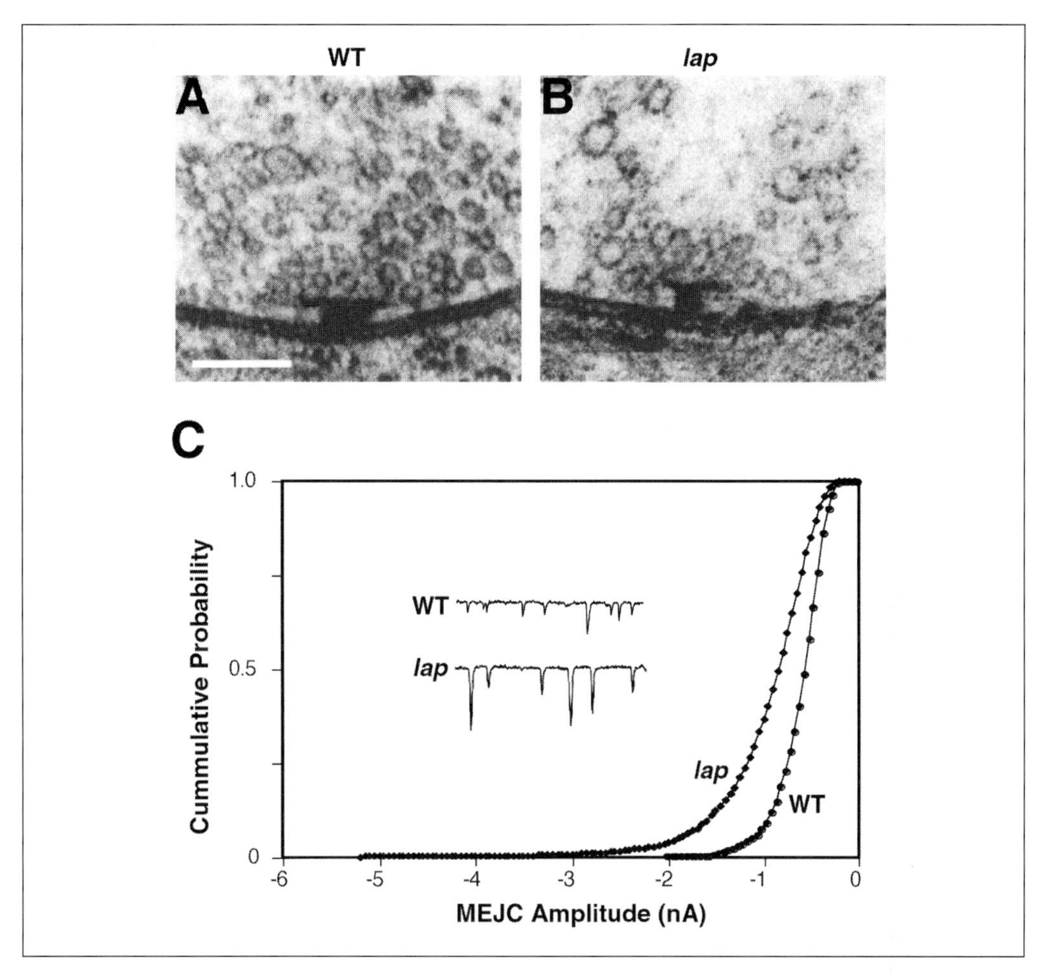

Fig. 8 The sizes of synaptic vesicles and quanta are increased in *lap* mutants. (A and B) Synaptic vesicles are shown to cluster around the dense body ('T'-shaped electron-dense substances) at the active zone in third instar larval NMJ synapses in *Drosophila*. Synaptic vesicles with a relatively uniform size are observed in wild-type. However, synaptic vesicles in *lap* mutants are variable and enlarged in size. In addition, the number of synaptic vesicles is reduced in *lap* mutants. Scale bar, 0.2 μm. (C) Spontaneous, miniature excitatory junction current (MEJC) recorded from voltage-clamped (at –80 mV) third larval body wall muscles and cumulative probability plot of MEJC amplitude. Note that MEJC amplitude in *lap* mutants is larger compared with that in wild-type. Furthermore, quantal events larger than ~ 2 nA are only found in *lap* mutants. Horizontal bar, 400 msec; vertical bar, 2.5 nA. (Reprinted with permission from ref. 52. Copyright 1998. Cell Press.)

In addition to altered synaptic vesicle sizes, *lap* mutants, but not *unc-11*, display a reduced efficiency of synaptic vesicle endocytosis, as judged by the severe loss of synaptic vesicles at nerve terminals. Consistently, FM1-43 uptake is impaired, and synaptic depression is more prominent upon high frequency nerve stimulation. In addition, AP180 also regulate the efficiency of synaptic vesicle endocytosis by recruiting clathrin to endocytic sites, as clathrin is mislocalized in *lap* synaptic boutons. These studies of *lap* mutants indicate that AP180 regulates both the efficiency and efficacy of synaptic vesicle reassembly during endocytosis in *Drosophila*.

Morphological analysis of *unc-11* and *lap* mutants indicates an additional unexpected role for AP180 in sorting specific synaptic vesicle proteins to synaptic vesicles (113, B. Zhang, unpublished). While synaptotagmin and most other synaptic vesicle proteins are restricted to synaptic sites, in these mutants synaptobrevin is no longer restricted to synaptic regions (varicosities) but along the axonal tracts where synaptic vesicles are not found. The mislocalization of synaptobrevin remains a puzzle, as UNC-11 does not bind synaptobrevin *in vitro* (113). Neither LAP nor mouse AP180 seems to interact with neuronal synaptobrevin (B. Zhang, E. Chapman, and B. Ganetzky, unpublished data). This suggests that AP180 recruits synaptic vesicle proteins, albeit indirectly, during coated synaptic vesicle reassembly.

Taken together, assays for synaptic vesicle recycling in fly and worm AP180 mutants indicate that it regulates sorting of synaptic vesicle proteins as well as the fine-tuning of vesicle formation. The combined weight of biochemical, morphological, genetic, and physiological evidence strongly support a role for clathrin and associated proteins in an essential pathway for internalization of synaptic vesicle membrane.

3.1.4 Clathrin-independent formation of synaptic vesicles

A clathrin-independent mechanism may also be involved in synaptic vesicle recycling. This has been suggested by a recent demonstration that synaptic vesicles form *in vitro* from donor membranes isolated from PC12 neuroendocrine cell homogenates (90, 114–116). When donor endosomal membranes are incubated with two cytoplasmic proteins ARF1 (a GTPase) and AP3, budding vesicles contain synaptic vesicle membrane proteins but not proteins such as the transferrin receptor, normally excluded from synaptic vesicles. This *in vitro* reconstituted sorting event appears not to require clathrin and dynamin, two proteins previously implicated in vesicle budding at the presynaptic plasma membrane. AP3 is ubiquitously expressed in all tissues (85), including nerve cells and terminals (117). However, it remains unclear whether it functions as a relay station for vesicle recycling or for *de novo* biosynthesis of vesicles in the soma.

3.2 Vesicle fission

Once synaptic vesicle proteins are sorted into a nascent endocytic vesicle, membrane fission at the vesicle neck is essential for vesicle budding. This energy-dependent process involves a host of molecules which may act together to achieve local membrane fission (118–120). Molecular mechanisms involved in vesicle fission are less well outlined than those involved in sorting membrane proteins. However, a critical role for the GTPase dynamin has been established by phenotypic analysis of *Drosophila* shi^{ts1} mutants (26, 30), by cell biological analysis of dynamin function in mammalian synaptosomes (69) and cultured cells (121), and by biochemical and structural studies of purified protein (122, 123). Proteins associated with dynamin, including amphiphysin (124–126), synaptojanin (127), DAP160 (49), may participate with dynamin in vesicle fission. However, direct evidence from perturbational studies only exists for amphiphysin (68).

In the following sections we will outline our current knowledge of the molecular functions of dynamin and associated proteins that are probably involved in vesicle fission. In our discussions of dynamin function, we will try and convey the difficulty in narrowing down the function of one protein to a very specific step in a vesicle cycle. Since proteins function normally as large complexes, recruitment of other components of the fission complex may be one role for each of these proteins. A highly precise dissection of the molecular mechanism of vesicle fission will eventually require information on co-ordinated molecular movements within a protein assembly.

3.2.1 Dynamin is required for vesicle fission

Dynamin is a 100 kDa GTPase originally isolated as a microtubule-bundling motor protein (128, 129). Electron micrographs of dynamin bound to microtubules reveal a regular helical decoration of 20 nm microtubules with dynamin. While the precise role of this microtubule binding *in vivo* remains unclear at present, the formation of these helices appears pertinent to dynamin function in membrane fission. Three different dynamin genes have been identified in vertebrates to date as well as several more distantly related dynamin-like proteins probably involved in membrane fission in other contexts (119). Most of our discussion will relate to vertebrate dynamin 1, specifically expressed in the nervous system and believed to participate in synaptic vesicle recycling, and to the closely related *Drosophila* shi^{ts1} (24, 28) and *C. elegans* dyn-1 genes (130).

The first evidence for dynamin's involvement in synaptic vesicle recycling came from the discovery that the *Drosophila* shi^{ts1} gene, essential for synaptic vesicle recycling, encodes dynamin (24, 28). Temperature-sensitive *Drosophila* shi^{ts1} mutants have conditional blocks in an essential function of dynamin's GTPase domain at elevated temperatures (28, 131). At elevated temperatures, synaptic vesicle recycling at nerve terminals in shi^{ts1} mutants is completely blocked leading to vesicle depletion and the accumulation of endocytic invaginations on plasma membrane. These endocytic invaginations are ringed by an electron dense 'collar' whose appearance under the electron microscope is consistent with a helical assembly of about 20 nm diameter (21, 26, 122). These 'collared pits' are likely to represent nascent endocytic vesicles trapped at a precise stage prior to vesicle fission, and the collars themselves are likely to contain mutant dynamin.

Several additional studies support the idea that dynamin rings at the neck of budding vesicles hydrolyse GTP and mechanically detach coated vesicles from the plasma membrane at the fission step (119, 120, 132). Dynamin assembles *in vitro* into rings and helical tubes whose pitch and diameter are very similar to the collars found at shi^{ts1} mutant nerve terminals (21, 30, 122). In synaptosomes incubated with GTPγS to inhibit GTP hydrolysis by dynamin (and other GTPases), nascent endocytic vesicles with exaggerated necks are formed on cisternae-like structures. These necks are ringed by an apparent helical coil of dynamin (33, 69) (Fig. 2). Thus, dynamin has the properties and is strategically located to play a role in vesicle fission. There is recent evidence that dynamin can generate a mechanochemical force to facilitate vesiculation and even provide the mechanical 'squeeze' necessary for vesicle fission (123). Purified dynamin binds to phosphatidylserine bilayers and forms long tubes. Treatment with GTP causes conformational changes in dynamin and leads to vesiculation of these dynamin-lined membrane tubes (123). Thus, both *in vivo* and *in vitro* studies demonstrate a direct role for dynamin in membrane fission.

Despite evidence that dynamin is involved in membrane fission, additional functions for dynamin may exist at synapses. Dynamin may play a role in the actual formation of invaginated vesicles. Such a function may explain the observation that the number of budding invaginations is far fewer than expected in vesicle-depleted *shi*^{ts1} nerve terminals (21) as well as in mammalian cells overexpressing a dominant negative form of dynamin (121). Significantly, dynamin alone can induce vesiculation of phosphatidylserine liposomes and form dynamin-coated tubules from protein-free liposomes (123, 133). Another proposed function for dynamin is tubule formation (134) to explain the observation that when *shi*^{ts1} synapses are restored to permissive conditions, tubular invaginations and cisternae are observed first rather than a wave of small vesicles. Despite these caveats, an essential function for dynamin in membrane fission appears proven beyond a reasonable doubt.

3.2.2 Molecular activities of dynamin

Apart from GTP hydrolysis, dynamin has several molecular activities presumably important for its function. It forms polymers with itself, alternates between membrane bound and non-membrane bound states, and associates with several accessory proteins also involved in vesicle budding. Different domains of the dynamin molecule may contribute to these diverse activities (135) (Fig. 7). While an N-terminal GTPase domain binds and hydrolyses GTP, other domains mediate protein–protein and potentially protein–lipid interactions: a middle domain (MD) of unknown function (131), the pleckstrin homology (PH) domain potentially involved in membrane binding (136), a coil-coiled domain possibly involved in multimerization, and a C-terminal proline/arginine-rich (PR) domain mediating several different protein–protein interactions. In addition, dynamin contains a small GTPase effector (GE) domain that is essential for GTP hydrolysis (137).

The GTPase activity of dynamin is regulated by interaction with other dynamin subunits or with other molecules via the PH and PR domain, by calcineurin-dependent dephosphorylation (138), and PKC-dependent phosphorylation (139). *In vitro*, the GTPase activity of dynamin is altered by binding to microtubules (140–142), phospholipids (143), phosphoinositol (144), SH3 domain-containing proteins (140), and the $\beta\gamma$ subunits of G-proteins (145). The implications for synaptic vesicle recycling by these regulations have not been elucidated. However, the *in vivo* roles of some of these dynamin-interacting proteins have not been studied.

3.2.3 Amphiphysin appears to recruit dynamin to the endocytic neck

Amphiphysin, a presynaptic protein that interacts with dynamin, is curiously an autoimmune antigen of Stiff-Man's syndrome (146, 147). Two amphiphysin genes have been identified. Amphiphysin 1 is neuronally expressed, and enriched at presynaptic terminals where it is co-localized with dynamin (124). Multiple alternatively spliced isoforms of amphiphysins are expressed in non-neuronal tissue where they may play a much broader role than endocytosis (148).

Amphiphysin 1 contains three domains: the N-terminal coil-coiled domain, a middle PR-rich domain, and a C-terminal SH3 domain (Fig. 7). These domains mediate direct protein–protein interactions that are potentially important for synaptic vesicle endocytosis. Amphiphysins directly bind to clathrin (149) and α-adaptin (124) via its N-terminus, and to dynamin (124–126), synaptojanin (127), and endophilin (150) through its SH3 domain. Amphiphysin may also indirectly bind AP180 (151).

An essential requirement for amphiphysin–dynamin interactions in synaptic vesicle endocytosis has been demonstrated in an elegant set of experiments at the lamprey reticulospinal synapse (68). Interactions between amphiphysin and dynamin occur via the SH3 domain of dynamin and a very limited region of dynamin — a consensus PSRPNR (proline, serine, arginine, proline, asparagine, and arginine) sequence in its proline-rich domain (126). Microinjection of either a synthetic human amphiphysin SH3 domain or a peptide containing the PSRPNR sequence results in a specific use-dependent accumulation of budding vesicles along the lamprey giant reticulospinal presynaptic plasmalemma (Fig. 9) (68). These budding vesicles are reminiscent of endocytic pits arrested in shi^{ts1} synapses, but remarkably, lack the dynamin collars apparent in shi^{ts1} preparations. This study suggests that amphiphysin is essential for membrane fission which occurs subsequent to coated pit formation. This study further suggests that amphiphysin is required to recruit dynamin molecules to the necks of coated vesicles (68, 152).

Amphiphysins also undergo rapid dephosphorylation and phosphorylation at nerve terminals and in synaptosomes, processes that may be critical to regulate the formation of protein complexes involved in endocytosis (62, 151, 153). This phenomenon is considered in more detail in our subsequent discussion of protein complex assembly and disassembly during regulation of synaptic vesicle endocytosis. Briefly, the phosphorylation of amphiphysin reduces binding to α- and β-adaptins and inhibits the *in vitro* formation of an endocytic core complex containing clathrin, adaptor proteins, and dynamin, as well as synaptojanin, EPS15, and two other SH3 domain-containing proteins (151).

3.2.4 Specificity of SH3 proteins in endocytosis

Many other SH3 domain-containing proteins are also found to bind to dynamin *in vitro*. However, these interactions appear to be mediated by different regions of the dynamin PR domain. Amphiphysins are the only SH3 domain proteins that bind dynamin at the PSRPNR consensus sequence (126, 154). Other SH3 domains, such as

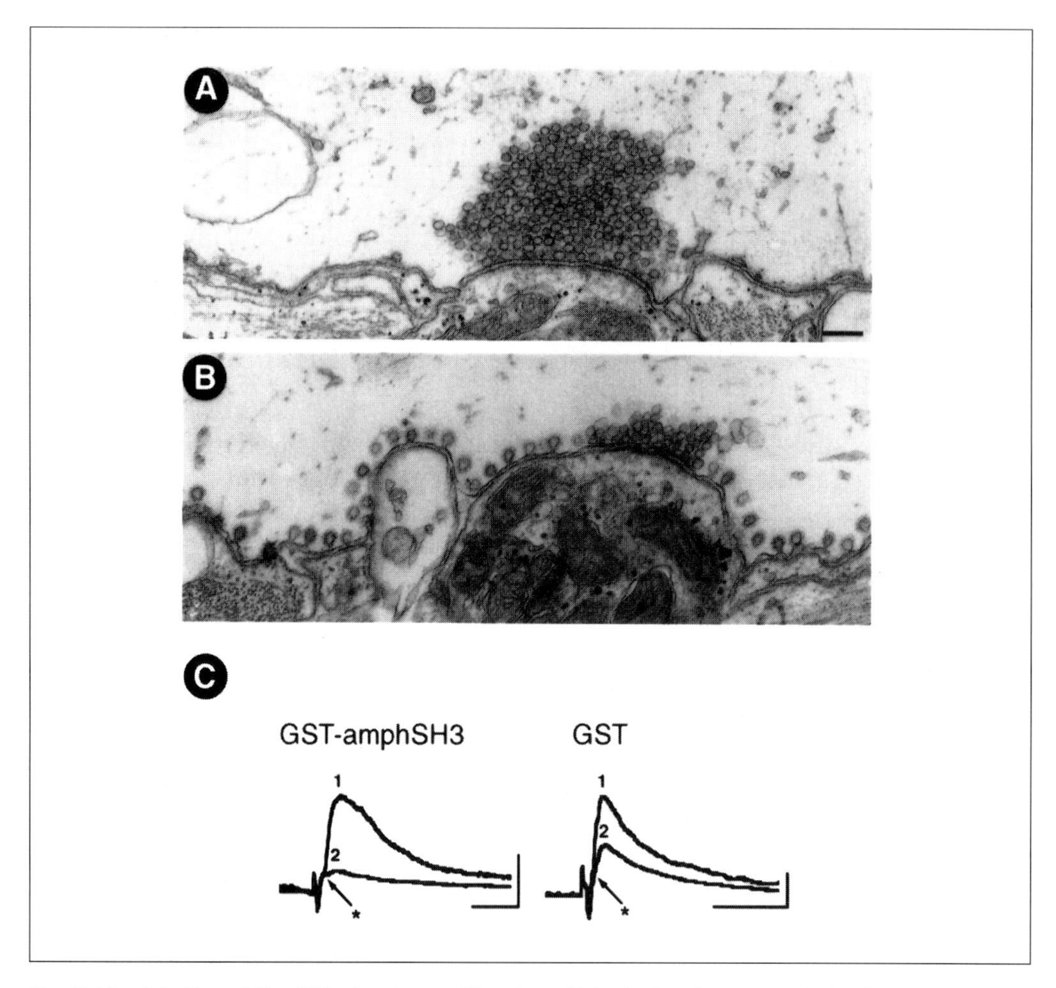

Fig. 9 Microinjection of the SH3 domain peptide of amphiphysin into lamprey reticulospinal axons blocks synaptic vesicle endocytosis. (A) Synaptic vesicles are clustered around the release site in an unstimulated synapse after injection of a GST–amphiphysin SH3 domain peptide. (B) Budding vesicles are accumulated on the plasma membrane after the axon in (A) has been stimulated at 0.2 Hz for 30 min. Note the expansion of axonal plasma membrane. Scale bar, 0.2 μm. (C) The amplitude of evoked excitatory postsynaptic potentials (EPSPs) in a spinal neuron postsynaptic to a SH3 domain peptide injected axon is significantly reduced or abolished, whereas GST-injected axon causes the same degree of EPSP reduction associated with prolonged stimulation at 5 Hz. Horizontal bars, 20 msec; vertical bar, 0.5 mV. (Reprinted with permission from ref. 68. Copyright 1997. American Association for the Advancement of Science.)

Grb2 and Src's, bind to the sequence PQVPSR (proline, glutamine, valine, proline, serine, and arginine) (155), a site overlapping but upstream of the consensus sequence for amphiphysin. Crystal structure analysis in combination with mutational expression in fibroblasts have shown that an unusual patch of acidic residues inserted in amphiphysin's SH3 domain is essential for efficient binding to the two arginine residues in PSRPNR and for endocytosis (154). In contrast, nonamphiphysin SH3 proteins such as Grb2 and Src's do not affect receptor-mediated endocytosis (154, 156), suggesting that they neither interact with dynamin *in vivo* nor affect synaptic vesicle recycling. This study demonstrates the importance and necessity for further *in vivo* validation of models proposed on the basis of *in vitro* data.

3.2.5 Clathrin-coated vesicle formation and fission may be co-ordinated

EPS15, initially identified as an epidermal growth factor receptor (tyrosine kinase) pathway substrate (clone) 15 (157), is a multi-domain protein which, *in vitro*, binds a large number of proteins associated with vesicle budding during endocytosis (Fig. 7). It contains three EPS15 homology (EH) domains at the N-terminus, which mediate molecular interactions by binding to NPF repeats (asparagine, proline, and phenylalanine) in target proteins (158). A central coiled-coil domain mediates homophilic oligomerization (158–160). 15 DPF (aspartic acid, proline, and phenylalanine) repeats are found in a C-terminal PR domain of EPS15, which mediate interaction with α-adaptin (161, 162). The interaction with α-adaptin is further enhanced by EPS15 oligomerization (160). An *in vivo* role for EPS15–AP2 interactions is suggested by studies of AP2-dependent transferrin receptor uptake in HeLa cells (163). In these cells, overexpression of a truncated EPS15 including the AP2 binding elements significantly reduces transferrin uptake at the plasma membrane. However, when the AP2 binding regions are deleted from this truncated protein, similar overexpression does not affect transferrin uptake (161, 163).

A direct role for EPS15 in membrane fission is suggested by its localization on the rim of the coated pits or nascent endocytic vesicles, where dynamin has also been localized (160, 164). In early stages of coated vesicle assembly, planar clathrin lattices formed on the plasma membrane include EPS15. However, at later stages when invaginated coated pits are clearly visible, EPS15 is found at the vesicle rim. Thus, it is likely that EPS15, initially part of a planar clathrin–AP2 lattice, is somehow expelled from the AP2–EPS15 complex during clathrin cage formation. During coated vesicle invagination, EPS15 is localized at vesicle necks, a location consistent with a role in membrane fission (164). Thus, it is conceivable that EPS15 functions mainly to co-ordinate coat formation and fission.

3.3 Post-fission events in synaptic vesicle recycling

The fate of synaptic vesicle proteins after they are internalized has been a source of many debates. It is conceivable that a rapid clathrin-independent mechanism for synaptic vesicle recycling exists and that vesicles internalized by this pathway are already competent to undergo regulated exocytosis. However, before they can release transmitter, vesicles are first acidified by a vesicle-associated ATP-driven proton pump, and this proton gradient is used to load vesicles with transmitter via a proton-coupled neurotransmitter transporter on synaptic vesicle membrane (see Chapter 5).

After synaptic vesicles are internalized by a clathrin-mediated pathway, coated vesicles must be uncoated prior to subsequent fusion events. The uncoating of vesicles has not been studied at nerve terminals *in vivo*. However, *in vitro*, two components of clathrin-coated vesicles, a DnaJ domain containing protein named auxillin and a DnaK domain containing heat shock cognate protein hsc70 serve to uncoat clathrin-coated vesicles (165–167). Uncoating occurs quickly after coated vesicles detach from the plasma membrane, presumably because the uncoating machinery is activated by an unknown regulatory mechanism immediately after vesicle fission. The hsc70 uncoating protein has ATPase activity that may provide the energy required for cage disassembly. A critical step in cage disassembly may be weakening of heavy chain–light chains interactions mediated through the hub domain of clathrin (168, 169).

After uncoating, internalized vesicles containing synaptic vesicle proteins may already be competent for transmitter uptake and regulated exocytosis (33). Alternatively, they might transit through an intermediate sorting station termed an endosome from which synaptic vesicles emerge following an additional vesicle budding event. Since a major function of sorting endosomes is to uncouple lysosomally-directed ligands from recycling receptors, it is conceivable that such a sorting station may simply be eliminated during synaptic vesicle recycling at nerve terminals. Studies of quantal FM1-43 labelled vesicle cycling in hippocampal synapses (170) as well as analysis of *lap* and *unc-11* mutants (52, 113) seem to support this notion. Most evidence for a second clathrin-independent vesicle budding event, or for participation of early endosomes in the traffic of synaptic vesicle proteins, comes from studies in PC12 cells, not from synapses (115, 116). With the exception of large cisternae that may arise from bulk internalization of plasma membrane, it is fair to state that, at nerve terminals molecular or structural components involved in a second vesicle budding event during vesicle recycling have not yet been described. However, if indeed an alternative ARF- and AP3-dependent pathway for vesicle recycling exists at some nerve terminals, we might gain insights into this issue from future studies of AP3 mutant mouse and *Drosophila* (171).

3.4 Protein–protein interactions in synaptic vesicle endocytosis

In the preceding section we have discussed molecular mechanisms of synaptic vesicle recycling in the context of molecules implicated in one of three broadly defined stages of synaptic vesicle endocytosis, sorting and vesicle formation, vesicle fission and uncoating. However, intensive and thorough analysis of clathrin, adaptin, dynamin, and amphiphysin binding molecules purified from brain homogenates has resulted in the identification of several other proteins very likely to participate in endocytosis. These molecules include synaptojanin, EPS15, EPSIN (172), endophilins (150), and DAP160 whose structures and binding properties *in vitro* are summarized in Fig. 7 and Table 1. The stages of synaptic vesicle endocytosis at which these molecules function remain unclear, and a major challenge in unravelling molecular mechanisms of vesicle recycling is to determine the nature of sequential and dynamic interactions among these proteins in the context of the synaptic vesicle cycle.

3.5 Molecular regulation during synaptic vesicle endocytosis

Synaptic vesicle recycling is regulated at several different levels. First, it is likely the molecular machinery for endocytosis is activated by Ca^{2+} influx into presynaptic terminals via the protein phosphatase calcineurin, which is present at very high levels at the nerve terminal. Such a mechanism would allow the temporal activation of endocytosis to be coupled precisely to vesicle exocytosis. Secondly, the assembly of clathrin-coated vesicles must be highly regulated to ensure quantal stability and precision in transmission. A potential regulator of this process is phosphoinositol lipids, which have been shown to affect clathrin caging by interacting with AP2 (173), AP180 (111, 174), and perhaps synaptotagmin (175–177). Thirdly, each step in the vesicle budding process must be regulated such that a strictly sequential series of events is maintained. For instance, uncoating must only be activated after vesicle fission. Details of these more subtle forms of regulation are yet unknown.

In vivo, Ca^{2+} is required, at low levels, for endocytosis of synaptic vesicles. However, the molecular mechanism by which they participate in vesicle recycling remains unclear. Recent biochemical reconstitution *in vitro* of complex formation among clathrin, adaptor proteins, amphiphysin, dynamin, synaptojanin, and endophilin revealed a dramatic dependence of complex stability on the phosphorylation state of these molecules (151). If brain homogenates were dephosphorylated, the complex formed efficiently. However, if phosphorylated, this multicomponent complex, presumed to represent a protein assembly involved in synaptic vesicle endocytosis, does not assemble (151). The likely targets for phosphorylation-dependent regulation include dynamin, amphiphysin, and synaptojanin. Phosphorylation of dynamin or synaptojanin inhibits their association with amphiphysin, while phosphorylation of amphiphysin inhibits its association with clathrin and AP2. Viewed simplistically, it is possible that Ca^{2+}-dependent dephosphorylation of these molecules would allow amphiphysin to bind clathrin and adaptor proteins in coated vesicles. Coat-associated amphiphysin could then recruit dynamin and synaptojanin to endocytic vesicles for vesicle fission. In addition to regulating binding between proteins, phosphorylation and dephosphorylation of dynamin regulates the GTPase activity of dynamin by altering the ratio of GTP-bound to GDP-bound dynamin. This could have profound effects on the localization and activity of dynamin and its associated molecules (178).

Ca^{2+} also has many other effects *in vitro*. Ca^{2+} binding to dynamin inhibits its GTPase activity and may, by increasing the stability of GTP-bound dynamin, mobilize cytosolic dynamin to the plasma membrane (138, 179). Ca^{2+} induces the oligomerization of synaptotagmin (180), a process which may play a role in facilitating clathrin coat formation around synaptic vesicle proteins. Ca^{2+} also binds directly, or via calmodulin, to clathrin light chains, and may thus regulate cage assembly or disassembly.

4. Conclusion, remarks and perspectives

A model for synaptic vesicle endocytosis and recycling is proposed in Fig. 10. As we know very little about the 'kiss-and-run' mechanism at nerve terminals, this model focuses primarily on several distinct steps and two independent pathways of clathrin-mediated endocytosis revealed by genetic, biochemical, and cell biological studies. The initiation of clathrin-coated synaptic vesicle formation is accomplished in part by clathrin and the clathrin adaptor proteins AP180 and AP2, presumably through interactions with the synaptic vesicle membrane protein synaptotagmin. Budding vesicles are either formed directly from the plasma membrane at the active zone or indirectly from the endocytic intermediate cisternae via a bulk uptake pathway. During the coated vesicle formation, AP180 and perhaps its interacting

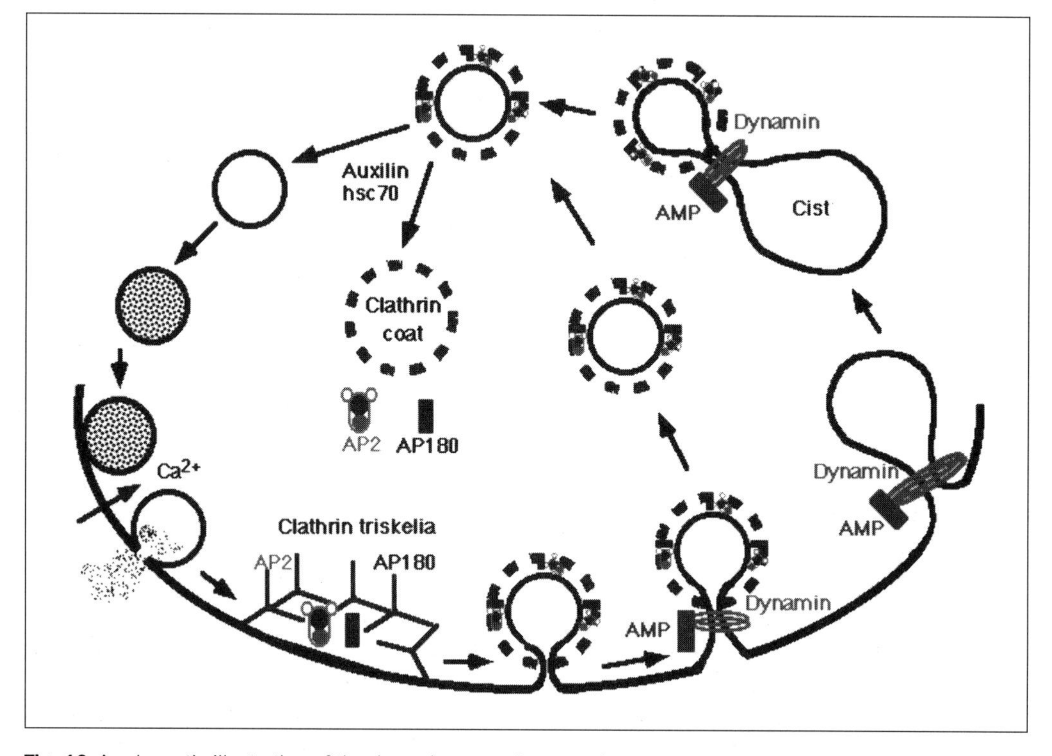

Fig. 10 A schematic illustration of the dynamic events in synaptic vesicle endocytosis and recycling. Exocytosis and transmitter release is induced by an influx of Ca^{2+} at the nerve terminal. Following exocytosis, synaptic vesicles may be recycled through two distinct pathways: a 'kiss-and-run' mechanism and a clathrin-mediated mechanism. In 'kiss-and-run', synaptic vesicles are thought to be simply snapped off from the plasma membrane and readily enter the releasable pool. Clathrin-mediated endocytosis involves several well co-ordinated steps, which begins with clathrin coat formation, a process assisted by AP2 and AP180, followed by vesicular invagination, fission, and uncoating. Other proteins, such as EPS15 and EPSIN, may also participate the formation of coat formation. Amphiphysin recruits dynamin to the neck of endocytic vesicles, where dynamin acts to severe coated vesicles from the plasma membrane. Coated vesicles also form from cisternae through a bulk uptake pathway, at sites away from the active zone. Auxilin and hsc70 are proposed to strip off the clathrin coats prior to 'naked' vesicles enter the releasable pool either directly or indirectly via early endosomes.

partners determine the size of the clathrin-coated synaptic vesicles through a direct regulation of clathrin cage size and by defining the amount of membrane to be retrieved into synaptic vesicles. After the formation of budding vesicles, the cytosolic protein amphiphysin is thought to recruit the dynamin GTPase to the endocytic vesicle neck, where it assists clathrin-coated synaptic vesicles to pinch off from the plasma membrane. EPS15 and EPSIN may also play a role in vesicle formation and fission. Once internalized, the clathrin coats are removed, presumably by auxilin and hsc70. Finally, newly recycled, 'naked' synaptic vesicles are directly translocated to either active zones or reserve pools for subsequent release.

It should be clear that many steps, links, and protein and lipid players essential for understanding synaptic vesicle endocytosis are still missing from our summary figure. A major task of future research in synaptic vesicle recycling is to discover the sequential order of interactions during vesicle formation and budding, and to lay out the regulatory mechanisms which maintain these ordered events. As more and more components of the endocytic machinery become identified, those interested in molecular mechanisms of vesicle recycling will face the challenge of understanding a multicomponent protein machine, a general problem of cell biologists today, perhaps better suited for analysis by biophysical rather than physiological methods.

Acknowledgements

In reviewing a very large body of literature, we have had to make difficult decisions about what to include and exclude. We offer our apologies to those whose work or opinions we may have overlooked. We are grateful to several colleagues for providing the original figures for reproduction: Jane Koenig and Kazuo Ikeda (Figs 1 and 5), Pietro De Camilli and Kohji Takei (Fig. 2), Lennard Brodin (Fig. 9), and to publishers of *Neuron*, *Journal of Cell Biology*, *Journal of Neuroscience*, *Science*, and *Nature* for permission to reproduce published figures. We wish to thank Pietro De Camilli, Hiroshi Kuromi, Yoshiaki Kidokoro, and Mike Nonet for providing preprints of their unpublished papers. We also thank Jane Koenig for constructive discussions, Bill Betz for a useful discussion on FM1-43 assays for endocytosis, Hugo Bellen, Gautam Bhave, Mark Wu, Lennard Brodin, and Regis Kelly for comments on the manuscript, and Salpy Sarikhanian for secretarial assistance. Bing Zhang is supported by a postdoctoral fellowship from the American Cancer Society. Mani Ramaswami's research is supported by grants from the NIH (RO1-NS34889 and KO2-NS02001) and from the McKnight and Sloan Foundations.

References

1. Betz, W. J. and Angleson, J. K. (1998) The synaptic vesicle cycle. *Annu. Rev. Physiol.*, **60**, 347.
2. Fesce, R., Grohovaz, R., Valtorta, F., and Meldolesi, J. (1994) Neurotransmitter release: fusion or 'kiss-and-run'. *Trends Cell Biol.*, **4**, 1.

3. Bittner, G. D. and Kennedy, D. (1970) Quantitative aspects of transmitter release. *J. Cell Biol.*, **47**, 585.
4. Ceccarelli, B., Grohovaz, F., and Hurlbut, W. P. (1979) Freeze-fracture studies of frog neuromuscular junctions during intense release of neurotransmitter. I. Effects of black widow spider venom and Ca^{2+}-free solutions on the structure of the active zone. *J. Cell Biol.*, **81**, 163.
5. Ceccarelli, B., Grohovaz, F., and Hurlbut, W. P. (1979) Freeze-fracture studies of frog neuromuscular junctions during intense release of neurotransmitter. II. Effects of electrical stimulation and high potassium. *J. Cell Biol.*, **81**, 178.
6. Ceccarelli, B., Hurlbut, W. P., and Mauro, A. (1972) Depletion of vesicles from frog neuromuscular junctions by tetanic stimulation. *J. Cell Biol.*, **54**, 30.
7. Clark, A. W., Hurlbut, W. P., and Mauro, A. (1972) Changes in the fine structure of the neuromuscular junction of the frog caused by black widow spider venom. *J. Cell Biol.*, **52**, 1.
8. Heuser, J. E. and Reese, T. S. (1973) Evidence for recycling of synaptic vesicle membrane during transmitter release at the frog neuromuscular junction. *J. Cell Biol.*, **57**, 315.
9. Heuser, J. E., Reese, T. S., Dennis, M. J., Jan, Y. N., Jan, L., and Evans, L. (1979) Synaptic vesicle exocytosis captured by quick freezing and correlated with quantal transmitter release. *J. Cell Biol.*, **81**, 275.
10. Heuser, J. E., Reese, T. S., and Landis, D. M. (1974) Functional changes in frog neuromuscular junctions studied with freeze-fracture. *J. Neurocytol.*, **3**, 109.
11. Kraszewski, K., Mundigl, O., Daniell, L., Verderio, C., Matteoli, M., and De Camilli, P. (1995) Synaptic vesicle dynamics in living cultured hippocampal neurons visualized with CY3-conjugated antibodies directed against the lumenal domain of synaptotagmin. *J. Neurosci.*, **15**, 4328.
12. Matteoli, M., Takei, K., Perin, M. S., Sudhof, T. C., and De Camilli, P. (1992) Exo-endocytotic recycling of synaptic vesicles in developing processes of cultured hippocampal neurons. *J. Cell Biol.*, **117**, 849.
13. Torri-Tarelli, F., Villa, A., Valtorta, F., De Camilli, P., Greengard, P., and Ceccarelli, B. (1990) Redistribution of synaptophysin and synapsin I during alpha-latrotoxin-induced release of neurotransmitter at the neuromuscular junction. *J. Cell Biol.*, **110**, 449.
14. Valtorta, F., Jahn, R., Fesce, R., Greengard, P., and Ceccarelli, B. (1988) Synaptophysin (p38) at the frog neuromuscular junction: its incorporation into the axolemma and recycling after intense quantal secretion. *J. Cell Biol.*, **107**, 2717.
15. von Wedel, R. J., Carlson, S. S., and Kelly, R. B. (1981) Transfer of synaptic vesicle antigens to the presynaptic plasma membrane during exocytosis. *Proc. Natl. Acad. Sci. USA*, **78**, 1014.
16. Betz, W. J. and Bewick, G. S. (1992) Optical analysis of synaptic vesicle recycling at the frog neuromuscular junction. *Science*, **255**, 200.
17. Betz, W. J., Mao, F., and Bewick, G. S. (1992) Activity-dependent fluorescent staining and destaining of living vertebrate motor nerve terminals. *J. Neurosci.*, **12**, 363.
18. Atwood, H. L., Lang, F., and Morin, W. A. (1972) Synaptic vesicles: selective depletion in crayfish excitatory and inhibitory axons. *Nature*, **176**, 1353.
19. Ryan, T. A., Reuter, H., Wendland, B., Schweizer, F. E., Tsien, R. W., and Smith, S. J. (1993) The kinetics of synaptic vesicle recycling measured at single pre-synaptic boutons. *Neuron*, **11**, 713.
20. Ryan, T. A. and Smith, S. J. (1995) Vesicle pool mobilization during action potential firing at hippocampal synapses. *Neuron*, **14**, 983.

21. Koenig, J. and Ikeda, K. (1989) Disappearance and reappearance of synaptic vesicle membrane upon transmitter release observed under reversible blockage of membrane retrieval. *J. Neurosci.*, **9**, 3844.
22. Ikeda, K., Ozawa, S., and Hagiwara, S. (1976) Synaptic transmission reversibly conditioned by single-gene mutation in *Drosophila melanogaster*. *Nature*, **259**, 489.
23. Ramaswami, M., Krishnan, K. S., and Kelly, R. B. (1994) Intermediates in synaptic vesicle recycling revealed by optical imaging of *Drosophila* neuromuscular junctions. *Neuron*, **13**, 363.
24. Chen, M. S., Obar, R. A., Schroeder, C. C., Austin, T. W., Poodry, C. A., Wadsworth, S. C., *et al.* (1991) Multiple forms of dynamin are encoded by *shibire*, a *Drosophila* gene involved in endocytosis. *Nature*, **351**, 583.
25. Grigliatti, T. A., Hall, L., Rosenbluth, R., and Suzuki, D. T. (1973) Temperature-sensitive mutations in *Drosophila melanogaster*. XIV. A selection of immobile adults. *Mol. Gen. Genet.*, **120**, 107.
26. Kosaka, T. and Ikeda, K. (1983) Possible temperature-dependent blockage of synaptic vesicle recycling induced by a single gene mutation in *Drosophila*. *J. Neurobiol.*, **14**, 207.
27. Poodry, C. A. and Edgar, L. (1979) Reversible alteration in the neuromuscular junctions of *Drosophila melanogaster* bearing a temperature-sensitive mutation, *shibire*. *J. Cell Biol.*, **81**, 520.
28. van der Bliek, A. M. and Meyerowitz, E. M. (1991) Dynamin-like protein encoded by the *Drosophila shibire* gene associated with vesicular traffic. *Nature*, **351**, 411.
29. Estes, P., Roos, J., van der Bliek, A., Kelly, R. B., Krishnan, K. S., and Ramaswami, M. (1996) Traffic of dynamin within single *Drosophila* synaptic boutons relative to compartment-specific presynaptic markers. *J. Neurosci.*, **16**, 5443.
30. Koenig, J. and Ikeda, K. (1996) Synaptic vesicles have two distinct recycling pathways. *J. Cell Biol.*, **135**, 797.
31. van de Goor, J., Ramaswami, M., and Kelly, R. (1995) Redistribution of synaptic vesicles and their proteins in temperature-sensitive *shibire*$^{(ts1)}$ mutant *Drosophila*. *Proc. Natl. Acad. Sci. USA*, **92**, 5739.
32. Maycox, P. R., Link, E., Reetz, A., Morris, S. A., and Jahn, R. (1992) Clathrin-coated vesicles in nervous tissue are involved primarily in synaptic vesicle recycling. *J. Cell Biol.*, **118**, 1379.
33. Takei, K., Mundigl, O., Daniell, L., and De Camilli, P. (1996) The synaptic vesicle cycle: a single vesicle budding step involving clathrin and dynamin. *J. Cell Biol.*, **133**, 1237.
34. Jorgensen, E. M., Hartwieg, E., Schuske, K., Nonet, M. L., Jin, Y., and Horvitz, H. R. (1995) Defective recycling of synaptic vesicles in synaptotagmin mutants of *Caenorhabditis elegans*. *Nature*, **378**, 196.
35. Zhang, J. Z., Davletov, B. A., Sudhof, T. C., and Anderson, R. G. (1994) Synaptotagmin I is a high affinity receptor for clathrin AP-2: implications for membrane recycling. *Cell*, **78**, 751.
36. Neher, E. and Marty, A. (1982) Discrete changes of cell membrane capacitance observed under conditions of enhanced secretion in bovine adrenal chromaffin cells. *Proc. Natl. Acad. Sci. USA*, **79**, 6712.
37. Wightman, R. M., Jankowski, J. A., Kennedy, R. T., Kawagoe, K. T., Schroeder, T. J., Leszczyszyn, D. J., *et al.* (1991) Temporally resolved catecholamine spikes correspond to single vesicle release from individual chromaffin cells. *Proc. Natl. Acad. Sci. USA*, **88**, 10754.
38. Zhou, Z. and Misler, S. (1995) Amperometric detection of stimulus-induced quantal

release of catecholamines from cultured superior cervical ganglion neurons. *Proc. Natl. Acad. Sci. USA*, **92**, 6938.

39. Alvarez de Toledo, G., Fernandez-Chacon, R., and Fernandez, J. M. (1993) Release of secretory products during transient vesicle fusion. *Nature*, **10**, 554.
40. Spruce, A. E., Breckenridge, L. J., Lee, A. K., and Almers, W. (1990) Properties of the fusion pore that forms during exocytosis of a mast cell secretory vesicle. *Neuron*, **4**, 643.
41. Oberhauser, A. F., Robinson, I. M., and Fernandez, J. M. (1996) Simultaneous capacitance and amperometric measurements of exocytosis: a comparison. *Biophys. J.*, **71**, 1131.
42. Betz, W. J. and Wu, L.-G. (1995) Kinetics of synaptic-vesicle recycling. *Curr. Biol.*, **5**, 1098.
43. von Gersdorff, H. and Matthews, G. (1994) Dynamics of synaptic vesicle fusion and membrane retrieval in synaptic terminals. *Nature*, **367**, 735.
44. Klingauf, J., Kavalali, E. T., and Tsien, R. W. (1998) Kinetics and regulation of fast endocytosis at hippocampal synapses. *Nature*, **394**, 581.
45. Henkel, A. and Betz, W. J. (1995) Staurosporine blocks evoked release of FM1-43 but not acetylcholine from frog motor nerve terminals. *J. Neurosci.*, **15**, 8246.
46. Heuser, J. H. and Reese, T. (1981) Structural changes after transmitter release at the frog neuromuscular junction. *J. Cell Biol.*, **88**, 564.
47. Ceccarelli, B. and Hurlbut, W. P. (1980) Vesicle hypothesis of release of quanta of acetylcholine. *Physiol. Rev.*, **60**, 396.
48. Gonzalez-Gaitan, M. and Jackle, H. (1997) Role of *Drosophila* alpha-adaptin in presynaptic vesicle recycling. *Cell*, **88**, 767.
49. Roos, J. and Kelly, R. B. (1998) Dap160, a neural-specific Eps15 homology and multiple SH3 domain-containing protein that interacts with *Drosophila* dynamin. *J. Biol. Chem.*, **273**, 19108.
50. Artalejo, C. R., Elhamdani, A., and Palfrey, H. C. (1996) Calmodulin is the divalent cation receptor for rapid endocytosis, but not exocytosis, in adrenal chromaffin cells. *Neuron*, **16**, 195.
51. Artalejo, C. R., Henley, J. R., McNiven, M. A., and Palfrey, H. C. (1995) Rapid endocytosis coupled to exocytosis in adrenal chromaffin cells involves Ca^{2+}, GTP, and dynamin but not clathrin. *Proc. Natl. Acad. Sci. USA*, **92**, 8328.
52. Zhang, B., Koh, Y. H., Beckstead, R. B., Budnik, B., Ganetzky, B., and Bellen, H. J. (1998) Synaptic vesicle size and number are regulated by a clathrin adaptor protein required for endocytosis. *Neuron*, **21**, 1465.
53. Gad, H., Low, P., Zotova, E., Brodin, L., and Shupliakov, O. (1998) Dissociation between Ca^{2+}-triggered synaptic vesicle exocytosis and clathrin-mediated endocytosis at a central synapse. *Neuron*, **21**, 607.
54. Kuromi, H. and Kidokoro, Y. (1998) Two distinct pools of synaptic vesicles in single presynaptic boutons in a temperature-sensitive *Drosophila* mutant, *shibire*. *Neuron*, **20**, 917.
55. Lagnado, L., Gomis, A., and Job, C. (1996) Continuous vesicle cycling in the synaptic terminal of retinal bipolar cells. *Neuron*, **17**, 957.
56. Pieribone, V. A., Shupliakov, O., Brodin, L., Hilfiker-Rothenfluh, S., Czernik, A. J., and Greengard, P. (1995) Distinct pools of synaptic vesicles in neurotransmitter release. *Nature*, **375**, 493.
57. Neher, E. (1998) Vesicle pools and Ca^{2+} microdomains: new tools for understanding their roles in neurotransmitter release. *Neuron*, **20**, 389.
58. Kuromi, H. and Kidojoro, Y. (1999) The optically dertermined size of exo/endo cycling vesicle pool correlates with the quantal content at the neuromuscular junction of *Drosophila* larvae. *J. Neurosci.*, **19**, 1557.

59. Stevens, C. F. and Sullivan, J. M. (1998) Regulation of the readily releasable vesicle pool by protein kinase C. *Neuron*, **21**, 885.
60. Wang, C. and Zucker, R. S. (1998) Regulation of synaptic vesicle recycling by calcium and serotonin. *Neuron*, **21**, 155.
61. Cousin, M. A. and Robinson, P. J. (1998) Ba^{2+} does not support synaptic vesicle retrieval in rat cerebrocortical synaptosomes. *Neurosci. Lett.*, **253**, 1.
62. Marks, B. and McMahon, H. T. (1998) Calcium triggers calcineurin-dependent synaptic vesicle recycling in mammalian nerve terminals. *Curr. Biol.*, **8**, 740.
63. von Gersdorff, H. and Matthews, G. (1994) Inhibition of endocytosis by elevated internal calcium in a synaptic terminal. *Nature*, **370**, 652.
64. Wu, L. G. and Betz, W. J. (1996) Nerve activity but not intracellular calcium determines the time course of endocytosis at the frog neuromuscular junction. *Neuron*, **17**, 769.
65. Wu, L. G. and Betz, W. J. (1998) Kinetics of synaptic depression and vesicle recycling after tetanic stimulation of frog motor nerve terminals. *Biophys. J.*, **74**, 3003.
66. Cremona, O. and De Camilli, P. (1997) Synaptic vesicle endocytosis. *Curr. Opin. Neurobiol.*, **7**, 323.
67. Schweizer, F. E., Betz, H., and Augustine, G. (1995) From vesicle docking to endocytosis: intermediate reactions of exocytosis. *Neuron*, **14**, 689.
68. Shupliakov, O., Low, P., Grabs, D., Gad, H., Chen, H., David, C., *et al.* (1997) Synaptic vesicle endocytosis impaired by disruption of dynamin-SH3 domain interactions. *Science*, **276**, 259.
69. Takei, K., McPherson, P. S., Schmid, S., and De Camilli, P. (1995) Tubular membrane invaginations coated by dynamin rings are induced by GTPγS in nerve terminals. *Nature*, **374**, 186.
70. Alder, J., Kanki, H., Valtorta, F., Greengard, P., and Poo, M. M. (1995) Overexpression of synaptophysin enhances neurotransmitter secretion at *Xenopus* neuromuscular synapses. *J. Neurosci.*, **15**, 511.
71. Betz, A., Ashery, U., Rickmann, M., Augustin, I., Neher, E., Sudhof, T. C., *et al.* (1998) Munc13-1 is a presynaptic phorbol ester receptor that enhances neurotransmitter release. *Neuron*, **21**, 123.
72. Bennett, M. K., Calakos, N., Kreiner, T., and Scheller, R. H. (1992) Synaptic vesicle membrane proteins interact to form a multimeric complex. *J. Cell Biol.*, **116**, 761.
73. Roth, T. F. and Porter, K. R. (1964) Yolk protein uptake in the oocyte of the mosquito *Aedes aegypti*. *J. Cell Biol.*, **20**, 313.
74. Douglas, W. W., Nagasawa, J., and Schulz, R. A. (1971) Coated microvesicles in neurosecretory terminals of posterior pituitary glands shed their coats to become smooth 'synaptic' vesicles. *Nature*, **232**, 340.
75. Nagasawa, J., Douglas, W. W., and Schulz, R. A. (1970) Ultrastructural evidence of secretion by exocytosis and of 'synaptic vesicle' formation in posterior pituitary glands. *Nature*, **227**, 407.
76. Pearse, B. M. F. and Crowther, R. A. (1987) Structure and assembly of coated vesicles. *Annu. Rev. Biophys. Biophys. Chem.*, **16**, 49.
77. Jackson, A. P. (1992) Endocytosis in the brain: the role of clathrin light-chains. *Biochem. Soc. Trans.*, **20**, 653.
78. Jackson, A. P., Seow, H.-F., Holmes, N., Drickamer, K., and Parham, P. (1987) Clathrin light chains contain brain-specific insertion sequences and a region of homology with intermediate filaments. *Nature*, **326**, 154.
79. Kirchhausen, T., Scarmato, P., Harrison, S. C., Monroe, J. J., Chow, E. P., Mattaliano, R. J.,

et al. (1987) Clathrin light chains LCA and LCB are similar, polymorphic, and share repeated heptad motifs. *Science*, **236**, 320.

80. Stamm, S., Casper, D., Dinsmore, J., Kaufmann, C. A., Brosius, J., and Helfman, D. M. (1992) Clathrin light chain B: gene structure and neuron-specific splicing. *Nucleic Acids Res.*, **20**, 5097.
81. Keen, J. H. (1987) Clathrin assembly proteins: affinity purification and a model for coat assembly. *J. Cell Biol.*, **105**, 1989.
82. Zaremba, S. and Keen, J. H. (1983) Assembly polypeptides from coated vesicles mediate reassembly of unique clathrin coats. *J. Cell Biol.*, **97**, 1339.
83. Simpson, F., Peden, A. A., Christopoulou, L., and Robinson, M. S. (1997) Characterization of the adaptor-related protein complex, AP-3. *J. Cell Biol.*, **137**, 835.
84. Dell'Angelica, E. C., Klumperman, J., Stoorvogel, W., and Bonifacino, J. S. (1998) Association of the AP-3 adaptor complex with clathrin. *Science*, **280**, 431.
85. Dell'Angelica, E. C., Ohno, H., Ooi, C. E., Rabinovich, E., Roche, K. W., and Bonifacino, J. S. (1997) AP-3: an adaptor-like protein complex with ubiquitous expression. *EMBO J.*, **16**, 917.
86. Page, L. J. and Robinson, M. S. (1995) Targeting signals and subunit interactions in coated vesicle adaptor complexes. *J. Cell Biol.*, **131**, 619.
87. Robinson, M. S. (1994) The role of clathrin, adaptors and dynamin in endocytosis. *Curr. Opin. Cell Biol.*, **6**, 538.
88. De Camilli, P. and Takei, K. (1996) Molecular mechanisms in synaptic vesicle endocytosis and recycling. *Neuron*, **16**, 481.
89. Cowles, C. R., Odorizzi, G., Payne, G. S., and Emr, S. D. (1997) The AP-3 adaptor complex is essential for cargo-selective transport to the yeast vacuole. *Cell*, **91**, 109.
90. Faundez, V., Horng, J.-T., and Kelly, R. B. (1998) A function for the AP3 coat complex in synaptic vesicle formation from endosomes. *Cell*, **93**, 423.
91. Kantheti, P., Qiao, X., Diaz, M. E., Peden, A. A., Meyer, G. E., Carskadon, S. L., *et al.* (1998) Mutation in AP-3 delta in the mocha mouse links endosomal transport to storage deficiency in platelets, melanosomes, and synaptic vesicles. *Neuron*, **21**, 111.
92. Ooi, C. E., Moreira, J. E., Dell'Angelica, E. C., Poy, G., Wassarman, D. A., and Bonifacino, J. S. (1997) Altered expression of a novel adaptin leads to defective pigment granule biogenesis in the *Drosophila* eye color mutant garnet. *EMBO J.*, **16**, 4508.
93. Stepp, J. D., Huang, K., and Lemmon, S. K. (1997) The yeast adaptor protein complex, AP-3, is essential for the efficient delivery of alkaline phosphatase by the alternate pathway to the vacuole. *J. Cell Biol.*, **139**, 1761.
94. West, M. A., Bright, N. A., and Robinson, M. S. (1997) The role of ADP-ribosylation factor and phospholipase D in adaptor recruitment. *J. Cell Biol.*, **138**, 1239.
95. Le Borgne, R. and Hoflack, B. (1998) Mechanisms of protein sorting and coat assembly: insights from the clathrin-coated vesicle pathway. *Curr. Opin. Cell Biol.*, **10**, 499.
96. Letourneur, R. and Klausner, R. D. (1992) A novel di-leucine motif and a tyrosine-based motif independently mediate lysosomal targeting and endocytosis of CD3 chains. *Cell*, **69**, 1143.
97. Robinson, M. S. (1989) Cloning of cDNAs encoding two related 100-kD coated vesicle proteins (alpha-adaptins). *J. Cell Biol.*, **108**, 833.
98. Rapoport, I., Miyazaki, M., Boll, W., Duckworth, B., Cantley, L. C., Shoelson, S., *et al.* (1997) Regulatory interactions in the recognition of endocytic sorting signals by AP-2 complexes. *EMBO J.*, **16**, 2240.
99. Gonzales-Gaitan, M. and Jackle, H. (1997) Role of *Drosophila* α-adaptin in presynaptic vesicle recycling. *Cell*, **88**, 767.

100. Bazinet, C., Katzen, A. L., Morgan, M., Mahowald, A. P., and Lemmon, S. K. (1993) The *Drosophila* clathrin heavy chain gene: clathrin function is essential in a multicellular organism. *Genetics*, **134**, 1119.
101. Reist, N. E., Buchanan, J., Li, J., DiAntonio, A., Buxton, E. M., and Schwarz, T. L. (1998) Morphologically docked synaptic vesicles are reduced in synaptotagmin mutants of *Drosophila*. *J. Neurosci.*, **18**, 7662.
102. Van der Kloot, W. (1991) The regulation of quantal size. *Prog. Neurobiol.*, **36**, 93.
103. Schikorski, T. and Stevens, C. F. (1997) Quantitative ultrastructural analysis of hippocampal excitatory synapses. *J. Neurosci.*, **17**, 5858.
104. Ahle, S. and Ungewickell, E. (1986) Purification and properties of a new clathrin assembly protein. *EMBO J.*, **5**, 3143.
105. Kohtz, D. S. and Puszkin, S. (1988) A neuronal protein (NP185) associated with clathrin-coated vesicles. Characterization of NP185 with monoclonal antibodies. *J. Biol. Chem.*, **263**, 7418.
106. Morris, S. A., Mann, A., and Ungewickell, E. (1990) Analysis of 100–180-kDa phosphoproteins in clathrin-coated vesicles from bovine brain. *J. Biol. Chem.*, **265**, 3354.
107. Murphy, J. E., Pleasure, I. T., Puszkin, S., Prasad, K., and Keen, J. H. (1991) Clathrin assembly protein AP-3. The identity of the 155K protein, AP180, and NP185 and demonstration of a clathrin binding domain. *J. Biol. Chem.*, **266**, 4401.
108. Sousa, R., Tannery, N. H., Zhou, S. B., and Lafer, E. M. (1992) Characterization of a novel synapse-specific protein. 1. Developmental expression and cellular localization of the F1-20 protein and messenger RNA. *J. Neurosci.*, **12**, 2130.
109. Zhou, S., Sousa, R., Tannery, N. H., and Lafer, E. M. (1992) Characterization of a novel synapse-specific protein. II. cDNA cloning and sequence analysis of the F1-20 protein. *J. Neurosci.*, **12**, 2144.
110. Lindner, R. and Ungewickell, E. (1992) Clathrin-associated proteins of bovine brain coated vesicles. *J. Biol. Chem.*, **267**, 16567.
111. Norris, F. A., Ungewickell, E., and Majerus, P. W. (1995) Inositol hexakisphosphate binds to clathrin assembly protein 3 (AP-3/AP180) and inhibits clathrin cage assembly *in vitro*. *J. Biol. Chem.*, **270**, 214.
112. Ye, W. and Lafer, E. M. (1995) Bacterially expressed F1-20/AP-3 assembles clathrin into cages with a narrow size distribution: implications for the regulation of quantal size during neurotransmission. *J. Neurosci. Res.*, **41**, 15.
113. Nonet, M. L., Holgado, A. M., Brewer, F., Serpe, C. J., Norbeck, B. A., Holleran, J., *et al.* (1999) UNC-11, a *C. elegans* AP180 homolog, regulates the size and protein composition of synaptic vesicles. *Mol. Cell. Biol.* (in press).
114. Clift-O'Grady, L., Desnos, C., Lichtenstein, Y., Faundez, V., Horng, J. T., and Kelly, R. B. (1998) Reconstitution of synaptic vesicle biogenesis from PC12 cell membranes. *Methods*, **16**, 150.
115. Faundez, V., Horng, J.-T., and Kelly, R. B. (1997) ADP ribosylation factor 1 is required for synaptic vesicle budding in PC12 cells. *J. Cell Biol.*, **138**, 505.
116. Lichtenstein, Y., Desnos, C., Faundez, V., Kelly, R. B., and Clift-O'Grady, L. (1998) Vesiculation and sorting from PC12-derived endosomes *in vitro*. *Proc. Natl. Acad. Sci. USA*, **95**, 11223.
117. Newman, L. S., McKeever, M. O., Okano, H. J., and Darnell, R. B. (1995) β-NAP, a cerebellar degeneration antigen, is a neuron-specific vesicle coat protein. *Cell*, **82**, 773.
118. Nemoto, Y., Arribas, M., Haffner, C., and DeCamilli, P. (1997) Synaptojanin 2, a novel

synaptojanin isoform with a distinct targeting domain and expression pattern. *J. Biol. Chem.*, **272**, 30817.

119. Schmid, S. L., McNiven, M. A., and De Camilli, P. (1998) Dynamin and its partners: a progress report. *Curr. Opin. Cell Biol.*, **10**, 504.
120. Warnock, D. E. and Schmid, S. L. (1996) Dynamin GTPase, a force-generating molecular switch. *BioEssays*, **18**, 885.
121. Damke, H., Baba, T., Warnock, D. E., and Schmid, S. L. (1994) Induction of mutant dynamin specifically blocks endocytic coated vesicle formation. *J. Cell Biol.*, **127**, 915.
122. Hinshaw, J. E. and Schmid, S. L. (1995) Dynamin self-assembles into rings suggesting a mechanism for coated vesicle budding. *Nature*, **374**, 190.
123. Sweitzer, S. M. and Hinshaw, J. E. (1998) Dynamin undergoes a GTP-dependent conformational change causing vesiculation. *Cell*, **93**, 1021.
124. David, C., McPherson, P. S., Mundigl, O., and De Camilli, P. (1996) A role of amphiphysin in synaptic vesicle endocytosis suggested by its binding to dynamin in nerve terminals. *Proc. Natl. Acad. Sci. USA*, **93**, 331.
125. Gout, I., Dhand, R., Hiles, I. D., Fry, M. J., Panayotou, G., Das, P., *et al.* (1993) The GTPase dynamin binds to and is activated by a subset of SH3 domains. *Cell*, **75**, 25.
126. Grabs, D., Slepnev, V. I., Songyang, Z., David, C., Lynch, M., Cantley, L. C., *et al.* (1997) The SH3 domain of amphiphysin binds the proline-rich domain of dynamin at a single site that defines a new SH3 binding consensus sequence. *J. Biol. Chem.*, **272**, 13419.
127. McPherson, P. S., Garcia, E. P., Slepnev, V. I., David, C., Zhang, X., Grabs, D., *et al.* (1996) A presynaptic inositol-5-phospatase. *Nature*, **379**, 353.
128. Scaife, R. and Margolis, R. L. (1990) Biochemical and immunochemical analysis of rat brain dynamin interaction with microtubules and organelles *in vivo* and *in vitro*. *J. Cell Biol.*, **111**, 3023.
129. Shpetner, H. S. and Vallee, R. B. (1989) Identification of dynamin, a novel mechanochemical enzyme that mediates interactions between microtubules. *Cell*, **59**, 421.
130. Clark, S. G., Shurland, D. L., Meyerowitz, E. M., Bargmann, C. I., and van der Bliek, A. M. (1997) A dynamin GTPase mutation causes a rapid and reversible temperature-inducible locomotion defect in *C. elegans*. *Proc. Natl. Acad. Sci. USA*, **94**, 10438.
131. Grant, D., Unadkat, S., Katzen, A., Krishnan, K. S., and Ramaswami, M. (1998) Probable mechanisms underlying interallelic complementation and temperature-sensitivity of mutations at the *shibire* locus of *Drosophila melanogaster*. *Genetics*, **149**, 1019.
132. Kelly, R. B. (1995) Endocytosis. Ringing necks with dynamin [news; comment]. *Nature*, **374**, 116.
133. Takei, K., Haucke, V., Slepnev, V., Farsad, K., Salazar, M., Chen, H., *et al.* (1998) Generation of coated intermediates of clathrin-mediated endocytosis on protein-free liposomes. *Cell*, **94**, 131.
134. Roos, J. and Kelly, R. B. (1997) Is dynamin really a 'pinchase'? *Trends Cell Biol.*, **7**, 157.
135. Muhlberg, A. B., Warnock, D. E., and Schmid, S. L. (1997) Domain structure and intramolecular regulation of dynamin GTPase. *EMBO J.*, **16**, 6676.
136. Shaw, G. (1995) The pleckstrin homology domain: an intriguing multifunctional protein module. *BioEssays*, **18**, 35.
137. Warnock, D. E., Hinshaw, J. E., and Schmid, S. L. (1996) Dynamin self-assembly stimulates its GTPase activity. *J. Biol. Chem.*, **271**, 22310.
138. Liu, J. P., Sim, A. T., and Robinson, P. J. (1994) Calcineurin inhibition of dynamin I GTPase activity coupled to nerve terminal depolarization. *Science*, **265**, 970.

139. Robinson, P. J., Sontag, J. M., Liu, J. P., Fykse, E. M., Slaughter, C., McMahon, H., *et al.* (1993) Dynamin GTPase regulated by protein kinase C phosphorylation in nerve terminals. *Nature*, **365**, 163.
140. Herskovits, J. S., Shpetner, H. S., Burgess, C. C., and Vallee, R. B. (1993) Microtubules and Src homology 3 domains stimulate the dynamin GTPase via its C-terminal domain. *Proc. Natl. Acad. Sci. USA*, **90**, 11468.
141. Maeda, K., Nakata, T., Noda, Y., Sato-Yoshitake, R., and Hirokawa, N. (1992) Interaction of dynamin with microtubules: its structure and GTPase activity investigated by using highly purified dynamin. *Mol. Biol. Cell*, **3**, 1181.
142. Shpetner, H. S. and Vallee, R. B. (1992) Dynamin is a GTPase stimulated to high levels of activity by microtubules. *Nature*, **355**, 733.
143. Tuma, P. L., Stachniak, M. C., and Collins, C. A. (1993) Activation of dynamin GTPase by acidic phospholipids and endogenous rat brain vesicles. *J. Biol. Chem.*, **268**, 17240.
144. Salim, K., Bottomley, M. J., Quenfurth, E., Zvelebil, J., Gout, I., Scaife, R., *et al.* (1996) Distinct specificity in the recognition of phosphoinositides by the pleckstrin homology domains of dynbamin and Brutons tyrosine kinase. *EMBO J.*, **15**, 6241.
145. Lin, H. C. and Gilman, A. G. (1996) Regulation of dynamin I GTPase activity by G protein betagamma subunits and phosphatidylinositol 4,5-bisphosphate. *J. Biol. Chem.*, **271**, 27979.
146. De Camilli, P., Thomas, A., Cofiell, R., Folli, F., Lichte, B., Piccolo, G., *et al.* (1993) The synaptic vesicle-associated protein amphiphysin is the 128-kD autoantigen of Stiff-Man syndrome with breast cancer. *J. Exp. Med.*, **178**, 2219.
147. Lichte, B., Veh, R. W., Meyer, H. E., and Kilimann, M. W. (1992) Amphiphysin, a novel protein associated with synaptic vesicles. *EMBO J.*, **11**, 2521.
148. Wigge, P. and McMahon, H. T. (1998) The amphiphysin family of proteins and their role in endocytosis at the synapse. *Trends Neurosci.*, **21**, 339.
149. McMahon, H. T., Wigge, P., and Smith, C. (1997) Clathrin interacts specifically with amphiphysin and is displaced by dynamin. *FEBS Lett.*, **413**, 319.
150. Micheva, K. D., Ramjaun, A. R., Kay, B. K., and McPherson, P. S. (1997) SH3 domain-dependent interactions of endophilin with amphiphysin. *FEBS Lett.*, **414**, 308.
151. Slepnev, V. I., Ochoa, G. C., Butler, M. H., Grabs, D., and De Camilli, P. (1998) Role of phosphorylation in regulation of the assembly of endocytic coat complexes. *Science*, **281**, 821.
152. Shpetner, H. S., Herskovits, J. S., and Vallee, R. B. (1996) A binding site for SH3 domains targets dynamin to coated pits. *J. Biol. Chem.*, **271**, 13.
153. Bauerfeind, R., Takei, K., and De Camilli, P. (1997) Amphiphysin I is associated with coated endocytic intermediates and undergoes stimulation-dependent dephosphorylation in nerve terminals. *J. Biol. Chem.*, **272**, 30984.
154. Owen, D. J., Wigge, P., Vallis, Y., Moore, J. D., Evans, P. R., and McMahon, H. T. (1998) Crystal structure of the amphiphysin-2 SH3 domain and its role in the prevention of dynamin ring formation. *EMBO J.*, **17**, 5273.
155. Sparks, A. B., Rider, J. E., Hoffman, N. G., Fowlkes, D. M., Quillam, L. A., and Kay, B. K. (1996) Distinct ligand preferences of Src homology 3 domains from Src, Yes, Abl, Cortactin, p53bp2, PLCgamma, Crk, and Grb2. *Proc. Natl. Acad. Sci. USA*, **93**, 1540.
156. Wigge, P., Vallis, Y., and McMahon, H. T. (1997) Inhibition of receptor-mediated endocytosis by the amphiphysin SH3 domain. *Curr. Biol.*, **7**, 554.
157. Fazioli, F., Minichiello, L., Matoskova, B., Wong, W. T., and Di Fiore, P. P. (1993) eps15, a novel tyrosine kinase substrate, exhibits transforming activity. *Mol. Cell Biol.*, **13**, 5814.

158. Wong, W. T., Schumacher, C., Salcini, A. E., Romano, A., Castagnino, P., Pelicci, P. G., *et al.* (1995) A protein-binding domain, EH, identified in the receptor tyrosine kinase substrate Eps15 and conserved in evolution. *Proc. Natl. Acad. Sci. USA*, **92**, 9530.
159. Cupers, P., ter Haar, E., Boll, W., and Kirchhausen, T. (1997) Parallel dimers and anti-parallel tetramers formed by epidermal growth factor receptor pathway substrate clone 15. *J. Biol. Chem.*, **272**, 33430.
160. Tebar, F., Sorkina, T., Sorkin, A., Ericsson, M., and Kirchhausen, T. (1996) Eps15 is a component of clathrin-coated pits and vesicles and is located at the rim of coated pits. *J. Biol. Chem.*, **271**, 28727.
161. Benmerah, A., Begue, B., Dautry-Varsat, A., and Cerf-Bensussan, N. (1996) The ear of alpha-adaptin interacts with the COOH-terminal domain of the Eps 15 protein. *J. Biol. Chem.*, **271**, 12111.
162. Benmerah, A., Gagnon, J., Begue, B., Megarbane, B., Dautry-Varsat, A., and Cerf-Bensussan, N. (1995) The tyrosine kinase substrate eps15 is constitutively associated with the plasma membrane adaptor AP-2. *J. Cell Biol.*, **131**, 1831.
163. Benmerah, A., Lamaze, C., Begue, B., Schmid, S. L., Dautry-Varsat, A., and Cerf-Bensussan, N. (1998) AP-2/Eps15 interaction is required for receptor-mediated endocytosis. *J. Cell Biol.*, **140**, 1055.
164. Cupers, P., Jadhav, A. P., and Kirchhausen, T. (1998) Assembly of clathrin coats disrupts the association between Eps15 and AP-2 adaptors. *J. Biol. Chem.*, **273**, 1847.
165. Ahle, S. and Ungewickell, E. (1990) Auxilin, a newly identified clathrin-associated protein in coated vesicles from bovine brain. *J. Cell Biol.*, **111**, 19.
166. Schlossman, D. M., Schmid, S. L., Braell, W. A., and Rothman, J. E. (1984) An enzyme that removes clathrin coats: purification of an uncoating ATPase. *J. Cell Biol.*, **99**, 723.
167. Ungewickell, E., Ungewickell, H., Holstein, S. E., Lindner, R., Prasad, K., Barouch, W., *et al.* (1995) Role of auxilin in uncoating clathrin-coated vesicles. *Nature*, **378**, 632.
168. Pishvaee, B., Munn, A., and Payne, G. S. (1997) A novel structural model for regulation of clathrin function. *EMBO J.*, **16**, 2227.
169. Ungewickell, E. and Ungewickell, H. (1991) Bovine brain clathrin light chains impede heavy chain assembly *in vitro*. *J. Biol. Chem.*, **266**, 12710.
170. Murthy, V. N. and Stevens, C. F. (1998) Synaptic vesicles retain their identity through the endocytic cycle. *Nature*, **392**, 497.
171. Lloyd, V., Ramaswami, M., and Krämer, H. (1998) Not just pretty eyes: *Drosophila* eye color mutations and lysosomal delivery. *Trends Cell Biol.*, **8**, 257.
172. Chen, H., Fre, S., Slepnev, V. I., Capua, M. R., Takei, K., Butler, M. H., *et al.* (1998) Epsin is an EH-domain-binding protein implicated in clathrin-mediated endocytosis. *Nature*, **394**, 793.
173. Gaidarov, I., Chen, Q., Falck, J. R., Reddy, K. K., and Keen, J. H. (1996) A functional phosphatidylinositol 3,4,5-trisphosphate/phosphoinositide binding domain in the clathrin adaptor AP-2 α subunit. *J. Biol. Chem.*, **271**, 20922.
174. Hao, W., Tan, Z., Prasad, K., Reddy, K. K., Chen, J., Prestwich, G. D., *et al.* (1997) Regulation of AP-3 function by inositides: identification of phosphatidylinositol 3,4,5-trisphosphate as a potent ligand. *J. Biol. Chem.*, **272**, 6393.
175. Fukuda, M., Aruga, J., Niinobe, M., Aimoto, S., and Mikoshiba, K. (1994) Inositol-1,3,4,5-tetrakisphosphate binding to C2B domain of IP4BP/synaptotagmin II. *J. Biol. Chem.*, **269**, 29206.
176. Llinas, R., Sugimori, M., Lang, E. J., Morita, M., Fukuda, M., Niinobe, M., *et al.* (1994) The

inositol high-polyphosphate series blocks synaptic transmission by preventing vesicular fusion: A squid giant synapse study. *Proc. Natl. Acad. Sci. USA*, **91**, 12990.

177. Mizutani, A., Fukuda, M., Niinobe, M., and Mikoshiba, K. (1997) Regulation of AP-2-synaptotagmin interaction by inositol high polyphosphates. *Biochem. Biophys. Res. Commun.*, **240**, 128.
178. Robinson, P. J., Liu, J.-P., Powell, K. A., Fykse, E. M., and Sudhof, T. C. (1994) Phosphorylation of dynamin I and synaptic-vesicle recycling. *Trends Neurosci.*, **17**, 348.
179. Liu, J. P., Zhang, Q. X., Baldwin, G., and Robinson, P. J. (1996) Calcium binds dynamin I and inhibits its GTPase activity. *J. Neurochem.*, **66**, 2074.
180. Chapman, E. R., An, S., Edwardson, J. M., and Jahn, R. (1996) A novel function for the second C2 domain of synaptotagmin. Ca^{2+}-triggered dimerization. *J. Biol. Chem.*, **271**, 5844.
181. Liu, S. H., Marks, M. S., and Brodsky, F. M. (1998) A dominant-negative clathrin mutant differentially affects trafficking of molecules with distinct sorting motifs in the class II major histocompatibility complex (MHC) pathway. *J. Cell Biol.*, **140**, 1023.
182. Wendland, B. and Emr, S. D. (1998) Pan1p, yeast eps15, functions as a multivalent adaptor that coordinates protein-protein interactions essential for endocytosis. *J. Cell Biol.*, **141**, 71.
183. Dreyling, M. H., Martinez-Climent, J. A., Zheng, M., Mao, J., Rowley, J. D., and Bohlander, S. K. (1996) The t(10;11) (p13;14) in the U937 cell line results in the fusion of the *AF10* gene and *CALM*, encoding a new member of the AP-3 clathrin assembly protein family. *Proc. Natl. Acad. Sci. USA*, **93**, 4804.
184. Volchuk, A., Narine, S., Foster, L. J., Grabs, D., De Camilli, P., and Klip, A. (1998) Perturbation of dynamin II with an amphiphysin SH3 domain increases GLUT4 glucose transporters at the plasma membrane in 3T3-L1 adipocytes. Dynamin II participates in GLUT4 endocytosis. *J. Biol. Chem.*, **273**, 8169.
185. Singer-Kruger, B., Nemoto, Y., Daniell, L., Ferro-Novick, S., and De Camilli, P. (1998) Synaptojanin family members are implicated in endocytic membrane traffic in yeast. *J. Cell Sci.*, **111**, 3347.
186. Novick, P., Osmond, B. C., and Botstein, D. (1989) Suppressors of yeast actin mutations. *Genetics*, **121**, 659.
187. Ringstad, N., Nemoto, Y., and De Camilli, P. (1997) The SH3p4/Sh3p8/SH3p13 protein family: binding partners for synaptojanin and dynamin via a Grb2-like Src homology 3 domain. *Proc. Natl. Acad. Sci. USA*, **94**, 8569.
188. Micheva, K. D., Kay, B. K., and McPherson, P. S. (1997) Synaptojanin forms two separate complexes in the nerve terminal. Interactions with endophilin and amphiphysin. *J. Biol. Chem.*, **272**, 27239.
189. Kane, N. S., Robichon, A., Dickinson, J. A., and Greenspan, R. J. (1997) Learning without performance in PKC-deficient *Drosophila*. *Neuron*, **18**, 307.
190. Kuromi, H., Yoshihara, M., and Kidokoro, Y. (1997) An inhibitory role of calcineurin in endocytosis of synaptic vesicles at nerve terminals of *Drosophila* larvae. *Neurosci. Res*, **27**, 101.
191. Engisch, K. L. and Nowycky, M. C. (1998) Compensatory and excess retrieval: two types of endocytosis following single step depolarizations in bovine adrenal chromaffin cells. *J. Physiol. (Lond)*, **506**, 591.

Index

4-aminopyridine 390
7S complex 96, 99, 102, 107, 212
17S complex 112
20S complex 92, 99, 104, 212

acetylcholine 34, 131, 170
acetylcholinesterase 175
actin 368
active zone 4, 5, 6, 24, 224, 237–8
activity dependent labelling 68
adaptins, *see* AP
adherens junction 20
adhesion 22, 23
Aequorea victoria 61
aequorin 61
aex mutants 268
aex-3 269, 287
agatoxins 223–4
Agelenopsis aperta 223
agrin 24
akt/Protein Kinase B (PKB) 134
aldicarb 175, 268–9
amino acid transporters 158
ammodytoxin 225
amperometry 35, 46–51
  and capacitance 49
  carbon electrode 48
  Faraday's law 46
  neurotransmitters 46, 48
  oxidizable compounds 46
  sensitivity 46
amphetamines 150, 174, 375
amphiphysin 272, 404, 406, 415
amyotrophic lateral sclerosis (ALS) 166
anticonvulsants
  tiagabine 148
antidepressant
  desipramine 149
  fluoxetine 149, 150
  paroxetine 149
  tricyclics 150
antihypertensive
  reserpine 172
AP1 408
AP2 17, 135, 139, 249, 257, 286, 289, 318, 338, 392, 394, 404–5, 408
AP3 17, 408, 412
AP50 338, 339
AP180 135, 249, 257, 272, 277, 289, 305, 392, 394, 404–5, 410
ARF1 17, 138, 412
ascorbate 177
ASCT 159, 163
ATP 91, 94, 96–100, 104, 139, 145, 177
ATPase 91, 96–9, 104, 145, 288, 418
ATPgamma-S 254
arachidonic acid 168
attention deficit-hyperactivity disorder (ADHD) 375
auxilin 404–5, 418
axon 24

bassoon 23
Bet1p 87
beta-alanine 148
Black Widow Spider Venom (BWSV) 330, 401, *see* latrotoxin
Bombesin 131
Bos1p 87
boutons 3
bradykinin 131
brain specific Na+-dependent phosphate transporter (BNPI) 166
bungarotoxin 225

C2 domains 94, 103, 285–6, 316, 360
cacophony 314
cadherins 20
Caenorhabditis elegans 265–93
  advantages 265
  behavioural assays 274
  biochemistry 276
  cell biology 276
  electrophysiology 277–81
  nervous system 266
  neurotransmitters 266
calcineurin 407
calcium 8–10, 110, 132
  channel 9, 20, 88, 110, 223–4, 239, 284–5, 313–14, 322, 325
  dependence 319, 322
  in endocytosis 401–3, 419
  imaging 35, 51
  and latrotoxin 222
  release from internal stores 132, 222
  sensing 101, 103, 316
  sensitivity 94
  signalling 312
  synapsin 368–71
calcium caged compounds 50, 60–1
  DM-nitrophen 60
  nitr family 60
  nitrophenyl EGTA 60
calcium indicator dyes 51, 56–60
  calcium green 57, 67
  calcium orange 57
  dextran conjugation 60
  dual wavelength indicators 59
  ester loading 59
  fluo-3 57
  fura-2 57–8, 70, 239
  furapta 59
  indo-1 59
  loading 59
  n-aequorin-J 58
  proteinaceous 61, *see* proteinacious calcium indicators
  rhod-2 57
  single wavelength indicators 57
calmodulin 312, 314, 400
CaM kinase 88, 245, 380
cAMP 25, 358, 366, 368
CAPS 135, 139, 272
capacitance measurement 36–42, 49–50, 94
  admittance 39
  and amperometry 49
  capacitor 38
  chromaffin cells 69
  endocytosis 395
  limits of measurement 38
  optimal frequency range 41
  phase tracking 40
  step increase 42
  unit/farads 38

CASK 20
catenins 20
CCD Camera 54
  pulsed laser imaging 63
  temporal resolution 54
  spatial resolution 55
  sensitivity 55
cell adhesion molecules (CAMs) 21
cha-1 269
chaperone 323
chemical crosslinkers 90
cholesterol 126
choline acetyltransferase 269, 288
circular dichroism 106
CIRL/latrophillin 221
cisternae 395, 400, 414
clathrin 17, 289, 392, 394, 400, 404–5
clathrin coated vesicles 16, 394, 405, 408
clostridial neurotoxins 208–20
  action 220, 251
  BoNT 209–13
  cleavage sites 212–13
  heavy chain 209
  internalization 211
  light chain 209–11
  specificity 216–17
  targets 212–13
  TeNT 209–13
Clostridium botulinum 209
Clostridium tetani 209
cocaine 145, 149, 150
coiled coil motifs 96
Colomboma mice 375
comatose 248, 306, 336
complexin 245, 248–9
confocal microscopy 64
conotoxins 223–4
core complex 101–2, 325, 327
CREB 25
Cre-recombinase 359
cystein string protein/CSP 88, 272, 306, 313, 322
Cystic Fibrosis Transport ion channel (CFTR) 378
cytochalasin 18
cytoskeleton 94, 107
cytosol 91

DAP160 23, 404, 406, 418
dendrite 5
dendritic spine 5, 6, 20
dense core vesicles 136
  membrane composition 127
depression 356
desipramine 149
diacyl glycerol (DAG) 132
diffusion of calcium 10
Discs Large (DLG) 21
DM-nitrophen 50
DNA J domain 323, 418
DOC2 94, 285, 314
dopamine 149, 171, 174, 275, 377
dopamine transporter (DAT) 149
  knock-out 150
Drosophila melanogaster
  reverse genetics 305–7
  neuromuscular junction 308–9
  electrophysiology 310–12
dynamin 17, 24, 135, 140, 272, 290, 391, 394, 404, 405, 412–14

early endosome 87
eat mutants 268
EEA1 135, 139
egl-8 288
egl-10 269, 286
egl-19 284–5
egl-30 269, 286
electrical capacitance 36
electron microscopy 98, 240, 254–8, 309, 361, 392
electropharyngeograms (EPGs) 277
electroretinograms 310
endocrine cells 3
endocrine communication 1
endocytosis 16–19, 22, 39, 69, 139, 226, 257, 288, 316, 318 *see* Chapter 11
  model 420
  phophorylation 415
  proteins 405
  regulation 419
endophilin 406, 418
endoplasmatic reticulum 87, 130
endosomes 16, 19, 81, 82, 418
epifluorescence 52
epinephrine 171, 174
eps15 404, 406, 417–18
epsin 24, 404, 406, 418
ethyl methane sulfonate (EMS) 269, 306
evanescent wave 72
evoked transmitter release 8, 310
excitatory amino acid transporters (EAAT) 158
  affinity 160
  exotoxicity 167
  glial 159
  glutamate binding 164
  knock-out mice 167
  localization 159, 165
  mechanisms 160
  neuronal 159
  physiological role 164
  postsynaptic 159
  presynaptic 159
  structure 162
  turnover rate 164
excitatory junctional current (EJC) 312
excitatory junctional potential (EJP) 312
excitatory postsynaptic current (EPSC) 361
excitatory postsynaptic potential (EPSP) 356
excitotoxicity 166
exocyst 374
exocytosis 16, 22, 39, 49, 69, 104
expressed sequence tags (EST) 271

Fasciclin II 21
facilitation 10, 356, 365
fluorescence 51
fluorophore 51
fluoxetine 149, 150
FM 1–43 18, 68, 70, 391, 395
FM 4–64 68
FRET 62, 98, 101
Frequenin 313–15
fusion pore 10–12, 42–5
  amperometry 48
  composition 10–12, 45
  conductance 45
  diameter 45
  rate of release 49
FYVE finger domain 136, 139

GABA 3, 146, 166, 171, 176
GABA transporter (GAT; VGAT) 146, 147–9, 176
GAP 43, 88
GDP dissociation inhibitor (GDI) 107
GFP 61, 276
glomus cell 48
glutamate 131, 158–69, 171, 308
glutamate receptor 277
  NMDA 358
glutaminase 165
glutamine 165
glutamate transporters (GLT) 158, *see* EAAT

glycine 158, 171
glycine transport (GLYT) 150
Golgi complex 87, 90, 91, 130
Gos1p 87
GOS-28 86
G-proteins 287
GTP 94
GTPase 95, 107, 413
  activating protein (GAP) 107, 368
GTP exchange factor (GEF) 107, 368
GTP-gammaS 41, 253, 413
Guanine nucleotide dissociation inhibitor (GDI) 368

halotane 283
hippocampal neurons 69
hippocampal slices 356, 365
histamine 131
horse raddish peroxidase (HRP) 395
hrs-2 88, 90, 96, 272, 307
HSC70 325, 405, 418
hsp70 325
hypertonic sucrose 327, 361

imaprime 275
inositol phospolipids 126
  in membrane trafficing 136
integrin 22, 26
internal reflection fluorescence microscopy 71
iontophoresis 311
IP3 132
  IP3 receptor 132
IP4 254
isoflurane 283

Katz's quantal hypothesis 8, 9
ketanserine 172, 179
kinesin like proteins 24
kiss-and-run hypothesis 12, 16, 18, 392

laminin 22, 25
lamprey reticulospinal synapse 415
lap 306, 410
large dense core vesicles 1, 139, 183, 267, 308
late endosome 87
Latrodectus mactans 221
latrophillin/CIRL 221, 373
latrotoxin 220–3, 251, 327, 361, 373
learning 25, 26
levamisole 275
lin-10 277
lipids, *see* phosphatidyl and cholesterol
lipid kinases 128–30
liposomes 106
Loligo peali 238
long term depression 13
long term potentiation (LTP) 13, 358
  postsynaptic 358
  presynaptic 358
  in rab3A mutants 366

mast cells 38, 48, 61
MDCK cells 112, 157
melanotrophs 61, 94
membrin 86
metalloproteases 212
methylphenidate 375
microtubules 24
miniature postsynaptic potentials/currents 241, 310
Mint-1 20, 272, 277, 283
monoamine transporters 149–50
  dopamine 149
  norepinephrine 149
  serotonin 149
mouse 352
  electrophysiology 355
  knock-outs 353
  phenotypic analyses 354
MPP+ 149, 169, 173
MPTP 173–4, 181
munc-18/n-sec1p/Rop 20, 86, 245, 249, 333
muscimol 275
MuSk 25
mvti1 86

Na+/K+ ATPase 145
nematode 265
NEM-sensitive factor, *see* NSF
N-ethylmaleimide (NEM) 88, 91, 254, 335
neurexins 20, 22, 88, 221, 373
neuromuscular junction 4, 25, 69, 280, 308
neuroligin 20, 22
neurotransmitter
  membrane transport 146
  transporters 145–84
  transport mechanisms 152–7
  vesicular transport 146
nipecotic acid 147–8
norepinephrine 131, 171, 174
norepinephrine transporter (NET) 146, 149
notexin 225
n-sec1/Munc18/Rop, *see* munc-18
NSF or NEM-sensitive factor 88, 91, 97 104–5, 212, 243, 245, 248–9, 272, 284, 335
  structure 98–100
numerical aperture 52
N-WASP 24
nystatin 60
Nyv1p 87

orphan transporters 151
ouabain 145

paracrine communication 1
Parkinson's disease 150, 173–4,
paroxetine 149
patch clamp 36
  whole cell configuration 39
PC12 cells 71, 72, 139
P-elements 306
pH 94
phosphoinosite-dependent protein kinase (PDK) 134
phosphoinositides 128, 130
phosphoinositide binding proteins 135
phosphoinositide transfer proteins (PITP) 136
phospolipase A (PLA) 225–6
phospolipase C (PLC) 130, 131, 135, 288
phospolipase D (PLD) 137
phospolipase toxins 225
phospholipids 126–40
  composition of membranes 127
  posphatidyl choline (PC) 126–7
  posphatidyl ethanolamine (PE) 126–7
  posphatidylserine (PS) 126–7, 414
  posphatidylinositol (PI) 126–40
    synthesis 130
photobleaching 52
PI- and PIP-kinases 128–40
piccolo 23
pixels 54
plekstrin homology domains (PH) 129, 133–5
polysialogangliosides 210

postsynaptic densities 6
post-tetanic potentiation 357
presynaptic dense bodies/active zone 4, 5, 224, 268
probability of release 9
proline transport 150
proteinacious calcium indicators 61
  Aequorin 61
  GFP 61
protein kinase A (PKA) 358, 366
protein kinase B (PKB) 134, 135
protein kinase C (PKC) 94, 131, 132, 157, 168, 380, 406
proton pump 88
psychostimulants
  amphetamines 150
  methylenedioxymetamphetamine (ecstasy) 150
pulsed laser imaging 63

quantal event 2, 7, 81
quantal size 180–1
quinidine 315

rab interacting molecule (RIM) 107, 108
rab proteins 86, 88, 90, 95, 108, 139, 245, 271, 272, 272, 285, 287, 363–8
rab3a cycle 107, 108, 287
rab3a knock-out 108, 110
rabphillin 88, 90, 94, 107, 108, 135, 139, 245, 249, 272, 285, 287, 314, 365
Rac 129
rbf-1 285, 287
readily releasable pool of vesicles 12, 94, 327
reserpine 172, 174, 222
retinal bipolar neurons 69, 70
retzius neuron 48
Rho 129
ric-4 269
rim 272, 285, 287, 314
RNA-mediated interference 274
Rop 313, 333
rotary shadowing electron microscopy 98, 101
ryanodine receptor 132

SCAMP 272
SDS-resistance 106
Sec1p 87, 93, 245
Sec3p 93
Sec4p 87, 93, 107
Sec5p 93
Sec6p 88, 93, 107, 109, 112, 374
Sec8p 88, 93, 107, 109, 112, 374
Sec9p 87, 93
Sec10p 93
Sec14p 136
Sec15p 93, 374
Sec17p 93
Sec18p 93
Sec22p 87, 93
Sed5p 87
serotonin 131, 171, 174, 275
serotonin transporter (SERT) 149
  knock-out 150
seven transmembrane receptors 130–1
  G-proteins 130
Sft1p 87
shibire 290, 391, 413
Sly1p 87
SNAP 25 86, 88, 90, 92, 96, 97, 100–7, 212, 219, 245, 249, 253, 282, 272, 325, 374
SNAPs (alpha, beta, gamma) 88, 91, 92, 97, 104, 243, 245, 249, 284
SNAREs 14, 92, 96–111, 213, 218–19, 256, 281
snb-1 269, 282
Snc1/2 87, 94, 332
snt-1 269, 285, 289
sphingomyelin (SM) 126
sphingolipids 126
spontaneous transmitter release or minis 8, 9
squid giant synapse 105, 238
Sso1/2p 87, 94, 332
stauroporin 18, 357
Stiff-Man's syndrome 415
stoned 289, 338
Stoned A 313, 338
styryl dyes 68
substance P, K 131
SV2 88 178, 290
SVOP 272, 290
synapse (CNS) 5, 6
  lamprey 23
  malleability 25
synapsin 88, 239, 243, 249, 272, 339, 368
synaptic depression 8, 12
synaptic vesicles 1, 6
  active pool 401
  antibodies 85
  brain 85
  C. elegans 267
  collapse 391
  composition 85
  distribution 254
  docking 6, 9, 253, 308, 316, 319, 332
  Drosophila 312
  electric organ 83
  endocytosis 16–19, *see* Chapter 11
  fusion 256, 316
  pools 400
  priming 255
  proteins 85
  purification 83–4
  rab3a knock-out mice 367
  reformation 391
  release 12
  reserve pool 6, 12, 255, 308, 369–71, 400
  size 5, 68, 289
  synapsin knock-out mice 369
  targeting 112
  transport 24, 81
  transporter targetting 183
synaptobrevin/VAMP 86, 88, 92, 96, 97, 100–7, 110, 112, 212, 219, 249, 253, 272, 277, 289, 306–7, 313, 325, 330, 372, 412
synaptogyrin 272
synaptojanin 140, 272, 289, 404, 406, 418
synaptophysin 88, 97, 272, 372–3
synaptososmes 6, 83, 89
synaptotagmin/p65 15, 21, 70, 86, 87, 88, 89, 94, 95, 101, 103, 110, 111, 135, 139, 211, 247, 249, 251, 257, 272, 285, 289, 306–7, 313–22, 360–3, 404, 407
syndapin 24
synprint 285
syntaxin 20, 86, 88, 89, 92, 96, 97, 100–7, 110, 212, 219, 223, 249, 253, 272, 276, 282, 306–7, 313, 325–6, 336, 377

taipoxin 225
temperature 94
tetrabenazine 172, 179
tiagabine 148
tight junctions 20, 112
Tlg1/2 87
tomosyn 89, 272
Torpedo californica 175–7, 214, 322
Tourette's syndrome 375
transcytosis 211

transglutaminase 216
trans-Golgi network 3, 87
trichlorfon 268
tricyclics 150
two photon excitation 67
tyrosine kinase receptors 133

Ufe1p 87
unc-2 269, 284
unc-10 269, 285, 287
unc-11 269, 277, 289, 410
unc-13 268, 272, 276, 285–6, 314
unc-17 175, 269, 288
unc-18 268, 272, 276, 283, 286
unc-26 269, 289
unc-31 269, 286–7
unc-32 *288*
unc-36 284
unc-41 269, 289, 338
unc-47 175, 288
unc-64 269
unc-101 289
unc-104 269

Vac1p 139
vacuolar ATPase 288
vacuolar fusion in yeast 104
Vacuolar H+-ATPase 169
vacuole 86
Vam3p 87
Vam7p 86, 87
VAMP, *see* synaptobrevin
varicosities 3, 24
vasopressin 131
vesamicol 182
vesicular neurotransmitter transporters 169–84
  acetylcholine (VAChT) 175–6, 181, 288
  amino acids 176–7
  mechanisms 178–9
  monoamines (VMATs) 171–5
    knock-out 181
  regulation 180–4
vesicular stomatitis virus assay 91
Vps15p 130
Vps27p 139
Vps33p 86, 87
Vps34p 130, 139
Vps45 86, 87
Vti1p 87

Williams syndrome 377

X-ray diffraction 102

yeast secretion 92–3
yeast two hybrid system 90
Ykt1p 87
Ypt1p 87, 107
Ypt6p 87
Ypt7p 87
Ypt31p 87
Ypt51p 87

Zinc 169, 212